How It's Done

AN INVITATION TO SOCIAL RESEARCH

Third Edition

Emily Stier Adler
Rhode Island College

Roger Clark
Rhode Island College

THOMSON

WADSWORTH

Australia • Brazil • Canada • Mexico • Singapore • Spain
United Kingdom • United States

THOMSON

★

WADSWORTH

***How It's Done: An Invitation to Social Research,* Third Edition**

Emily Stier Adler and Roger Clark

Acquisitions Editor: Chris Caldeira
Assistant Editor: Christina Ho
Editorial Assistant: Tali Beesley
Technology Project Manager: David Lionetti
Marketing Manager: Michelle Williams
Marketing Assistant: Jaren Boland
Marketing Communications Manager: Linda Yip
Project Manager, Editorial Production: Matt Ballantyne
Creative Director: Rob Hugel
Art Director: John Walker
Print Buyer: Linda Hsu

Permissions Editor: Bob Kauser
Production Service: Sonia Taneja, ICC Macmillan Inc.
Photo Researcher: Terri Wright
Copy Editor: Ivan Weiss
Cover Designer: Yvo Riezebos
Cover Images (From Left): Spencer Grant/Photo Edit;
    Richard Lord/Photo Edit; Sparky/The Image Bank/
    Getty Images, Inc.; Michael Newman/Photo Edit
Compositor: ICC Macmillan Inc.
Text and Cover Printer: West Group

Library of Congress Control Number: 2006937606

ISBN-13: 978-0-495-09338-1
ISBN-10: 0-495-09338-6

Thomson Higher Education
10 Davis Drive
Belmont, CA 94002-3098
USA

For more information about our products, contact us at:
Thomson Learning Academic Resource Center
1-800-423-0563
For permission to use material from this text or
product, submit a request online at
http://www.thomsonrights.com.
Any additional questions about permissions can be
submitted by e-mail to thomsonrights@thomson.com.

Brief Contents

Contents

11 Observational Techniques / 297

12 Using Available Data / 333

We dedicate this book to
George L. Adler and Beverly Lyon Clark

Preface

We'd like to invite you to participate in one of the most exciting, exhilarating, and sometimes exasperating activities we know: social science research. We extend the invitation not only because we know, from personal experience, how rewarding and useful research can be, but also because we've seen what pleasure it can bring other students of the social world. Our invitation comes with some words of reassurance, especially for those of you who entertain a little self-doubt about your ability to do research. First, we think you'll be glad to discover, as you read *How It's Done,* how much you already know about how social research is done. If you're like most people, native curiosity has been pushing you to do social research for much of your life. This book is meant simply to assist you in this natural activity by showing you some tried-and-true ways to enlightening and plausible insights about the social world.

Special Features

Active Engagement in Research

Our second word of reassurance is that we've done everything we can to minimize your chances for exasperation and maximize your opportunities for excitement and exhilaration. Our philosophy is simple. We believe that honing one's skill in doing social research is analogous to honing one's skills in other enjoyable and rewarding human endeavors, like sport, art, or dance. The best way isn't simply to read about it. It's to do it and to watch experts do it. So, just as you'd hesitate to teach yourself tennis, ballet, or painting only by reading about them, we won't ask you to try learning the fine points of research methodology by reading alone. We'll encourage you to get out and practice the techniques we describe. We've designed exercises at the end of each chapter to help you work on the "ground strokes," "serve," "volleys," and "overheads" of social research. We don't think you'll need to do all of the exercises at home. Your instructor might ask you to do some in class and might want you to ignore some altogether. In any case, we think that, by book's end, you should have enough control of the fundamentals to do the kind of on-the-job research that social science majors are increasingly asked to do, whether they find themselves in social service agencies, the justice system, business and industry, government, or graduate school.

The exercises reflect our conviction that we all learn best when we're actively engaged. Other features of the text also encourage such active engagement, including the Stop and Think questions that run through each chapter, which encourage you to actively respond to what you're reading.

Engaging Examples of Actual Research

Moreover, just as you might wish to gain inspiration and technical insight for ballet by studying the work of Anna Pavlova or Mikhail Baryshnikov, we'll encourage you to study the work of some accomplished researchers.

Thus, we build most of our chapters around a research essay, what we call focal research, usually previously published, that is intended to make the research process transparent, rather than opaque. We have chosen these essays for their appeal and accessibility, and to tap what we hope are some of your varied interests: for instance, crime, gender, defining oneself as an adult, attitudes toward rape, immigrants' lives, and others.

Behind-the-Scene Glimpses of the Research Process

These focal research pieces are themselves a defining feature of our book. In addition to such exemplary "performances," however, we've included behind-the-scenes glimpses of the research process. We're able to provide these glimpses because many researchers have given generously of their time to answer our questions about what they've done, the special problems they've encountered, and the ways they've dealt with these problems. The glimpses should give you an idea of the kinds of choices and situations the researchers faced, where often the "real" is far from the "ideal." You'll see how they handled the choices and situations and hear them present their current thinking about the compromises they made. In short, we think you'll discover that good research is an achievable goal, and a very human enterprise.

Clear and Inviting Writing

We've also tried to minimize your chances for exasperation by writing as clearly as we can. A goal of all social science is to interpret social life, something you've all been doing for quite a while. We want to assist you in this endeavor, and we believe that an understanding of social science research methods can help. But unless we're clear in our presentation of those methods, your chances of gaining that understanding are not great. There are, of course, times when we'll introduce you to concepts that are commonly used in social science research which might be new to you. When we do, however, we will try to provide definitions to make the concepts as clear as possible. The definitions are highlighted in the margin of the text and in the glossary at the end of the text.

Balance Between Quantitative and Qualitative Approaches

We think you'll also appreciate the balance between quantitative and qualitative research methods presented here. Quantitative methods focus on things that are measured numerically. ("He glanced at her 42 times during the performance.") Qualitative methods focus on descriptions of the essence of things. ("She appeared annoyed at his constant glances.") We believe both methodological approaches are too useful to ignore. Emblematic of this belief is the inclusion of a chapter (Chapter 15), which devotes about as much space to the discussion of qualitative data analysis as it does to quantitative data analysis. The presence of such a chapter is another defining feature of the book.

Moreover, in addition to more conventional strategies, we will introduce you to some relatively new research strategies, such as using the Internet to refine ideas and collect data. We cover the link between theory and research, compare research to other ways of knowing, and focus on basic and applied research.

Our aims, then, in writing this book have been (1) to give you first-hand experiences with the research process, (2) to provide you with engaging examples of social science research, (3) to offer behind-the-scenes glimpses of how professional researchers have done their work, (4) to keep our own presentation of the "nuts-and-bolts" of social science research as clear and inviting as possible, (5) to give a balanced presentation of qualitative and quantitative research methods, and (6) to introduce recent technological innovations. Whether we succeed in these goals, and in the more important one of sharing our excitement about social research, remains to be seen. Be assured, however, of our conviction that there is excitement to be had.

What Is New in the Third Edition

The third edition represents a substantial revision of the second. Once again, we've rewritten major sections of every chapter to clarify the process of social research and to provide up-to-date material from the social research literature. In doing so, however, we've tried to focus our presentation on the essentials of social research and are therefore able to offer a shorter book than we have before. We have, for instance, combined our data analysis chapters, Chapters 15 and 16 in the second edition, into one chapter on data analysis (Chapter 15), deleting or shortening sections on quantitative data analysis that readers have told us are unnecessary for an introductory text: for instance, a section on multivariate analysis. We have retained in our new Chapter 15, however, a thorough introduction to both quantitative and qualitative data analyses. Our interest in focusing and condensing our presentation is also reflected in the presence of fewer appendices, even though we have added a new appendix on writing research proposals, and in the absence of any "summing up" sections outside of the book's 15 chapters. At the same time we've added a substantial section to Chapter 13 that compares various data collection techniques and advocates the use of multiple techniques.

Our data analysis chapter (Chapter 15) also reflects, as does the rest of the current text, our belief that research, as practiced by social (and all other) scientists, is increasingly computer-assisted and Internet based. So, for instance, in the data analysis chapter, we now introduce students to data that they can analyze online. In other chapters we also present ways of finding research reports and data that can be accessed quickly online.

Themes from the first and second editions have been retained here. This edition has five new focal research pieces and one updated one. While incorporating the new pieces, we have enhanced the balance between qualitative and quantitative research in the book. In Chapter 7, for instance, we present a new piece on how young people "invent" adulthoods, based on qualitative interviews done with young people in England over time. New focal research pieces also appear in Chapters 3, 5, 6, and 13, and the focal research piece in Chapter 4 has been revised substantially. In all cases, our new contributors have volunteered new "behind-the-scenes" insights into the research process, insights that we gratefully share here.

Acknowledgments

We cannot possibly thank all those who have contributed to the completion of this edition, but we can try to thank those whose help has been most indispensable and hope that others will forgive our neglect. We'd first like to thank all the students who have taken research methods courses with us at Rhode Island College for their general good-naturedness and patience as we've worked out ideas that are crystallized here, and then worked them out some more. We'd also like to thank our colleagues in the Sociology Department and the administration of the college, for many acts of encouragement and more tangible assistance, including released time.

We'd like to thank colleagues, near and far, who have permitted us to incorporate their writing as focal research and, in many cases, then read how we've incorporated it, and told us how we might do better. They are Patricia A. Adler, Peter Adler, Richard Aniskiewicz, Benjamin Bates, Erica Chito Childs, Desirée Ciambrone, Sandra Enos, Joseph R. Ferrari, Paula Foster, Norma B. Gray, Jessica Guilmain, Sheila Henderson, Michele Hoffnung, Janet Holland, G. David Johnson, Matthew T. Lee, Kristy Maher, Ramiro Martinez, Sheena McGrellis, Jr., Donald C. Naylor, Gloria J. Palileo, Matthew M. Reavy, Paul Khalil Saucier, Sue Sharpe, Jessica Holden Sherwood, Jocelyn Tavarez, Rachel Thomson, David Wright, and Earl Wysong. These researchers include those who've given us the behind-the-scenes glimpses of the research process that we think distinguishes our book. We'd like to extend special thanks to Leslie Hossfeld, University of North Carolina—Wilmington, who composed the excellent instructor's manual accompanying the book.

We're very grateful to Chris Caldeira, our editor at Wadsworth, and to Ivan Weiss for their help with the third edition. We've benefited greatly from the comments of the social science colleagues who have reviewed our manuscripts at various stages of development. Mary Archibald; Cynthia Beall, Case Western Reserve University; Daniel Berg, Humboldt State University; Susan Brinkley, University of Tampa; Jeffrey A. Burr, State University of New York, Buffalo; Richard Butler, Benedict College; Daniel Cervi, University of New Hampshire; Charles Corley, Michigan State University; Norman Dolch, Louisiana State University, Shreveport; Craig Eckert, Eastern Illinois University; Pamela I. Jackson, Rhode Island College; Alice Kemp, University of New Orleans; Michael Kleinman, University of South Florida; James Marquart, Sam Houston State University; James Mathieu, Loyola Marymount University; Steven Meier, University of Idaho; Deborah Merrill, Clark University; Lawrence Rosen, Temple University; Josephine Ruggiero, Providence College; Kevin Thompson, University of North Dakota; Steven Vassar,

Mankato State University; and Gennaro Vito, University of Louisville, offered important assistance for the first edition. For the second edition, Gai Berlage, Iowa College; Susan Chase, University of Tulsa; Dan Cooper, Hardin-Simmons University; William Faulkner, Western Illinois University; Lawrence Hazelrigg, Florida State University; Josephine Ruggiero, Providence College; Jeffrey Will, University of North Florida; and Don Williams, Hudsonville College, have been most helpful. For the third edition we would like to thank Ronda Priest, University of Southern Indiana; Dennis Downey, University of Utah; Scott M. Myers, Montana State University; William Wagner, California State University, Bakersfield; and John Mitrano, Central Connecticut State University, for their valuable comments.

Finally, we'd like to thank our spouses, George L. Adler and Beverly Lyon Clark, for providing many of the resources we've thanked others for: general good-naturedness and patience, editorial assistance, encouragement, and other tangible aids, including released time.

The Uses of Social Research

© Emily Stier Adler

Introduction

How do you see yourself 10 years from now? Will you be single? Will you be a parent? What about people you know? Does it seem to you that almost everyone with whom you've talked about such things is either planning to marry and have children or is already married with children? Or does it seem to be the other way around? Does everyone seem to be planning to avoid marriage, perhaps to live alone, at least for extended periods of his or her life? What's the reality? Studying something systematically, with method, means not relying on your impressions, or on anyone else's, either. It means checking our impressions against the facts. It also means being sure about what one is looking for. One way researchers achieve this level of definiteness is to begin their research with a **research question,** or a question about one or more topics that can be answered with research. One such focusing question about our initial concern, for instance, might be: Is almost everyone in the country married with children or are they living alone?

A recent U.S. Census Bureau report (Hobbs, 2005) provides one apparently relevant set of facts. It shows that in the year 2000, for the first time in U.S. history, the number of Americans living alone, about 25.8 percent of all households, surpassed the proportion of American households that are made up of a married couple with children, about 25.5 percent. This could be the starting point of your investigation.

research question, a question about one or more topics or concepts that can be answered through research.

STOP AND THINK *This would only be a start, however. What about our initial interest in the future of you and people like you is not addressed by the "facts" of the Census report?*

Although the Census report does tell us something that's relevant to our main research—whether young people today are planning to marry and have children or planning not to marry—it requires a little interpreting, doesn't it? At first glance, its facts seem almost to address this question. They suggest that in 2000, more households in the United States were made up of people living alone than of couples with children, and so seem to imply that more people will end up alone than in families with children. Or do they?

We've made up the pie chart in Figure 1.1 to help us think about the question of whether the facts of paragraph two really settle the issue addressed in our research question. The chart indicates one thing that we already knew: More American households are made up of people living alone than of couples with children. But, perhaps, looking at the chart will remind you that all households made up of married couples with children had at least *twice as many adults living in them* as households made up of people living alone. So, in fact, there actually were more adults living as part of couples with children in 2000 than there were adults living alone. It also reminds us that we should consider other kinds of households. One such group is married couples without children, a group that once again is almost as numerous as households made up of people living alone, and, again, would be therefore made up of households that have at least twice as many adults in them as households made up of people living alone. The presence of this large group might make us want to revise our original question to

FIGURE 1.1

Total Households in 2000: 105,480,101

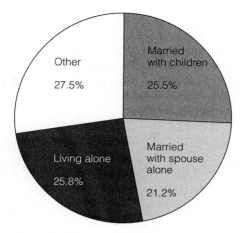

Source: Data from Hobbs, 2005: 6.

include something about the number of people who plan to live with another adult (or adults), but without children. Moreover, we suspect that more of these people, people in couples without children, "plan" to become couples with children than "plan" to live alone. In effect, then, the chart also leads us to a second research question, a question that, in some ways, might even better embody the concern expressed at the beginning of paragraph one: Are more young people planning to marry without having children, to marry and have children, or to live alone?

STOP AND THINK   *Can you think of a better way of finding out how young people plan to live in the future than with the Census data about the year 2000?*

We suspect that you can think of such a way. Perhaps you've thought of doing some kind of questionnaire survey, the nature of which we talk more about in Chapter 9. Perhaps you've also realized that you'd want to survey some kind of representative sample of young people. We talk about questions of gathering such samples in Chapter 5. Perhaps you've thought a little about the kinds of questions you might want to ask. We deal with those kinds of questions in Chapter 6. Perhaps you've even thought about examining what others have had to say about the issue. You might, for instance, be interested in Michele Hoffnung's research, the focal piece of Chapter 4. Maybe someone's already collected just the information you need and would be willing to share it with you. We examine this possibility in Chapter 12, on finding and using available data.

Whatever you've thought, you've clearly begun to engage the question, as we (Emily and Roger) have done with questions we've studied, as a mystery to be solved. Learning about social research is a lot like learning to solve mysteries. It's about challenge, frustration, excitement, and exhilaration. We, Emily and Roger, are addicted to social research. Even as we are writing about research, Emily's working on a study of how people make the transition to retirement. She's passionately interested in this mystery, partly because she's begun the process of retirement herself. Roger's continuing his work on gender in children's picture books. He's been consumed by this topic ever since his children were young and choosing good books for them was personally important to him. The current chapter, however, is about something a little more mundane, perhaps even practical: the reasons why we do research and the purposes to which that research can be put.

Research versus Other Ways of Knowing

Knowledge from Authorities

The new data about households in America—that slightly more of them are made up of people living alone than of people living with a spouse and children—are fascinating. But perhaps they're not as fascinating as what we, Emily and Roger, learned many years ago, literally at our mothers' knees, that "In fourteen hundred and ninety-two, Columbus sailed the ocean blue," despite nearly everyone's belief that he was sailing in a direction that jeopardized his very existence.

Now, why do we think we "know" these things? Basically, it's because some authority told us so. In the first case, we read it in a Census Bureau report. In the second case, we relied on our moms, who were reporting a commonly accepted version of America's "discovery." **Authorities,** like the Census Bureau and our moms, are among the most common sources of knowledge for most of us. Authorities are socially defined sources of knowledge. Many social institutions, like religion, news media, government agencies, and schools, are authorities, and individuals within them are often seen as having superior access to relevant knowledge. In modern societies like ours, we often attribute authority to what we see and hear on the television from, say, newscasters. Sometimes authorities use research as their basis for knowledge but we usually don't evaluate their sources. For most of us, most of the time, learning from authorities is good enough, and it certainly helps keep life simpler than it would be otherwise. Life would be ridiculously difficult and problematic if, for instance, we who live in the Western world had to reinvent a "proper" way of greeting people each time we met them. At some point, we're told about the customs of shaking or slapping hands, or saying "Hi," and we move on from there.

authorities, socially defined sources of knowledge.

STOP AND THINK

What, do you think, are the major disadvantages of receiving our "knowledge" from authorities?

Although life is made simpler by "knowledge" from authorities, sometimes such knowledge is inappropriate, misleading, or downright incorrect. Very few people today take seriously the "flat-earth" theories that have been attributed to Columbus' social world. More interesting, perhaps, is an increasingly accepted view that, our mothers' teachings notwithstanding, very few people in Columbus' social world took the flat-earth view seriously either; that, in fact, this view of the world was wrongly attributed to them by late-nineteenth-century historians to demonstrate how misguided people who accepted religious over scientific authority could be (for example, Gould, 1995: 38–50). In fact, we mean no offense to our moms when we say they represent a whole category of authorities that can mislead: authorities in one area of expertise (in the case of our moms, us) who try to speak authoritatively on subjects in which they are not experts (in the case of our moms, the world view of people in Columbus' society).

Knowledge from Personal Inquiry

But if we can't always trust authorities, like our moms, or even experts, like those late-nineteenth-century historians (and, by extension, even our teachers or the books they assign) for a completely truthful view of the world, who or what can we trust? For some of us, the answer is that we can trust the evidence of our own senses. **Personal inquiry,** or inquiry that employs the senses' evidence for arriving at knowledge, is another common way of knowing.

personal inquiry, inquiry that employs the senses' evidence.

STOP AND THINK *Can you think of any disadvantages in trusting personal inquiry alone as a source of "knowledge"?*

The problem with personal inquiry is that it, like the pronouncements of authorities, can lead to misleading, even false, conclusions. As geometricians like to say, "Seeing is deceiving." This caution is actually as appropriate for students of the social world as it is for students of regular polygons because the evidence of our senses can be distorted. Most of us, for instance, developed our early ideas of what a "family" was by observing our own families closely. By the time we were six or seven, each of us (Emily and Roger) had observed that our own families consisted of two biological parents and their children. As a result, we concluded that all families were made up of two parents and their biological children.[1]

There's obviously nothing wrong with personal inquiry . . . except that it frequently leads to "knowledge" that's pretty half-baked. There are many reasons for this problem. One of them is obvious from our example: Humans tend to <u>overgeneralize</u> from a limited number of cases. Both of us had experienced one type of family and, in a very human way, assumed that what was true of our families was true of all human families. Another barrier to discovering the truth is the human tendency to <u>perceive selectively</u> what we've been conditioned to perceive. Thus, even as 10-year-olds, we might have walked into a commune, with lots of adults and many children, and not entertained the possibility that this group considered itself a family. We just hadn't had the kind of experience that made such an observation possible. A third problem with knowledge from personal inquiry is that it often suffers from <u>premature closure</u>—our tendency to stop searching once we

---

[1] Of course, by the time that Roger, in his early forties, had adopted two children from another country, his ideas of "family" had changed many times.

think we have an answer. At 10, Emily and Roger thought they knew what a family was, so they simply didn't pursue the issue further.

So, neither relying on authorities nor relying on one's own personal inquiry is a foolproof way to the truth. In fact, there might not be such a way. But authors of research methods books (ourselves included) tend to value a way that's made some pretty astounding contributions to the human condition: the scientific method. In the next two subsections, we'll first, discuss some relative strengths of the scientific method for knowledge acquisition and, second, give you some idea about what this knowledge is intended to do.

The Scientific Method and Its Strengths

scientific method, a way of conducting empirical research following rules that specify objectivity, logic, and communication among a community of knowledge seekers, and the connection between research and theory.

Specifying precise procedures that constitute the **scientific method** is a dicey business at best, but we'd like to suggest four steps that are often involved, to a greater or lesser degree, and that suggest the relative emphasis placed on *care* and *community* that distinguishes science from other modes of knowing. An early step is to specify the goals or objectives that distinguish a particular inquiry (here care is paramount). In our first example, we eventually specified the goal of answering the research question "Are more young people planning to marry and have children or to live alone?" A subsequent step involves reviewing a literature, or reading what's been published about a topic (here learning what a relevant community thinks is the goal). We could do worse, in the pursuit of that early research question—about whether young people are planning to marry and have children or live alone, for instance, than read Michele Hoffnung's work (see Chapter 4) on what an earlier generation of "young people" planned to do. At some point, it becomes important to specify what is actually observed (care again). We'd want to define, for instance, what we mean by "young people" and how we plan to measure what their plans are. A later step is to share one's findings with others in a relevant community so that they can scrutinize what's been done (community again). We might want, for instance, to prepare a paper for a conference of family sociologists. These steps or procedures will come up again throughout this book, but we'd now like to stress some of the strengths that accrue to the scientific method because of their use.

The Promotion of Skepticism and Intersubjectivity

One great strength of the scientific method, over modes that rely on authorities and personal inquiry, is that, ideally, it promotes skepticism about its own knowledge claims. Perhaps you looked at our presentation of the data about today's households (e.g., that fewer are made up of couples with children than are made up of people living alone) and said, "Hey. Those facts alone don't tell us much about people's intentions. And they don't even indicate that more people are living alone than living in couples today." If so, we applaud your skepticism. One way in which healthy skepticism is generated is through the communities of knowledge seekers. Each member of

objectivity, the ability to see the world as it really is.

intersubjectivity, agreements about reality that result from comparing the observations of more than one observer.

these communities has a legitimate claim to being a knowledge producer, as long as she or he conforms to other standards of the method (mentioned later). It is sometimes said that one major benefit of having groups of relatively equal knowledge producers working in the same general area is that their joint efforts contribute to the greater **objectivity** of the group. Objectivity refers to the ability to see the world as clearly as possible, free from personal feelings, opinions, or prejudices about what it is or what it should be. We're frankly not completely convinced by the argument that a community of relatively equal knowledge producers necessarily enhances objectivity,[2] but we do think it increases the chances of what is sometimes called **intersubjectivity.** Intersubjectivity refers to agreements about reality that come from the practice of comparing one's results with those of others and discovering that the results are consistent with one another. Equally important, when intersubjective agreement eludes a community of scientists because various members get substantially different results, it's an important clue that knowledge remains elusive and that knowledge claims should be viewed with skepticism. Until the 1980s, for instance, the medical community believed that stomach ulcers were caused by stress. Then a couple of Australian scientists, Barry Marshall and Robin Warren, found, through biopsies, that people with ulcerous stomachs often had the *H. pylori* bacterium lurking nearby and theorized that the bacterium had caused the ulcers. Marshall and Warren thus became skeptical of the medical community's consensus about the causes of ulcers. Few members of that community took them seriously, however, until Marshall, experimenting on himself, swallowed *H. pylori* and developed pre-ulcerous symptoms. Subsequently, Marshall and Warren found that most stomach ulcers could be successfully treated with antibiotics. For their work in discovering the bacterium that causes stomach inflammation, ulcers, and cancer, Marshall and Warren won the 2005 Nobel Prize for Physiology or Medicine (Altman, 2005, D3).

The Extensive Use of Communication

Another related ideal of the scientific method is *adequate communication* within the community of knowledge seekers, implicit in the scientific procedures of referring to previous published accounts in an area and of sharing findings with others. Unlike insights that come through personal inquiry, scientific insights are supposed to be subjected to the scrutiny of the larger community, and therefore need to be as broadly publicized as possible. Communication of scientific findings can be done through oral presentations (as at conferences) or written ones (especially through publication of articles and books). For example, Hoffnung has presented her findings about college students' plans for the future at scientific conferences and in scholarly articles. In Chapter 4, she offers you a glimpse of her most recent

---

[2] After all, just because most Western biologists were creationists (believing that each species was independently created by a supreme being) before Darwin published *The Origin of Species*, doesn't mean that creationism was objectively true.

findings. Increasingly, the computer revolution has provided new media (discussed in Chapter 12) for communication about research and the exchange of data. You, for instance, can use a variety of print and online resources that your college library may make available to you to find references to, and even copies of, research articles about your topic. We did, and found research by Orrange (2003) that supplements Hoffnung's work, and our interests in the kinds of family lives that young people envision for themselves. Once findings are communicated, they then become grist for a critical mill. Others are thereby invited to question (or openly admire) the particular approach that's been reported or to try to reproduce (or *replicate*) the findings using other approaches or other circumstances. Adequate communication thus facilitates the ideal of reaching intersubjective "truths."

Testing Ideas Factually

These communal aspects of the scientific method are complemented by at least three other goals: that "knowledge" be *factually testable,* be *logical,* and be *explicable through scientific theory.* Factual, or empirical, testability means that scientific knowledge, like personal inquiry, must be supported by observation. We began that kind of "observation" in relation to our initial question about plans for future living by gathering some facts about American households from the U.S. Census. But rather than simply using evidence to support a particular view, as we sometimes do in personal inquiry, scientific observation also includes trying to imagine the kinds of observations that would undermine the view and then pursuing those observations. Confronted with the Census data about, say, the increase in the number of households with people living alone, we questioned whether those data were in fact the best data for answering our question and imagined other possible sources (e.g., a questionnaire survey of our own). Similarly, when confronted with the idea that people of Columbus' day held a "flat-earth" view of the world, recent historians have consulted the writings of scientists of that day and earlier and found evidence of a pretty widespread belief that the earth was spherical—surprisingly similar to our beliefs today (Gould, 1995: 38–50). The pursuit of counter-examples, and the parallel belief that one can never fully prove that something is true (but that one can show that something is not true), is a key element of the scientific method.

The Use of Logic

The *Star Trek* character Spock embodies the desirability of logical reasoning in science. Logical reasoning involves making sure that conclusions follow from premises. One particular model of logical reasoning is called the *syllogism,* or reasoning from general propositions to more particular ones. Syllogisms start with one premise that is a statement about a general class of items (for example, "All social scientists are human"). Syllogisms proceed to a premise that is a statement about particular items (for example, "Roger and Emily are social scientists"). Syllogisms then move to conclusions about the particular items, based on the previous two premises (for example,

"Roger and Emily are human"). Spock, like most scientists, approved of syllogistic and other forms of logical reasoning. You used logical reasoning, too, if when confronted with the Census data about the compositions of households in America in 2000, you wondered if those data were adequately related to the question about young people's plans to address our research question about their plans.

Like many scientists, Spock was always quick to point out illogical reasoning. If young Roger or Emily had proposed the notion that all families consisted of two biological parents and their offspring and presented the reasoning—"We belong to such families. Therefore everyone must."—Spock might have responded, "That's illogical." One canon of the scientific method is that one must adhere, as closely as possible, to the rigors of logical thinking. Few practicing scientists invoke the scientific standard of logic as frequently as Spock did, but fewer still would wish to appear as illogical as Roger and Emily were in their reasoning.

The relative strengths of the scientific method, then, derive from several attributes of its practice. Ideally, the method involves communities of relatively equal knowledge seekers (scientists) among whom findings are communicated freely for careful scrutiny. Knowledge claims ideally are subjected to factual tests and to tests of logical reasoning. They're also supposed to be explicable in terms of scientific theory, something we'll discuss more as we pursue some of the major purposes of scientific research, especially in the section on explanatory research.

STOP AND THINK *Suppose I submit a research report to a journal and the journal's editor writes back that s/he can't publish my findings because expert reviewers don't find them persuasive. Which of the strengths of the scientific method is the editor relying on to make his or her judgment?*

The Uses and Purposes of Social Research

By now, you might be saying: "OK, research methods may be useful for finding out about the world for solving mysteries. But aren't there any more practical reasons for learning them?" We think there are.

We hope, for instance, that quite a few of you will go on and apply your knowledge of research methods as part of your professional lives. We know that many of our students have done do so in a variety of ways: as graduate students and professors in the social sciences, as social workers, as police or correctional officers, as analysts in state agencies, as advocates for specific groups or policies, as community organizers, or as family counselors, to name but a few. These students, like ourselves, have tended to direct their research toward two different audiences: toward the scientific community in general, on the one hand, and toward people interested in specific institutions or programs, on the other. When they've engaged the scientific community in general, as Sandra Enos does in the research, reported in Chapter 10, on how women in prison mother their children, they've tended

> ## BOX 1.1
>
> ## Examples of Applied Research
>
> *Here are just a few examples of applied research some of our graduates have done or are planning to do:*
>
> "When I worked for the Department of Children and Youth and their families, we conducted a survey of foster parents to see what they thought of foster care and the agency's services. The parents' responses provided us with very useful information about the needs of foster families, their intentions for the future, and the kinds of agency support that they felt would be appropriate."
> *Graduate employed by a state Department of Children, Youth and Their Families*
>
> "As the Department of Corrections was under a court order because of crowding and prison conditions, it was important that we plan for the future. We needed to project inmate populations and did so using a variety of data sources and existing statistics. In fact, we were accurate in our projections."
> *Graduate employed by a state Department of Corrections*
>
> "I'm working at a literacy program designed to help children in poverty by providing books for the preschoolers and information and support for their parents. I've realized that while the staff all think this is a great program, we've never really determined how effective it is. It would be wonderful if we could see how well the program is working and what we could do to make it even better. I plan on talking to the director about the possibility of doing evaluation research."
> *Graduate employed by a private pediatric early literacy program*

basic research, research designed to add to our fundamental understanding and knowledge of the social world regardless of practical or immediate implications.

applied research, research intended to be useful in the immediate future and to suggest action or increase effectiveness in some area.

to engage in what is sometimes called **basic research.** Enos reports to other scientists, for instance, that, while incarcerated, white women tend to place their children in the care of their husbands or the state (e.g., foster care), and African American women tend to place their children in the care of their own mothers. Basic research is designed to add to our knowledge and understanding of the social world for the sake of that knowledge and understanding. Much of the focal research in this book, including both Enos' work on women in prison (Chapter 10) and Hoffnung's work on college students' plans (and achievements) about their futures (Chapter 4), is basic research.

When our former students have addressed a clientele with interests in particular institutions or programs, like the students quoted in Box 1.1, they've tended to engage in what is called applied research. **Applied research** aims to have practical results and produce work that is intended to be useful in the immediate future. Schools, legislatures, government and social service agencies, health care institutions, corporations, and the like all have specific purposes and ways of "doing business." Applied research, including evaluation research and action-oriented research, is designed to provide information that is immediately useful to those participating in institutions or programs. Such research can be done for or with organizations and communities, and can include a focus on the action implications of the research. Evaluation research, for example, can be designed to assess the impact of a specific program, policy, or legal change. It often focuses on whether a program or policy has succeeded in effecting intentional or planned changes. Participatory action research, which we'll discuss in Chapter 14, is done jointly by researchers and community members. It often has an emancipatory purpose. Participatory action research and related research strategies are the cornerstone of what Feagin and Vera call "liberation sociology" and social science, a social science that focuses on "alleviating or eliminating various social oppressions" and on "creating societies that are more just and egalitarian" (Feagin and Vera, 2001, 1). You'll find an example of applied research in Chapter 4, where Wysong, Aniskiewicz, and Wright evaluate the long-term effects of a school-based drug education program (DARE) on adolescent drug use.

Even if you don't enter a profession in which you'll do research of the sort we discuss in this book, we still think learning something about research methods can be one of the most useful things you do in college. Why? Oddly, perhaps, our answer implicates another apparently esoteric subject: social theory.

social theory, an explanation about how and why people behave and interact in the ways that they do.

When we speak of **social theory,** we're not only referring to the kinds of things you study in specialized social theory courses, although we do include those things. In our view, social theories, like all other theories, are explanations about how and why things are as they are. In the case of social theories, the explanations are about why people "behave, interact, and organize themselves in certain ways" (Turner, 1991: 1). Such explanations are useful, we feel, not only because they affect how we act as citizens—as when, for instance, we inform, or fail to inform, elected representatives of our feelings about matters like welfare, joblessness, crime, and domestic violence—but also because we believe that Charles Lemert (1993: 1) is right when he argues that "social theory is a basic survival skill." Individuals survive in society to the extent that they can say plausible and coherent things about that society.

Useful social theory, in our view, concerns itself with those things in our everyday lives that can and do affect us profoundly, even if we are not aware of them. We believe that once we can name and create explanations (or create theories) about these things, we have that much more control over them. At the very least, the inability to name and create such explanations leaves us powerless to do anything. These explanations can be about why some people live alone and some don't, why some are homeless and some aren't, why some commit crimes and some don't, why some do housework and others don't, why some people live to be adults and others don't. These explanations can come from people who are paid to produce them, like social scientists, or from people who are simply trying to make sense of their lives. Lemert (1993) reminds us that the title for Alex Kotlowitz's (1991) *There Are No Children Here* was first uttered by the mother of a 10-year-old boy, Lafeyette, who lived in one of Chicago's most dangerous public housing projects. This mother observed, "But you know, there are no children here. They've seen too much to be children" (Kotlowitz, 1991: 10). Hers is eloquent social theory, with serious survival implications for those living in a social world where nighttime gunfire is commonplace.

We'll have more to say about social theory and its connection to research methods in the next chapter. But, for now, we'll simply say that we believe the most significant value of knowledge of research methods is that it permits a critical evaluation of what others tell us when we and others develop social theory. This critical capacity should, among other things, enable us to interpret and use the research findings produced by others. Our simple answer, then, to the question about the value of a research methods course is not that it adds to your stock of knowledge about the world, but that it adds to your knowledge of *how* you know the things you know, *how* others know what they know, and, ultimately, *how* this knowledge can be used to construct and evaluate the theories by which we live our lives.

The major purposes of scientific research, in many ways overlapping with the "uses" (of, say, supplying other scientists with basic information, supplying interested persons with information about programs, or developing theories) we've mentioned, include *exploration, description,* and *explanation.* Although any research project can have more than one purpose, let's look at the purposes individually.

Exploratory Research

exploratory research, ground-breaking research on a relatively unstudied topic or in a new area.

In research with an **exploratory** purpose, the investigator works on a relatively unstudied topic or in a new area, to become familiar with this area, to develop some general ideas about it, and perhaps even to generate some theoretical perspectives on it. Exploratory research is almost always inductive in nature, as the researcher starts with observations about the subject and tries to develop tentative generalizations about it (see Chapter 2).

An example of exploratory research is the Adlers' study of children's adult-organized afterschool activities, presented in Chapter 11. The Adlers began by noting that such activities are a relatively new phenomenon: Until the 1970s, most kids would come home from school and engage in unorganized play with other kids, usually from their own neighborhood. The Adlers wondered whether they'd be able to spot patterns in the kinds of adult-organized activities for children that have subsequently emerged. For about six years they observed and interacted with children, including their own, both inside and outside their schools. They noticed that children who are involved in adult-organized afterschool activities typically experience "a passage through an 'extracurricular career' that begins with a recreational ambiance but progresses into competitive and finally elite activities as they grow older and become more skilled" (Adler and Adler, 1994a: 309). They derive a theory of afterschool activities that, among other things, suggests that these activities socialize young people to the work values of their surrounding society by eventually focusing their leisure activities into almost highly specialized ways. Similarly, Orrange (2002) felt that little was known of how today's pre-professional women and men were planning their work and family lives, so he did 43 in-depth interviews of advanced professional students in business and law. He found that the men and women he interviewed had different orientations toward family life. Almost all men saw families in their futures, with themselves being either primary providers or egalitarian providers. He found that women foresaw themselves as being either egalitarian providers in families or as being single. Both the Adlers and Orrange, then, did exploratory research: research into a relatively new subject (children's after-school activities and the family and work plans of pre-professional graduate students), collected data through observations or in-depth interviews of a relatively few cases, and tried to spot themes that emerged from their data. Although exploratory analyses, with their focus on relatively unexplored areas of research, do not always employ this kind of thematic analysis of data on relatively few cases, when they do they involve what is called **qualitative data analysis,** or analysis that tends to involve the interpretation of actions or the representations of meanings in words (see Chapter 15).

qualitative data analysis, analysis that results in the interpretation of action or representation of meanings in the researcher's own words.

descriptive study, research designed to describe groups, activities, situations, or events.

Descriptive Research

In a **descriptive study,** a researcher describes groups, activities, situations, or events, with a focus on structure, attitudes, or behavior. Researchers who do descriptive studies typically know something about the topic under study before they collect their data, so the intended outcome is a relatively accurate and precise picture. Examples of descriptive studies include the kinds of polls done during political election campaigns, which are intended to describe how voters intend to vote, and the U.S. Census, which is designed to describe the U.S. population on a variety of characteristics. The Census Bureau description of the makeup of households in the United States (2005) is just that: a description, in this case of the whole U.S. population. As the pie chart in Figure 1.1 indicates, about 25.8 percent of American households in 2000 were made up of people living alone. About 21.2 percent were made up of people living with a spouse, and so forth. Unlike exploratory studies, which might help readers become familiar with a new topic, descriptive studies can provide a very detailed and precise idea of the way things are. They can also provide a sense of how things have changed over time. Thus, a study by Clark, Folgo, and Pichette (2005), of the visibility of women artists in art history books took as its starting point an assertion, by White (1973), that no women artists had been included in college art texts prior to 1973. Clark et al. found that this assertion wasn't quite true, that there had been in fact two women artists mentioned in one art history book published before 1974, but that it was essentially true: the ratio of women artists to men artists mentioned in those early texts was about 0.2:100. Clark et al. found that by 2004 the ratio of women artists to men artists in existing art history texts had substantially increased—to about 10.1:100—but that it still showed men artists being much more visible than women artists.

Descriptive research, in its search for a picture of how the land lies, can be based upon data from surveys (as the Census report is), or on content analysis (as Clark et al.'s report is). In fact, it can be based upon most of the kinds of data-gathering techniques you'll learn about in this text, except perhaps experimental techniques, which are generally focused on explanatory research. Often, as in the case of the Census Bureau's study of American households, descriptive research generates data about a large number of cases: there were actually 105,480,101 households enumerated by the census of 2000. To analyze these data meaningfully, descriptive researchers frequently use **quantitative data analysis,** or analysis that is based on the statistical summary of data (see Chapter 15).

quantitative data analysis, analysis based on the statistical summary of data.

Explanatory Research

explanatory research, research designed to explain why subjects vary in one way or another.

Unlike descriptive research, which tends to focus on how things are, the goal of **explanatory research** is to explain why things are the way they are. Explanatory research looks for causes and reasons. Unlike exploratory research, which tends to be inductive, building theoretical perspectives from data, explanatory research tends to be deductive, moving from more general to less general statements. Thus, for instance, explanatory research often uses preexisting theories to decide what kinds of data should be collected (see Chapter 2). In a study that we present in Chapter 9, researchers Norma Gray, Gloria Palileo, and David Johnson ask themselves the explanatory question: Why are some people more likely than others to blame the victims of rape for the rape? At the individual level, the researchers knew, the tendency to blame rape victims might be explained by having personally known someone who had been raped, having personally experienced rape (or rape-like) assaults, having committed rape (or rape-like) assaults, and so on. But these researchers also believed that such intimate experience with rape situations couldn't possibly explain all the variation in people's willingness to attribute blame to rape victims. Even people who hadn't had such experiences, they knew from previous research, would vary in how much they attributed blame to victims.

Gray and her associates did what many social scientists do: They consulted a theory, or an explanation of how and why things are the way they are. Gray and her associates consulted a theory called "attribution theory," one that tries to describe why people attribute any characteristics to other people.

In Chapter 2, you will read about attribution theory in the researchers' own words, but for now we'll just mention a key element of the theory: It expects an observer, when deciding whether to attribute responsibility to the victim of any event, to compare his or her own social characteristics to those of the victim. The theory suggests that an observer is less likely to attribute responsibility to the victim if the observer's own social characteristics are similar to the victim's than if they are not. From attribution theory, Gray, Palileo, and Johnson derived an expectation about the way that gender and the willingness to blame rape victims go together. After surveying students in college classes, the researchers used statistical analysis of data from a large number of cases to find that the women students in their study attributed responsibility to rape victims less often than the men did. In their work, Gray, Palileo, and Johnson describe something (rape blame attribution) that "goes with" something else (gender) and explain this connection through a theory (attribution theory). In doing so, they explain their knowledge claim through scientific theory, one of the ideals, you will recall, of the scientific method.

Explanatory analyses, with their focus on areas upon which a researcher might be able to shed theoretical light in advance of collecting data, may, like descriptive analyses, generate data about relatively large numbers of cases and employ statistical analyses to make sense of these cases. When they do, they, like many descriptive analyses, involve quantitative data analysis, or analysis based on the statistical summary of data (see Chapter 15).

STOP AND THINK   *Suppose you've been asked to learn something about the new kinds of communities that have arisen out of young people's use of instant messaging. Of the three kinds of research outlined above (exploratory, descriptive, explanatory), what kind of study have you been asked to do?*

Summary

We've argued here that, at its best, research is like trying to solve a mystery, using the scientific method as your guide. We've distinguished the scientific approach of social research methods from two other approaches to knowledge about the social world: a reliance on the word of "authorities" and an exclusive dependence on "personal inquiry." We've suggested that the scientific method compensates for the shortcomings of these two other approaches in several ways. First, science emphasizes the value of communities of relatively equal knowledge seekers who are expected to be critical of one another's work. Next, science stresses the simultaneous importance of empirical testing, logical reasoning, and the development or testing of theories that make sense of the social world. Two communities that researchers report to are the community of other scientists (basic research) and communities of people interested in particular institutions or programs (applied research). We've argued that knowing research methods may have both professional and other practical benefits, not the least of which is the creation of usable theories about our social world. We suggest that social research methods can help us explore, describe, and explain aspects of the social world.

EXERCISE 1.1

Ways of Knowing About Social Behavior

This exercise compares our recollections of our everyday world with what we find out when we start with a research question and collect data.

Part 1: Our Everyday Ways of Knowing

Pick any *two* of the following questions, and answer them based on your recollections of past behavior.

1. What does the "typical" student wear as footwear to class? (Will the majority wear shoes, boots, athletic shoes, sandals, and so on?)

2. While eating in a school cafeteria, do most people sit alone or in groups?

3. Of those sitting in groups in a cafeteria, are most of the groups composed of people of the same gender or are most mixed-gender groups?

4. Of your professors this semester who have regularly scheduled office hours, how many of them will be in their offices and available to meet with you during their next scheduled office hour?

Based on your recollection of prior personal inquiry, describe your expectations of social behavior.

Part 2: Collecting Data Based on a Research Question

Use the two questions you picked for Part 1 as "research questions." With these questions in mind, collect data by carefully making observations that you think are appropriate. Then answer the same two questions, but this time base your answers on the observations you made.

Part 3: Comparing the Ways of Knowing

Write a paragraph comparing the two ways of knowing you used. (For example, was there any difference in accuracy? In ease of collecting data? Which method do you have more confidence in?)

EXERCISE 1.2

Social Science as a Community Endeavor

This exercise is meant to reinforce your appreciation of how important the notion of a community of relatively equal knowledge seekers is to social research. We'd like you to read any one of the "focal research" articles in subsequent chapters and simply analyze that article as a "conversation" between the author and one or two authors who have gone before. In particular, summarize the major point of the article, as you see it, and see how the author uses that point to criticize, support, or amend a point made by some other author in the past.

1. What is the name of the focal research article you chose to read?

2. Who is (are) the author(s) of this article?

3. What is the main point of this article?

4. Does the author seem to be criticizing, supporting, or amending the ideas of any previous authors to whom she or he refers?

5. If so, which author or authors is/are being criticized, supported, or amended in the focal research?

6. Write the title of one of the articles or books with which the author of the focal research seems to be engaged.

7. Describe how you figured out your answer to the previous question.

8. What, according to the author of the focal research, is the central idea of this book or article?

9. Does the author of the focal research finally take a critical or supportive position in relation to this idea?

10. Explain your previous answer.

EXERCISE 1.3

Ways of Knowing About the Weather

This exercise is designed to compare three ways of knowing about the weather.

Part 1:

Knowledge from Authorities.   The night before: See what the experts have to say about the weather for tomorrow by watching a TV report or listening to a radio newscast. Write what the experts said about tomorrow's weather (including the temperature, the chances of precipitation, and the amount of wind).

Knowledge from Casual Personal Inquiry.   That day, before you go outside, look through only one window for a few seconds but don't look at a thermometer. After taking a quick glance, turn away, and then write down your perceptions of the weather outside (including the temperature, the amount and kind of precipitation, and the amount of wind).

Knowledge from Research.   Although we're not asking you to approximate the entire scientific method (such as reviewing the literature and sharing your findings with others in a research community), you can use some aspects of the method: specifying the goals of your inquiry and making and recording careful observations.

Your research question is "What is the weather like outside?" To answer the question, use a method of collecting data (detailed observation of the outside environment) and any tools at your disposal (thermometer, barometer, and so on). Go outside for at least five minutes and make observations. Then come inside and write down your observations of the weather outside (including the temperature, the amount and kind of precipitation, and the amount of wind).

Part 2: Comparing the Methods

Write a paragraph comparing the information you obtained using each of the ways of knowing. (For example, was there any difference in accuracy? In ease of collecting data? Which method do you have the most confidence in?)

2

© Emily Stier Adler

Theory and Research

19

Introduction

Concepts, Variables, and Hypotheses

We imagine that you've given some thought as to whether abortion is moral or not and therefore whether it should be legal or not. But have you ever considered the ways in which legalized abortion might be associated with legal (or illegal) behaviors of other kinds? In this chapter, on the relationship between theory and research, we will begin with an example that forces us to think about such an association. Let's examine Box 2.1, a summary of a theory by John Donohue and Steven Levitt, that connects the legalization of abortion in the early 1970s and a decrease in crime rates in the U.S. in the early 1990s. You'll recall from Chapter 1 that a **theory** is an explanation of the "hows" and "whys" of social life.

Donohue and Levitt's theory about the connection between abortion legalization and crime rates, like all theories, is formulated in terms of **concepts,** which are words or signs that refer to phenomena that share common characteristics. Some of these concepts refer to phenomena that are, at least for the purposes of the theory, relatively fixed or invariable in nature. Donohue and Levitt, for example, refer to the "high-crime late adolescent years" and "birth cohort." These concepts require clarification for the purposes of research, or what researchers call **conceptualization,** but, once defined, they do not vary. For Donohue and Levitt, the high-crime late adolescent years begin at about 17 years of age, and a birth cohort is the people born in a calendar year. The first birth cohort in the U.S. that could have been affected by the 1973 *Roe v. Wade* decision would have been born in 1974, therefore Donohue and Levitt expect that there should have been a diminution in this cohort's crime rate in comparison with previous cohorts 17 years later, or about 1991.

theory, an explanation about how and why something is as it is.

concepts, words or signs that refer to phenomena that share common characteristics.

conceptualization, the process of clarifying what we mean by a concept.

STOP AND THINK *Do you see how the concept of a "high-crime late adolescent years" is relatively fixed? How its definition does not admit of much variation? (It begins at about 17 and ends, one guesses, at about 19 years of age.) Can you see any concepts used by Donohue and Levitt that could vary from one person or birth cohort to another?*

In addition to using concepts that identify phenomena that are relatively fixed in nature, that don't vary for the purposes of the research, like "high-crime late adolescent years," Donohue and Levitt use concepts such as "nurturing home environment" and "criminal activity" that actually might vary from one person to another or from one birth cohort to another. The individual person might experience a nurturing home or not and might engage in criminal activity later in life, or not. A higher percentage of the children in one birth cohort, for instance, might experience a nurturing home environment than the children in another birth cohort. A higher percentage of the people in one birth cohort might engage in criminal activity when they are 17 years of age than people in another birth cohort. A concept that varies is called a **variable.**

variable, a characteristic that may vary from one subject to another or for one subject over time. A concept that varies.

> **BOX 2.1**
>
> ## A Theoretical Statement: Why the Legalization of Abortion Contributed to the Decline of Crime Rates
>
> In 2001, John Donohue and Steven Levitt published a highly controversial research report, "The Impact of Legalized Abortion on Crime," in which they claimed to show that the introduction of legalized abortion in the U.S. in the early 1970s led to a drop in crime in the early 1990s. The key theoretical argument in Donohue and Levitt's (2001) paper was that abortion "is more frequent among parents who are least willing or able to provide a nurturing home environment" (386). Moreover, children from less nurturing home environments are more likely than others to engage in criminal activity, once they reach the "high-crime late adolescent years" (386). Therefore, to the extent that the legalization of abortion in the early 1970s enabled parents who would provide the least nurturing home environments to avoid having children, it also led to a lower crime rate when the affected birth cohort reached its late adolescent years.

STOP AND THINK   *The lowest possible value for the percentage of a birth cohort that has experienced a nurturing home environment is zero percent. What's the highest possible value? What are the lowest and highest possible values for the percentage of a birth cohort that engage in criminal activity when they are 17 years old?*

Another way to think of a variable is as a characteristic that can vary from one subject to another or for one subject over time. To be a variable, a concept must have at least two categories. For the variable "gender," for instance, we can use the categories "female" and "male." Gender is most frequently used to describe individual people. The variable of "criminal activity" used by Donohue and Levitt, however, can be meant, in the theoretical formulation above, to describe a birth cohort. However, getting information about a birth cohort is very difficult, so Donohue and Levitt use substitutes for birth cohort, focusing instead on the crime rates for the whole U.S. population in different years. They end up measuring the concept of "criminal activity" in a variety of ways. One of these ways is as a characteristic of the whole nation and in terms of the number of crimes committed in a year per 100,000 people (see Table 2.1). This variable could have many more than two, perhaps even an infinite number, of categories. Scientists can use variables to describe the social world, just as you might use the variable "gender" to describe students in your research methods class.

STOP AND THINK   *Notice that the crime rate is a variable characteristic of the United States because it can vary over time. Notice also that this variable could take on many, rather than just two, values. (Actually, in principle it could vary from zero to infinity. Right?)*

TABLE 2.1   **What Happened to the Crime Rate in the United States Between 1985 and 2002**

The World Almanac (2005: 162) reports the following information about the crime rate in the United States.

Year	Total Crime Rate (Number of Crimes per 100,000 people)
1985	5,207.1
1986	5,480.4
1987	5,550.0
1988	5,664.2
1989	5,741.0
1990	5,820.3
1991	5,897.8
1992	5,660.2
1993	5,484.4
1994	5,373.9
1995	5,275.9
1996	5,086.6
1997	4,930.0
1998	4,615.5
1999	4,266.5
2000	4,124.8
2001	4,162.6
2002	4,118.8

What pattern in Table 2.1 do you spot in the U.S. crime rate between 1985 and 2002? In what year did a reversal of the previous pattern of increases begin? When was the first cohort affected by Roe v. Wade *born? (1974) How many years was this before the downturn in the national crime rate? How might this be significant for the thesis advanced by Donohue and Levitt?*

When we think of "crime activity" as the crime rate of the country, then, we see that it does decline at about the time that Donohue and Levitt's theory predicts: about 17 years (in 1991) after *Roe v. Wade* would have had its initial effects (in 1974). This illustrates a frequent use of theory for research: to point to variable characteristics of subjects that can be expected to be associated with each other. Thus, Donohue and Levitt are claiming that, among other things, they expect that a downturn in the crime rate should occur in jurisdictions where abortion became legal sometime earlier. Similarly, Gray, Palileo, and Johnson, in the research mentioned in Chapter 1 and in the next section of this chapter, tell us that the female students in their sample were less likely to attribute responsibility to rape victims than were the male students. Using the categories of two variables, "gender" and "rape blaming behavior," they found that "females" were more likely to do "less blaming of rape victims," and "males" were more likely to do "more blaming of rape victims."

The finding supported an expectation that Gray and her associates had derived from attribution theory about the way the two variables would be associated. They stated this expectation in the form of a **hypothesis.** A hypothesis is a statement about how two or more variables are expected to relate to one another. The hypothesis that Gray and her associates developed is this:

> Females will attribute less rape victim responsibility than will males (Gray, Palileo, and Johnson, Chapter 9).

Note that this hypothesis links two variable characteristics of people: gender and the attribution of responsibility to rape victims. You might also note that, like many other social science hypotheses, this one doesn't refer to lead-pipe certainties (such as "If a person is a female, she will never attribute responsibility to rape victims, and if a person is male, he always will."). Rather, this hypothesis speaks of tendencies.

The business of research, you've probably surmised, can be complex and can involve some fairly sophisticated ideas (theory, hypotheses, concepts, variables—to name but four), all of which we'll discuss more later. Meanwhile, we'd like to introduce you to one more fairly sophisticated idea before moving on to a more general consideration of the relationship between research and theory. This is the idea that explanations involve, at minimum, the notion that change in one variable can affect or influence change in another. Not surprisingly, scientists have special names for the "affecting" and "affected" variables. Variables that are affected by change in other variables are called **dependent variables**—dependent because they depend on change in the first variable. **Independent variables,** on the other hand, are variables that affect change in dependent variables. Thus, when Donohue and Levitt hypothesize that making abortion legal leads to declines in the crime rate, they are implying that a change in the independent variable (whether or not abortions are legal) will cause change in the dependent variable (the crime rate). And, when Gray and her associates hypothesize that females are less likely than males to attribute responsibility to rape victims, they are suggesting that a change in an independent variable (gender) will cause change in a dependent variable (attribution of responsibility). Having an arrow stand for something like "is associated with," hypotheses can be depicted using diagrams like the one in Figure 2.1. The arrows in this diagram suggest that being female "is associated with" being less likely to blame the victim, while being male "is associated with" being more likely to blame the victim. And, in general, when one or more categories of one variable are expected to be associated with one or more categories of another variable, the two variables are expected to be associated with each other.

hypothesis, a testable statement about how two or more variables are expected to be related to one another.

dependent variable, a variable that a researcher sees as being affected or influenced by another variable (contrast with independent variable).

independent variable, a variable that a researcher sees as affecting or influencing another variable (contrast with dependent variable).

FIGURE 2.1

A Diagram of the Hypothe-
sized Relationship Between
Gender and Attribution of
Responsibility

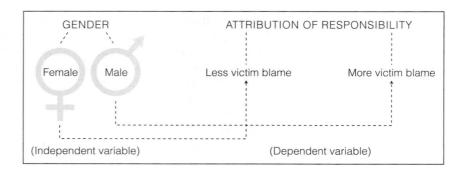

STOP AND THINK   *Can you draw a diagram of Donohue and Levitt's hypothesis linking the legalization of abortion to declines in criminal activity? (Hint: It might be easiest to think in terms of individuals, not cohorts or countries. What kind of person, according to Donohue and Levitt's thesis, is more likely to engage in criminal activity: someone born when abortion is legal or someone who is not?)*

We want to assure you that at this point we don't necessarily expect you to feel altogether comfortable with all of the ideas we've introduced so far. They'll all come up again in this book and, by the end, we expect you'll feel quite comfortable with them. For now, we want to emphasize a major function of theory for research: to help lead us to hypotheses about the relationship between or among variables. We'd like soon to turn to a more general consideration of the relationship between social research and social theory. But first let us say a little more, by way of a word of caution, about the difficulty of establishing causality in social science.

Social Science and Causality: A Word of Caution

The fact that two variables are associated with each other doesn't necessarily mean that change in one variable causes change in another variable. Table 2.1 shows, generally, what Donohue and Levitt, in their much more sophisticated analysis (Donohue and Levitt, 2001), contend: When abortion rates (the independent variable) go up, societies are likely to experience a decline in crime rates later (the dependent variable), when the more "wanted" generation grows up. But does this mean that making abortion legal causes lower crime rates? Donohue and Levitt suggest that it does. Others argue that it doesn't.

In the social sciences, we have difficulty establishing causality for several reasons. Some reasons are purely technical, and we'll discuss them in greater detail in Chapter 8. One reason establishing causality is difficult is that to show that change in one variable causes change in another, we want to be sure that the "cause" comes before, or at least not after, the "effect." But many of our most cherished methods of collecting information in the social sciences just don't permit us to be sure which variable comes first. Cramer, Gallant, and Langlois (2005), for instance, found that students who gave indications, in a questionnaire survey, of being inclined toward silence, rather than expressiveness, were more likely than others to give indications of psychological depression. The association of these two variables, however, doesn't prove that self-silencing causes depression because, given the nature of questionnaire surveys, Cramer et al. had to collect information about the two variables at the same time. They couldn't tell whether self-silencing came before depression or depression came before self-silencing. It would seem that Donohue and Levitt's research at least passes the temporal-sequencing test: that the "cause" (abortions, subsequent to *Roe v. Wade*) comes before the "effect" (a decline in crime rates in the 1990s). In this case, however, there can remain doubt about whether the "right" time "before" has been specified. Donohue and Levitt, in their sophisticated statistical analysis, make very plausible guesses about appropriate time "lags," or periods over which abortion will have its greatest effects on crime, but even they can't be sure that their "lags" are exactly right.

STOP AND THINK *Donohue and Levitt (2001: 382) think there was something pretty significant about the 17 years between when* Roe v. Wade *would have affected its first birth cohort (1974) and the downturn, beginning in 1991, in the overall crime rate in the United States. Can you see any reasons why 17 years might be a significant length of time? Can you imagine, assuming that abortion has any effect on crime, why one might expect a shorter or longer period of time between the legalization of abortion and the onset of significant declines?*

Another technical problem with establishing causality is that even if we can hypothesize the correct causes of something, we usually can't demonstrate that another factor isn't the reason why the "cause" and "effect" are associated. We might be able to show, for instance, that members of street gangs are more likely to come from single-parent households than non-members. However, because we're not likely to want or be able to create street gangs in a laboratory, we can't test whether other things (like family poverty) might create the association between "family structure" (an independent variable) and "street gang membership" (a dependent variable). Similarly, it's possible that some "other thing" (like swings in the economic cycle) might create conditions that make "abortion rates" (people may be more likely to have more abortions during economic downturns) and "subsequent crime rates" (people may be less likely to commit crimes during economic upturns) go together. Doing experiments (discussed in Chapter 8) is most useful in demonstrating causality, but this strategy frequently doesn't lend itself well (for ethical and practical reasons, discussed in the next two chapters) to social science investigations.

antecedent variable, a variable that comes before both an independent variable and a dependent variable.

The idea of a third variable, sometimes called an **antecedent variable,** that comes before, and is actually responsible for, the association between an independent variable and a dependent variable, is so important that we'd like to give you another example to help you remember it. Firefighters themselves will tell you of the association between the number of firefighters at a fire (an independent variable) and the damage done at the fire (a dependent variable): that the more firefighters at a fire, the more damage occurs.

Can you think of an antecedent variable that explains why fires that draw more firefighters are more likely to do more damage than fires that draw fewer firefighters?

Perhaps it occurred to you that a characteristic of fires that would account both for the number of firefighters drawn to them and the amount of damage done that occurs is the size of the fire. Smaller fires tend to draw fewer firefighters and do less damage than larger fires. In this case, the antecedent variable, size of fire, explains why the independent variable and the dependent variable are associated. When this happens, when an antecedent variable provides such an explanation, the original association between the independent variable and the dependent variable is said to be **spurious,** or non-causal. Neither one causes the other; their association is due to the presence of an antecedent variable that creates the association. It is only when the association between an independent variable and a dependent variable is non-spurious (not generated by some third variable acting on the two) that we can conclude that it is causal. But this is hard to demonstrate, because it means taking into account *all possible* antecedent variables, of which there are an infinite number.

spurious, non-causal.

The three conditions that must exist before we can say an independent variable "causes" and dependent variable are, then:

(a) that the two variables are associated in fact. In the case of the fires, this would mean showing that fires that drew many firefighters were associated with more damage than other fires. This is the condition of *empirical association.*

(b) the independent variable, in fact, precedes, or at least doesn't come after, the dependent variable. In the case of the fires, this would mean showing that fire damage never came before the arrival of the firefighters. This is the condition of *temporal precedence* or *time order.*

(c) there is no third variable, antecedent to the independent variable and the dependent variable, that is responsible for their association. In the case of the fires, this would mean showing that there is no variable, like the size of the fire, that led to the association between the number of firefighters at the fire and the damage done at the fire. This is the condition of *elimination of alternative explanations* (or *demonstrating non-spuriousness*).

Most social science research designs (as you'll see in Chapter 7 and 8) make it difficult to establish the condition of temporal precedence and make it impossible to establish the condition of non-spuriousness, so we rarely feel sure enough to say, for sure, that one thing causes another. In fact, only the experimental design is meant to do so. We often settle for establishing the first condition, the condition of empirical association, knowing that it is a necessary condition for, but not a completely satisfactory demonstration of, causality.

STOP AND THINK *Think of what you know about the association that interests Donohue and Levitt: the one between the legality of abortion and criminal activity. Which condition(s) for establishing causality has/have been at least plausibly met for you? Which condition(s) has/have not been at least plausibly met?*

The Relationship Between Theory and Research

If social theories are explanations of the way people act, interact, or organize themselves, then good social theories are those that enable people to articulate and understand something about everyday features of social life that had previously been hidden from their notice. We begin this section with two examples of social theories used by the social-scientist authors of focal research pieces that we use later in this text. Although the complete articles are included in later chapters, here we will excerpt the authors' descriptions of a theory they've employed or developed through their research. The first excerpt is from Norma Gray, Gloria Palileo, and G. David Johnson's essay (see Chapter 9) entitled "Explaining Rape Victim Blame: A Test of Attribution Theory."

STOP AND THINK *As you read each excerpt, see if you can guess whether it appears before or after the authors' presentation of their own research.*

FOCAL RESEARCH # Excerpt from "Explaining Rape Victim Blame": Attribution Theory

by Norma B. Gray, Gloria J. Palileo, and G. David Johnson

We test hypotheses derived from attribution theory . . . in order to explain variations in rape myth acceptance. A basic assumption of attribution theory is that individuals will attribute causality for observed behavior, i.e., they will make decisions about who or what is responsible for behavior they observe. Individuals will attribute causality and responsibility either to the actor's "personal disposition" or to external (e.g., other persons, the environment, chance) causes. . . . The attributions an individual makes about an actor's responsibility for his or her own behavior are influenced by (1) the social location of the observer and the observed (e.g., their class position, gender, and race); (2) the observer's previous experience (particularly in related activities); (3) the setting; and (4) the potential psychological benefit to be derived from alternative attributions (Shaver, 1975). . . .

[In particular] [w]e examine the association between the social characteristics of the observers and the attribution of responsibility from the . . .

perspective . . . of the "defensive attribution" (Shaver, 1975) . . . model . . . The defensive attribution model states that the greater the similarity in social characteristics or experiences between observer and observed [victim] (e.g., if they are the same gender and/or the same race), the greater the likelihood that the observer will attribute responsibility to someone or something other than the observed victim. This attribution is motivated by the need for the observer to protect his or her self-esteem and to avoid self-blame if he or she should be similarly victimized in the future.

REFERENCE

Shaver, Kelly. 1975. *An introduction to attribution processes.* Cambridge, MA: Winthrop.

Our second excerpt is from Patricia and Peter Adler's "The Institutionalization of Afterschool Activities" (see Chapter 11 for full article).

FOCAL RESEARCH

Excerpt from "The Institutionalization of Afterschool Activities": Functions and Dysfunctions

by Patricia A. Adler and Peter Adler

In considering the benefits or harm generated by the infusion of adult control and the associated adult-oriented structures and values into the leisure of children and youth, we see elements of both. Our observations suggest that in progressing through the stages of afterschool activities, children learn several important norms and values about the nature of adult society. They discover the importance placed by adults on rules, regulations, and order. Creativity is encouraged, but acknowledged within the boundaries of certain well-defined parameters. Obedience, discipline, sacrifice, seriousness, and focused attention are valued; deviance, dabbling, and self-indulgence are not. Coordination with others, the organic model of working toward challenging and complex goals, stands as the ultimate model toward which young people are directed. . . .

At the same time, the earlier imposition of adult norms and values onto childhood may rob children of developmentally valuable play, unchannelled and pursued for merely expressive rather than instrumental purposes. By participating in increasing amounts of adult-organized activities, children are steered away from goal-setting, negotiation, improvisation, and self-reliance toward acceptance of adult authority and adult pre-set goals. While these afterschool activities may prepare children for the formally rational, hierarchical, and, in Foucault's (1977) words, disciplined adult world, children's spontaneous play may teach different but important social lessons. Adult-organized activities may prepare children for passively accepting the adult world as given; the activities that children organize themselves may prepare them for creatively constructing alternative worlds.

REFERENCE

Foucault, Michel. 1977. *Discipline and punishment.* New York: Pantheon.

Deductive Reasoning

Gray, Palileo, and Johnson's description of "attribution theory" comes at the beginning of their article, before they present their own research; the Adlers' theory of what might be called the "functions and dysfunctions" of adult-organized afterschool activities for children comes at the end of their article, after the presentation of their findings. Note, however, that both theories attempt to shed light on what might have been a previously hidden aspect of social life. They do so with concepts (such as "social characteristics" and "self-blame" or "obedience" and "self-reliance") that are linked, explicitly or implicitly, with one another. These concepts are neither so personal as the one espoused by Lafeyette's mother at the beginning of *There Are No Children Here* (remember the boy mentioned in Chapter 1?) nor as grandiose as some theories you might have come across in other social science courses. But they are theories, nonetheless.

What distinguishes these theories, and many others presented by scientists, from those presented by pseudoscientists (astrologers and the like) is that they are offered in a way that invites skepticism.[1] Neither Gray, Palileo, and Johnson nor the Adlers are 100 percent sure they've got it right. Nor would they want a theory that didn't "fit" with the observable world. For both sets of researchers, a major value of research methods is that they offer a way to test or build social theories that fit the observable social world. We'll spend the rest of this chapter discussing how research can act as a kind of "go-between" for theory and the observable world.

Many social scientists engage in research to see how well existing theory stands up against the real world. That is, they want to *test* at least parts of the theory. Gray, Palileo, and Johnson engaged in their research because of their interest in explaining why some people attribute blame to rape victims and others don't. They started, as we've suggested, with a general theory of why people attribute characteristics to others: attribution theory. This theory, like many others, is made up of very general, though not necessarily testable, propositions. One of these propositions is that the greater the similarity in social characteristics or experiences between an observer and an observed victim is, the greater is the likelihood that the observer will attribute

---

[1] The implicit or explicit invitation to skepticism is what distinguishes scientific theories from pseudoscientific ones, according to Carl Sagan (1995) in *The Demon-Haunted World.*

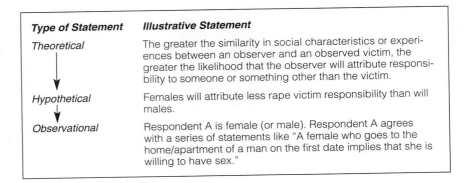

Type of Statement	Illustrative Statement
Theoretical	The greater the similarity in social characteristics or experiences between an observer and an observed victim, the greater the likelihood that the observer will attribute responsibility to someone or something other than the victim.
Hypothetical	Females will attribute less rape victim responsibility than will males.
Observational	Respondent A is female (or male). Respondent A agrees with a series of statements like "A female who goes to the home/apartment of a man on the first date implies that she is willing to have sex."

responsibility to someone or something other than the victim. From this theory, they deduced, among others, the testable hypothesis: females would be less likely than males would to attribute responsibility to rape victims.

Having made this deduction, Gray, Palileo, and Johnson then needed to decide what kind of observations would count as support for the hypothesis and what would count as nonsupport. In so doing, they devised a way of measuring the variables of their hypothesis for the purpose of their study. **Measurement,** as we will suggest in Chapter 6, involves devising strategies for classifying subjects (in this case, students) by categories (say, female or male) to represent variable concepts (say, gender). To measure things such as the gender of their respondents, Gray, Palileo, and Johnson could have asked questions like "What is your sex?" Instead, they used visual clues to give males and females different questionnaires. To measure whether the students might attribute blame to rape victims, they asked students to express their level of agreement with a series of prepared statements, such as "A female who goes to the home/apartment of a man on the first date implies that she is willing to have sex." Direct observation and responses to questions, then, were the ways in which Gray, Palileo, and Johnson observed the social world.[2]

Once they'd decided how they would measure important variables, Gray and her colleagues were prepared to make relevant observations to ascertain how particular students responded to questions. Here we'd like to emphasize that in the process of testing attribution theory, Gray and her associates did what is typical of most such exercises: They engaged in a largely **deductive** process in which more general statements were used as a basis for deriving less general, more specific statements. Figure 2.2 summarizes this process.

Theory testing rarely ends with the kinds of observational statements suggested in Figure 2.2. These statements usually are summarized with statistics, some of which we'll introduce in Chapter 15, "Quantitative and Qualitative Data Analysis." The resulting summaries are often **empirical generalizations,** or statements that summarize a set of individual observations. Gray and her associates, for instance, report the empirical

measurement, the process of devising strategies for classifying subjects by categories to represent variable concepts.

deductive reasoning, reasoning that moves from more general to less general statements.

empirical generalization, a statement that summarizes a set of individual observations.

---

[2] Answers to questions on a questionnaire might seem to be *indirect* ways of discerning things like attitudes. But much scientific observation is indirect. An x-ray of your tooth doesn't provide direct evidence of the presence or absence of a cavity, but, when read by a specialist, it can provide pretty conclusive *indirect* evidence.

generalization that females were, indeed, less likely than males were to attribute responsibility to rape victims.

As with Gray, Palileo, and Johnson's study, research is frequently used as a means of testing theory: to hold it up against the observable world and see if the theory remains plausible or becomes largely unbelievable. When researchers use their work to test theory in this way, they typically spend some of their effort deducing hypothetical statements from more theoretical ones, and then making observational statements that, when summarized, support or fail to support their hypotheses. In general, they engage in deductive processes, moving from more general to less general kinds of statements.

Inductive Reasoning

In contrast with the more or less deductive variety of research we've been discussing is a kind that emphasizes moving from more specific kinds of statements (usually about observations) to more general ones and is, therefore, a process called **inductive reasoning.** Many social scientists engage in research to develop or build theories about some aspect of social life that has previously been inadequately understood. Theory that is derived from data in this fashion is sometimes called **grounded theory** (Glaser, 2005; Glaser, 1993; Glaser and Strauss, 1967).

The process of building grounded theory from data is illustrated by Patricia and Peter Adler's research reported in Chapter 11. The Adlers had observed their own children's involvement in afterschool activities and wanted a deeper understanding of those activities. Unlike Gray and her colleagues, who started with a particular theory, Adler and Adler focused initially on a series of individual observations (their own and those of other participants) about afterschool activities. They quote in the article from which Chapter 11's focal research is derived, for instance, a 12-year-old girl who claims to like her dance company because less capable dancers aren't admitted and therefore don't slow her progress:

> I really like being in the performing company. Having tryouts cut out all of the people who can't really dance, who aren't coordinated, and who just don't pay attention. For so many years my dance classes were filled with those kids and they dragged the class down. . . . Now we learn much more because we can move faster, the classes are tough and the rehearsals are serious. (Adler and Adler, 1994a, page 318)

Having immersed themselves in many such observations, the Adlers then developed statements at a higher level of generality, statements that summarize a much larger set of observations and therefore can be seen as empirical generalizations. One generalization suggests that children who participate in adult-organized afterschool activities tend to experience predictable developmental changes that involve increasing involvement, commitment, and fervor with age:

> A broad range of adult-organized afterschool activities are available to youth. Three categories emerge, varying in their degree of organization, rationalization, competition, commitment, and professionalization:

inductive reasoning, reasoning that moves from less general to more general statements.

grounded theory, theory derived from data in the course of a study.

Type of Statement	Illustrative Statement
Theoretical	Adult-organized afterschool activities prepare young people for the adult corporate world.
↑	
Empirical Generalization	Young people tend to follow extracurricular careers that move from purely recreational to more competitive.
↑	
Observational	A 12-year-old girl has a preference for a selective dance company.

recreational, competitive, and elite. . . . Not all young people follow the progression of the extracurricular career; some remain at more recreational levels or retreat from intense types of afterschool participation to more moderate and less demanding activities. However, these are exceptions to the norm, and [this] depiction represents the path followed by most young people. (Adler and Adler, in the Focal Research in Chapter 11, page 298)

The Adlers might easily have left their study here, providing us with a kind of descriptive knowledge about children's participation in afterschool activities that we'd not had before. Instead, they offered a theory of afterschool activities that is illuminating and, perhaps, disturbing. On the one hand, they point out how modern versions of afterschool activities foster values (obedience, discipline, sacrifice, and seriousness among them) that prepare boys *and* girls for futures in the adult corporate world. On the other hand, they suggest that these values are taught at the expense of others (such as goal-setting, negotiation, improvisation, and self-reliance). As a result, the Adlers made subtle and informed contributions to existing theories of socialization and stratification. The Adlers' movement from specific observations to more general theoretical ones suggests a different approach to relating theory and research than the one used by Gray and her associates. Figure 2.3 summarizes this alternate approach.

STOP AND THINK *Consider the Donohue and Levitt research on the relationship between abortion and crime. Does that research, as we've described it, seem to rely more on deductive or inductive reasoning? What makes you say so?*

The Cyclical Model of Science

A division exists among social scientists who prefer to describe the appropriate relationship between theory and research in terms of theory-testing and those who prefer theory-building. Nonetheless, as Walter Wallace (1971) has suggested, it is probably most useful to think of the real interaction between theory and research as involving a perpetual flow of theory-building into theory-testing, and back again. Wallace captured this circular or cyclical model of how science works in his now famous representation, shown in Figure 2.4.

The significance of Wallace's representation of science is manifold. First, it implies that if we look carefully at any individual piece of research, we might

FIGURE 2.4

Wallace's Cyclical Model of Science Theories

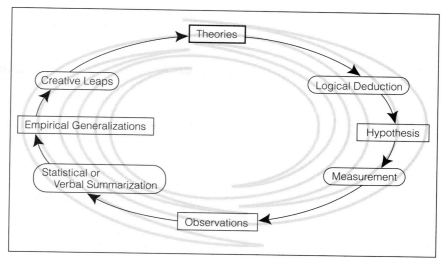

Source: Adapted from Walter Wallace (1971). Boxed items represent the kinds of statements generated by researchers during their research. Items in ovals represent the typical ways in which researchers move from one type of statement to the next. Reprinted with permission from: Wallace, Walter L. *The logic of science in sociology* (Chicago: Aldine). Copyright ©1971 by Walter L. Wallace.

see more than the merely deductive or inductive approach we've outlined. We've already indicated, for instance, that Gray and her associates' research goes well beyond the purely deductive work implied by the "right half" of the Wallace model. Gray, Palileo, and Johnson do use statistical procedures that lead to empirical generalizations about their observations. (One such generalization: that females are less likely than males are to attribute responsibility to rape victims.) This led, in turn, to a renewed affirmation of their original theoretical perspective—what they call "attribution theory." Moreover, although much of the Adlers' work is explicitly inductive, involving verbal summaries of observations and creative leaps from empirical generalizations to theoretical statements shown on the "left half" of Wallace's model, the Adlers actually did quite a bit of hypothesis testing as well. Thus, in arriving at the inference that children who participate in afterschool activities tend to experience predictable developmental stages (from recreational to competitive to elite activities), the Adlers first entertained hypotheses, based on a few observations, that children typically move from, say, recreational to competitive activities, and then cross-checked these hypotheses against observations they made about other children. This kind of cross-checking of hypotheses against observations is actually an important element in the development of grounded theory (Glaser, 2005; Glaser, 1993; Glaser and Strauss).

A second implication of Wallace's model is that, although no individual piece of research need touch on all elements of the model, a scientific community as a whole might very well do so. Thus, for instance, some researchers might focus on the problem of measuring key concepts, whereas others use those measurement techniques in testing critical hypotheses.

Gray, Palileo, and Johnson wanted to test the hypothesis that females are less likely than males are to attribute responsibility to rape victims. In creating such a test, Gray, Palileo, and Johnson used measurement strategies developed by several other researchers, notably Martha Burt's (1980) technique for measuring "rape myth acceptance" and the Koss, Gidycz, and Wisniewski (1987) measures of sexual aggression and victimization.

Another example of how one might use Wallace's model to analyze the workings of a scientific community comes out of the controversy stirred by Donohue and Levitt's work on abortion and crime. Levitt told how he became interested in the connection (Stille, 2001). It began, Levitt says, with a simple observation: that before *Roe v. Wade* in 1973, about 700,000 abortions were performed each year; the number reached 1.6 million in 1978, remained fairly constant through the 1980s, and had begun to decline during the 1990s. "When I first learned of the magnitude of abortions performed in the late 1970s, I was staggered by the number and thought, 'How could this not have had some dramatic social impact?'" (Stille, 2001, *Arts and Ideas*). From this simple fact, Levitt reports, and the observation that "the five states that legalized abortion . . . [before *Roe v. Wade*] saw drops in crime before the other 45 states" (which methodology students will recognize as an empirical generalization) (Stille, 2001), he and Donohue built their theory that increased abortion rates lead, via the reduction of unwanted births, to decreased crime rates (general proposition). They then came up with a series of hypotheses, suggesting, among other things, that various crime rates in different states should drop at particular times after the rise of abortion rates in those states (hypotheses). They collected appropriate data to test these hypotheses (observations) and found that the resulting analyses (giving them empirical generalizations) affirmed those hypotheses (Donohue and Levitt, 2001). But, their research drew a critical firestorm, one that led some of the rivals of Donohue and Levitt already to offer alternative theoretical explanations. For instance, Ted Joyce, an economist, suggested that in states where abortion was made legal before 1973, when *Roe v. Wade* was decided, there was not the decline in crime rates that one would have expected, given Donohue and Levitt's thesis, before 1991. Joyce (2003) collected data (observations) and published his own counterstudy (empirical generalizations and theory). This in turn led to Donohue and Levitt's (2003) research, which challenged Joyce's study with the claim that the nationwide crack cocaine epidemic hit the high-abortion early legalizing states harder and earlier (in the late 1980s) than it did other states and that taking this epidemic into account would explain away Joyce's findings (theory and hypotheses). Donohue and Levitt collected the appropriate data (observations) and published their own study (with theory, hypotheses, and empirical generalizations). And, more recently, Levitt (2004) again collected data to test a theory that suggests the 1990s downturn in criminality reflected four factors: increase in the numbers of police, a rising prison population, a waning crack epidemic, and the legalization of abortion.

Most important, Wallace's model reinforces the basic point that theory and research are integrally related to each other in the sciences. Whether the scientist uses research for the purpose of theory-testing, theory-building, or both, the practitioner learns to think of theory and research as necessarily intertwined. Perhaps even more than the "chicken and the egg" of the old riddle or the "love and marriage" of the old song, when it comes to theory and research, "you can't have one without the other."

Summary

In this chapter, we added to our discussion of theory by introducing the terminology of concept, variable, hypothesis, and empirical generalization. Using these terms and several examples of actual research, we illustrated the relationships that most scientists, including social scientists, see existing between research and theory. Many of the nuances of the connections between theory and research have been captured by Walter Wallace's cyclical model of how science works. Wallace's model suggests, on the one hand, how research can be used to test theories or explanations of the way the world really is. On the other, the model suggests how research can be used as a tool to discover new theories. Moreover, Wallace's model suggests that an appropriate model of science not only encompasses both of the previous views of the relationship between theory and research, but also the view that research and theory "speak" to each other in a never-ending dialogue.

EXERCISE 2.1

Hypotheses and Variables

1. Name a subfield in the social sciences that holds some interest for you (for example, marriage and family, crime, aging, stratification, gender studies, and so on).

2. Name a social problem that you associate with this subfield (for example, divorce, domestic violence, violent crime, medical care, the cycle of poverty, sharing of domestic work).

3. Form a hypothesis about the social problem you've named that involves a guess about the relationship between an independent variable and a dependent variable (for example, and this is just an example—you have to come up with your own hypothesis—"children whose parents have higher status occupations will begin their work lives as adults in higher status occupations than will children whose parents have low status occupations").

4. What, in your opinion, is the independent variable in your hypothesis (for example, "parents' occupation")?

5. What is the dependent variable of your hypothesis (for example, "child's occupation")?

6. Rewrite your hypothesis, this time in the form of a diagram like the one in Figure 2.1, a diagram that connects the categories of your variables.

EXERCISE 2.2

The Relationship Between Theory and Research

Using your campus library, find a recent research article in one of the following journals: *Social Science Quarterly, American Sociological Review, American Journal of Sociology, Journal of Marriage and the Family, Criminology, The Gerontologist, Social Forces, Qualitative Sociology, The Journal of Contemporary Ethnography.*

Or find an article in any other social science research journal that interests you. Read the article and determine how the research that is reported is related to theory.

1. What is the article's title?

2. Who is (are) the author(s)?

3. What is the main finding of the research that is reported?

4. Describe how the author(s) relate(s) the research to theory.

   Hint: do(es) the author(s) begin with a theory, draw hypotheses for testing, and test them? Do(es) the author(s) start with observations and build to a theory or a modification of some theory? Do(es) the author(s) work mostly inductively or deductively? Do(es) the author(s) use theory in any way?

5. Give support for your answer to question 4 with examples of the use of theory and its connection to observations from the research article.

© Susan Van Etten/Index Stock Imagery

3

Ethics and Social Research

Introduction

Have you been a victim of a crime—even a relatively small one, such as having had some personal property stolen, or a more serious crime, such as assault? If so, would you be willing to fill out a questionnaire or talk to a researcher who asks about your experiences? Would you have any concerns about participating in such a study?

Researchers have responsibilities to those who participate in their studies. Criminologist Peter Manning (2002: 546) believes "We owe them, I think, succor, aid in times of trouble, protection against personal loss or embarrassment, some distance that permits their growth . . . and perhaps feedback on the results of research if requested or appropriate." Although not every social scientist offers succor and aid in times of trouble, most aim to provide the other items on Manning's list. That is, they strive to act *ethically* as they plan and conduct studies and disseminate their findings. In the next chapters we'll be talking about evaluating studies on practical and methodological grounds, but first we want to talk about evaluating research using ethical criteria. In this chapter we'll consider **ethical principles in research,** the abstract set of standards and principles used to determine appropriate and acceptable research conduct. Determining appropriate research behavior is a complex topic and a source of heated debate. One issue is whether there is a set of universally applicable ethical principles, or if it's better to take a case-by-case approach and consider each project's specific circumstances when deciding what is, and what is not, ethical. A second issue is that unforeseen circumstances and unintended consequences at every stage of research—from planning to publication—can create ethical problems. Finally, there is the issue of whether the ethical *costs* of the study should be measured against the study's *benefits* for the participants or others.

ethical principles in research, the set of values, standards, and principles used to determine appropriate and acceptable conduct at all stages of the research process.

STOP AND THINK

Do you know about any research project in the social sciences, medicine, or biology where the researcher did something that was ethically questionable? What was it about the study that raised questions for you?

A Historical Perspective on Research Ethics

Individual scholars might have been interested in defining appropriate research conduct earlier, but ethics did not become a concern of social science organizations as a whole until the middle of the twentieth century. The first ad hoc committee of the American Psychological Association (APA) established in the late 1930s to discuss the creation of a code of ethics did not support developing one. The APA did not develop and adopt a code of ethics until the 1950s (Hamnett, Porter, Singh, and Kumar, 1984: 18).

There are similar trends in other social science organizations. The British Sociological Association approved a code of ethics in 1973, the American Anthropological Association first drafted its statement in 1967, and the members of the American Sociological Association (ASA) were so divided about creating a statement on ethics that it took more than eight years from the proposal of the code to its adoption in 1971 (Hamnett et al., 1984). The

BOX 3.1

The American Sociological Association's Code of Ethics

The American Sociological Association asks that sociologists demonstrate professional competence, integrity, professional, social, and scientific responsibility, and respect for people's rights, dignity, and diversity. A 30-page document posted on the association's website covers issues including informed consent, confidentiality, and plagiarism.

Review the complete document at (http://www.asanet.org/galleries/default-file/Code%20of%20Ethics.pdf).

most recent version of the ASA Code of Ethics presents a set of values for professional and scientific work, and also presents the rules and procedures of the ASA committee charged with investigating complaints of unethical behavior. (See Box 3.1).

The relatively recent impetus for disciplines to consider ethical issues stems from several factors. One was the reaction to the total disregard for human dignity perpetrated during World War II by researchers in concentration camps controlled by Nazi Germany. After the Nuremberg trials, the United Nations General Assembly adopted a set of principles about research on human beings (Seidman, 1991: 47). The first principle of the Nuremberg Code is that it is absolutely essential to have the voluntary consent of human subjects (Glantz, 1996).

There has also been increasing awareness of and distress among scientists, policymakers, and the general public concerning problematic biomedical research involving human subjects. For example, government-sponsored research conducted during the 1950s studied the effects of radiation on human subjects. Funded by the Atomic Energy Commission, this research included medical research on using radioisotopes to diagnose or cure disease. However, some experiments were gruesome, such as one where people with "relatively short life expectancies," such as semicomatose cancer patients, were injected with plutonium to determine how much uranium was needed to produce kidney damage (Budiansky, 1994). In a study of cancer done in 1963, three physicians at the Brooklyn Jewish Chronic Disease Hospital injected live cancer cells into 22 elderly patients without informing them that live cells were being used or that the procedure was unrelated to therapy (Katz, 1972).

One long-lasting research scandal ended as the result of public outrage. The Tuskegee study (labeled as such because the project was conducted in Tuskegee, Alabama) was conducted by the United States Public Health Service to study the effect of untreated syphilis to determine the natural history of the disease (Jones, 1981). The 40-year research project was started in 1932, with a sample of 399 poor black men with late-stage syphilis. The study ultimately included 600 men, all of whom were offered free medical care in exchange for their medical data, and none of whom were told that they had syphilis. More than 400 of the men were not offered the standard (and often fatal) treatment for syphilis in the 1930s. In later years, they were not provided with penicillin even though it was available by the late 1940s (Jones, 1981). Many of the men died of the disease, and some of them unknowingly transmitted it to their wives and children (*The Economist* 1997: 27). As late as 1965, an outsider's questioning of the morality of the researchers was dismissed as eccentric (Kirp, 1997: 50). The project ended in 1972, but not until 1997 was an official governmental apology offered by President Clinton to the surviving victims and their families (Mitchell, 1997: 10).

The disgrace of the Tuskegee Syphilis Study was one of the impetuses for the passage of National Research Act of 1974. This act created a commission to identify principles for conducting biomedical and behavioral research ethically and prompted the development of regulations requiring the establishment of committees explicitly designed to protect the rights of research subjects at institutions that receive federal funds (Singer and Levine, 2003). By the 1970s, social scientists could look to federal guidelines and their professional associations for help in conducting ethical research. (See, for example, the current National Science Foundation statement about federal policy in Box 3.2).

The committee charged with evaluating the ethics of research proposals is called an **institutional review board (IRB)** or a committee on human subjects research. Such committees exist at almost every college, university, and research center. Before any federal, state, or institutional monies can be spent on research with human subjects, the research plan must be approved by the committee, at least one of whose members is unaffiliated with the institution. For the past 30 years, IRBs have attempted to verify that the benefits of proposed research outweigh the risks, and to determine if subjects have been informed of participation risks. The need to seek approval from such a committee varies depending on the topic, the source of funding, and the specific institutions involved. The Office of Budget, Finance and Award Management of the National Science Foundation, for example, notes that IRBs can determine if research projects, such as ones involving anonymous surveys and observations of public behavior, may be exempted (NSF, 2006). Typically, research projects are subject to review when they deal with sensitive issues or can cause damage to participants, or when subjects can be identified. Each researcher should be familiar with appropriate ethical codes and the need to submit proposals to appropriate review committees. Ultimately, the responsibility for ethical research behavior lies with the individual researcher.

There is sometimes controversy about how IRBs deal with social research. The impetus for IRB regulation came from abuses in biomedical research, and the membership of many boards is composed mainly of those in the medical and biological sciences. With few social scientists on IRBs, some feel that social research is not understood and that the methodology of informed consent is based what works best in clinical studies rather than in

institutional review board (IRB), the committee at a college, university, or research center responsible for evaluating the ethics of proposed research.

BOX 3.2

The National Science Foundation's Ethical Standards

What are the overall goals of the federal policy (the Common Rule)?

The policy is designed to ensure minimal standards for the ethical treatment of research subjects. The major goal is to limit harm to participants in research. That means that no one should suffer harm just because they became involved as subjects or respondents in a research project. Institutions engaged in research should foster a culture of ethical research.

ETHICAL RESEARCH rests on three principles:

1. RESPECT for persons, meaning the researcher gives adequate and comprehensive information about the research and any risks likely to occur, understandable to the participant, and allows them to voluntarily decide whether to participate.

2. BENEFICENCE, meaning the research is designed to maximize benefits and minimize risks to subjects and society.

3. JUSTICE, meaning that the research is fair to individual subjects and does not exploit or ignore one group (e.g., the poor) to benefit another group (e.g., the wealthy).

Research produces benefits valued by society. Regulatory oversight seeks to ensure that any potential harm of the research is balanced by its potential benefits.

Source: National Science Foundation website http://www.nsf.gov/bfa/dias/policy/hsfaqs.jsp#relation

---

social research. In addition, some complain that IRB procedures result in needless delay (Bosk and deVries, 2004). However, in general, researchers support the IRB process. In a survey of researchers at the University of North Dakota, the overwhelming majority of faculty and graduate students said that IRB approval is an important protection for human subjects; all of the graduate students and 88 percent of the faculty felt that they had been treated fairly by the committee, and very few complained that they could not carry out a research project because of problems with the IRB (Ferraro, Szigeti, Dawes, and Pan, 1999).

The IRB at Brown University was very helpful to Desirée Ciambrone, a sociologist whose work we'll read in the following focal research section. While taking a graduate seminar in medical sociology at Brown University, Ciambrone became interested in the topic of women and AIDS. Conversations about AIDS in Africa with a fellow graduate student who had worked on public health and policy issues convinced Ciambrone that focusing on

American women's experiences with HIV/AIDS was a worthwhile and doable project. As this was her first totally independent research project, she learned a great deal from the experience of conducting a study from start to finish. She learned about the impact of illness on people's lives and how an HIV diagnosis affects women's social and familial roles. She also learned that submitting a research proposal to an IRB can bring to your attention issues that you might not have considered.

FOCAL **RESEARCH**

Ethical Concerns and Researcher Responsibilities in Studying Women with HIV/AIDS

By Desirée Ciambrone[1]

Conducting one's first independent research project is a daunting task; eager to immerse oneself in the project, there is a great deal of preparatory work to be done. I chose to study women with HIV infection and found that the anxiety and complexity of conducting social science research are heightened when people facing difficult circumstances are involved.

Studying Women with HIV: Getting Ready

Respondent Recruitment

Finding women with HIV to speak with was no easy task. I sought respondents from health and social service agencies located primarily in the Northeastern United States. Some agencies requested copies of my interview guide, my research proposal, and/or the Brown University Institutional Review Board approval. I discussed the goal of my study with my contacts and urged them to post/distribute a letter describing the study and how to get in touch with me if they had any questions about the study or wished to participate. Better yet, I urged that agency contacts speak with women directly and encourage their participation.

Many organizations I contacted were justifiably protective of their clients and appeared unwilling to assist in respondent recruitment despite their verbal acquiescence. Others were too busy to help me with my study; they posted my letter among a myriad of other brochures and leaflets but did little to personally encourage women to participate. Several key contacts in local hospitals were very helpful and referred most of the 37 HIV-positive women who volunteered to be interviewed for my study.

---

[1] This article was written for this text by Desirée Ciambrone and is published with permission. A complete description of her study and findings can be found in D. Ciambrone, *Women's Experiences with HIV/AIDS Mending Fractured Selves,* New York: Haworth Press, 2003.

Informed Consent

The protection of human subjects is of utmost importance when conducting any type of research. Thus although my study would require "only an interview" (as opposed to a clinical trial, for example, where drugs and other treatments may be involved), obtaining *informed consent* from my participants was crucial. The Institutional Review Board (IRB) at Brown University provided the guidelines that helped me carry out my research in an ethical way, including providing potential participants with an understanding of the potential risks and benefits of participation. In my initial proposal, I was quick to note the benign nature of my study. I described the benefits of being able to share one's story with an interested listener and the potential impact of my results on social policy and medical care provision. I noted that I wanted to talk to women about their lives and about living with HIV/AIDS. Initially, I did not see any risks that could result from this. Comments from the IRB prompted a more careful reflection of the *potential* risks of participation in my study. Indeed, I realized my questions might elicit a negative emotional response; such uneasiness or distress is, in fact, a risk. Thus, my informed consent was to be rewritten to include the possibility of emotional discomfort (See Appendix A).

Informed consent means that respondents demonstrate that they understand the study, what is expected of them, and how the information will be used. It also requires that participants *voluntarily* enter into a research project; that they haven't been coerced or duped into participation. Given previous research in which human subjects were misinformed and/or forced to participate in dangerous and life threatening experiments under the guise of science, the apprehensions of vulnerable populations are justified.

Women with HIV, while not unintelligent, are less likely to be highly educated and may not be knowledgeable about research; thus, it was my responsibility to explain the study in a clear, straightforward manner. I explained that I was a graduate student in medical sociology at Brown University who was conducting research in order to learn about the daily lives of women with HIV/AIDS. I told them that I wanted to interview them about their experiences in whatever setting was comfortable for them, their homes or a more neutral setting. I said I would tape record the interviews in order to have an accurate, complete record of their stories but that, in order to protect their privacy, I would not use their real names on the tapes, transcripts or in reports, and would keep the information confidential. I needed to describe what the study was, and perhaps more importantly, *what it was not.* I had to be clear that I was not a health care provider; I could not offer medical advice or prescribe medications. I was also not a social worker; I could not act as their case manager linking them to needed services. On the other hand, I felt I owed women answers to their questions. So I became educated about local services for persons with HIV/AIDS so that if a woman had a concern I might say, "Well, perhaps X organization can help. I know they deal with those types of issues." I was careful to make certain my respondents understood that while I was knowledgeable about many services, I did not know of *all* that was available nor could I guarantee the organization would be able to help. It was a challenge to be frank, honest, and share whatever I could *and* be sure that women did not perceive me as someone who would solve their problems; even with the best of intensions, that was not in my power.

An IRB follows research from its conception through its completion. Brown's IRB requires annual reports in which researchers describe progress, and explain any changes made to protocols or informed consent forms. This mechanism ensures that changes made are warranted and ethical. As an example, one of the things I reported in my annual progress report was my decision to offer respondents compensation for participation. I felt that this would attract those less accustomed to sharing their stories and offer all the participants some compensation for their time. To this end, I offered each participant a $10.00 gift certificate from a local supermarket chain, a decision which was approved by the IRB.

The IRB protocol outlines the appropriate storage of data. For example, I had to ensure that my respondents' personal identifying information was not attached to the data. Thus, instead of using the real names on the transcripts, pseudonyms were used. In accordance with federal regulations, transcripts, audio tapes, and other materials were stored under lock and key in a limited-access location.

The Interview Experience

Women with HIV/AIDS, particularly those with substance abuse histories, face unique challenges including poverty, caring for young children and other adults, reproductive decisions, access to health care and social services, inadequate housing, negotiating condom use, sexual undesirability, and abandonment. Race and class inequities exacerbate these problems. I knew all of this from my review of the literature, but had no *personal experience* with this population.

Being able to talk to the women in person was essential as I was able to listen to them speak, to follow up on key issues, to observe body language, and to see where they lived. It was also important for me to learn that these women are knowledgeable, articulate people—experts on their experiences. I developed a newfound respect for these women's agency; in the face of poverty, drug use, and illness these women managed to live productive, meaningful lives. Several did public speaking and worked to help others with HIV/AIDS in their community.

Connecting with Women with HIV

In graduate school, I had learned about the importance of the researcher's role in the research process. Thus, at every stage of my research I made a conscious effort to empower the study participants and present them as experts on living with HIV. Prior to interviewing, I contemplated the worse case interview scenarios, particularly access and rapport problems stemming from differential social positions. I did not encounter skepticism or scrutiny; none of my respondents questioned my interest in HIV/AIDS, nor did they appear uncomfortable sharing their stories with someone unlike themselves. This was probably due to the fact that many of these women were

accustomed to talking to "outsiders" about their experiences via their participation in clinical trials of medical treatment at local health care centers.
Women who had not participated in such projects may well have been skeptical of my intentions and less frank about their lives.

In addition to social distance and nondisclosure issues, I had also prepared myself for other possible problematic interview situations, such as lack
of privacy and dire living circumstances. For most of the interviews, respondents and I had adequate privacy. Despite occasional commonplace interruptions (e.g., phone calls, demands of young children), the interviewees
tried to create conducive interview situations.

The living conditions of some women were more troubling than relational issues. Several women lived in poor, traditionally high crime neighborhoods or public housing developments; neighborhoods I was familiar with
from a young age as "places not to go." In order to feel as safe as possible, I
scheduled most of my interviews, especially those in "tough" neighborhoods, during the day. Entering questionable dwellings was difficult. By and
large, the women's homes were adequately furnished and relatively well
kept. It was distressing to witness the apparent poverty and the destitute
living conditions of some of my respondents. For example, one woman's
home was filthy—the floors appeared as if they had not been swept or
washed for quite some time, there were dirty dishes and pans overflowing in
the sink, and ashtrays were filled to the rim with cigarette butts. I felt somewhat uneasy in homes like this, and less accepting of women's hospitality
(e.g., a cup of coffee). Not accepting food and beverages made me feel
rude and I never wanted to appear judgmental of women's homes. And although I had no intention of verbalizing my thoughts, I wondered how I
could explain that it wasn't the HIV status that made me uncomfortable at
times; it was the unsanitary home!

Eliciting Women's Stories and Confidentiality

Given the sensitive nature of my questions (e.g., about sexual relations and
illicit drug use), I took great pains to make sure that the women knew the
information provided was completely *confidential* and would not interfere
with the receipt of medical/social services. I reiterated that they did not
have to answer any questions that made them feel uncomfortable. While
there were a couple of women who did not want to discuss certain topics,
overall, the women I interviewed were incredibly open and often offered
more intimate details than I had expected.

Many women were very interested in the purpose of the study. Often,
women wanted to know how *their* responses compared to those of other
women. Many also asked questions about my academic experience and personal life (e.g., What are you studying? Are you married? Do you have
kids?). Recognizing the importance of establishing trust and minimizing
power differentials between us, I answered the questions and told them
about myself. When asked about the other interviewees, I was concerned
that answering questions about the study would lead respondents to feel
there was a normative or desired response to my questions that they should
confirm. As I was willing to discuss my research, but didn't want women's

narratives to be influenced by those of other respondents, I suggested that we talk about the study after the interview so as to be sure we had ample time to discuss their experiences and views. Evading interviewees' questions hinders efforts to establish trust and build rapport with respondents. My self-disclosure put my respondents at ease by reassuring them that the interview was a guided conversation and that their experiences and opinions were what was important, not getting the "right" answer. The open-ended nature of my questions allowed women to highlight the issues that were important to them and explore areas that I may not have inquired about.

Interviewing women who were active drug users posed a challenge. I felt most uncomfortable around these women as their experiences were so far removed from mine. In addition to striving to appear calm and comfortable, I had to remind myself that my role was that of researcher, not social worker or care provider. Thus, I could not offer unsolicited advice about the dangers of drug use on their weakened immune systems, the potential ramifications of nonadherence to their medications, or provide solutions for the apparent disrespect the women were shown by their partners or children. Furthermore, when women did ask me for my advice I was careful to make it clear to them that I was a sociologist, not a therapist or a counselor. While I answered their questions about HIV and services available to the best of my ability, I didn't feel it was appropriate to offer unsolicited advice even when I had strong feelings about their choices. For example, one woman was married to a devout Christian who had convinced her to leave her health in God's hands and forgo her treatment regimen. Sheila felt torn between her husband's religious beliefs and her health care providers' recommendations.[2] Despite a dropping T-cell count and increasing viral load, Sheila stopped taking her medications at her husband's request. I found her situation troubling, and while I worried about Sheila's health, I felt it was not my role to encourage medication use or discredit her faith in religion.

Lessons Learned

In designing and conducting this study I learned a great deal about the *process* of conducting research, including the importance of following ethical practices, using informed consent procedures, protecting my respondents' privacy and confidentiality, and preserving the integrity of their narratives. My understanding of how women cope with HIV/AIDS was enhanced as a result.

Appendix A: Informed Consent

Informed Consent for Participation in Women with HIV Infection Study

You are being asked to participate in an important study to learn more about women's experiences with HIV infection. This study is being conducted by

---

[2] Sheila is a pseudonym to protect the respondent's confidentiality.

Desirée Ciambrone, a graduate student in the Sociology Department at Brown University.

During this interview, I will ask you about your illness, how HIV infection affects your daily life, and about the people who help you if, and/or when, you need assistance.

- This interview will be tape recorded.
- In order to participate in this project I must be 18 years old or over.
- I may be contacted for further questions regarding this study.
- My participation in this project is completely voluntary; the decision to participate is up to me and no one else.
- I do not have to answer every question; I may choose not to answer certain questions.
- I can ask the interviewer to leave at any time. Even if I sign this form, I can change my mind later and not participate in the study.
- The information I provide will be kept confidential to the extent of the law.
- The only known risk to me is the possibility of some emotional discomfort.
- A direct benefit of participation in this study is the opportunity to talk with an interested listener and to think reflexively about my own experiences. I understand that the information obtained by this study may help other women who have had similar life experiences and may help professionals understand and treat women with HIV infection.

If I have any questions about this study, I may contact Desirée Ciambrone at (401) 863-1275 at any time. Or, I may contact Alice A. Tangredi-Hannon or Dorinda Williams at the Office of Research Administration at Brown University (401) 863-2777 with questions about my rights as a participant in this study.

I have read the consent form and fully understand it. All my questions have been answered. I agree to take part in this study, and I will receive a copy of this completed form.

Signature of Participant   Date

Signature of Interviewer   Date

Principles for Doing Ethical Research

Ethics in research is a set of moral and social standards that includes both *prohibitions against* and *prescriptions for* specific kinds of behavior in research. Applying ethical principles gives a researcher a moral compass for making decisions about their research. One interesting thing to note is that although researchers can scientifically study people's values and their ethical standards, researchers cannot use scientific methods to determine how to behave ethically in their own work. Instead, they must examine their behavior using other frames of reference. To act ethically, each researcher must consider many factors, including safeguarding the welfare of those who participate in their studies, protecting the rights of colleagues, and serving the larger society. As we consider specific ethical principles, we'll examine these issues.

STOP AND THINK

What ethical standards do you think are most important when doing research? For example, if you were asked to observe friends at a party to study social interaction among college students for a class assignment, what would you do? Would you do the observation without first asking your friends for their permission?

Principles Regarding Participants in Research

Protect Participants from Harm

protecting study participants from harm, the principle that participants in studies are not harmed, physically, psychologically, emotionally, legally, socially, or financially as a result of their participation in a study.

First and foremost among ethical standards is the obligation to **protect study participants from harm.** This means that the physical, emotional, social, financial, legal, and psychological well-being of those who participate in a research project must be protected, both during the time that the data are collected and after the conclusion of the study. Overall, there should not be negative consequences as a result of being in a study.

Social science data are frequently collected by asking study participants questions. In the next chapter, we'll read about a research project in which study participants were asked about their career and family goals. Answering such questions did not put people at risk and represented a minimal intrusion on their privacy. In addition, their answers were held in confidence. In contrast, in studies like Ciambrone's, where participants are vulnerable in some way, answering questions is more likely to cause emotional or psychological discomfort. In Ciambrone's case, the IRB requirement that she inform potential interviewees about the emotional risk of participation was appropriate.

One study that has been criticized as violating the "no harm" principle and other ethical standards is the Milgram experiment on obedience to authority figures. In this famous experiment, psychologist Stanley Milgram deceived his subjects by telling them that they were participating in a study on the effects of punishment on learning. The subjects, men aged 20 to 50, were told that they would be "teachers," who, after watching a "learner" being strapped into place, were taken into the main experimental room and seated before an impressive shock generator with 30 switches ranging from 15 volts to 450 volts. The subjects were ordered by the experimenter to administer increasingly severe shocks to the "learner" whenever he gave an incorrect answer. They were pressed to do so even when the "learner" complained verbally, demanded release, or screamed in agony. The "learner," or victim, was played by someone who was helping the researcher and who actually received no shock at all; the purpose of the experiment was to see at what point, and under what conditions, ordinary people would refuse to obey an experimenter ordering them to "torture" others (Milgram, 1974).

The results of the experiment were surprising to Milgram and others. At another time, Milgram asked groups of psychiatrists, college students, and middle-class adults to predict the results. Not one person in these groups predicted that any subject would give high-voltage shocks (Milgram, cited in Korn, 1997: 100). Milgram was caught unawares when he found that although many of the hundreds of participants exhibited doubt and anxiety about administering the shocks, with "numbing regularity good people were seen to knuckle under to the demands of authority and perform actions that were callous and severe" (Milgram, 1974: 123). After the data were collected, there were post-experimental debriefings where the participants were told that the "learners" had not received any shocks.

It might surprise you that Milgram's work was consistent with the ethical principles for psychological research at the time. The American Psychological Association's 1953 guidelines allowed for deception by withholding information from or giving misinformation to subjects when such deception was, in the researcher's judgment, required by the research problem (Herrera, 1997: 29). Not until the 1992 revision of the APA's statement on ethical principles were additional restrictions placed on the use of deception (Korn, 1997). Despite the APA guidelines, when the results of the experiment were published, Milgram found himself in the middle of a controversy over ethics. Critics blamed him for deceiving his subjects about the purpose of the study and for causing his subjects a great deal of emotional stress both during and after the study (Baumrind, 1964).

Milgram's study raises the issue of whether the end can justify the means. Thinking about Milgram's work, you might feel that the study was acceptable because it made an important contribution to our understanding of the dangers of obedience. Or, you might agree with those who have vilified Milgram's work because it violated the ethical principle of protecting subjects from harm.

Whatever your perspective on Milgram's study, his research points out that in many projects, the line between ethical and unethical methods is not always obvious. Often, it is difficult to design a study in which there is absolutely no risk of psychological or emotional distress to participants. For example, for at least a few people, being interviewed about their marriages or their relationships with siblings can dredge up unhappy memories. Similarly, for others, talking about experiences with sexual assault, divorce, or the death of a loved one can produce adverse effects or create difficulties for them in groups to which they belong. It can be impossible to anticipate each and every consequence of participation for each and every person.

Foreseeing possible negative reactions by subjects does *not* mean a study should be abandoned. Instead, a realistic and achievable ethical goal is for studies to create, *on balance,* more positive than negative results and to give participants the right to make informed decisions about that balance.

Offering help is one way to attract participants, keep their cooperation, and reciprocate for their contributions. For example, in a study on sexual assault, Bart and O'Brien (1985) offered a psychotherapy session at the research project's expense after each interview and provided information about self-defense courses and legal assistance. Ciambrone did tell her participants about the possible emotional consequences of participating, but found that most seemed to enjoy the interview and liked the idea of telling their stories to help others with HIV. They said they appreciated her interest and in having the opportunity to share their views. A main theme was that the interview made them feel "normal," accepted, and valued rather than being defined exclusively by their HIV status.

Voluntary Participation and Informed Consent

voluntary participation, the principle that study participants choose to participate of their own free will.

informed consent, the principle that potential participants are given adequate and accurate information about a study before they are asked to agree to participate.

informed consent form, a statement that describes the study and the researcher and formally requests participation.

The ethical strengths of Ciambrone's study include the fact that all the interviewees volunteered to participate after being informed of the researcher's identity and affiliation, and the study's purpose, risks, and topics. The ethical principles being followed were **voluntary participation** and **informed consent.** These principles require that, *before participating,* all potential participants be told that they are not obliged to participate, that they can discontinue their participation at any time, that they be given accurate information about the study's purpose, the method of data collection, whether the information obtained will be collected anonymously or kept confidential, and the availability of results. Violations of ethical standards include not asking for consent, presenting inadequate information or misinformation, and using pressure or threats to obtain participation.

One way to provide potential respondents with information about the study is through a written **informed consent form,** like the ones included in the Appendix of Ciambrone's article above and in Chapter 10. Anonymous questionnaires that pose few risks for participants may be exempted from the requirement for informed consent, and some researchers rely on verbal rather than written permission as illustrated in Box 3.3. In recent years, it is more common to use written statements to provide potential interviewees with sufficient information to decide about participation. Informed consent should specify the purpose of the study, the benefits to participants and to others, the identity and affiliations of the researcher, the nature and likelihood of risks for participants, a statement of how results will be disseminated, an account of who will have access to the study's records, and a statement about the sponsor or study's funding source.

STOP AND THINK   *Who should give informed consent in a study involving high school students under 18? The students? Their parents? Both?*

There is some debate about the issue of who can appropriately give informed consent. Of particular concern are children, adolescents, and individuals unable to act in an informed manner. Most agree that, for minors, parental permission is important. However, there is debate about the age at which someone can decide for himself or herself to grant or withhold consent and who should be authorized to give consent for those judged

BOX 3.3

An Example of Verbal Parental Consent

A research project by the National Alliance for Caregiving (2005: 45) was conducted to determine the impact of children's participation in caregiving within households. First an adult in the household was interviewed by phone and then asked for permission to speak directly to his/her child. A verbal consent statement was read that began: "Before we proceed, I would like to tell you about the interview so you feel comfortable having me talk to your child. I will be asking him/her about the kinds of things s/he typically does, including his/her responsibilities and free time activities. We will also talk about the kinds of things s/he does for the person she/he helps care for and how the responsibilities affect him/her. Please be assured that your responses and your child's will be kept strictly confidential. Individual results will not be released to anyone. . . . If you'd like, you may listen to the interview."

Source: http://www.uhfnyc.org/usr_doc/Young_Caregivers_Study_083105.pdf

unable to decide for themselves (Ringheim, 1995). Studying anorexic teenagers, Halse and Honey (2005: 2152) note that difficult family relationships or histories could make parental consent a "double-edged sword, protecting some girls and erasing other girls' potential for agency by increasing the opportunity for parental coercion." Sometimes researchers are granted approval from IRBs to waive parental permission. Diviak et al. (2004), noting that teen smokers might not be willing to admit their smoking habit to parents, surveyed 21 researchers engaged in smoking cessation research with adolescents and found that 13 had asked for parental permission to be waived. Four of the researchers' projects were authorized to collect teen permission only.

passive consent, when no response is considered an affirmative consent to participate in research; also called "opt out informed consent," this is sometimes used for parental consent for children's participation in school-based research.

Methods of obtaining permission vary. In school settings, for example, active consent requires someone, such as a parent or guardian, to sign and return a form to give permission for their child to participate in a study, whereas **passive consent** requires them to sign and return a form if they do *not* want their child to participate. Getting consent forms returned can be quite difficult, even if parents are willing to have their children participate. In a study of high school students in Australia, de Meyrick (2005) sent 2,500 consent letters to parents but only 200, or less than 8 percent, were returned. Things such as response rate and research costs may be why Morris and Jacobs (2000) found in a survey of researchers that only 69 percent of them agreed that passive consent was definitely or probably ethically problematic.

STOP AND THINK

Have you ever been asked by a faculty member while you were his or her student to participate in a study that he or she was conducting? Were you ever asked while in class to complete a questionnaire or participate in an interview? Did you feel any pressure or did you feel you could decline easily? Do you think that students in this kind of situation who agree to participate are true volunteers?

As a way of ascertaining if participants are really *volunteers*, researchers should ask the following questions: Do potential participants have enough information to make a judgment about participating? Do they feel pressured by payments, gifts, or authority figures? Are they competent to give consent? Are they confused or mentally ill? Do they believe that they have the right to stop participating at any stage of process? Those who are dependent on others, such as those in shelters or in nursing homes, might be especially vulnerable to pressure to cooperate. For example, in her discussion of research with homeless women, Emily Paradis (2000) notes the possibility of a dependent or coercive relationship when the institution in which the research takes place is the major source of subsistence, advocacy, or emotional support for the participants, and when the participants are used to making accommodations to survive. Such relationships might make it harder for those approached to decline participation in a study.

Other studies raise different questions about consent. For a study on police interrogations, for example, researcher Richard Leo was required by his university's institutional review board to secure informed consent from the police officers, but not from the criminal suspects (Leo, 1995). In another research project, Julia O'Connell Davidson obtained informed consent from a prostitute and her receptionist, but did not ask for permission from the prostitute's clients. Davidson says that the clients did not know a sociologist was listening to their conversations in the hall and watching them leave the house, nor did they consent to having the prostitute share information about their sexual preferences or physical and psychological defects (Davidson and Layder, 1994). Davidson was not troubled by the uninvited intrusion into the clients' world, arguing that the prostitute had willingly provided knowledge that belonged to her.

New technologies can lead to new ethical dilemmas. For example, what about using e-mail material, such as comments posted on an e-mail forum? Although some consider such postings to be in the public domain and therefore available without consent for research purposes, others believe that permission should be obtained from participants (Sixsmith and Murray, 2001). Similar differences of opinion can be found about other sources of data such as field research, which is where the researcher observes daily life in groups, communities, or organizations.

STOP AND THINK　*How about observing people in public, such as watching crowd behavior at a sporting event? What research procedures would be ethical? Would you or could you ask for permission? What would you do if you saw illegal behavior? How about if law enforcement personnel asked you for the notes you took while watching the crowd?*

Most researchers would argue that it's unnecessary, impractical, or impossible to get informed consent when observing in public places. Sometimes there is no simple distinction between "public" and "private" settings. For the study described in *My Freshman Year: What a Professor Learned by Becoming a Student*, 52-year-old anthropologist Rebekah Nathan obtained approval from her university's IRB and followed the ethical protocol of informed consent when conducting formal interviews with students (Hoover, 2005: 36). However, in day-to-day interactions with students, Nathan let students assume she was one of them, which some might argue is unethical. How about attending a meeting of Alcoholics Anonymous? Maurice Punch (1994: 92), a qualitative researcher, asks "Can we assume that alcoholics are too distressed to worry about someone observing their predicament (or that their appearance at A.A. meetings signal their willingness to be open about their problem in the company of others)?" He argues that observation in many public and semi-public places is acceptable even if subjects are unaware of being observed. The Code of Ethics developed by the American Sociological Association (1999: 12) supports observing without permission in public places, but suggests consultation with an IRB if there is any doubt about the need for informed consent.

Anonymity and Confidentiality

Even if informed consent is not obtained, the ethical principles of anonymity and confidentiality do apply. **Anonymity** is when it is impossible for anyone, including the researcher, to connect specific data to any particular member of the sample. To collect data anonymously, data must be collected without names, personal identification numbers, or information that could identify subjects.

Although it is not always possible or practical to collect data anonymously, research ethics dictate keeping them confidential. **Confidentiality** is keeping the information disclosed by study participants, including their identities, from *all* other parties, including parents, teachers, and school administrators, and others who have given permission or aided participation in the research. To keep material confidential, the data collected from all respondents must be kept secure. Identifying respondents by code numbers or pseudonyms are useful ways to keep identities and data separate. When results are made public, confidentiality can be achieved by not using real names or by grouping all the data together and reporting summary statistics for the whole sample. In addition, researchers can change identifiers like specific occupations or industries and names of cities, counties, states, and organizations when presenting study results.

The issues of anonymity and confidentiality are at the very heart of ethical research. One way to ensure the security of participant data is to destroy all identifying information as soon as possible. Such a step, however, eliminates the possibility of doing future research on that sample because collecting data from the same participants more than once means being able to identify them. If a researcher wants to keep in touch with study participants over a number of years, then records of names, addresses, and phone numbers must be kept. However, keeping identifying information means that confidentiality could be at risk because it is possible through subpoena to link a given data record to a specific person (Laumann et al., 1994: 72).

One very controversial study, *Tearoom Trade,* by Laud Humphreys (1975), illustrates several ethical principles in doing research with human

anonymity, when no one, including the researcher, knows the identities of research participants.

confidentiality, when no third party knows the identities of the research participants.

subjects—perhaps more by violation of than by compliance with the principles. This well-known study gives an account of brief, impersonal, homosexual acts in public restrooms (known as "tearooms"). Humphreys was convinced that the only way to understand what he called "highly discreditable behavior" without distortion was to observe while pretending to "be in the same boat as those engaging in it" (Humphreys, 1975: 25). For that reason, he took the role of a lookout (called the "watchqueen"), a man who watches the interaction and is situated so as to warn participants of the approach of anyone else.

Humphreys observed in restrooms for more than 120 hours. Once, when picked up by the police at a restroom, he allowed himself to be jailed rather than alert anyone to the nature of his research. He also interviewed 12 men to whom he disclosed the nature of his research, and then decided to obtain information about a larger, more representative sample. For this purpose, he followed men from the restrooms and recorded the license plate numbers of the cars of more than 100 men who had been involved in homosexual encounters. Humphreys obtained the names and addresses of the men from state officials who were willing to give him access to license registers when he told them he was doing "market research." Waiting a year, and using a social health survey interview that he had developed for another study, Humphreys and a graduate student approached and interviewed 50 men in their homes as if they were part of a regular survey. Using this deception, Humphreys collected information about the men and their backgrounds.

Humphreys viewed his primary ethical concern as the safeguarding of respondents from public exposure. For this reason, he did not identify his subjects by name and presented his published data in two ways: as aggregated, statistical analyses and as vignettes serving as case studies. The publication of Humphreys' study received front-page publicity, and because homosexual acts were against state law at the time, he worried that a grand jury investigation would result. Humphreys spent weeks burning tapes, deleting passages from transcripts, and feeding material into a shredder (1975: 229), and resolved to go to prison rather than betray his research subjects. Much to his relief, no criminal investigation took place.

STOP AND THINK *Review the principles of ethical research that have been discussed in this chapter. Before we tell you what we think, decide which of the principles you believe Humphreys violated. Make some guesses about why he behaved the way he did.*

Two principles that Humphreys violated are voluntary participation and informed consent. By watching the men covertly, tracking them through their license plates, asking them to participate in a survey without telling them its true purpose, and lying to the men about how they were selected for the study, Humphreys violated these principles several times over. After the study's publication, critics took him to task for these actions.

Humphreys' work provided insights about sexual behavior that had received little scholarly attention and told about groups that were often targets of hostile acts. Because of the usefulness of the work and his observational setting, Humphreys maintained that he suffered minimal doubt or hesitation. He argued that these were *public* restrooms and that his role, a

> **BOX 3.4**
>
> ## Another Point of View
>
> In "The Ethics of Deception in Social Research: A Case Study," Erich Goode argues a different point of view:
>
> "I intend to argue that it *is* ethical to engage in *certain kinds* of deception . . . the ethics of disguised observation [should be] evaluated on a situational, case-by-case basis. . . . In *specific* social settings, some kinds of deception should be seen as entirely consistent with good ethics. . . .
>
> I strongly believe every society *deserves* a skeptical, inquiring, tough-minded, challenging sociological community. It is one of the *obligations* of such a society to nurture such a community and to tolerate occasional intrusions by its members into the private lives of citizens. . . . I have no problem with the *ethics* of the occasional deceptive prying and intruding that some sociologists do. I approve of it as long as no one's safety is threatened." (1996: 14 and 32)

natural one in the setting, actually provided extra protection for the participants (Humphreys, 1975: 227). But, reevaluating his work in the years after the study's publication, Humphreys came to agree with critics about the use of license plate numbers to locate respondents and approach them for interviews at their homes. He came to feel that he had put the men in danger of public exposure and legal repercussions. We will never know how many sleepless nights and worry-filled days the uninformed and unwilling research subjects had after the publication of and publicity surrounding Humphreys' book. However, no subpoena was ever issued; the data remained confidential, and no legal action was taken against the subjects.

Decades later, the Humphreys study continues to be a source of debate about where to draw the line between ethical and unethical behavior. Although many of its substantive findings have been forgotten, Humphreys' work has become a classic because of the intense debate about ethics that it generated. Discussing the choices that researchers like Humphreys made can help others to understand more clearly the legal and ethical issues implicit in doing social research. Debate about the costs and benefits of deception continues. For a less common point of view, see Box 3.4.

Ethical Issues Concerning Colleagues and the General Public

STOP AND THINK   *Assume that you have designed a study, and you feel confident that the participants will face minimal risks. Can you think of any other ethical responsibilities that should concern you?*

In addition to treating research participants ethically, researchers have responsibilities to colleagues and to the public at large. The actions of an unethical researcher can affect the reputations of other social scientists and the willingness of potential research participants in the future, so the appropriate treatment of research participants is part of the responsibility to colleagues. Soliciting input from groups involved in the research topic, such as local advocacy groups or community leaders, is an important part of ethical behavior (Dickert and Sugarman, 2005).

honest reporting, the ethical responsibility to produce and report accurate data.

A central ethical principle is **honest reporting,** which is the responsibility to produce accurate data, report honestly, and disseminate it in both professional and community forums. Recent scandals in the biological sciences and a survey of scientists that found a significant number anonymously admitting to having committed some form of scientific misbehavior (Singer, 2005) indicate that dishonest reporting might be more widespread than previously thought. We want to emphasize that the presentation of data without fraud, distortion, or omission; disclosure of the research's limitations; and a willingness to admit when hypotheses are not supported by the data are important responsibilities to colleagues and the general public.

If a sponsor funds a study, the researcher must be aware of the possibility of losing control of the study. Funding agencies and sponsors can influence research agendas and methodology by making awards on the basis of certain criteria. In addition, they can limit the freedom of researchers by imposing controls on the release of research reports. A researcher hired and paid by an organization will be expected by that organization to work in its best interest. That organization might attempt to keep data that are gathered from public disclosure (Homan, 1991, cited by Davidson and Layder, 1994: 56).

Sometimes a conflict of interest exists between the aims of the sponsor of the study and those of the larger community, especially if the sponsor is interested in influencing or manipulating people, such as consumers, workers, taxpayers, or clients (Gorden, 1975: 142). For example, in one city, a team of researchers representing a city planning commission, a community welfare council, a college, and several cooperating organizations, including local Boy Scout officials, studied the metropolitan area's social problem rate. At the conclusion of the study, the planning commission, the welfare council, and a local council of churches attempted to suppress the research findings that did not fit their groups' expectations or images. In each case, "it was one of the front-line field workers, viewed as 'hired hands' by the sponsoring agencies, who had to press for full dissemination of the findings to the public at large" (Gorden, 1975: 143–144).

Ethical Conflicts and Dilemmas

Who owns the data? . . . Whose interpretation counts? Who has veto power? . . . What are the researcher's obligations after the data are collected? Can the data be used against the participants? Will the data be used on their behalf? Do researchers have an obligation to protect the communities and the social groups they study or just to guard the rights of individuals? (Marecek, Fine, and Kidder, 1997: 639)

Most examples of unethical behavior in the social and behavioral sciences research literature are rooted in ethical dilemmas rather than in the machinations of unethical people. Many of the dilemmas involve *conflicts*— conflicts between ethical principles, between ethical principles and legal pressures, or between ethical principles and research interests.

Conflict Between Ethical Principles

In some research situations, researchers encounter behavior or attitudes that are problematic or morally repugnant. Although ethical responsibilities to subjects might prohibit us from reporting or negatively evaluating such behavior, other ethical principles hold us responsible for the welfare of others and the larger society. James Ptacek encountered an example of this ethical dilemma in his study of wife batterers. He recruited his sample from a counseling organization devoted to helping such men, paid each participant a nominal fee, and told the men that participation would help the organization and that the interview was not a formal counseling session (1988: 135). He used a typical sociological interviewing style—staying dispassionately composed and nonjudgmental even when the men described bloody assaults on women. Even though Ptacek's goal was to facilitate their talking rather than to continually challenge the men, he did worry about the moral dimension of his impartial approach. In the three interviews where men reported ongoing if sporadic violence, Ptacek switched from an interviewer role to a confrontational counselor role after the formal questions had been completed. He noted, "At the very least, such confrontation ensures that among these batterers' contacts with various professionals, there is at least one place where the violence, in and of itself, is made a serious matter" (Ptacek, 1988: 138). Facing a conflict between ethical principles, Ptacek gave a higher priority to the elimination of abuse than to the psychological comfort of study participants.

When the method of presenting findings can result in participants being identified, the researcher faces another dilemma. Conflict between reporting results and maintaining confidentiality confronted the authors of *Small Town in Mass Society: Class, Power and Religion in a Rural Community*. Arthur Vidich and Phillip Bensman have described their community study of a town in upstate New York as unintentional and unplanned, a by-product of a separate organized and formal research project (1964: 314). One of the major problems they faced occurred because, during the data collection stage of the research, the townspeople of "Springdale" had been promised confidentiality and purely statistical analyses of the data. As the study progressed, the researchers came to focus on the town's political power structure and decided that the case study approach would be the most appropriate method of analysis. Identity concealment of townspeople described in the case study proved impossible. The use of pseudonyms and changing some personal and social characteristics to describe the members of the power hierarchy was not sufficient to keep their identities hidden, and, with the book's publication, the identities of those in the town's "invisible government" were recognizable to residents. The community leaders were distressed to be identified as part of such a government. Townspeople felt humiliated and were upset that their trust in the researchers had been betrayed (Vidich and Bensman, 1964: 338). Vidich and Bensman recognized that by publishing their analysis, they gave greater priority to the ethic of scientific inquiry than they did to the ethic of confidentiality. They came down on the side of honest reporting to colleagues and the public, but other researchers might have made different choices.

What if you wanted to study powerful organizations? And what if you wanted to study something that organizations did not want studied? Would you be tempted to be deceptive to open closed doors? Robert Jackall (1988) was turned down by 36 corporations when he asked for permission to study how bureaucracy shapes moral consciousness and managerial decisions. His analysis of the occupational ethics of corporate managers was made possible only when he used his personal contacts to gain access and recast his proposal using "bland, euphemistic language" (Jackall, 1988: 14). He was able to do field work and interviews in a chemical company, a defense contractor, and a public relations firm because the companies thought his interest was in issues of public relations. Jackall completed an analysis of the "moral mazes" navigated by executives only after he made a few twists and turns of his own. His work answers some questions and raises one: Can the knowledge gained by a study outweigh the ethical costs?

STOP AND THINK   *What would you do if someone you were interviewing told you that they were planning to engage in self-destructive behavior? Would you keep the information confidential or report it to someone?*

Although ethical responsibilities to subjects might prohibit us from reporting what study participants tell us, other ethical principles hold us responsible for the welfare of others and to the larger society. An uncommon but possible scenario is when a researcher obtains unanticipated information that a research participant intends to harm him/herself or a third party (Palys and Lowman, 2001). Don't we as researchers have the responsibility of trying to intervene when someone is in serious danger? The American Sociological Association's Committee on Professional Ethics has considered the choice that must be made if, for example, a researcher who gets permission from a family to observe their interaction discovers child abuse. Should the child abuse be reported and the confidentiality that might have been promised to the parent be breached, or should confidentiality be maintained and risk future harm to the child? On behalf of the ASA, Iutocovich, Hoppe, Kennedy, and Levine (1999: 5) argue that in such unanticipated situations, the social researcher must balance the importance of guarantees of confidentiality with other ethical principles. In Canada, the federal government's council on research recently gave similar advice noting that "The values underlying the respect of the protection of privacy and confidentiality are not absolute. . . . Compelling and specifically identifiable public interests, for example, the protection of health, life and safety, may justify infringement of privacy and confidentiality" (Tri-Council statement, cited by Palys and Lowman, 2001: 260).

STOP AND THINK   *What about a situation where an interviewee described having participated in illegal behavior? Would you keep that information confidential?*

Conflict Between Ethical Concerns and Legal Matters

Researchers may face legal pressures to violate confidentiality. For example, in 22 states, studying a topic such as child abuse or neglect means a researcher will face the legal requirement that anyone having reason to suspect maltreatment must report it to the authorities (Socolar et al., 1995: 580). This has significant implications for the informed consent process, as there are consequences of participating, including legal risks for subjects and legal liabilities for investigators (Socolar et al., 1995: 580). Some researchers have tried to avoid a legal obligation to report suspected child maltreatment by collecting data anonymously, but others would argue that ethical obligations remain.

There is also the issue of the legal limits of confidentiality. The only researcher in Canada to face a charge of contempt for refusing to divulge confidential information to a court successfully defended himself using a common law test to establish researcher-participant privilege on a case-by-case basis (Lowman and Palys, 1999: 5). But in the United States, under current rules of evidence, a social scientist has no legal right to refuse to turn over materials by claiming a privileged relationship with the participants (Leo, 1995: 124). For example, following the 1989 environmental disaster created when an oil tanker, the Exxon *Valdez*, ran aground off the coast of Alaska, research projects were conducted to assess the impact of the spill, including the social and psychological consequences. After the data were collected, some of the researchers and their associates were subpoenaed and deposed by Exxon, and one researcher, J. Steven Picou, was ordered to turn over all the information he had gathered (McNabb, 1995: 331). Although the case was settled before going to trial, the researchers learned that anything (files, agreements, tapes, diaries, etc.) can be subpoenaed and that criminal prosecution can result if the material is not released (McNabb, 1995: 332).

Researcher Richard Leo learned the same lesson on a much smaller scale. Leo had spent more than 500 hours "hanging out" inside the criminal investigation division of a large urban police department, observing police interrogation practices with departmental permission. Leo was subpoenaed by a suspect's public defender as a witness in a felony trial. Facing the threat of incarceration, Leo (1995: 128–130) complied and testified at the preliminary hearing but ultimately felt that he had betrayed confidentiality and spoiled the field for future police researchers because, as a result of his testimony, the suspect's confession was excluded from the jury.

In a case with a different outcome, sociologist Rik Scarce (1995) went to jail to protect the confidentiality of participants in his research. After completing a study of radical environmental activists, Scarce remained in contact with some of them. At one point, one activist, Rod Coronado, stayed at Scarce's house for several weeks while Scarce was away. The time in question included the evening that the Animal Liberation Front took an action that resulted in the release of 23 animals and $150,000 in damages at a local research facility. Scarce was subpoenaed to testify about Coronado before a grand jury. Even after being granted personal immunity, Scarce (2005: 16) refused to answer some questions because they called for information he had by virtue of confidential disclosure given in the course of his research. The district court judge held Scarce in contempt of court, ruling that if the government has a legitimate reason for demanding cooperation with a grand jury, a social scientist can not claim academic privilege; the appeals court upheld the ruling. After spending 159 days in jail, the judge decided that further incarceration was not likely to pressure Scarce into giving grand jury testimony and released him (Scarce, 2005: 207). The Supreme Court never heard Scarce's case and the precedent his case set stands. As sociologists Patricia and Peter Adler (2005: xii–xiii) note in the Forward to *Contempt of Court*, the book Scarce wrote chronicling his case and his time in jail, "Rik Scarce has laid it on the line for the rest of the community of scholars to ponder. He has not merely paid lip service to the ideals he cherishes, but he has served his time to honor them. . . . We can only hope that we are moving toward a time when the ethical becomes both respectable and legal."

Although these situations are rare, and most researchers do not usually face such difficult dilemmas, it is important to be aware of the legal limits of confidentiality. The American Sociological Association's statement (1999) takes that into account when it specifies that sociologists have an obligation to protect confidential information, while also noting that ownership of information is governed by law. Similarly, many institutional review boards ask researchers to specify that confidentiality is limited by the law. Ciambrone included in her informed consent form the statement that information provided by interviewees will be kept confidential to the extent of the law at the request of the IRB at her university. She tells us that she feels fortunate that no interviewee told her anything of concern, so that she did not have to consider breaking confidentiality.

Conflict Between Ethical Principles and Research Interests

STOP AND THINK   *You're thinking of a research project that involves studying people's behavior by observing them. You wonder what the effect of asking for permission will be. Could there be situations where you think it might be better for the outcome of the study if researchers **don't** ask for informed consent?*

There can be conflicts between following ethical principles and the methodological goals of data collection. Sometimes the choice is between using ethically compromised methods and not doing the research. We must ask at what point the researcher should refrain from doing a study because of the ethical problems. Did Milgram's work contribute so much to our understanding of human nature that the violations of ethics are worthwhile? How important were Humphreys' contributions? At what point should the Tuskegee study have been stopped?

At other times, the choice will be between methodologically valid data and ethically appropriate methods. We will explore some of these issues in future chapters as we discuss various study designs and methods of data collection. At this point, however, we'll present some examples of this conflict.

Dilemmas can occur when recruiting a sample of participants for a survey because telling every potential interviewee a lot about a study can lead to a greater number of refusals to participate than will a more cursory description of a study. Ruth Frankenberg (1993), for example, originally told people, accurately, that she was "doing research on white women and race," and was greeted with suspicion and hostility. When she reformulated her approach and told white women that she was interested in their interactions with people of different racial and cultural groups, she was more successful in obtaining interviews.

In observing groups or organizations who meet in private settings, it is ethical to ask for permission from those you'll be observing. If, however, those being observed are likely to change their behavior if they know about the observation, then you will need to make a choice. Consider, for example, studies that have used "pseudo-patients," research confederates that go into mental hospitals feigning symptoms. Some argue that such research creates high-quality data and yields new knowledge by observing true-to-life situations that would be distorted if informed consent were requested (Bulmer, 1982: 629–630). Judith Rollins (1985) faced this issue and decided to not disclose her identity as a graduate student researcher when she also worked as a domestic worker and collected data about her interactions with her employers. Rollins acknowledged feeling deceitful and guilty, especially when her employers were nice. However, she feels that her behavior was acceptable because there was no risk of harm or trauma to the subjects.

Matthew Lauder also feels that deception can be ethically defensible. In a study of a Canadian neo-National Socialist organization that had a history of racism and using violence, Lauder (2003) first tried collecting information by interviewing people using informed consent. But many group members refused to sign informed consent forms and his formal interviews produced little useful information about the anti-Semitic and religious character of the organization's ideology. Rather than abandoning the project, Lauder decided to change tactics. Using deception, he moved from an overt to a covert role by pretending to have converted to the group's worldview (Lauder, 2003: 190–191) and then by talking to members informally. He collecting a great deal of information covertly and protected confidentiality when using the data. Lauder feels that he balanced the needs of society against the needs of individuals and concluded that he was morally obligated to conduct the study.

Ward and Henderson (2003) came down on the side of ethics over research considerations as they conducted a long-term study of young people leaving the care of state service agencies. Noting that study participants had faced painful childhood experiences and emotional upset, they decided that some were too vulnerable to participate in the follow-up interview and others needed it postponed. They concluded that it was their responsibility to avoid placing participants in situations where it was very difficult for them to discuss sensitive issues (Ward and Henderson, 2003: 258).

Making Decisions to Maximize Benefit and Minimize Risk

Although most social scientific research does not place subjects in situations that directly and overtly jeopardize their health and well-being, social research does involve some risk. Discomfort, anxiety, reduced self-esteem, and revelation of intimate secrets are all possible costs to subjects who become involved in a research project. No investigator should think it possible to design a risk-free study, nor is this expected. Rather, the ethics of human subject research require that investigators calculate the risk-benefit equation and seek to balance the risks of a subject's involvement in the research against the possible benefits of a study for the individual and for the larger society.

Many colleges and universities have research review boards and most professional organizations have codes of ethics. Examining their guidelines can make us aware of the ethical issues involved in conducting a study and provide guidelines for appropriate behavior. Ultimately, however, the responsibility for ethical research lies with each individual researcher.

To make research decisions, it's wise to use separate but interconnected criteria focusing on the ethics, practicality, and methodological appropriateness of the proposed research. In selecting a research strategy, we should ask these questions: Is this strategy practical? Is it methodologically appropriate? Is it ethically acceptable? When the answer to all three questions is yes, there is no dilemma. But, if we feel that it will be necessary to sacrifice one goal to meet another, we must determine our priorities.

Each researcher must think about the consequences of doing a given study as opposed to *not* doing the study and must consider *all* options and methods to find a research strategy that balances being ethical, being practical, and the likelihood of obtaining good quality data. In doing research, each of us must recognize that we are balancing the rights of study participants against our desire for research conclusions in which we have confidence and that we can share with the public.

> **BOX 3.5**
>
> ## Summing Up Ethical Principles
>
> - Protect research participants from harm
> - Get informed consent
> - Be sure that study participants have not been pressured into volunteering
> - Collect data anonymously or keep data confidential
> - Submit the research proposal to a review board
> - Provide accurate research findings
> - Consider responsibilities to research colleagues and the general public
> - Maximize benefits and minimize risks

Summary

Even though researchers often face unanticipated situations, most aspire to conduct their studies ethically and to apply the values and standards of ethical behavior to all stages of the research process. Recognizing the need to use feasible and methodologically acceptable strategies, we should strive for appropriate and acceptable conduct during the planning, data collection, data analysis, and publication stages of research, as summarized in Box 3.5.

Being ethical in the treatment of study participants means collecting data anonymously or keeping identities confidential, providing adequate information about the study to potential participants, and, most important, working toward keeping negative consequences of participation to a minimum. Many ethical dilemmas can be anticipated and considered when planning a research strategy. Institutional review boards and the codes of ethics of professional associations can provide guidelines for individual researchers. When researchers face unusual ethical and legal dilemmas, they need to take special care in planning and conducting their studies.

In addition to the treatment of participants, ethical principles require researchers to conduct themselves in ways that demonstrate that they are responsible to other scholars and to society as a whole. Scholars contribute an important service to their fields and the general public by providing and disseminating accurate information and careful analyses. Researchers will confront difficult situations if ethical principles conflict with each other, when legal responsibilities and ethical standards clash, or when collecting high-quality data means violating ethical principles. These situations must involve careful decision making. The best way to prepare for ethical dilemmas is to try to anticipate possible ethical difficulties and to engage in open debate of the issues.

Knowing Your Responsibilities

All institutions that receive federal funds and most that conduct research have a committee or official group to review the ethical issues involved in studies done on campus or by members of the campus community. Called an Institutional Review Board (IRB), a Human Subjects Committee, or something similar, the group must approve research involving human subjects before such research begins. Your college or university almost certainly has such a committee.

Find out if there is a committee on your campus that evaluates and approves proposed research. If there is no such group, check with the chairperson of the department that is offering this course, and find out if there are any plans to develop such a committee and describe those plans. If there is an institutional review board, contact the person chairing this committee, and get a copy of the policies for research with human subjects. Answer the following questions based on your school's policies. Attach a copy of the policies and guidelines to your exercise.

1. What is the name of the review committee? Who is the committee's chairperson?

2. Based on either written material or information obtained from a member of the committee, specify the rules that apply to research conducted by *students* as part of course requirements. For example, does each project need to be reviewed in advance by the committee or is it sufficient for student research for courses to be reviewed by the faculty member teaching the course? If the committee needs to review all student projects, how much time would you need in advance of the project's start date to allot to the review process?

3. What are some advantages and disadvantages of your institution's procedures for reviewing the ethics of student research?

4. Based on the material you've received from the committee, describe at least three of the ethical principles that are included in the standards at your school.

Ethical Concerns: What Would You Do?

In the following examples, the researcher has made or will need to make a choice about ethical principles. For each situation, identify the ethical principle(s), speculate about why the researchers made the decisions they did, then comment whether or not you agree with their decisions.

A. Professor Ludwig wants to see the connection between experiencing family violence and academic achievement among students. He contacts the principal of a local high school who agrees to the school's participation. At a school assembly, the principal hands out Professor Ludwig's questionnaire, which includes questions about family behavior including violence between family members. The principal requests that the students fill out the survey and put their names on it. The following day, Professor Ludwig uses school records to obtain information about each student's level of academic achievement. He keeps all information that he obtains confidential.

1. What ethical principle(s) do you think was (were) violated in this research?

2. Describe probable reason(s) Professor Ludwig violated the principle(s) you noted in your answer to question 1.

3. Describe your reactions to the way this study was done.

B. Professor Hawkins receives funding from both a state agency and a national feminist organization to study the connection between marital power, spouses' financial resources, and marital violence. Deciding to interview both spouses from at least 100 married couples, she contacts potential participants by mail. Because she is a university professor, she writes to each couple on university letterhead stationery and asks each couple to participate in her study. She specifies that she will be interviewing them about their marriage, but does not specify the specific topics that the interview will cover or the source of her funding. She does specify that everything they tell her will be kept confidential.

1. What ethical principle(s) do you think was (were) violated in this research?

2. Describe the probable reasons Professor Hawkins violated the principle(s) you noted in your answer in question 1.

3. Describe your reactions to the way this study was done.

C. To study a radical "right-to-life" group, Professor Jenkins decides to join the group without telling the members that he is secretly studying it. After each meeting, he returns home and takes field notes about the members and their discussion. After several months of attending meetings, he is invited to become a member of the group's board of directors and accepts. One evening at the group's general membership meeting, one member of the board becomes enraged during the discussion about abortion information and procedures available at a local women's health clinic. He shouts that the clinic's staff deserve to die. Later that week, someone fires a gun through several windows of the clinic. Although no one is killed, several clinic staff members and one patient are injured. The police find out that Dr. Jenkins has been studying the local group and subpoena his field notes. Professor Jenkins complies with the subpoena, and the board member is arrested.

1. What ethical principle(s) do you think was (were) violated in this research?

2. Describe the probable reasons Professor Jenkins violated the principles you noted in your answer to question 1.

3. Describe your reactions to the way this study was done and the researcher's response to the subpoena.

4

Selecting Researchable
Topics and Questions

© Royalty-Free/CORBIS

Introduction

Do you know people who lived through a natural disaster such as a hurricane, a tornado, a tsunami, or the like? In what ways have their lives changed? Do you think their experiences are typical or unusual? Based on your current understanding of their experiences, could you create a research question about the social impact of living through a natural disaster?

Creating research questions by noticing patterns of behavior is a common way to begin the research process. In 1886, for example, after widespread rioting in London, sociologist Charles Booth published a paper reporting the results of his survey of poverty in London's East End (Bales, 1999). The press at the time was warning of revolution in the streets, but Booth argued that the working class was not well organized and did not pose a significant threat to the social order. In addition, he argued for social reform to address the poverty he documented. Booth's analyses were widely reported around the world in many newspapers, making him perhaps the first sociological "household name" (Bales, 1999). In Booth's work, we find an early example of how the events can shape research and, conversely, how research can shape the social and political responses to events and underlying social problems.

In 2006, four months after Hurricane Katrina, sociologist Ronald Kessler, director of the Hurricane Katrina Advisory Group Initiative, announced the start of two-year study of 2,000 survivors of the hurricane to see how they are coping. Kessler told the media, "The situation is worse here than in a lot of other disasters . . . [survivors] are spread out all over God's creation. There are people in Los Angeles and Oshkosh and in Florida and in New Jersey, so . . . we have to beat the bush all over the country" (Allen, 2006: A4). The study proposed by Kessler and his colleagues is another illustration of the connection between social problems and social research.

In Chapters 1 and 2, we talked about the building blocks of research in our discussion of concepts, variables, research questions, and hypotheses. Reading those chapters, you might have asked yourself, "Questions and hypotheses *about what?* Where does the idea for a research project come from?" We'll admit that we skipped over the issue of selecting research topics and questions. Now that you're acquainted with some aspects of research, including the ethical issues involved in pursuing the answers to research questions, we'll begin at the beginning.

In Chapter 2, we defined **concepts** as words that refer to phenomena that share common characteristics. Over the years, social scientists have studied many thousands of concepts, such as poverty, crime, social class, career success, unemployment, social support, and parental divorce. In this chapter, we'll add that a **research topic** is a concept, subject, or issue that can be studied through research. Because all the concepts we listed can be studied through research, they are also research topics. Interest in a research topic can emerge from a variety of sources, including personal experiences and the social world. In 1896, the public reaction to the riots that gripped his city was the impetus for Booth's study of poverty; in 2005, the suffering

concepts, words or signs that refer to phenomena that share common characteristics.

research topic, a concept, subject, or issue that can be studied through research.

caused by Hurricane Katrina and the national attention to the survivors motivated Kessler and his colleagues. As we'll read in this chapter, developmental psychologist Michele Hoffnung had both personal experiences and professional interests that set the stage for her study of career expectations. The ongoing work by Hoffnung and by Kessler and associates will yield important findings: *basic* information about the social world and understandings that can be *applied* to creating social policy.

research question, a question about one or more topics or concepts that can be answered through research.

Once a research topic is selected, researchers typically will do some reading or consult with others in the field. This is one way to figure out what they want to know more about. The next step is to develop a **research question,** which is a question about one or more topics that can be answered through research. Booth (Bales, 1999) ultimately asked and answered three questions: How many people were living in poverty? What brought them to poverty and kept them impoverished? What can be done to alleviate the poverty? The Hurricane Katrina Advisory Group research will answer questions about the numbers of people going hungry or not getting adequate medical care, and questions about their long-term mental health and the extent to which the survivors feel overwhelmed by the task of rebuilding (Allen, 2006). We'll read about Hoffnung's research questions in this chapter.

STOP AND THINK   *Think about your life and the issues you're facing and the concerns you have. Are there any questions you'd like answers to? This is one way to create a research question.*

Research questions can be about local or global phenomena; can be about individuals, organizations, or entire societies; can be very specific or more general; can focus on the past, the present, or the future, and can ask basic questions about social reality or seek solutions to social problems. Some examples of research questions can be found in Box 4.1.

hypothesis, a testable statement about how two or more variables are expected to be related to one another.

Research questions are similar to hypotheses, except that a **hypothesis** presents an *expectation* about the way two or more variables are related, but a research question does not. Like hypotheses, research questions can be arrived at inductively by generating a general question from specific observations, or by deducing a specific question from a more general understanding. Both research questions and hypotheses can be "cutting edge" and explore new areas of study, can seek to fill gaps in existing knowledge, or can involve rechecking things that we already have evidence of. Whereas research projects that have explanatory or evaluation purposes typically begin with one or more hypotheses, most exploratory and some descriptive projects start with research questions.

STOP AND THINK   *We're willing to bet that by this point in your college career you've read dozens of articles (and most likely some books) that report the results of research. You can probably identify the research questions or hypotheses that the studies started with. But, think about some articles you've read in the recent past, and try to recall if there was any information about the actual genesis of each project. That is, do you know why those particular social or behavioral scientists did those particular research projects at that point in time?*

> **BOX 4.1**
>
> ## Examples of Research Questions
>
> - Do students who take internships while in school get better jobs upon graduation than those that don't take internships?
> - What kinds of jobs do college graduates with degrees in sociology get after they graduate?
> - Does parental divorce have long-term effects on the children? If so, what are they?
> - Has there been a change in the number of unemployed people living in my city or town in the past decade? If so, what federal, state, and local policies have had an impact on unemployment?
> - Which regions of the United States have experienced the greatest population growth in the past decade?
> - Does the death penalty have an impact on the crime rate?
> - Do neighborhoods that differ in social class also differ in neighborliness?
> - What are the differences in the experiences of immigrants in the United States when compared to those of immigrants to European nations?

If the articles you've read are like most journal articles, they probably included a review of the literature on the research topic and made connections between the research question and at least some of the previous work in the field. In journal articles, research questions and hypotheses are presented as if they've emerged dispassionately (and, most often, deductively) simply as the result of a careful and thorough review of the prior work on a given topic.

As a result of the formal and impersonal writing style that is often used in academic writing, we often don't learn much about the beginnings and the middles of many projects. Although real research is "messy"—full of false starts, dead ends, and circuitous routes—we frequently get sanitized versions of the process. There are some wonderful exceptions,[1] but the "inside story" of the research process is rarely told publicly. More typically, written accounts of research projects are presented in formalized ways. Articles often include clearly articulated hypotheses, summaries of the work in the field that the study builds on, descriptions of carefully planned and flawlessly executed methods of collecting data, and analyses that support at least some of the hypotheses. Researchers will frequently comment on the limitations of their work, but only sometimes on problems encountered, mistakes made, and pits fallen into. Most articles published in scholarly journals present the research process as a seamless whole. There is little sense that, to paraphrase John Lennon, real research is what happens to you

---

[1] See, for example, some of the classics in social research including *Street Corner Society* (Whyte, 1955), *Sociologists at Work* (Hammond, 1964), and *Tell Them Who I Am* (Liebow, 1993) as well the more recent *Moving Up and Out: Poverty, Education and the Single Parent Family* (Holyfield, 2002).

when you're busy making other plans. Nor is there a sense that researchers are multidimensional people with passions, concerns, and lives beyond that of "researcher."

Because our primary goal is to explain research techniques and principles, we've tried to do more than reproduce conventional social science writing in the excerpts of research that we've included. To give you a realistic sense of the research process, we've focused on small, manageable research projects of the kind being done by many social scientists. In each case, we wanted to get "the story behind the story." We have been most fortunate to have found a wonderful group of researchers, each of whom was willing to give us a candid account. From them, we learned about the interests, insights, or experiences that were the seeds from which their projects grew, the kinds of practical and ethical concerns they had at the start, and how "real life" shaped their data collections and analyses. As a result, in each chapter, we are able to share at least a little of "the inside scoop."

As you read this, researcher Hoffnung is likely still working on her study of women's lives. Although the "end of the story" has yet to be written, in the following article, we'll read about the beginning and the middle.

FOCAL RESEARCH

Studying Women's Lives: From College into Their 30s

by Michele Hoffnung[2]

The Research Questions

In 1992, I began a study of women's work and family choices that will take fifteen or more years to complete. Several factors led me to undertake this time-consuming research project. First and foremost, my personal experience as a woman committed both to career and motherhood gave me an understanding of what such a commitment entails. Second, my concern about the unequal status of women in the labor force led me to focus on the costs of the conflicting demands of work and family for women. Third, the scarcity of theories about women's adult development and longitudinal research using female samples presented a challenge to me as a feminist researcher.

Even as a girl, I always knew I wanted children. I loved babies. My mother was one of eleven children; my father one of eight. While I was one of three children, I wanted to have *at least* four. I also expected to go to college. During my college years, I became very interested in psychology.

[2] This article was written for this text by Michele Hoffnung, Ph.D., Quinnipiac University, and is published with permission.

My success as a student led my professors to suggest I go to graduate school. I did and loved it. During my years at the University of Michigan, while I committed myself to an academic career, I also got married. My desire for motherhood remained unchanged; with my course work completed, I had my first child at the end of my third year of graduate school.

The first year of motherhood, I worked on my dissertation. Since my son was a good sleeper, since my husband was willing and able to share infant care except for nursing, and since we were able to hire a student to help us for a couple of hours a day, I managed. But, the next year was far harder than I ever imagined. Starting as a new assistant professor was more taxing and less self-scheduled than researching and writing a dissertation. I found balancing a new career and caring for a baby extremely difficult. I spent three days at school and four days at home with my son, writing lectures during nap time and late into the night. By mid-year I was exhausted. Luckily, I had an intersession in which to partially recover, so I survived.

What, I wondered, did women do who had less flexible jobs, husbands with less time or willingness to share family responsibilities, less money for some outside help, and no intersessions? The social science literature I read was proscriptive (saying what women should do or feel) rather than descriptive (describing what they *actually* did and felt). When my second child was out of infancy, I finally found the time to engage this question as a researcher. I conducted in-depth interviews with forty mothers of at least one preschool child to find out about their lives and discovered that mothers who were pursuing their careers were likely to carefully plan their families. This showed up particularly in their rigorous use of birth control, their systematic preparation for childbirth, and their careful selection of child care. They formed marital relationships with a less gender-stereotyped division of labor than the noncareer mothers; their husbands shared parenting. I also found that all of the women were grappling with the issue of employment; even at-home mothers were concerned with when and whether they would be able to work again (Hoffnung, 1992).

In my first study of women's lives, the women I interviewed were products of the fifties and sixties, when career aspirations were less fashionable for women than they are now. Elizabeth Almquist, Shirley Angrist, and Richard Mickelsen (1980) found that, in 1968, as seniors in college, 40 percent of the women in their sample wanted both career and family while the rest wanted family. Seven years later, they found that the women had underestimated the importance of work in their lives as most of them had exceeded their earlier work plans.

What, I wondered, were the realities for women growing up in more recent decades? Studies done during the 1980s and 1990s indicated that a vast majority of college women plan to combine family and career. When asked which roles they plan to have, the great majority of women said they want it all: marriage, motherhood, and career. While we know that most college women say that they plan to have a career as well as a family, we have little data on the kinds of families they want or how they are going to balance career and family. When Kristine Baber and Patricia Monaghan (1988) questioned 250 college women, they found that these women presented a contemporary vision for combining career and family. All planned careers, 99 percent planned to marry as well, and 98 percent also planned to be mothers. When asked how they would accomplish their goals, 71 percent said they planned to establish their careers before having their first child and the more career-oriented the women, the later they projected their first birth. The most common plan was to return to work part time after the child was at least a year old. Such plans for "combining," if implemented, can have a negative impact on a woman's career development. We also know that, whether they originally planned to or not, most women are employed these days, even mothers of preschool children. Many women have jobs rather than careers; others are underemployed given their talents and training, often because of their family commitments.

My experiences, previous research, and current interests led me to plan a long-term study of a group of young women. I thought that starting at a time when they had plans to "do it all," and continuing on as they made life choices, would reveal patterns in their lives and the efficacy of various choices. For example, does the level of career commitment affect family choices? Does it affect work satisfaction? My research goal is to see how women's ideas and expectations change as they become more educated, face the world of graduate and professional school and employment, and make the choices of whether and when to marry, whether and when to have children, and whether and how to combine career and motherhood. These are important questions for our time, as most mothers work outside the home, even though the family continues to have sole responsibility for infant and preschool care.

The Study Design

My earlier study was retrospective, with adult respondents recalling their childhood hopes and dreams; this one was to be prospective, asking young adults still in the relatively sheltered environment of college to look toward their futures. Since I am following the women over time, I can find out about the present rather than the past and watch them change and grow. My strategy is designed to avoid the problem of having respondents' memories altered by the choices they've actually made, but it requires a commitment of many years and creates a range of other methodological challenges.

In 1992–1993, I interviewed 200 seniors at five New England colleges about their expectations and plans for the future regarding employment, marriage, and motherhood. Since then, I have followed their progress, most years with a brief mailed questionnaire. During the seventh year of the study, I interviewed the participants by telephone. This regular contact has helped me maintain a high response rate over the years. In 2004, 164 women, 82 percent of the sample, returned their questionnaires.

Collecting Data

Because my study requires that respondents participate over a period of many years, I have had to interest the women enough in my project to commit to responding annually. Because American young adults move frequently, I have devised ways to contact them as they travel and relocate. In this over-time process, I see my relationships with the women as more ongoing, personal, and collaborative than is typical of more traditional and impersonal survey procedures. At the same time, I strive to maintain a value-free, open-ended environment in which they can honestly tell about their individual lives (Hoffnung, 2001).

A major methodological concern was to select a sample that included white women and racial/ethnic women from a wide variety of backgrounds. This was important because until recently, most social science research, even research on women, has focused on members of the white middle class. I wanted to recruit a sample that would be varied and accessible to me. I decided to select a stratified random sample of female college seniors from five different colleges and universities. While my strategy excludes the lower ranges of the socioeconomic spectrum, by including a small private college of mostly first-generation college students, a large state university, two women's colleges, and an elite private university, I reached women with a wide range of family, educational, and economic backgrounds. My proposal was approved by my university's IRB; each college required me to follow its own procedures to gain access to the senior class lists which would allow me to pick random samples of both white women and racial/ethnic women. Once obtained, the college permission gave me credibility when I called to ask the women to participate, and meant that alumni associations would help me locate students in the years after graduation.

Since in the initial in-person interview I could expect students to come to me only if I made it convenient, I needed to find an office in which to do my interviews in private and a phone for people to call if they couldn't come. One college was my home institution, where I already had an office. At another college, I applied for and received a fellowship and office space at a research institute housed on campus. For the other three campuses, I used strategies based on the contacts I had. At one school, I called three feminist psychologists whom I had met professionally; one offered me her own (but phoneless) research space. At the second, I called the chairman of the psychology department and he provided me with unused space reserved for student projects. A campus phone in the hall enabled me to call interviewees, although I couldn't receive calls. In the third, I called the chairman of the history department, who knew my husband, and asked for help. His secretary provided me with the office and phone of a faculty member on sabbatical.

Even with the office problem solved, students needed to be called, recruited, scheduled, reminded, and if they did not show, called again and rescheduled. Each interview took about an hour, but many more hours went into phone calls and waiting for those who forgot or skipped their appointments. I quickly learned to make reminder phone calls, and I always called and rescheduled interviews for those who missed an appointment. As a result, most of the random sample I contacted completed the interview. I suspect a couple of women finally showed up to stop my phone calls!

In the initial interview, I used a set of questions that was consistent from respondent to respondent, yet was flexible enough to allow respondents to explain themselves. I asked questions about their families of origin, their attitudes about work, marriage, motherhood, and their short- and long-term plans for the future. In the mailed surveys I've used in subsequent years, I've asked questions about relationships, work, school, and a few open-ended questions about aspirations, feelings, and concerns. In the recent phone interview, I focused on work experience and aspirations, personal histories and expectations, and experiences and expectations concerning motherhood. Because some of the women are married and some are not and only some are mothers, this set of interviews took different forms with different participants. Notes and comments from the women during each data collection have helped to shape the next year's questionnaire or interview.

Findings

What have I found so far? As I anticipated, in their senior year, while only one woman was married and another one was pregnant, virtually all planned to have a career (96 percent) and to have children (99 percent). Only 87 percent expected to marry—less than in earlier studies of young women. Since white, non-Hispanic women were significantly more likely to expect to marry than other women, the smaller percentage may be due to my more culturally diverse sample of women (118 white, 25 African American, 20 Hispanic, and 36 Asian American). This appears to be a realistic appraisal, since other data indicates that African American women have the highest rates of nonmarriage; Chinese and Japanese American women the next highest rates, and European Americans have the lowest nonmarriage rates (Ferguson, 2000).

For these 1993 graduates, career was the primary focus of their 20s. Eleven years after college graduation, more than half had obtained graduate degrees. One third earned a masters' degree, 21 percent a doctorate (including MD, JD, and PhD), and 8 percent are still students. In contrast, their spouses are not as well educated: 21 percent have not earned a bachelors degrees, 34 percent have just a bachelors degree, 20 percent a masters' degree, and 25 percent a doctorate.

By 2004, almost three quarters (73 percent) are married, 3 percent are engaged, and 6 percent are in committed relationships. Another 4 percent are separated or divorced. Although 4 percent either said they are not dating or did not specify the sex of their partner or partners, 5 percent have women partners (always or sometimes) and 91 percent have men partners exclusively.

As seniors in college virtually all of these women wanted children. By year eleven, 39 percent are mothers. Of those, 38 percent have one child; 49 percent have two; 14 percent have three. Another 4 percent were pregnant with their first child. Now in their early 30s, family life has begun to have an impact on their careers. In 2004, 67 percent of the women were employed full-time and 22 percent were employed part-time. Of the mothers, 42 percent worked full time and 54 percent had reduced or relinquished employment. A quarter (26 percent) of the mothers were home full-time caring for children, whereas only 3 percent of their spouses were. In the years to come, as more of the women have children and the children grow beyond infancy, watching the interaction of family and career responsibilities will be on my research agenda.

REFERENCES

Almquist, E. M., Angrist, S. S., and Mickelsen, R. (1980). Women's career aspirations and achievements: College and seven years after. *Sociology of Work and Occupations, 7,* 367–384.

Baber, K. M., and Monaghan, P. (1988). College women's career and motherhood expectations: New options, old dilemmas. *Sex Roles, 19,* 189–203.

Ferguson, Susan J. (2000). Challenging traditional marriage: Never married Chinese American and Japanese American women. *Gender & Society, 14,* 136–159.

Hoffnung, M. (1992). *What's a mother to do? Conversations on work and family.* Pasadena, CA: Trilogy.

Hoffnung, M. (2001, March). *Maintaining participation in feminist longitudinal research: Studying young women's lives over time.* Paper presented as part of the Feminist Research Methods Symposium at the National Conference of the Association for Women in Psychology, Los Angeles, CA.

Sources of Research Questions

STOP AND THINK   *Chances are that recent newspapers or local newscasts covered stories about social phenomena. Perhaps there's a story about a family coping with the loss of a loved one, a graph showing changes in the crime rate, an editorial about a proposed change to state laws to allow gay and lesbian couples to marry, or an article on the job market for new college graduates. Do you have a personal interest in any of these topics? If you needed to design a research project for a class assignment, which of these or other topics might you select to study?*

In the focal research, Hoffnung tells us the "story behind the story." Her account points out that the selection of a research question is often the result of many factors, including personal interests, values, and passions; the desire to satisfy scientific curiosity; previous work on a topic (by the researcher in question or by others); the current political, economic, and social climates; being able to get access to data; and having a way to fund a study. Although each research project will have a unique history, some factors are common to most.

Values and Science

During the nineteenth century and for a good part of the twentieth, it was commonly thought that all science was "value free," but today it is more widely believed that values, both social and personal, are part of all human endeavors, including science. Science and even the most objective scientific products are socially situated (McCorkel and Myers, 2003: 201). Group interests and values can subtly influence scientific investigations, and they are especially influential during the creation and evaluation of hypotheses. Values come into play because, "At any given time, several competing hypotheses may explain the facts equally well, and each may suggest an alternate route for further research" (National Academy of Sciences, 1993: 342). However, social and personal values do not necessarily harm science.

The desire to produce knowledge is a social value. So is the desire to do accurate work and the belief that such work can ultimately benefit rather than harm humankind. One must simply acknowledge that values contribute to the motivations and conceptual outlook of scientists. The danger comes when scientists allow their values to introduce biases into their work that distort the results of scientific investigation (National Academy of Sciences, 1993: 343). Articulating values can help others evaluate a researcher's work, and such expression can make the researcher more aware and perhaps more objective as a result.

Personal Factors

STOP AND THINK *Compare your life today to your parents' lives when they were your age. How did they keep in touch with their friends? What are some of the differences that you can identify (such as cell phones, e-mail, instant messaging, blogging, and the Internet)? Think about the social and psychological impacts of the widespread use of a technological innovation, such as the cell phone, and create a research question that's of interest to you.*

Researchers often undertake studies to satisfy intellectual curiosity and to contribute to society and to social science. However, personal interests and experiences can influence the selection of a specific research topic or question. It shouldn't surprise us that researchers often study topics with special significance for them. After all, research projects can take months or years to complete; having a strong personal interest can lead to the willingness to make the necessary investment of time and energy.

Hoffnung tells us how her personal interest in career and motherhood influenced her work. Karen Norman describes the connection between her study of homeless youth and her own experience of having been homeless, pregnant, and in an abusive relationship (2000: 334); Jane Bock writes of conducting a study of single mothers by choice as a "midlife single woman who had long considered single motherhood but had not yet taken any action" (2000: 66). William Foote Whyte recalled the impetus for his classic study, *Street Corner Society,* as resulting from a day spent visiting the slums of Philadelphia with a group of other college students. He remembered the day "not only for the images of dilapidated buildings and crowded people but also for the sense of embarrassment I felt as a tourist in the district. I had the common young man's urge to do good to these people . . . I began to think sometimes about going back to such a district and really learning to know the people and the conditions of their lives" (Whyte, 1955: 281–282). Several years later, having been given a fellowship to pursue any line of research, Whyte's memory of the vague notion of studying a slum district became the genesis of a well-known research project.

Another important study came about through much more difficult personal circumstances. In 1984, anthropologist Elliot Liebow learned that he had cancer and a limited life expectancy. Not wanting to spend what he thought would be the last months of his life working at his government job, Liebow retired on disability, but found himself with a great deal of time on his hands (1993: vii). He became a volunteer at a soup kitchen and later at an emergency shelter for homeless women and found he enjoyed meeting the residents. Liebow was struck by the enormous efforts that most of the women made to secure even elementary necessities and he appreciated their humor and lack of self-pity (1993: viii). Liebow volunteered for two years, but then, out of either habit or training, during the last three years of his work at the shelter, he collected data as a participant observer—talking to, observing, and taking notes about the women and their lives. Before his death, Liebow completed an important contribution to helping others, the thoughtful and sympathetic analysis, *Tell Them Who I Am: The Lives of Homeless Women* (1993).

Other, much less dramatic examples of personal interests and experiences can be found in many research projects. Emily's first major study, for example, was a study of the division of labor and the division of power in marriage. She began the study in the first few years of her own marriage, as half of a two-career couple. And, perhaps because both of us, Emily and Roger, are parents who have read hundreds of books to children over the years, we've produced, separately and together, more than a dozen studies on children's literature (two of which we've included as focal research sections in this text). In Box 4.2, Jessica Holden Sherwood recounts her search for a research topic and how she came to study members of exclusive country clubs, a social world that she had some connection to.

Research and the Social, Political, and Economic World

Personal interests alone rarely account for the selection of a research topic. Knowledge is socially constructed and socially situated. Social relations produce individual research projects and the enterprise of research itself. In fact, all knowledge claims "bear the fingerprints of the communities that produce them" (Harding, 1993: 57). As social, political, and economic climates change, the kinds of research questions that social scientists pose

BOX 4.2

Reflections on "Studying Up" by Jessica Holden Sherwood[3]

I wasn't one of those people who had a plan right from the beginning. Approaching my doctoral dissertation, I was like any student in need of a paper topic. At my university, graduate students must choose specialty areas within sociology. For one of mine, I picked social inequality, as in race, class, and gender. Often, studying inequality means studying minorities, the poor, and women. Missing from this picture is the other side of each equation: the white, the rich, and men. Sometimes I would read or hear a social scientist point out the importance of "studying up" as well as "down" regarding inequality. While these groups are not what we usually think of when we think of race, class, and gender, that's part of the point. If they *don't* get studied, we'll never have a complete understanding of inequalities—who maintains them, why, and how.

The call to "study up" or study the "unmarked" (the categories that seem standard and go unnoticed) resonated with me. My graduate experience at a large state university has been eye-opening in terms of my own position on the socioeconomic scale. Coming from an upper-middle-class background, my socioeconomic position had been unmarked for me. Now I recognize its significance: first, that as a member of the unmarked group (like "white"), one of my privileges has been never to think about that group membership[4]; and second, that my life was organized in such a way (such as attending an expensive private college) as to preserve that obliviousness well into my twenties.

One reason that "studying up" is rare is that it's considered difficult. Especially when it comes to class, advantaged group members are thought of as elusive subjects of study. I thought that my own social location—being white, knowing some people that are rich and/or powerful, and having the cultural capital[5] to interact with them effectively and comfortably—might enable me to overcome that elusiveness.

[3] Jessica Holden Sherwood wrote this while she was a graduate student in the Department of Sociology, North Carolina State University. Her dissertation, *Talk About Country Clubs: Ideology and the Reproduction of Privilege* can be found at http://www.lib.ncsu.edu/theses/available/etd-04062004-083555. This article is published with permission.

[4] McIntosh, Peggy. 1992. "White Privilege and Male Privilege: A Personal Account of Coming to See Correspondences Through Work in Women's Studies," in *Race, Class and Gender: An Anthology,* ed. Margaret Anderson and Patricia Hill Collins. Belmont, CA: Wadsworth.

[5] Cultural capital is an acquired cluster of habits, tastes, manners and skills, whose social display often sends a message about one's class standing. From Bourdieu, Pierre. 1977. "Cultural Reproduction and Social Reproduction," in *Power and Ideology in Education,* eds. Jerome Karabel and A. H. Halsey. New York: Oxford University Press.

(continued)

> ### BOX 4.2 (*continued*)
>
> Another facet of my biography is that I have one Jewish parent and one WASPy parent. (WASP—White Anglo-Saxon Protestant—is a common label in the Northeast.) So, I have a tenuous relationship with the mythical "old-WASPy-Yankee" peoples. It is not hopelessly foreign and distant, but it's also not a circle I firmly and wholly belong to. So I am faced with the option of making efforts to belong to that circle, and also asking myself why I even want to. This tenuous relationship is probably one reason that I care about, and have chosen to study, this particular slice of social life.
>
> Next, I needed a place to find these people, a context in which to study them. I attended an elite boarding school, and I considered doing research at that school or a similar one. But, I was more interested in the doings of adults than teenagers. I'm familiar with a small island nearby that's a destination for wealthy vacationers, and I considered studying its social life. But, anybody can visit that island, and the wealthy vacationers don't all know each other, so it's not exactly a coherent community. I've settled on something that's convenient and concrete: a study of country clubs and other exclusive social clubs in one section of the Northeast. These private social clubs provide a specific context for the social lives of people who are "unmarked" and advantaged rather than disadvantaged.
>
> I'm finding a unique blend of pros and cons as I do this research. On one hand, I'm having a perfectly smooth time securing interviews with the people who are members of the clubs. On the other hand, though I expect my dissertation to be critical, I'll hesitate to condemn the people I've studied, because some of them are neighbors and friends. But in any case, I'm pleased with the research topic I've settled on and feel it's enriched by my consideration of its connection to my biography.

and the kinds of questions that are acceptable and of interest to the research community and the larger society change with them.

STOP AND THINK *Some topics have been the focus of considerable research in the social sciences for a century or more, but others have received attention only within the past decade or two. Can you think of examples of topics that have been studied only recently?*

Some topics have been largely invisible to the general public and generally ignored by social scientists until recently. In Chapter 12, we'll read about an interesting study that compares homicide rates of different groups. While in graduate school, one of the study's authors, Ramiro Martinez, was told that homicide was the leading cause of death among young African American males. But when he asked about the Latino-specific rate, he was told that this information was not available. This omission became one of the reasons he began his work (Martinez, personal communication).

Before the 1970s, few studies focused on women and, when considered at all, data about women were analyzed using male frames of reference[6] (Glazer, 1977; Westkott, 1979). However, as a result of the women's movement, which first developed outside of the university in the late 1960s and early 1970s, scholars began to study women and their lives. Hoffnung, for example, noticed that few of the long-term studies in her specialty, adult development, included women as subjects. She found herself perplexed that the theories that she had studied as a student focused on only men and their lives. In her professional work, she sought to redress the imbalance and began using new theories being developed by feminist thinkers such as Nancy Chodorow, Carol Gilligan, and Jean Baker Miller (Hoffnung, personal communication).

Critiques of gender imbalance are found frequently in social science. For example, in the mid-1990s, when three out of five American women were employed, Weaver (1994) noted that the number of studies focusing on women's retirement were quite small. Studies of the homeless have followed a similar pattern, with only the more recent studies paying attention to women's experiences (Liebow, 1993; Lindsey, 1997). Given the recent changes in the social climate, current research is much more likely to focus on women and men.

Change in society also has influenced the amount of research on other topics. In 1995, Allen and Demo argued that sexual orientation was understudied, especially in research on families. Analyzing 9 journals that published more than 8,000 articles on family research between 1980 and 1993, they found that only 27 articles focused explicitly on gay men, lesbian women, or their families (Allen and Demo, 1995). The data led them to conclude that sexist and heterosexist assumptions continued to underlie most of the research on families, with the research focusing mainly on heterosexual partnerships and parenthood. Allen and Demo felt that studying lesbians and gay men as individuals, but not as family members, reflected a societywide belief that "gayness" and family are mutually exclusive concepts (1995: 112). To see if this criticism was still valid, Emily did a search of *Sociological Abstracts,* a database of scholarly sources that we'll talk about later in this chapter. Her search found citations for only 20 articles about gay men and lesbians as family members published between 1996 and 2000, and more than 200 published between 2001 and 2005. If this particular omission has been rectified, there will always be topics that need more research. Chito Childs (2005), for example, noticing that black women's voices were overlooked in research on interracial relationships, designed a study focusing on them; Hagestad and Uhlenberg (2005) point out that while ageism has been studied extensively, age segregation is a neglected topic; Szinovacz and

---

[6] The natural, no less than the social, sciences offer many interesting examples of such male bias. In one federally funded study examining the effects of diet on breast cancer, only men were used as sample subjects (Tavris, 1996).

Davey (2006) note that research on grandchild care has ignored the impact of other simultaneous life transitions, such as retirement.

When researchers decide to break new ground, they may find a lack of support for their projects. Helena Lopata, a well-established sociologist, encountered negative reactions from many in her social and professional circles when she decided to study the social role of widows. She pursued her interest despite the view that the subject was judged "depressing and as unworthy of serious sociological research while 'more important' scientific problems remained unstudied" (Lopata, 1980: 69).

Topics and issues become the focus of research when they become visible, more common, and/or a source of concern to the public or the scholarly community. We've seen, for example, more studies on the elderly as the proportion of the population older than 70 has increased. Similarly, as the numbers of children living in single-parent and in two-worker households has risen, the impacts on children of family structure and workplace demands have received increasing scholarly attention. In Great Britain, Rachel Thomson and colleagues planned a study that we'll read about in Chapter 7 when they noticed that many British policy and curriculum decisions were being made because of untested assumptions about ethical and moral decline among young people. Finally, consider the annual global costs of public expenditures on criminal justice—which in the late 1990s was $360 billion (Farrell and Clark, 2004)—and then note how many hundreds of articles on crime and criminal behavior are published annually.

Sometimes current events focus increased attention on an existing research topic. Although the social and psychological aspects of natural disasters have been studied for decades, recent events are probably the reason why attention to the social dimensions of disasters has increased. A quick search of articles indexed in *Sociological Abstracts* finds the number of articles presenting research on the social impact of hurricanes, earthquakes, and tsunamis to have tripled over a five-year period—from 12 articles between 1996 and 2000 to 36 articles between 2001 and 2005. Social scientists are concerned about contemporary social problems and often select research topics to suggest solutions.

STOP AND THINK *Let's say you're reading a newspaper and come across a column that argues against a sportswear company producing TV shows with rap stars. Derrick Jackson (2005) writes in one such article that "It is tragic enough that black rappers and hip-hop moguls prostitute themselves to the Fortune 500 with the very stereotypes about violence, stupidity, and sexual drive that white society used to justify slavery, colonization, segregation, and lynching. After slave rebellions, the Underground Railroad, patriotism in world wars, marches on Washington, and murders of civil rights workers, Jay-Z makes millions saying, 'I take and rape villages.' African-Americans can no longer afford to coddle these people." Think about the issues being raised, including the impact of songs and media, attitudes about profit-making companies and rappers, and how one might go about creating social change. If you had the resources to do a research project in this area, what would your research question be?*

Research Funding

As one set of researchers has noted, "It is always difficult to raise money for social science research; after all, it neither directly saves thousands of lives nor enables one to kill thousands of people" (Fischman et al., 2004: x). Even a small study can be expensive and the cost of doing research can run into hundreds, thousands, or millions of dollars, so funding is an important consideration. Many research projects are funded by private foundations, government agencies, local and state institutions, or corporate sponsors. The United States has the largest number of research foundations among the Western industrialized nations. Many of these foundations offer support for the social and behavioral sciences. In 1981, for example, the four largest contributors to social science research—the Carnegie, Ford, Rockefeller, and Sloan foundations—contributed 17 percent of the total of $160 million granted to all social science research that year (Hewa, 1993: 71). Some foundations support specific kinds of social research. In 2005, the Robert Wood Johnson Foundation gave over $100 million in grant awards to support 329 projects involving research and evaluation in the areas of health and heath care (Knickman, 2005).

National funding for behavioral and social science research has been very important since the middle of the twentieth century. In the United States, social and behavioral research is funded by, among others, the National Institutes of Health; the National Science Foundation (NSF); the National Institute of Justice, the Department of Education, and the National Institute on Aging. Although the amount of funding decreased in the early 1990s, with the share of total federal research funding for the social and behavioral sciences falling from 8 to 4.5 percent (Smith and Torrey, 1996: 611), government support remains important. One branch of the NSF, the Division of Social and Economic Sciences, received $92 million in funding in fiscal year 2005 and supports research in areas that include sociology, economics, law and social science, political science, methodology, and the societal and ethical dimensions of science and technology (COSSA, 2005: 45). Within months of Hurricane Katrina, a project on the mental health of the survivors of the hurricane was funded by a $1 million grant from the National Institute of Mental Health (Allen, 2006). In the United Kingdom, the government-supported Economic and Social Research Council, the UK's leading social and economic research and training agency, allocates £46 million per year to social research in the United Kingdom (HERO, 2006).

Funding research expresses a value choice—the judgment not just that knowledge is preferable to ignorance, but that "it is worth allocating resources to research against the other directions in which modern states expend, or could expend, resources" (Hammersley, 1995: 110). The particular values associated with specific projects affect funding as the appropriations process for research is part of a larger political process. Each year, for example, after the president submits the budget, constituent institutes appear before appropriations committees to defend it. Interest groups, such as the Consortium of Social Science Associations, the umbrella organization of the professional societies in the behavioral and social sciences, testify as well. Research that is perceived to be threatening to those making decisions might have a more difficult time getting financial support (Auerbach, 2000).

Funding can be in the form of grants or in the form of a contract for an agency or sponsor. Grants typically allow the researcher more control of the topic, methods of data collection, and analysis techniques, whereas contracts usually commit the researcher to a specific topic, timetable, and methods. Although some sources are more restrictive than others, the availability of funding and economic support can influence a study, specifically the questions asked, the amount and kinds of data collected, and the availability of the resulting research report.

A fascinating example of the effect of funding on research is the story behind one of the largest study of sexual practices and beliefs in the United States (Laumann, Gagnon, Michael, and Michaels, 1994). The project was influenced by a combination of factors including a major health crisis, public values, politics, and funding. The initial motivation for the project was a growing AIDS epidemic and a public health community with little data on sexual behavior because of the controversial nature of doing such research in our society (Laumann et al., 1994: xxvii). By the mid-1980s, coalitions of federal agencies had expressed support for a national survey of sexual practices. In 1987, the National Institute of Child Health and Human Development requested proposals to design studies—one on adult sexual behavior and the other on adolescent sexuality (Laumann et al., 1994: 39). Researchers at the National Opinion Research Center in Chicago won the contract and designed the adult study, but Reagan and Bush political appointees at the highest administrative levels of the Department of Health and Human Services refused to allow approval of even a narrowly focused survey of sexual practices. When it became clear that public funding would not be forthcoming, the researchers found financial support from a consortium of private foundations. With the funding in place, the researchers were able to complete the first large, nationally representative study of human sexuality in decades. It is ironic that a *refusal* to sponsor research ultimately led to a much more comprehensive study than had been proposed originally (Laumann et al., 1994: 41). Another example of the political nature of research can be found in the 2003 action of U.S. Representative Patrick Toomey, who sponsored an amendment to a bill which, had it succeeded, would have rescinded National Institute of Health funding for five already approved grants on sexual behavior and health (Silver, 2006: 5).

Developing a Researchable Question

Regardless of the source of an idea for a project—scientific curiosity, personal experiences, social or political climate, the priorities of funding agencies, or even chance factors—once a researcher has a topic, there's still work ahead.

STOP AND THINK   *Let's say you did get interested in the topic mentioned earlier—the social and psychological impacts of the use of the cell phone. Now, where would you go from here?*

researchable question, a question that can be answered with research that is feasible.

Several steps are needed to turn a research question into a **researchable question,** a question that is feasible to answer with research. The first step is to narrow down the broad area of interest into something that's manageable. You can't study *everything* connected to cell phones, for example, but you could study the effect of these phones on family relationships. You can't study *all* age groups, but you can study a few. You might not be able to study people in *many* communities, but you could study one or two. You might not be able to study *dozens* of behaviors and attitudes and how they change over time, but you could study some current attitudes and behaviors. While there are many research questions that *could* be asked, one possible researchable question is: In the community in which I live, how does cell phone use affect parent-child relationships; more specifically, how does the use of cell phones affect parents' and adolescents' attempts to maintain and resist parental authority?

Reviewing the Literature

STOP AND THINK   *Let's assume that we've selected the impact of cell phones on family relationships as our research topic, and we want to begin a review of the literature. Where should we start?*

The accumulation of scientific wisdom is a slow, gradual process. Each researcher builds on the work others have done and offers his or her findings as a starting point for new research. Reviewing previous work helps us to figure out what has already been studied, the conclusions that were reached, and the remaining unanswered questions. Although a thorough review is time consuming, it can generate a more cumulative social science, with each research project becoming a building block in the construction of an overall understanding of social reality. Reading, summarizing, and synthesizing at least a significant portion of the work that has preceded our own is an important step in every project. We can look for warnings of possible pitfalls and reap the benefits of others' insights.

Because both the *process* of reviewing previous work and the resulting *written summary* of the work that has been reviewed are both called a **literature review,** we'll discuss both usages of that term in the sections that follow.

literature review, the process of searching for, reading, summarizing, and synthesizing existing work on a topic or the resulting written summary of a search.

Academic Sources

To start the *process* of reviewing the literature means figuring out what literature or sources you want to search. Books, articles, and government documents are among the most commonly used sources for materials that describe research. Popular literature, including newspapers and magazines, might be good sources of research ideas, but academic journals will be more useful for your literature review.

A good place to start looking for books and published government documents is in your university's library catalog. Many libraries allow others in addition to their patrons to search their catalogs electronically. If you find a book, article, or government publication that you want to use at a library other than your own, you can often request it through an interlibrary loan system. In addition, most government agencies have a great deal of material that can be downloaded from their websites. In addition to the sites of specific agencies, www.fedstats.gov links to information from 100 government agencies.

Hundreds of print and online journals publish research in the social sciences and related fields. Many of them are peer-reviewed journals, which means that the articles have been evaluated by several experts in the field before publication. Almost every university library has electronic databases that allow you to search many journals simultaneously. Some databases list citations and abstracts and others provide the full text of the articles. *Sociological Abstracts,* for example, is a database that provides citations and abstracts of journal articles, book reviews, book chapters, dissertations, and conference papers. It draws from more than 1,800 serial publications in sociology and related disciplines in the social and behavioral sciences. Other useful databases that many libraries have access to include *Academic Search, ERIC, Criminal Justice Abstracts, PsycArticles, PsycInfo, Social Sciences, Social Services,* and *Worldwide Political Science Abstracts.* You might also want to check out Google Scholar at http://scholar.google.com/, a free search engine that covers peer-reviewed papers, books, and articles from academic publishers, professional societies, and universities.

keywords, the terms used to search for sources in a literature review.

Although use of the different databases might vary, most allow you to search titles and abstracts only, or to search the entire article using one or more **keywords,** which are the terms and concepts of interest. You'll generate numerous citations (title of the article, name of the author, name of the journal, volume, date of publication, and pages) with commonly used keyword(s). You can limit the search to titles and abstracts only. With common keywords, narrow the focus by using multiple keywords—such as by including "and" between the terms. For more unusual topics, think of broader terms or synonyms that could be used as keywords in additional searches. If you want articles that focus on research findings, use the word "research" and/or "study" as a key term in the search.

STOP AND THINK   *If you were going to use a database to look for scholarly articles on the impact of the cell phone on family relationships, what keywords might you use?*

Using the electronic *Sociological Abstract* as our database, we searched for articles published between 2000 and 2005 for using the following terms: "mobile" or "cell," "phone or phones," and "research" or "study" searching the entire article record. We received 89 results, 55 of which were published in peer-reviewed journal articles, but when we added other search words, such as parent, family, or adolescent, the number of results dropped significantly.

Once a list of citations is generated, the titles and abstracts can help you decide if the book or article is relevant to your work. Most university libraries have many thousands of books, documents, and academic journals (in print, in microforms, or online). If a library doesn't subscribe to a

> **BOX 4.3**
>
> # Parts of a Research Article
>
> *The title* tells what the research is about. It usually includes the names of at least a few of the major concepts.
>
> *The abstract* is a short account of what the research is about including a statement of the paper's major argument and its methods.
>
> *Introduction* will usually introduce the topic and give a brief summary or review of the literature in the field, sometimes including relevant theories. In the introductory sections, the author usually justifies the need for the research, describes the work that will be added to, and gives credit to those who have contributed ideas, theories, conceptual definitions, or measurement strategies.
>
> *Data and Methods* sections describe and explain the research strategies and usually cover the population, sample, sampling technique, study design, method of data collection and measurement techniques that were used, and any practical and ethical considerations that influenced the choices.
>
> *Results or Findings* are usually the point of the paper. In findings sections, researchers incorporate data or summaries of data.
>
> *Discussion* section highlights the major findings and typically connects the findings to the literature review.
>
> *Conclusions* usually sum up the current study, the directions for future work, and may caution the reader as to the study's limitations.
>
> *References* list the books and articles referred to in the article.

specific journal or own the book, you can usually request a copy through interlibrary loan. When reading a journal article, you'll find that it will usually include a literature review with references to additional sources that you might want to check out. Box 4.3 describes the sections in a typical journal article, including the review of the literature.

A thorough literature search is a process of discovery. In addition to providing useful information, each article, book, or other item can direct you to new ideas and sources. Just as when you're on a hike in the woods with only a general idea of the direction that you want to travel, you'll need a good compass and some common sense to find your way.

Using the Literature in a Study

When you read the focal research sections in the chapters to come, you'll find that most of the researchers have used the existing literature to develop their own research questions or hypotheses. Some researchers, however, postpone a complete literature review until after data collection so they can develop new concepts and theories without being influenced by prior work in the field. In both approaches, prior work in the field is used at *some point* in the research process.

Existing research helps put new work into context. What comes before determines if a study is breaking new ground, using an existing concept in a new way, expanding a theory, or replicating a study in a new setting or with a new sample. When writing your own literature review, you will want to be up to date and thorough, but you will have to choose which sources to use and *give credit to,* using an appropriate method of citing sources. In writing a literature review for a research proposal or project, you should pick the materials that focus on the most relevant concepts, theories, and findings to provide a context for your own work.

You can use the literature review sections of published articles and research monographs as guides for writing your own literature reviews. A good literature review provides a perspective for the study you are proposing, in much the same way that a frame helps to focus attention on a painting. They can serve many purposes, among them a discussion of the significance of a research question, providing an overview of the thinking about a research question as it has developed over time, and presenting one or more theoretical perspectives on the topic. One endpoint of a literature review is one or more research questions that can be answered with data. In addition, a literature review can include definitions of useful concepts and variables, identify testable hypotheses deduced from prior research, and suggest some useful strategies and methodologies. Desirée Ciambrone, in her study of women with HIV/AIDS described in Chapter 3, found that doing a thorough literature review was one of the most important steps in the research process, as it honed her thinking on the topics she selected for her own interviews. For a discussion of writing a research report including some tips on writing the literature review, see Appendix A of this text.

STOP AND THINK *Let's assume that we've picked a topic and developed a research question about cell phones and family relationships. Let's also assume that as a result of the literature review, we have some ideas for a research strategy. Now, we need to see if our study is "doable."*

Practical Matters

Even with a general idea of a research strategy, it's important to make sure that the strategy is feasible before beginning the actual data collection. No matter how important or interesting a question is, we don't have a researchable question unless it is feasible or practical to answer the question with research. **Feasibility** or practicality means considering three separate but related concerns: access, time, and money.

feasibility, whether it is practical to complete a study in terms of access, time, and money.

access, the ability to obtain the information needed to answer a research question.

Access

Access is the ability to obtain the information needed to answer a research question. Although it's very easy to get some data, other kinds are much

BOX 4.4

Practical Matters

Think About It

If you wanted to do a study, would it be practical?

Will you be able to get access to the data?

Do you have the time to do this research?

Do you have the funds to do this research?

more difficult to secure. In her study of college seniors, Hoffnung had relatively easy access. Once the ethical issues of her proposed research had been reviewed and approved by the Institutional Review Board, Hoffnung was permitted to obtain a random sample of names and phone numbers of seniors. Hoffnung was able to cite the university's "stamp of approval," when she called the students to introduce herself, describe her project, and request participation. Relatively easy access and Hoffnung's personal and professional skills all contributed to the success of the project, as most of the women selected agreed to participate in the initial and later phases of the study.

STOP AND THINK   *College students are the subjects of many studies in the social and behavioral sciences, partly because of convenient access. Are there other groups or situations that you think are also easily accessible? Any that you think are particularly difficult to study?*

People and information that are connected in some way to organizations and institutions are fairly accessible. Examples of groups that are usually accessible include employees of businesses or organizations, members of professional unions or associations, patients in a hospital, residents of a nursing home, and families whose children attend a school. In each case, we have access to a place that allows us to easily meet or contact people to ask for their participation and an organizational structure that could facilitate the interaction. Ciambrone notes that finding women with HIV/AIDS to interview was the most difficult part of her study. She found enough study participants only when her contacts at a local hospital started referring women who were participating in an ongoing clinical trial being conducted to evaluate medical treatment.

Most public information is easy to obtain and therefore practical to analyze. Typically, materials produced by government or public agencies and most media, such as newspapers, children's books, and television commercials, are quite easy to access. For example, if you were interested in comparing states by the percentage of residents who classify themselves as Hispanic or Latino, you would find this information very easy to obtain. (See, for example, http://quickfacts.census.gov/qfd/states/.)

On the other hand, sometimes gaining access to records or to potential study participants can be quite challenging. As we noted in the previous chapter, Robert Jackall (1988) was turned down when correctly identifying his topic as a study of how bureaucracy shapes moral consciousness and managerial decisions, and was given access by corporations only when he recast his topic as a technical one. In a study of commercial fishing families, researchers mailed out 2,000 surveys but only had 43 returned. Phone calls inquiring about reasons for refusal prompted fishermen to reply in anger that the study was a waste of taxpayers' money (McGraw, Zvonkovic, and Walker, 2000: 69). Interested in the health of women moving from welfare to employment, Kneipp, Castleman, and Gailor (2004) were granted access to the Florida Department of Children and Family's database. They randomly selected the names of 150 women from one county who had received welfare at some time in the past year and had left welfare for employment, but found that with the phone numbers provided, they could not locate two thirds of the women. In their work on attitudes about DNA testing, Turney and Pocknee (2005) found it difficult to recruit men involved in father's rights groups and, as a result, employed a variety of strategies, including e-mail, telephone, and asking participants for the names of other fathers. (See Box 4.5 for examples of difficult access groups and records.)

In some situations, gaining access is an ongoing process as people "check out" researchers, find out if they are to be trusted, and then gradually let them into various parts of their lives and organizations. Jessica Holden Sherwood, in the research described in this chapter, was interested in an elite group: members of exclusive country clubs. Beginning with a few personal acquaintances who were club members, she asked them to identify others and then had no trouble securing interviews, most likely because she always introduced herself using the names of others who were participating (Sherwood, 2004).

As you read the focal research selections in this text, you'll notice variations in the accessibility of data. In the next chapter, for example, you'll see that Matthew Reavy had very easy access to his data, which were *USA Today* tracking polls. In Chapter 12, you'll learn about a study of homicide by Ramiro Martinez and Matthew Lee. Martinez was introduced to the chief of police for Miami through a colleague of a colleague and eventually received permission to do the study (Martinez, personal communication). By spending countless hours accompanying detectives on calls and demonstrating that he was trustworthy, Martinez was able to gain their confidence and that of the administrators and, as a result, access to the data (Lee, personal communication).

BOX 4.5

Groups or Records with Difficult Access

Group or Record	*Examples*
People without an institutional connection	Children 2 years of age
	Homeless people
	Those not registered to vote
	Atheists
People who are not easily identifiable	Undocumented immigrants
	Italian Americans
	HIV-positive individuals
People who don't want attention from outsiders	Members of gangs
	Anonymous philanthropists
	The Amish
	Child abusers
People who are very busy or who guard their privacy	U.S. Senators
	Celebrities
	Chief Executive Officers
Records that are private or closed	Diaries
	Tax records
	Census data for individual households

Time and Money

Research can be expensive and time consuming (see Box 4.6). In a study where subjects will be interviewed, salaries for interviewers and transcribers need to be budgeted. If the interviewees are to be given small gifts, like the $10 gift certificates that Ciambrone gave the women in her study, the study's cost increases. If mailed questionnaires are planned, costs include paper, printing, and postage. If historical or archival documents are used, there could be copying costs or time expenditures. In the study using homicide records that you'll read in Chapter 12, the researchers needed to fly to where the records were, rent a car, and stay in hotels in that city. They even ended up buying their own copy machine because they found the one in the police department frequently out of order or unavailable (Martinez, personal communication).

Some projects can be completed quickly and very inexpensively. For example, the data collection costs might be negligible when using easily obtained public documents or books that can be borrowed from libraries. Similarly, doing surveys by collecting data through questionnaires, especially if they can be handed out in groups, can be relatively inexpensive. Using survey data that someone else has already administered and that is

---

BOX 4.6

How Much Did Sex in America Cost?

Funded by nine private foundations, the survey of Americans' sexual practices and beliefs was an expensive and time-consuming project. It took 20 interviewers seven months to complete the 3,432 in-person interviews that lasted, on average, 1½ hours. Michael, Gagnon, Laumann, and Kolata (1994: 33) figured that the interviews with a random sample of adult Americans cost an average of $450 each when the interviewer training, trips to the interviewees' residences, and the data entry for analysis is included in the estimate.

---

archived on a database such as the ones maintained by the Inter-university Consortium for Political and Social Research (ICPSR) are also inexpensive. Typically, most Web-based information is easily accessible. To cite an example, the American Family Immigration History Center recently made documents from their archives about passengers who came to America through Ellis Island from 1892 to 1924 accessible through their site (www.ellisislandrecords.org). Although there is no charge for the data, it can take many hours to find information about specific individuals.

Projects that collect data over time can be much more costly than are those with only one data collection. Similarly, projects that study large numbers of people or groups are more expensive than are those that study smaller samples. Researchers can sometimes rely on donated materials and volunteer workers to make a study feasible. Other times, unanticipated costs and events can lead to a scaled-down project.

research costs, all monetary expenditures needed for planning, executing, and reporting research.

Although each method of data collection has unique costs and time expenditures, we can identify some general categories of expense. **Research costs** include the salaries needed for planning the study; any costs involved in pre-testing the data collection instruments; the actual costs of data collection and data analysis such as salaries of staff, payments to subjects for their time, rent for office space or other facilities; the cost of equipment (such as recording devices and computers) and supplies (such as paper or postage); and other operating expenses. **Time expenditures** include the time to plan the study, to complete the data collection process, and to organize and analyze the data.

time expenditures, the time it takes to complete all activities of a research project from the planning stage to the final report.

The focal research in this chapter is an example of an ambitious and a "doable" study. Hoffnung has invested a great deal of her own time in the project for many years and expects to do so for some years to come. By doing the interviewing herself, she is able to control the quality of the interviews and avoid the expense of hiring interviewers. She has obtained small research grants to cover costs of postage, tapes, recorders, and, the largest expense, interview transcription. She has kept costs to a minimum by employing work-study students as assistants for the follow-up questionnaires, by using her existing and donated office space, and by working on her personal computer.

Summary

Research questions can vary in scope and purpose. Some projects are designed to address very broad and basic questions whose answers will help us describe and understand our social world more completely. Other projects work toward solving social problems or creating information that can be used directly by members of communities and participants in programs or organizations. Questions can focus on one or more concepts; can be narrowly or broadly defined; concentrate on the present or changes over time; and are developed inductively or deductively. Questions can be generated from a number of sources, such as personal interests and experiences; the social, political, economic, and intellectual climates; previous research; chance factors; or the ability to get funding. The availability of funding can determine not only the size and scope of a project, but also the specific question, the timetable, and the methodology.

A review of the literature is essential at some point in the research process. By reviewing the work that other researchers have done, each new study builds upon the previous work and provides a foundation for the next wave of data collection and analyses.

Planning a study means considering the practical matters of time, money, and access to data. It's necessary to plan a feasible study. No matter how important or interesting a research question is, unless the research plan is practical, the study will not be completed. Hoffnung gained access to a sample of women to interview and she has been able to keep in contact with most of them since 1992. She has had sufficient resources of time and funding to be able to continue to collect and analyze data so that she will be able to answer the research questions she set for herself more than a decade ago.

EXERCISE 4.1

Selecting a Topic

1. Get a copy of a local or regional newspaper or go to a newspaper's website. Find three articles that deal with social behavior or characteristics. List the name of the paper, the date of publication, and the headline for each of these articles.

2. Select one of the stories and make a copy to turn in with this exercise.

3. Create one research question on the topic of the article that could be answered by social research and state it.

4. Assuming you could get enough funding and had the time to do research to answer the question you stated, would you do the research? Describe why you would or would not do the research by focusing on your personal interests, today's social and political climate, and the contribution you think the research would make to social science or society.

Starting a Literature Review

Select a topic you are interested in or one of the topics from Hoffnung's study, such as career aspirations, self-concept, two-career family, family planning attitudes, or role conflicts. Using your library's resources, find five citations of *original social research from social science journals* on this topic that are no more than five years old. (Be careful not to get an article that *summarizes* other research or is solely a discussion of theory.) Complete the following exercise:

1. Write down the topic you selected and the database(s) you used for your search.

2. List the five citations giving full bibliographic information for each including name(s) of the author(s), title of the article, name of the journal, date of publication, and pages of the article.

3. Select one of the articles from your five citations that is available to you either online or in print and read it. Put an asterisk (*) next to the name of the article that you've read.

4. Describe the review of the literature that the researcher(s) presented in the article. For example, is the literature review extensive? To what extent does it focus on theories and theoretical perspectives? To what extent does it present the research findings from previous research? How adequate is the literature review in covering previous work on the topic?

5. Does the researcher(s) start with a research question or hypothesis? If so, list one research question or hypothesis. Otherwise, list the topic of the study. Note the page number of the article where you found the research question, hypothesis, or the topic.

Using the Internet and the World Wide Web

1. Select one of the following topics: two-career family, death penalty legislation, or immigrant assimilation. Write down the name of your topic.

2. Using a search engine on the Web, such as Google, Yahoo, Altavista or Ask.com, do a search for research on this topic by searching for an exact phrase that's of interest (such as "death penalty legislation") *and* the word **research.**

3. How many results did you obtain in your search? If you ended up with more than a few hundred items, narrow down your topic in some way. Describe what you did to narrow the search. Describe the number and quality of the results you obtained.

4. Check out at least three of the sites that you found in your search. Identify the website, describe the information on the site, and evaluate the usefulness of this information for you and/or other researchers.

EXERCISE 4.4

Summing Up Practical Matters

Think about the following research questions and the suggested ways that a researcher might collect data. For each, give a cost and time estimate and your evaluation of whether access will be easy or hard.

Research Question	Will access be easy or hard?	Time consuming or not?	What are some expenses?
1. Do teachers' characteristics and teaching styles have an impact on amount of student participation in class? (Suggested method: Observe 50 classes for several hours each.)			
2. What kinds of countries have the highest infant mortality rates? (Suggested method: Use government documents for a large number of countries.)			
3. Is the psychological impact of unemployment greater for people in their 20s or 50s? (Suggested method: Do a survey using a questionnaire mailed to 300 people who have contacted the unemployment office in one state.)			

Research Question	Will access be easy or hard?	Time consuming or not?	What are some expenses?
4. Which regions of the United States have experienced the greatest population growth over the past 20 years? (Suggested method: Use data from the U.S Bureau of the Census.)			
5. Are people who are more religiously observant more satisfied with their lives than those who are less observant? (Suggested method: Interview 100 people for approximately 1 hour each.)			

© Mark Richards/PhotoEdit, Inc.

5

Sampling

Introduction

Have you ever asked yourself, "How do pollsters figure out what Americans feel about things like the war in Iraq, gay marriages, presidential candidates, and all those other things they tell us about on the news? They surely don't ask *all* Americans about their feelings, do they?" An important part of the answer to this question involves the topic of this chapter: sampling.

sampling, the process of drawing a number of individual cases from a larger population.

Sampling is the process of drawing a number of individual cases from a larger population. Researchers sample in the hopes that they can gain insight into a larger population without studying each member of the population. Presidential polls are based upon samples of the population that might vote in an election. We'd like to begin this chapter on sampling with three stories about presidential election polls (see Box 5.1), two that are examples of how misleading sampling can be, the other of how informative it can be.

STOP AND THINK   *Are there any of the 2004 poll results that, even when you take into consideration the margin of error, don't seem to embrace the final election results?*

Why did the *Literary Digest* poll, with two million respondents, do so poorly in 1936? Why did Gallup, with his apparently superior technique, miss the mark in 1948? And what was it about the 2004 polls that permitted them to adopt a sensibly cautious approach to the results of the 2004 election? The main reason has to do with the topic of this chapter: sampling. The science of sampling has progressed dramatically since the 1930s, to the point where even a basic introduction must touch on some relatively technical matters. But the logic of sampling can be easily grasped, and we think you'll actually enjoy learning something about it. Let's start at the beginning by addressing a question that has probably already occurred to you: why sample?

Why Sample?

Sampling is a means to an end: to learn something about a large group without having to study every member of that group. The *Literary Digest*, George Gallup, and the authors of the 2004 polls mentioned in Box 5.1 wanted to know how Americans were going to vote for president. Why didn't the *Literary Digest*, Gallup, and the 2004 poll takers simply survey the whole of the American voting-age population? Put this way, one answer to the question, "Why sample?" is obvious: We frequently sample because studying every single instance of a thing is impractical or too expensive. The *Literary Digest*, Gallup, and 2004 poll takers couldn't afford to contact every American of voting age.

We sample because studying every single instance is beyond our means. Suppose, for example, you wanted to know how all Americans felt about something? Why not ask them all? The U.S. Census Bureau is, in fact, obliged by law to learn something about every American every 10 years. But it's very expensive. The 2000 Census, during which 281,421,906 Americans were enumerated, cost in excess of $4.5 billion (Gauthier, 2002), maybe 60 percent of which was spent to employ the half million temporary

Three Different Sets of Pre-Presidential Election Polling Results

1936 *Literary Digest* Poll Predicts Alf Landon Will Beat Franklin Delano Roosevelt, 57% to 43%

Based on a survey of over 2 million people, a survey that had predicted with "uncanny accuracy" (its own words), the outcome of the 1924, 1928, and 1932 presidential elections, the *Literary Digest* predicted that Alf Landon would win the 1936 election by a margin of 14 percentage points over then-President Franklin Delano Roosevelt: 57% to 43%. Imagine the *Digest's* surprise when Roosevelt actually won and won big: 61% to 39%.

1948 Gallup Poll Predicts that Thomas Dewey Will Beat Harry Truman

Using a technique called quota sampling, George Gallup correctly predicted that Roosevelt would win the 1936 election, even if his prediction that Roosevelt would receive 54% of the vote was somewhat wide of the 61% Roosevelt actually received. Gallup got it completely wrong in 1948, however, when he predicted that Thomas Dewey would beat Harry Truman in that year's presidential election. Truman actually won. It was a very good day for Truman, but a very bad day for Gallup.

2004 Six Pre-Election Polls Say Presidential Election Is Statistical Dead Heat

Based on surveys of anywhere between 350 and 1,258 likely voters, six polls saw the presidential election as a statistical dead heat (CBS News, Nov. 1, 2004). All polls were completed on either October 31st or November 1st, the day before the election:

Poll	Bush	Kerry	Margin of Error
Newsweek	50%	44%	+/−4%
CBS News	47%	46%	+/−3%
Fox News	46%	48%	+/−3%
Marist University	48%	49%	+/−3.5%
Zogby	48%	47%	+/−2.9%
Washington Post	49%	48%	+/−3%

The margin of error means that the percentage of likely voters favoring each candidate is the estimated percentage plus or minus the number of percentage points indicated by the margin of error. Thus, given the *Newsweek* poll's estimate of 50% of likely voters favoring Bush and its margin of error of +/−4%, one can expect the true percentage of likely votes to favor Bush to lie between about 54% and 46%. Doing the same thing with Kerry's percentage, one finds that the percentage of likely

(continued)

> ### BOX 5.1 (*continued*)
>
> voters favoring him was somewhere between 48% and 40%. Because there was an overlap of two percent (from 46% to 48%) in these two estimates, one could fairly claim that, statistically speaking, they are tied (because each could, for instance, have 47% of the likely voters favoring him).
>
> In fact, you may recall, the election went to Bush, who won about 51.0% of the votes cast, compared to Kerry's 48.2%.
>
> Considering the margin of error for each of the pre-election polls, all of them were far better at predicting the actual election results than the *Literary Digest* poll had been in 1936 or the Gallup poll had been in 1948.

workers used to follow up on the (34 percent of) households that failed to return their forms by mail (U.S. Bureau of the Census, 2000). Such costs, in terms of money and people, are beyond the means of even the wealthiest research organizations.

So we sample to reduce costs. But we also sample because sampling can improve data quality. Sandra Enos, whose study of how imprisoned women handle mothering we present in Chapter 10, wanted to understand women's approaches to mothering in an in-depth fashion, with a real appreciation of each individual woman's social context. Enos intensively interviewed 25 incarcerated women, not only because studying the whole population of such women would have been expensive, but also because she wanted in-depth information about each of her subjects, rather than more superficial data on all.

 In short, we sample because we want to minimize the number (or quantity) of things we examine or maximize the quality of our examination of those things we do examine.

STOP AND THINK *Using the logic of this section, when do you think sampling is unnecessary?*

element, a kind of thing a researcher wants to sample. Also called a sampling unit.

units of analysis, the units about which information is collected, and the kind of thing a researcher wants to analyze. The term is used at the analysis stage, whereas the terms *element* or *sampling unit* can be used at the sampling stage.

We don't need to sample, and probably shouldn't, when the number of things we want to examine is small, when data are easily accessible, and when data quality is unaffected by the number of things we look at. For instance, suppose you were interested in the relationship between team batting average and winning percentage of major league baseball teams last year. Because there are only 30 major league teams, and because data on team batting averages and winning percentages are readily available (see any of this year's almanacs), you'd simply be jeopardizing the credibility of your study by using, say, 10 of the teams. So why not examine all 30 of them?

We need to define a few terms before proceeding much further. First, we need a term to describe the kind of thing about which we want information—for example, a major league baseball team, an adult American of voting age, all Americans. Social scientists define such things as **elements** or **sampling units** when they're engaged in sampling, and as **units of analysis** when they're engaged in data analysis. Sampling units or units of analysis can be people, organizations, institutions, collectivities, and so on.

In a public opinion survey, the element or unit of analysis would be individual people. In the aforementioned study of major league baseball teams, it would be individual teams. In a study concerned with how the population size of a city is related to its expenditures on education, the element or unit of analysis would be the individual city. A crucial question to ask of any study that you read about or propose to do is this: What is the unit of analysis of this study or what kind of sampling unit do I want to get information about?

STOP AND THINK

See if you can identify the units of analysis for each of the following studies: (a) Norgaard and York's (2005) study that found that nations with a higher proportion of women in Parliament were more likely to have ratified environmental treaties than other nations. (b) Lee and Ousey's (2005) study that found that black homicide rates are lower in urban areas where blacks have access to mainstream integrated institutions such as churches. (c) Eliassen, Taylor, and Lloyd's (2005) study that found that moderately religious young adults were more likely to have high levels of repression than either very religious or nonreligious young adults.

 population, the group of elements from which a researcher samples and to which she or he might like to generalize.

sample, a number of individual cases drawn from a larger population.

Next, we need a term for the entire group of elements that interests us theoretically—for example, all major league baseball teams last year, American adults who will vote on election day, or people in the United States. Social scientists define such a group as a **population.** Finally, we can formally define a **sample** as a subset of a population—used typically to gain information about or insight into the entire population.

STOP AND THINK

Suppose you're interested in describing the nationality of Nobel prize-winning scientists. What would an element in your study be? What would the population be?

Sampling Frames, Probability versus Nonprobability Samples

We can now begin to address the question of why the *Literary Digest* poll did so poorly in 1936, why Gallup did badly in 1948, and why the 2004 election eve polls did relatively well at predicting the outcome of the 2004 election. All of these polls sought information about the same population: people who would vote on election day. But the concept of "people who would vote on election day" is elusive, just as many target populations are, especially *before* an election, when even the most well-intentioned citizens can't be sure what might keep them from voting on election day (e.g., weather, natural or social disaster, death, etc.). One thing most researchers must do, then, is come up with a list of elements that approximates the elements of the population of interest. This list is referred to as a **sampling frame** or **study population.** The list that the *Literary Digest* used for its 1936 poll consisted of names from automobile registrations and telephone directories. But, in 1936, at the height of the Great Depression, substantial numbers of people voted who owned neither a car nor a telephone. So, even though the *Literary Digest* received two million responses to the (ten million) ballots it sent out, the best it could hope to do was get responses from people who owned cars or telephones—people who, because of their wealth, were more

sampling frame or study population, the group of sampling units or elements from which a sample is actually selected.

likely to vote for the Republican Landon than the Democrat Roosevelt. All of the 2004 polls, however, were of people who had telephones but, because the percentage of U.S. households with telephones had reached 94.6 percent by 2000 (Weisberg, 2005: 209), this sampling frame provided a much greater chance of reaching the desired population than the one used by the *Literary Digest* in 1936 had. Modern surveys conducted over the Internet, when intended to provide insights about the total American population, are a more recent example of surveys that are subject to inaccuracy because of a discrepancy between the target population and the sampling frame. As of 2005, no more than two thirds of the U.S. population used the Internet (Weisberg, 2005: 207).

So, the main reason the *Literary Digest* poll did so poorly is that it employed a relatively poor sampling frame. The explanation for Gallup's failure in 1948 is different, and has to do with the fact that Gallup employed a nonprobability, as opposed to a probability, sample. **Nonprobability samples** are those in which members of the population have an unknown (or perhaps no) chance of being included. Gallup used a fairly sophisticated type of nonprobability sampling, called quota sampling. Quota sampling begins with a description of the target population: its proportion of males and females, of people in different age groups, of people in different income levels, of people in different racial groups, and so on. Based on this description, which in Gallup's case was from the census, the researcher draws a sample that represents the target population in all ways that are deemed important (for example, gender, age, income levels, race). The problem is that the researchers who use quota sampling are satisfied once they find a sample that meets the description they are looking for. Hence they do not give everyone in the population a chance of appearing in their sample.

Probability samples, on the other hand, are selected in a way that gives every element of a population a known (and nonzero) chance, or probability, of being included. The samples used to predict the most recent presidential race were probability samples. They were all designed to give every likely voter of the United States a (known) chance of being part of their sample.

The advantage of probability over nonprobability samples is not, as you might hope, that they ensure that a sample is representative or typical of the population from which it is drawn. Still, probability samples are *usually* more representative than nonprobability samples of the populations from which they are drawn. To use the jargon of statisticians, probability samples are less likely to be **biased samples** than are nonprobability samples. Biased means that the cases that are chosen are unrepresentative of the population from which they've been drawn. And because probability samples increase the chances that samples are representative of the populations from which they are drawn, they tend to maximize the **generalizability** of results. Generalizability here refers to the ability to apply the results of a study to groups or situations beyond those actually studied.

This chapter's focal research piece, by Matthew Reavy, focuses on another attribute of probability samples: the fact that their estimates of what is

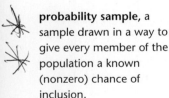
nonprobability sample, a sample that has been drawn in a way that doesn't give every member of the population a known chance of being selected.

probability sample, a sample drawn in a way to give every member of the population a known (nonzero) chance of inclusion.

biased sample, a sample that is not representative of the population from which it is drawn.

generalizability, the ability to apply the results of a study to groups or situations beyond those actually studied.

true of the larger population should be couched in terms of ranges rather than razor-sharp estimates. They come with those peculiar-looking margin of error qualifications that need to be taken seriously. Thus, when the *Newsweek* 2004 election eve poll claimed that Bush enjoyed the favor of 50 percent of likely voters, with a margin of error of plus or minus 4 percent, it was really saying that Bush was favored by somewhere between 46 percent to 54 percent of the likely voters. Sometimes, however, news media accounts of polls fail to mention anything about margins of error and this, as Reavy points out, can be terribly misleading to a public that relies upon them.

FOCAL RESEARCH

USA Today Reports of Tracking Polls Sometimes Ignore Sampling Error

by Matthew M. Reavy[1]

The media's use of pre-election polls has long been a subject of concern among journalists and media critics. One of the most serious objections is that newspapers misrepresent data through poor reporting. This paper examines how *USA Today* reported its own daily tracking poll for the 2000 U.S. presidential election.

Reports about daily tracking polls were criticized during the 1992 and 2000 presidential election campaigns. Rhee (1996), for instance, cited an incident in which *USA Today* reported a substantial change in the 1992 race between George Bush and Bill Clinton, even though that change fell within the poll's margin of error. Reavy (2002) found similar problems in CNN's reporting of Gallup poll results on its Web site during the 2000 presidential campaign. It is this problem, of how journalists handle sampling error, that forms the basis of the present research.

Research Questions

This article focuses on *USA Today*'s handling of change and difference in reporting results of its daily tracking poll. Specific attention is given to how the newspaper dealt with sampling error.

Research Question 1: Did *USA Today* report change from one tracking poll to another when statistically no change took place?

Research Question 2: Did *USA Today* report a difference between the top two candidates in a tracking poll when there was no statistical difference between those candidates?

[1] This is an abridged version of an article that appeared in the *Newspaper Research Journal* and is reprinted by permission from the *Newspaper Research Journal,* Spring 2004, Vol. 25, No. 2, published by the Newspaper Division of the Association for Education in Journalism and Mass Communication.

Method

This study employed a content analysis of *USA Today*'s coverage of its daily tracking poll results between Sept. 8 and Nov. 7, 2000—the day before the election.

The Lexis-Nexis database was searched for articles in *USA Today* that reported results of its tracking poll. The final universe included 33 articles. Because Lexis-Nexis cautioned subscribers to check microfiche records for text within graphics, issues of *USA Today* on microfiche were also searched for articles that appeared with the daily tracking poll graphic. An additional 17 articles were identified in this manner, bringing the total to 50 articles.

Two coders were used for reliability. A change in the poll was defined as any alteration in the lead—a narrowing or widening of the margin between the top two candidates—either from the previous poll or from an unspecified earlier time, taken to mean the previous poll. A difference between the two candidates was defined as any indication that one candidate or the other was ahead in the race. Any caveats included by the writer (e.g., "the race may be tightening") resulted in the body of the article being coded as showing no substantial difference or change in the poll.

Analysis was performed according to the National Council on Public Polls' (NCPP) recommendations for handling sampling error when reporting poll results (Gawaiser and Witt, 2003). So, for instance, in a poll that showed Bush at 44 and Gore at 47, based on a sample of 777, the adjusted sample size would be 707 (number voting for Bush or Gore). According to NCPP recommendations, the sampling error would be 3.66 for Bush and 3.68 for Gore. Combining those two sampling errors, we get 7.3. Because the actual difference between the two candidates (3) falls within the combined sampling error (7.3), the difference is not significant.

Results

Of the 50 articles in this study, 46 claimed a difference between the candidates (i.e., that one candidate or the other was leading the race) within the body of the article itself. Of those, 15 (32.6 percent) had headlines that suggested that one or the other candidate was leading the race. An additional 19 articles suggested that the poll had changed in some way. Of those, seven (36.8 percent) had headlines suggesting a change in the poll.

Research Question 1: Did *USA Today* report change from one tracking poll to another when statistically no change took place?

During the course of reporting on its daily tracking poll, *USA Today* published 19 articles suggesting that the poll had changed in some way. In 18 of these, the change fell within the sampling error of the poll. Ten of these articles (55.6 percent) failed to note that the results fell within the poll's margin of error; five (27.8 percent) made no mention of sampling error at all.

This led to possible inaccuracies in reporting poll results. For example, a Nov. 6 article indicated that "Republican presidential candidate George W. Bush's lead over Democrat Al Gore shrank to 2 points, 47%-45%" (*USA Today*, 2000a, 7)—this despite the fact that the gap between the two candidates had changed only two percentage points, well within the poll's margin of error for either candidate.

The situation was predictably more common in headlines. The newspaper ran seven headlines that claimed a change in the daily tracking poll. Of these, six reported changes that were within the sampling error of the poll. All six headlines (100 percent) failed to mention that the results fell within the margin of error.

Research Question 2: Did *USA Today* report a difference between the top candidates in a tracking poll when there was no statistical difference between those candidates?

Although the daily tracking poll indicated that the 2000 U.S. presidential race remained tight throughout, 46 of the polls (92 percent) suggested that one candidate or the other was in the lead. Of these, 34 "differences" were within the sampling error. The newspaper failed to cite this in 16 (47.1 percent) of those articles and failed to include any mention of sampling error in 8 (23.6 percent) of them.

The newspaper at times declared one candidate or the other to be in the lead even when that lead was not statistically significant. This practice continued right up to the last day of the poll when one article stated "The USA Today/CNN/Gallup Poll's final projection of the outcome for the 2000 presidential election has Bush beating Gore 48%-46%" (*USA Today*, 2000b, 1).

Again, headlines tended to emphasize the solidity of the lead more than did the articles. Of the 15 headlines that suggested a difference between the candidates, 11 dealt with poll results that fell within the margin of error. All of these (100 percent) failed to note this fact.

Discussion

The results of this study indicate that *USA Today* had a tendency to discuss change and difference in its Election 2000 daily tracking poll even when that change or difference could be accounted for by sampling error.

Journalists can learn three lessons from this study:

- Reporters should not say that one candidate is ahead of the other if the gap between the two candidates is less than the error margin. They can say the race is "close," the race is "roughly even" or even "there is little difference between the candidates."

- Journalists handling daily polls should discuss change over the life of the poll rather than from one poll to the next.

- Extra care needs to be taken with the headlines that appear over poll stories. Headlines rarely have enough space to include the caveats that can be included within the text of an article, so their authors should be especially careful not to mislead readers.

REFERENCES

Gawaiser, Sheldon R. and G. Evans Witt. 2003. 20 questions a journalist should ask about poll results. NCPP. Retrieved November 13, 2003. http://www.ncpp.org/qajsa.htm

Reavy, Matthew M. 2002. CNN's emphasis of change and difference in Web reporting of its daily tracking poll during the 2000 U.S. Presidential Campaign. Paper presented to the Communication and Policy division, AEJMC, Miami, FL.

Rhee, June Wong. 1996. How polls drive campaign coverage: The Gallup/CNN/USA Today tracking poll in USA Today's coverage the 1992 presidential campaign. *Political Communication* 13: 214.

USA Today. 2000a. New national polls show tighter race. November 6: 7.

USA Today. 2000b. Election focus: Turnout verdict today for Bush, Gore. November 7: 1.

Sources of Error Associated with Sampling

Was the last 2000 *USA Today* poll report, the one that predicted Bush would beat Gore, 48 percent to 46 percent, right or wrong? Well, because Gore actually outpolled Bush in the election, 48.4 percent to 47.9 percent, the *report* was wrong. But the *poll* itself, with its acknowledgement of a 3 percent margin of error, was right! Reavy's clearly on to something.

STOP AND THINK

Reavy's focus on the margin of error of polls reminds us that the even the most scientific samples are likely to involve error. What do you think is the most basic reason why samples involve error?

The overarching reason why samples fail to represent accurately the populations we're interested in is that populations themselves aren't uniform or *homogeneous* in their characteristics. If, for instance, all human beings were identical in every way, any sample of them, however small (say, even one person) would give us an adequate picture of the whole. The problem is that real populations tend to be heterogeneous, or varied in their composition—so any given sample might not be representative of the whole. Obviously, not all eligible voters in the U.S. vote for the same candidate in every election. If they did, and did so reliably, any sample, however selected, would perfectly predict the election's outcome. Weisberg (2005) lists three sources of survey error that are associated with sampling for populations that are not homogeneous, however: **coverage error,** nonresponse error, and **sampling error** (see Figure 5.1). The last of these, sampling error, will help make sense of why all those scientifically conducted polls conducted by *USA Today*, and reported about by Reavy, needed to be qualified by reference to a "margin of error." We'll be telling you about sampling error after briefly touching upon coverage and nonresponse errors.

coverage errors, errors that result from differences between the sampling frame and the target population.

sampling error, any difference between the characteristics of a sample and the characteristics of the population from which the sample is drawn.

Coverage Errors

Coverage errors are attributable to the difference between a sampling frame and a target population. It's useful to introduce a little more terminology

FIGURE 5.1

Sources of Error
Associated with
Sampling

Coverage errors	Nonresponse errors	Sampling error

Adapted from Weisberg (2005).

parameter, a summary of
a variable characteristic in
a population.

statistic, a summary of a
variable in a sample.

here. A **parameter** is a summary of a variable *in a population.* In the 1936 election, about 61 percent of the voting public favored Roosevelt over Landon for president. This percentage is a parameter of the population that voted for president in 1936. A **statistic,** on the other hand, is a summary of a variable *in a sample.* The best the *Literary Digest* could do was come up with a statistic, that Roosevelt had received 43 percent of the ballots cast in their poll. But the largest problem was that the sample with which that statistic was calculated was drawn from a sampling frame, lists of telephone and car owners, that bore only a remote resemblance to the target population. Hence the coverage error, defined precisely as the difference between the parameter for the sampling frame, or study population, and the parameter for the target population, was necessarily large.

STOP AND THINK *Can you imagine any sources of coverage error in the 2004 election eve polls?*

The 2004 polls also involved some coverage error. After all, they were based upon sampling frames that importantly included only people with telephones and then only people who said they were "likely voters" in those households. When we recall that even in 2000 the percentage of U.S. households with phones was 94.6 percent, not 100 percent, and that even people who thought they were "likely" voters couldn't be sure they'd vote (and those who thought they weren't likely voters couldn't be sure they wouldn't), we can see that the sampling frames for these polls were imperfect, if pretty good, matches for their target populations (people who did vote in the election).

STOP AND THINK *What do you think survey organizations do about the problem of unlisted telephone numbers?*

One of the great potential sources of coverage errors is that many telephone numbers are not listed in telephone books. Survey organizations have gotten around this problem through a process called random-digit dialing. Through random-digit dialing, interviewers are commonly put in telephone contact with interviewees by a computer that, first, randomly selects three-digit local exchanges and, then, randomly selects the next four digits. National surveys, of course, tend to employ the step of randomly selecting national area codes. We'll be saying more about random selection a little later. For now, notice that one of the great advantages of random digit-dialing is that it avoids the problem of unlisted telephone numbers, even if it brings into play numbers, like those of businesses, that are best left out of surveys of households.

Most of the time, when we think of coverage errors, we think of the eligible elements of a target population that are excluded in the sampling frame. The eligibles who were excluded by the *Literary Digest* poll, for instance, were voters who didn't own a car and didn't own a telephone. But coverage errors also might occur when elements that are ineligible for the target population are included in the sampling frame. Thus, a voting survey of households with telephones could contact noncitizens—ineligibles in U.S. elections. Similarly, a household survey, especially one based, as many large-scale surveys are today, on random-digit dialing, could contact some people at non-residential (perhaps business) numbers. The problem of ineligibles, however, can, in most cases, be dealt with by the use of screening questions ("Are you an eligible voter?" "Is this your home phone?"). The problem of excluded eligibles is more challenging, but not always insurmountable (see, for instance, Weisberg, 2005: 215–216).

The issue of excluded eligibles has become particularly problematic with the increased popularity of cell phones. Increasingly cell phone users possess only a cell phone, with no land line. The figure was only two percent in 2001 (Weisberg, 2005: 210), but it is increasing. Because most cell phone plans charge users for calls they receive, survey organizations avoid calling numbers assigned to cell phones. In response to concerns about excluding people who rely primarily on cell phones, one public opinion pollster has announced that it will be exploring the use of text-message surveys (Zogby, 2004). The list of technological innovations that currently frustrate survey research does not stop at cell phones, however. Answering machines and caller ID and other devices used to screen calls are increasingly used by "eligibles" to shield themselves from participation in surveys. And one can imagine that the increasing use of Internet technology as a substitute for traditional land line telephone connections will, in future, create a new coverage problem for telephone surveys.

Nonresponse Errors

Another source of error in both nonprobability and probability samples is **nonresponse error.** Nonresponse refers to the observations that cannot be made because some potential respondents refuse to answer, weren't "there" when contacted, and so forth. Nonresponse error is the error that results from differences between those who participate in a survey and those who don't. Nonresponse can involve large amounts of bias, as it was suspected to do in the *Literary Digest*'s 1936 presidential poll, when about 8 million of the *Digest*'s 10 million contacts simply didn't return their ballots. Travis Hirschi (1969) has argued that nonresponse bias is also enormous in questionnaire studies of delinquency among school-aged adolescents given in school because delinquents are much more likely than are their nondelinquent counterparts to have dropped out of school or to skip sessions when questionnaires are administered.

Nonresponse plagues carefully constructed probability sample surveys, as well as others. The various techniques, mentioned above, used to screen telephones generates a significant kind of nonresponse. And election

nonresponse errors, errors that result from differences between nonresponders and responders to a survey.

pollsters, along with many other researchers, frequently run into people who refuse to answer their questions for one reason or another, and they're doing so at ever increasing rates. Curtin, Presser, and Singer (2005) report, for instance, that response rates to the University of Michigan's telephone Survey of Consumer Attitudes dropped from about 72 percent in 1979 to about 48 percent in 2003.

STOP AND THINK   *What kinds of people might not be home to pick up the phone in the early evening when most survey organizations make their calls? What kinds of people might refuse to respond to telephone polls, even if they were contacted?*

If sampling is done from lists, as is frequently done in probability sampling, it is sometimes possible to study the nature of nonresponse bias, especially if the lists include some information about potential sampling units. Thus, in a study of married people, Benjamin Karney and his colleagues (1995) were able to determine that people who did not respond to their questionnaire survey generally had less education and were employed in lower-status jobs than were those who did respond. Karney and his colleagues used information from the marriage certificates of those that constituted their sampling frame. Weisberg (2005) points out that nonresponse can come at the unit level (when there is no response at all from the respondent) or the item level (when a respondent fails to answer a particular question), and that the former can be due to an inability to contact the respondent, an incapacity on the part of the respondent to be interviewed, or noncooperation on the part of the respondent. The latter, noncooperation, is most likely to be associated with error, especially insofar as respondents' lack of cooperation is systematically related to their attitudes toward the subject of the survey, as, for instance, when people with conservative social values refuse to answer a survey on sexual behavior (Weisberg, 2005: 160).

Sampling Error

Now we come to the type of error that speaks directly to Reavy's concern about tracking polls: sampling error. Sampling error is the error that arises when sample characteristics are used to estimate the characteristics of a study population or sampling frame. Here the question isn't whether there's a perfect match between the sampling frame and the target population, or whether non-responders are different from responders. Theoretically, if not practically, one can imagine no coverage errors (due to differences between the sampling frame and the target population) or nonresponse errors (due to differences between responders and non-responders), and still imagine that a sample does not perfectly represent the population from which it is drawn. In fact, because the election eve polls conducted by *Newsweek, CBS News, Fox News, Marist University, Zogby,* and *Washington Post* in 2004 found that 44 percent, 46 percent, 48 percent, 49 percent, 47 percent, and 48 percent of the respondents, respectively, who said they were likely voters expressed a preference for Kerry (see Box 5.1), it is clear that at least some of the samples on which the polls were based did not give a perfectly accurate picture of the percentage of all likely voters (the sampling frame for these polls) who preferred Kerry. If they all did, they'd all be the same.

Most students of sampling believe that probability samples, with their insistence that every element of a study population or sampling frame have some (known) chance of appearing in the sample, are likely to have smaller sampling errors than nonprobability samples. But all such students know that the sampling error of nonprobability samples cannot be estimated, whereas the sampling error of probability samples can be. To give some sense of how sampling error is, in fact, estimable with probability samples, we'd like to introduce another term: **sampling variability,** or the variability in sample statistics that can occur when different samples are drawn from the same study population or sampling frame.

Even probability samples entail sampling variability and, therefore, frequently involve sampling error. To demonstrate, we'd like you to imagine drawing random samples of two elements from a population of four unicorns: a population with four members named, let us say, Leopold, Frances, Germaine, and Quiggers. Let us also say that these unicorns were one, two, three, and four years of age, respectively.

sampling variability, the variability in sample statistics that occurs when different samples are drawn from the same population.

STOP AND THINK

How could you draw a sample of two unicorns, giving each unicorn an equal chance of appearing in the sample?

simple random sample, a probability sample in which every member of a study population has been given an equal chance of selection.

A sample in which every member of the population has an equal chance of being selected is called a **simple random sample.** Simple random samples, because they give every member of the population an equal chance of being selected, also give every member a known chance, and are therefore probability samples. Perhaps we should emphasize that the word "random" in the phrase "simple random sample" has a special meaning. It does not mean, as it can in other contexts, haphazard, erratic, or arbitrary; here it connotes a "blind" choice, almost lottery-like, of equally likely elements from a larger population. One way of drawing (selecting) a simple random sample of two of our unicorns would be to write the four unicorns' names on a piece of paper, cut the paper into equal-sized slips, each with one unicorn's name on it, drop the slips into a hat, shake well, and, being careful not to peek, pick two slips out of the hat. Let's say we have just done so and drawn the names of Frances and Quiggers (ages two and four). We thereby achieved our goal of a sample of unicorns whose average age ([2 years + 4 years]/2 = 3 years) was *not* the same as that of the population (2.5), hence proving that it is possible to draw a probability sample that misrepresents the larger population, and therefore entails sampling error.

Of course, if we had selected a sample of Frances and Germaine, the average age of sample members ([2 years + 3 years]/2 = 2.5 years) would have been the same as the average age for the whole population of unicorns. But we would also have proven that it is not only possible, as we did in our first attempt, to draw a sample that misrepresents the larger population, but also

FIGURE 5.2

The Sampling Distribution of Unique Samples of Two Unicorns

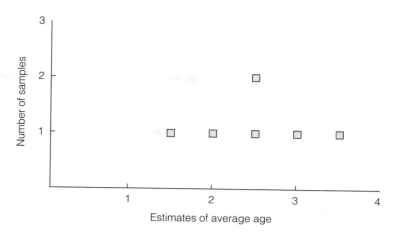

to draw probability samples from the same population that have different statistics (and, so, show sampling variability).

STOP AND THINK *How many possible unique samples of two unicorns could be drawn from our population?*

You could randomly select six possible unique samples of two unicorns from our population. These would be: Leopold and Frances (average age = 1.5), Leopold and Germaine (2.0), Leopold and Quiggers (2.5), Frances and Germaine (2.5), Frances and Quiggers (3.0) and Germaine and Quiggers (3.5).

sampling distribution, the distribution of a sample statistic (such as the average) computed from many samples.

We could make a visual display of these samples, called a **sampling distribution,** by placing a dot, representing each of the sample averages, on a graph big enough to accommodate all possible sample averages. A sampling distribution is a distribution of a sample statistic (like the average) computed from more than one sample. The sampling distribution of any statistic, according to probability theory, tends to hover around the population parameter, especially as the number of samples gets very large. Figure 5.2, a sampling distribution for the possible unique samples of two unicorns, suggests this property by showing that the sampling distribution of the average age for each possible pair of unicorns is centered around the average age for unicorns in the population (2.5 years).

STOP AND THINK *You're obviously pretty imaginative if you've followed us up to here. But, can you imagine the sampling distributions of all possible unique samples of one unicorn? Of four unicorns? What, if anything, do you notice about the possible error in your samples as the number of cases in the samples grow?*

The sampling distribution of the average ages of samples of one unicorn would have four distinct points, representing 1, 2, 3, and 4 years, all entailing sampling error. The sampling distribution of the average age for a sample of four unicorns would have one point, representing 2.5 years, involving no sampling error at all. Figure 5.2 suggests that the sampling distribution of samples with two cases generally has less average sampling error than that

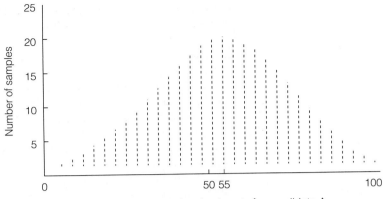

Percent of voters planning to vote for candidate A

for samples with one case, but generally more average sampling error than the sampling distribution with four cases. This illustrates a very important point about probability sampling: Increases in sample size, referred to as *n*, improve the likelihood that sampling statistics will accurately estimate population parameters.

Another, more realistic, example of a sampling distribution would be the samples you might draw of your town's voters to see what percentage of them plan to vote for Candidate A. Suppose you wanted to estimate how well this candidate will do in the upcoming election for mayor. You might use your town's list of registered voters (if it were available) as a study population, select a random sample of 100 registered voters, and ask members of the sample whether they plan to vote for Candidate A or not. The possible results of your survey would be, of course, that anywhere from zero to 100 percent of your sample say they plan to vote for Candidate A. Suppose your first sample shows that 52 percent of your respondents say they will vote for Candidate A. Suppose your second sample shows that 57 percent say they will vote for Candidate A. Suppose you keep drawing additional samples of 100 respondents and decide to plot the new sample statistics on a summary graph, like the one shown in Figure 5.3. This would be one way to display the sampling distribution of the variable "percent of registered voters planning to vote for Candidate A" for a large number of samples. Notice that, in this case, the sampling distribution seems to be centered around samples suggesting that 55 percent of the registered voters plan to vote for Candidate A.

A major point of this section has been to demonstrate that even when one employs a type of probability sampling (in this case, simple random sampling), sampling error can and frequently does occur. This means, unfortunately, that no sample statistic can be relied on to give a perfect estimate of a population parameter. Another point of this section has been to suggest something about the nature of sampling distributions (for example, that they tend to be centered around the population parameter when the number of samples is large), a topic to which mathematicians have given extensive consideration. The practical offshoot of this consideration is a branch of mathematics known as probability theory.

A tremendous advantage of probability samples is that they enable their users to use probability theory to estimate population parameters with a certain degree of confidence. Suppose, for instance, you drew a sample of 100 residents that showed that 55 percent planned to vote for Candidate A. Based on our previous discussion, you'd also know that there was a good chance that the population parameter isn't necessarily 55 percent. But you could strive to say, thanks to probability theory and to calculations like those made by Reavy for some of the *USA Today* polls, something like this: There is a 95 percent chance that between 45 to 65 percent of all voters plan to vote for Candidate A. Then you'd be announcing what is known as a **confidence interval,** or range of values in which the population parameter is expected to fall, of 20 percent (between 45 and 65 percent). You'd also be announcing a **confidence level,** or an estimate of the probability that the population parameter falls in this range (in this case, 95 percent). Frequently, "confidence level" and "confidence interval" are combined into **margin of error,** which is almost always a statement of what the 95 percent confidence interval is. Thus, when the *Newsweek* poll told us on the eve of the 2004 election that the margin of error of its survey, suggesting that Bush was favored by 50 percent of likely voters, was 4 percent, it meant to say that the 95 percent confidence interval extended from 46 percent (50% − 4%) to 54 percent (50% + 4%).

We haven't shown you how to calculate margins of error here, but they are not difficult to do and perhaps you'll learn how to do them in a statistics class. With the right formulas in hand, Reavy was able to do the calculations when they weren't reported, just by finding a footnote that mentioned the number of poll respondents, for the 2000 tracking polls he discusses in the focal research. More important, perhaps, is the ability to use the margin of error to read research, and polling reports, critically. Reavy's inspiration for his article came, in fact, when he applied his knowledge of margin of errors to do such a critical reading. "I became interested in the topic [of my article]," he reports, "while reading *USA Today*'s reporting of the 1996 Presidential election. In doing so, I was struck by the fact that, despite the newspaper's rather sensational headlines about leads gained and lost, the actual poll numbers could be interpreted as showing no change at all through the pre-election period. . . . The polls and the newspaper's coverage of the polls told two very different stories" (personal communication).

As our unicorn example suggests, though, the accuracy with which sample statistics estimate population parameters increases with the size of the sample. Probability theory tells us that if we want to achieve a margin of error of 10, we need a sample of about 100—and it doesn't matter what the size of the population is. If you'd like to reduce the margin of error from 10 percent to 5 percent, you'd need a sample of 400. To achieve its 4 percent margin of error, *Newsweek* talked with 882 likely voters; *Fox News* sampled 1200 likely voters to achieve a 3 percent margin of error. You'd need a sample of about 1600 to achieve a margin of error of about 2.5 percent. And, *in general, one needs to increase the sample size by about four times to achieve a halving of the margin of error.* Because of the costs involved in sampling more people, Weisberg (2005: 229) sees little reason for trying to achieve, say, a

---

confidence interval, a range of values within which the population parameter is expected to lie.

confidence level, the estimated probability that a population parameter will fall within a given confidence interval.

reduction in the margin of error from 3 percent to 1.5 percent by surveying 4800 respondents rather than 1200. It's far better, once sampling error has been reduced to, say, 3 percent or even 5 percent, to spend time and energy trying to reduce other sources of error, such as coverage, nonresponse, and measurement error, a kind of error we'll be focusing on in Chapters 6 (on measurement), 9 and 10 (on questionnaires and interviews). These kinds of error are normally far greater contributors to what Weisberg calls "total survey error" than can possibly be offset by increases in sample size beyond, say, 1200.

Sampling error, then, is an error that is introduced whether one uses a probability or nonprobability sample. A great advantage of probability samples, though, is that one can estimate the effect of sampling error and, for instance, produce a statement of a survey's margin of error. Margin of error, as Reavy suggests, is such a valuable and important by-product of probability sampling that it is not only wasteful, but also misleading, not to mention it in reports of surveys that employ probability sampling.

Types of Probability Sampling

In the previous section, we introduced you to simple random sampling, a type of probability sampling that is appropriate for many statistical procedures used by social scientists. Here, we'd like to introduce you to other ways of doing simple random sampling, and to other kinds of probability sampling and their relative advantages.

Simple Random Sampling

Simple random samples are selected so that each member of a study population has an equal probability of selection. To select a simple random sample, you need to have a list of all members of your population. Then, you randomly select elements until you have the number you want—perhaps using the blind drawing technique we mentioned earlier.

To show you another way of drawing a simple random sample, we'd like you to consider the list (or sampling frame) of Pulitzer Prize winners in Journalism for Commentary. The list of such award winners, given only since 1970, is relatively short, and so is useful for illustrative purposes (though if, in fact, your study population included so few elements, you would probably want to use the entire population). Until recently, there were only 24 members of this population:

01. Marquis Childs	09. William Safire	17. Jimmy Breslin
02. William Caldwell	10. Russell Baker	18. Charles Krauthammer
03. Mike Royko	11. **Ellen Goodman**	19. Dave Barry
04. David Broder	12. Dave Anderson	20. Clarence Page
05. Edwin Roberts	13. Art Buchwald	21. Jim Murray
06. **Mary McGrory**	14. Claude Sitton	22. Jim Hoagland
07. Red Smith	15. Vermont Royster	23. **Anna Quindlen**
08. George Will	16. Murray Kempton	24. **Liz Baalmaseda**

FIGURE 5.4

Random Number Table
Excerpt

5334	5795	2896	3019	7747	0140	7607	8145	7090	0454	4140
8626	7905	3735	9620	8714	0562	9496	3640	5249	7671	0535
5925	4687	2982	6227	6478	2638	2793	8298	8246	5892	9861
9110	2269	3789	2897	9194	6317	6276	4285	0980	5610	6945
9137	8348	0226	5434	9162	4303	6779	5025	5137	4630	3535
4048	2697	0556	2438	9791	0609	3903	3650	4899	1557	4745
2573	6288	5421	1563	9385	6545	5061	3905	1074	7840	4596
7537	5961	8327	0188	2104	0740	1055	3317	1282	0002	5368
6571	5440	8274	0819	1919	6789	4542	3570	1500	7044	9288
5302	0896	7577	4018	4619	4922	3297	0954	5898	1699	9276

STOP AND THINK *We have shown the female members of the population in boldface print. Because we've presented the population in this way, it is possible to calculate one of its parameters: the fraction of award winners who have been female. What is that fraction?*

Let us draw a simple random sample of six members from the list of award winners (that is, we want an *n* of 6). If we don't want to use the "blind" drawing technique, we could use a random number table. To do so, we would, first, need to locate a random number table, such as the one we've provided in Figure 5.4, a larger version of which exists in Appendix B of this book. (Such tables, incidentally, are generally produced by computers in a way that gives every number an equal chance of appearing in a given spot in the table.) Second, we'd need to assign an identification number to each member of our population, as we have already done by assigning the numbers 1 through 24 to the list of award winners. Now we need to randomly select a starting number whose number of digits is equal to the digits in your highest identification number. The highest identification number in our list was 24 (with two digits), so we need to select (randomly) a number with two digits in it. Let's say you did this by closing your eyes, pointing toward the page, and (after opening your eyes again) underlining the closest two-digit number—perhaps, the number 96 (see the "96") underlined in Figure 5.4.

Then, proceeding in any direction in the table, you could select numbers, discarding any random number that does not have a corresponding number in the population, until you'd chosen the number of random numbers you desire. Thus, for instance, you could start with 96 and continue across its row, picking off two-digit numbers, then down to the next row, and so forth, until you'd located six of them (remember our desired *n* numbers).

STOP AND THINK *What numbers would you select if you followed the procedure described in the previous sentence?*

The numbers you'd select, in this instance, would be the following:

96, 30, **19,** 77, 47, **01,** 40, 76, **07,** 81, 45, 70, 90, **04,** 54, 41, 40, 86, 26, 79, **05,** 37, 35, 96, **20**

Of course, you'd keep only six of these: those in bold print in the previous array: 19, 01, 07, 04, 05, 20. Incidentally, you'd be working with a definition of simple random selection that assumes that once an element (corresponding to, say, the number 19) is selected, it is removed from the pool eligible for future selection. So if, say, during your selection number "07" had appeared more than once, you would have discarded the second "07." We'll also be using this definition later on in our discussions.

Your resulting sample of six journalists would be: Marquis Childs (case 01), David Broder (04), Edwin Roberts (05), Red Smith (07), Dave Barry (19), and Clarence Page (20).

STOP AND THINK   *What fraction of this sample is female? Does this sample contain sampling error relative to its gender makeup?*

STOP AND THINK AGAIN   *Pick a different simple random sample of six journalists? (Hint: Select another starting number in the random number table and go from there.) Who appears in your new sample? Does this sample contain sampling error relative to gender makeup?*

A quicker way of drawing a random sample is to use Urbaniak and Plous' (2005) website, the Research Randomizer (http://randomizer.org). To random sample the same Pulitzer Prize winners, you'd simply tell the *Randomizer* you wanted one set of six numbers from the numbers 1 through 24 and, presto, you'd have your set of random numbers.

Systematic Sampling

systematic sampling, a probability sampling procedure that involves selecting every *k*th element from a list of population elements, after the first element has been randomly selected.

Even with a list of population elements, simple random sampling, done manually, can be tedious. It is rarely used in practice. Only one article abstracted in *Sociological Abstracts* between 2000 and 2005 inclusive, for instance, employed it as a sampling technique. If one does in fact have a list of sampling-frame elements, a much less tedious probability sampling technique is **systematic sampling.** In systematic sampling, every *k*th element on a list is (systematically) selected for inclusion. If, as in our previous example, the population list has 24 elements in it, and we wanted six elements in our sample, we'd select every fourth element (because 24/6 = 4) on the list. (If the division doesn't yield a whole number, you should round to the nearest whole number.) To make sure that human bias wasn't involved in the selection and that every element actually has a chance of appearing, we'd select the first element randomly from elements 1 through *k*. Thus, in the example of the Pulitzer Award winners, you'd randomly select a number between one and four, then pick the case associated with that number and with every fourth case after that one on the list.

STOP AND THINK   *Try to draw a sample of six award winners using systematic sampling. What members appear in your sample?*

In general, we call the distance between elements selected in a sample the selection interval (4 in the award-winner example but this is generally referred to by the letter "k"), which is calculated using the formula:

$$\text{selection interval (k)} = \frac{\text{population size}}{\text{sample size}}$$

We've mentioned the relative ease of selecting a systematic sample as one of the procedure's advantages. There is at least one other advantage worth mentioning: If your population is physically present (say, file folders in an organization's filing cabinets), you don't need to make a list. You need only determine the population size (that is, the number of folders) and the desired sample size. Then, you'd just need to look at every *k*th element (or folder) after you'd randomly selected the first folder. Ian Rockett, a colleague in epidemiology, tells us that he has used a modified form of systematic sampling in a hospital emergency use study. A potential interviewee was the first eligible patient entering the emergency room after the hour. Eligibility criteria included being 18 years of age or older and able to give informed consent to participate in the study. In each hospital, interviewing lasted three weeks and covered each of three daily eight-hour shifts twice during the three-week period.

The only real disadvantage of systematic sampling over simple random sampling, in fact, is that systematic sampling can introduce an error if the sampling frame is cyclical in nature. If, for instance, you want to sample months of several years (say, 1900 through 2006) and your months were listed, as they frequently are, January, February, March, and so on for each year, you'd get a very flawed sample if your selection interval turned out to be 12 and all the months in your sample turned out to be Januarys. Imagine the error in your estimates of average annual snowfalls in North Dakota if you just collected data about Januarys and multiplied by 12! A word to the wise: When you're tempted to use systematic sampling, check your population lists for cyclicality.

Stratified Sampling

One disadvantage of simple random sampling is that it can yield samples that are flawed in ways you want to avoid and can do something about. Take the earlier example of the Pulitzer Prize winners. Suppose you knew that approximately one-sixth of that population was female and you wanted a sample that was one-sixth female (unlike the one that we drew in the example). We could use a procedure called **stratified random sampling.** To draw a stratified random sample, we group the study population into strata (groups that share a given characteristic) and randomly sample within each stratum. Thus, to draw such a sample of the prize winners, stratified by gender, we'd need to separate out the 4 female winners and the 20 male winners, enumerate each group separately, and random sample, say, a quarter of the members from each group.

stratified random sampling, a probability sampling procedure that involves dividing the population in groups or strata defined by the presence of certain characteristics and then random sampling from each stratum.

S T O P   A N D   T H I N K *See if you can draw a stratified random sample of the prizewinners, such that one female and five males appear in the sample.*

More complexity can be added to the basic stratifying technique. If, for instance, you were drawing a sample of students at your college, using a roster provided to you by the registrar, and the roster provided information not only about each student's gender but also about his or her year in school, you could stratify by both gender and year in school to ensure that your sample wasn't biased by gender or year in school.

Stratified sampling can be done proportionately and disproportionately. The Pulitzer Prize sample you drew for the *Stop and Think* exercise was a proportionately stratified random sample because its proportion of females and males is the same as the proportion of females and males in the population. However, if you wanted to have more female prizewinners to analyze (maybe all of them), you could divide your population by gender and make sure your sample included a number of females that was disproportionate to their number in the population.

S T O P   A N D   T H I N K *How would you draw a disproportionate stratified random sample of eight prizewinners, with four females and four males in it?*

Disproportionate stratified sampling makes sense when the given strata, with relatively small representation in the population, are likely to display characteristics that are crucial to a study, and when proportionate sampling might yield uncharacteristic elements from those strata because too few elements were selected. Suppose, for instance, you had a population of 100, 10 of whom were women. You probably wouldn't want to take a 10 percent proportionate sample from men and women because you'd then be expecting one woman to "represent" all ten in the sample. You could oversample the women and then make an adjustment for their over-representation in the sample during your analysis. This is what Hoffnung did in her study of student views of their futures focused on in Chapter 4: she oversampled African American, Hispanic, and Asian American women to have large numbers of women of color in the study (personal communication).

It's also what Karl Pillemer and David Finkelhor (1988) did when they wanted to estimate the prevalence of elder abuse in the Boston metropolitan area. They drew a disproportionate random sample of elderly persons living in private residences. They knew that elder abuse was more common in households where the elderly live together with their children, even though the elderly in such households constitute no more than 10 percent of all elderly. Therefore, they oversampled elderly living with persons of a younger generation. Similarly, Perrin (2005), in his study of letters to the editor in 17 major newspapers, wished to determine whether there were changes in the expression of authoritarianism (and antiauthoritarianism) after September 11, 2001 (9/11). He therefore wanted the same number of letters to the editor after 9/11 as before 9/11, even though there were many more such letters from before 9/11. He therefore stratified the letters by whether they were published before or after 9/11 and oversampled the latter group.

Stratified random sampling has major advantages over simple random and systematic sampling, including its capacity to deal with otherwise underrepresented groups and thereby reduce sampling error with respect to those groups. For this reason, it is used more frequently by researchers. A survey of *Sociological Abstracts* (using the key words "simple random sampling," "systematic sampling," and "stratified random sampling"), for instance, indicates that while only one article between 2000 and 2005, inclusive, was based on research that used simple random sampling, and only five employed systematic sampling, more than 30 used stratified random sampling.

Cluster Sampling

cluster sampling, a probability sampling procedure that involves randomly selecting clusters of elements from a population and subsequently selecting every element in each selected cluster for inclusion in the sample.

Simple random, systematic, and stratified sampling all assume that you have a list of all population elements easily at hand and that the potential distances between potential respondents aren't likely to result in unreasonable expense. If, however, data collection involves visits to sites that are at some distance from one another, you might consider cluster sampling. **Cluster sampling** involves the random selection of groupings, known as clusters, from which all members are chosen for study. If, for instance, your study population is high school students in a state, you might start with a list of high schools, random sample schools from the list, ask for student rosters from each of the selected schools, and then contact each of those students.

STOP AND THINK   *How might you cluster sample students who are involved in student organizations that are supported by student activity fees at your college?*

Multistage Sampling

multistage sampling, a probability sampling procedure that involves several stages, such as randomly selecting clusters from a population, then randomly selecting elements from each of the clusters.

Cluster sampling is used much less frequently than is **multistage sampling,** which frequently involves a cluster sampling stage. Two-stage sampling, a simple version of the multistage design, usually involves randomly sampling clusters and then randomly sampling members of the selected clusters to produce a final sample. All random-digit dialing, of course, is a form of multistage sampling, with area codes being the first cluster that is random sampled, and the three-digit local exchanges being the second, etc., the last four-digit number being the third. Assuming that you reach someone at the resulting number, of course, you might need another stage, unless you want, say, the first person who answers. You might be tempted to ask your questions, for instance, of the person who answered the phone, especially if that person were an adult, but such a procedure could engender a kind of coverage error.

What coverage error might result from questioning the first adult to answer the phone?

Some people who are more likely than others to answer phones are the able-bodied, those who have friends they like to chat with on the phone, those who aren't afraid that creditors might be calling, and so forth. Any number of biases might be engendered if you queried only those who answered the phone. But if you, say, asked to speak to the adult who had the most recent birthday (as Bates and Harmon, 1993, did), you'd be much more likely to get a representative sample of the adults living in the residences you contacted. Weisberg (2005: 245–248) provides an excellent discussion of this and other methods for selecting random respondents in households during random-digit dialing surveys.

Amadou Diallo and colleagues (2001) used a slightly different kind of two-stage sampling (two-stage cluster sampling) in their effort to assess the ability of mothers in a rural area of Guinea to estimate fever in their children and to see which children had received antimalarial drugs. Diallo and colleagues first random sampled villages in the area (stage one), then random sampled households within those villages (stage two), and then asked mothers questions about all of their children and administered brief medical exams to all the children (stage three).

The complexity of sampling designs tends to increase as the target population becomes more geographically dispersed and as lists become harder to find. Irene Hess (1985) has provided a detailed overview of procedures used by the Social Research Center for these and similar national household polls, involving six stages of sampling. First, the Center first divides the country into 2,700 primary units (clusters of counties or metropolitan areas), allocates these units into 74 strata, and takes a disproportionate stratified sample of units from each of these strata. Second, the Center divides each of the primary units into sub-units (cities, towns, and rural areas), allocates these sub-units into strata, and takes disproportionate stratified samples from each of the strata. Third, the Center divides each sub-unit into sub-sub-units (usually blocks or "chunks") and systematically samples these sub-sub-units. Fourth, representatives of the Center "scout" these sub-sub-units to find all households in them, divide the sub-sub-units into sub-sub-sub-units and random sample these clusters (called sampling segments). Fifth, households (usually about four) in the sampling segments are randomly sampled, and sixth, the Center decides which member of each particular household should be interviewed to avoid bias.

You're probably as impressed as we are with the complexity of a sampling procedure like the one described in the previous paragraph. But, we'd also like you to consider what it does for researchers. One big advantage is that it permits the researchers to focus their investigation on relatively well-defined areas. If, for instance, your goal were to reach 2400 households, this procedure would enable you to focus your investigation on (2400 households total/4 households per sampling segment =) 600 sampling segments in the whole country. Moreover, you wouldn't need to start with a list of the whole population. In fact, in the previous example, you only need to start drawing up your list in the fourth stage (or once you've reached the block or "chunk" level).

The Social Research Center multistage design, like all probability sampling designs, is much better at ensuring representativeness (and generalizability) than any nonprobability sampling design of a comparable population is. There are times, however, when nonprobability sampling is necessary or desirable, so we now turn our attention to some nonprobability sampling designs.

Types of Nonprobability Sampling

We've already suggested that, although nonprobability samples are less likely than are probability samples to represent the populations they're meant to represent, they are useful when the researcher has limited resources or an inability to identify members of the population (say, underworld or upperworld criminals), or when one is doing exploratory research (say, about whether a problem exists or about the nature of a problem, assuming it exists). Let us describe each of these advantages as we describe some ways of doing nonprobability sampling.

Purposive Sampling

purposive sampling, a nonprobability sampling procedure that involves selecting elements based on the researcher's judgment about which elements will facilitate his or her investigation.

For many exploratory studies and much qualitative research, **purposive sampling,** in any of a number of forms, is desirable. In purposive sampling, the researcher selects sampling units based on his or her judgment of what units will facilitate an investigation. Especially early on in their study of afterschool activities, Adler and Adler (see Chapter 11) sought activities that they viewed as typical so that they could gain a basic understanding of them. They went to t-ball games, baseball games, dance classes, and soccer games, to name just a few, because they believed that these were illustrative of venues where children participated in adult-organized activities.

Quota Sampling

quota sampling, a nonprobability sampling procedure that involves describing the target population in terms of what are thought to be relevant criteria and then selecting sample elements to represent the "relevant" subgroups in proportion to their presence in the target population.

Quota sampling was the type of sampling used by Gallup, and others, so well in presidential preference polls until 1948. (You might remember that Gallup's projection, picking Roosevelt over Landon, was better for the 1936 election than was the *Literary Digest*'s, which depended on surveying owners of cars and telephones.) Quota sampling begins with a description of the target population: its proportion of males and females, of people in different age groups, of people in different income levels, of people in different racial groups, and so on.

STOP AND THINK

Where do you think Gallup and other pollsters might have gotten such descriptions of the American public?

Until 1948, Gallup relied on the decennial censuses of the American population for his descriptions of the American voting public. Based on these descriptions, a researcher who used quota sampling drew a sample that "represented" the target population in all the relevant descriptors (for example, gender, age, income levels, race, and so on).

Quota sampling at least attempts to deal with the issue of representativeness (in a way that neither convenience nor snowball sampling, discussed later, do). In addition, quota samples are frequently cheaper than probability samples are. Quota sampling is, however, subject to some major drawbacks. First is the difficulty of guaranteeing that the description of the target population is an accurate one. One explanation given for the inability of Gallup and other pollsters to accurately predict Truman's victory in the 1948 presidential election is that they were using a population description that was eight years old (the decennial census having been taken in 1940), and much had occurred during the forties to invalidate it as a description of the American population in 1948. Another drawback of quota sampling is how the final case selection is made. Thus, the researcher might know that she or he needs, say, 18 white males, aged 20 to 30 years, and select these final 18 by how pleasant they look (in the process, avoiding the less pleasant-looking ones). Pleasantness of appearance is likely to be one source of bias in the final sample.

Snowball Sampling

snowball sampling, a nonprobability sampling procedure that involves using members of the group of interest to identify other members of the group.

When population listings are unavailable (as in the case of "community elites," gay couples, upperworld or underworld criminals, and so forth), **snowball sampling** can be very useful. Snowball sampling involves using some members of the group of interest to identify other members. One of the earliest and most influential field studies in American sociology, William Foote Whyte's (1955) *Street-Corner Society: The Social Structure of an Italian Slum,* was based on a snowball sample: Whyte talked with a social worker, who introduced him to Doc, his main informant, who then introduced him to other "street corner boys," and so on. Durst and Lange (2005) wanted to find how war refugees from El Salvador were adjusting to North America, and so found a few Salvadoran refugees and asked them if they knew another, and so forth. With snowball sampling, the disadvantage of not being able to generalize can be more than compensated for by the capacity to provide an understanding of a subject that had hitherto been concealed by a veil of ignorance.

Convenience Sampling

convenience sampling, a nonprobability sampling procedure that involves selecting elements that are readily accessible to the researcher.

A **convenience sample** (sometimes called an available-subjects sample) is a group of elements (often people) that are readily accessible to, and therefore convenient for, the researcher. Psychology and sociology professors sometimes use their students as participants in their studies. Chase, Cornille, and English (2000) made very creative use of convenience sampling in their study of life satisfaction among people with spinal cord injuries. They sent letters requesting participation in their study to five list-servs related to disability, to a website bulletin board, and to organizations that focused on spinal cord injuries and disabilities. They also sent the letter to various organizations that were likely to deal with people with spinal cord injuries. Eighty-nine percent of their 158 respondents chose to use a Web-based form. Chase and colleagues argue that their own convenience in using the Web was matched by the convenience of using the Web for their specific population, people with spinal cord injuries, many of whom would have had much more difficulty responding through other media. Weisberg (2005: 213), in fact, makes the Internet sound like a pretty good medium for convenience samples when he points out that, in 2001, 54 percent of U.S. households were using the Internet.

Convenience samples are relatively inexpensive, and they can yield results that are wonderfully provocative and plausible. If, for instance, a student handed out questionnaires to 100 students sitting in the dining room of his university, he'd be collecting a convenience sample. Nevertheless, the temptation to generalize beyond the samples involved is dangerous and should be tempered with considerable caution.

STOP AND THINK *The student who handed out those 100 questionnaires in the dining room might be tempted to say he'd collected a "random" sample. Why is his really not a random sample?*

Choosing a Sampling Technique

Choosing an appropriate sampling technique, like making any number of other decisions in the research process, involves a balancing act, often among methodological, practical, theoretical, and ethical considerations. First, one needs to consider whether it's desirable to sample at all, or if the whole population can be used. Then, if it is methodologically important to be able to generalize to some larger population, as in political preference polls, choosing some sort of probability sample is of overwhelming importance. The methodological importance of being able to generalize from such a poll can clash with the practical difficulty of finding no available list of the total population and, so, dictate the use of some sort of multistage cluster sampling, rather than some other form of probability sampling (for example, simple random, systematic, or stratified) that depends on the presence of such a list. Or, however desirable it might be to find a probability sample of corporate embezzlers (a theoretical concern), your lack of access to the information you'd need to perform probability sampling (a practical concern) could force you to use a nonprobability sampling technique—and perhaps finally settle for those you can identify as corporate embezzlers and who volunteer to come clean (ethical).

You might find that your study of international espionage would be enhanced by a probability sample of the list of CIA agents your roommate has downloaded after illegally tapping into a government computer (methodological and practical considerations), but that your concern for going to prison (a practical consideration) and for jeopardizing the cover of these agents (practical and ethical) leads you to sample passages in the autobiographies of ex-spies (see Chapter 13 on content analysis) rather than to survey the spies themselves. In another study, your exploratory (theoretical) interest on the psychological consequences of adoption for non-adopted adult siblings could dovetail perfectly with your recent introduction to three such siblings (practical) and their obvious desire to talk to you about their experiences (practical and ethical).

Method, theory, practicality, and ethics—the touchstones of any research project—can become important, then, as you begin to think about an appropriate sampling strategy.

Summary

Sampling is a means to an end. We sample because studying every element in our population is frequently beyond our means or would jeopardize the quality of our data—by, for instance, preventing our close scrutiny of individual cases. On the other hand, we don't need to sample when studying every member of our population is feasible.

Probability samples give every member of a population a chance of being selected; nonprobability samples do not. Probability sampling is generally believed to yield more representative samples than does nonprobability sampling. Both probability and nonprobability sampling can be associated with coverage errors, nonresponse errors, and sampling error itself. Coverage errors are caused by differences between the sampling frame, or study population, and the target population. Nonresponse errors are caused by differences between nonresponders and responders to a survey. Sampling error is caused be differences between the sample and the study population, or sampling frame, from which it is drawn. A great advantage of probability samples, however, is that one can use them to estimate the amount of sampling error and thereby generate estimates of things like margins of error. Reavy's article reminds journalists that they ignore margins of error in their reports of political polls at their peril. It can also stand as a reminder to us to read research, and those polls, carefully for what they say about margins of error.

As researchers, we want to remember that we can reduce sampling variability, and margins of error, by taking larger samples. However, because it normally takes a quadrupling of a sample to halve sampling error, and margin of error, it eventually makes more sense to try to reduce other sources of survey error (e.g., coverage, nonresponse, measurement) than to spend resources on reducing sampling error.

We looked at five strategies for selecting probability samples: simple random sampling, systematic random sampling, stratified random sampling, cluster sampling, and multistage sampling. We then looked at four common strategies for selecting nonprobability samples: purposive sampling, quota sampling, snowball sampling, and convenience sampling. We emphasized that the appropriate choice of a sampling strategy always involves balancing theoretical, methodological, practical, and ethical concerns.

EXERCISE 5.1

Practice Sampling Problems

This exercise gives you an opportunity to practice some of the sampling techniques described in the chapter.

The following is a list of the states in the United States (abbreviated) and their populations in 2005 to the nearest tenth of a million.

State	Population	State	Population	State	Population	State	Population
AL	4.5	IN	6.2	NE	1.7	RI	1.1
AK	0.6	IA	2.9	NV	2.2	SC	4.1
AZ	5.5	KS	2.7	NH	1.3	SD	0.8
AR	2.7	KY	4.1	NJ	8.6	TN	5.8
CA	35.5	LA	4.5	NM	1.9	TX	22.1
CO	4.5	ME	1.3	NY	19.2	UT	2.4
CT	3.5	MD	5.5	NC	8.4	VT	0.6
DE	0.8	MA	6.4	ND	0.6	VA	7.4
FL	17.0	MI	10.0	OH	11.4	WA	6.1
GA	8.6	MN	5.1	OK	3.5	WV	1.8
HI	1.3	MS	2.9	OR	3.6	WI	5.4
ID	1.4	MO	5.7	PA	12.4	WY	0.5
IL	12.7	MT	0.9				

1. Number the states from 01 to 50, entering the numbers next to the abbreviated name on the list.

2. Use the random number table in Appendix B and select enough two-digit numbers to provide a sample of 12 states. Write all the numbers and cross out the ones you don't use.

3. List the 12 states that make it into your random sample.

4. Now, if you have easy access to the Internet, locate the *Research Randomizer* (http://randomizer.org/) and draw one set of 12 numbers from 1 to 50 from it. Then list the 12 states that would make it into your random sample this time.

5. This time, take a stratified random sample of 10 states, one of which has a population of 10 million or more and 9 of which have populations of less than 10 million. List the states you chose.

6. How might you draw a quota sample of 10 states, one of which has a population of 10 million or more and 9 of which have populations of less than 10 million?

   a. Describe one way of doing this.

   b. Describe, in your own words, the most important differences between the sampling procedures used in question 5 and question 6a.

7. Generate a sample of 15 telephone numbers using the random number table in Appendix B. Use the following set of prefixes in equal proportions: 456-, 831-, 751-.

EXERCISE 5.2

Examining Sampling in a Research Article

Using your campus library (or some other source), pick a research article from any professional journal that presents the results of a research project that involves some sort of sampling. Answer the following questions:

1. Name of author(s):
   Title of article:
   Journal name:
   Date of publication:
   Pages of article:

2. What kind of sampling unit or element was used?

3. Was the population from which these elements were selected described? If so, describe the population.

4. How was the sample drawn?

5. What do you think of this sampling strategy as a way of creating a representative sample?

6. Do(es) the author(s) attempt to generalize from the sample to a population?

7. Did the author(s) have any other purposes for sampling besides trying to enable generalization to a population? If so, describe this (these) purpose(s).

8. What other comments do you have about the use of sampling in this article?

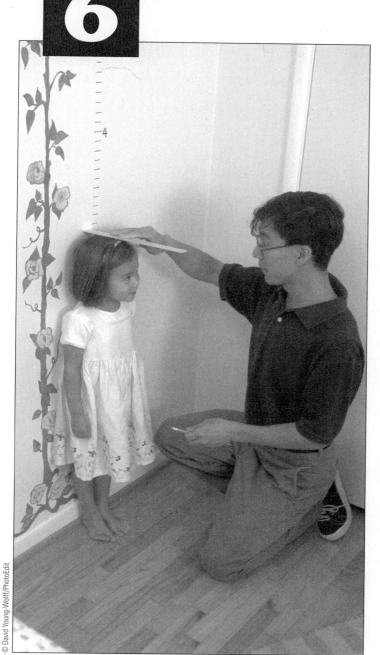
© David Young-Wolff/PhotoEdit

6

Measurement

Introduction

If you see a good friend today and he asks you how you are, what will you say? What will you think your friend means by his question? What kinds of things might come to mind as you think about the answer? What would you mean by your answer? These kinds of questions are like the ones researchers ask themselves when they try to decide how to measure concepts.

measurement, classifying units of analysis by categories to represent variable concepts.

Measurement generally means classifying units of analysis (say, individuals) into categories (say, satisfied with life or not satisfied with life) to represent variable concepts (say, life satisfaction). Like other issues we've considered, measurement is a process that frequently requires considerable care during the preparation of research, and attention when we read a research report. The building blocks of research are concepts, or ideas (like "impostor," "fatherhood," "satisfaction with life," "suicide," and "power") about which we've developed words or symbols. Such concepts are almost always abstractions that we make from experience. Concepts cannot often be observed directly. What we, as researchers, do during the measurement process is to define the concept, for the purposes of the research, in progressively precise ways so that we can take indirect measurements of it.

Suppose we want to determine (or measure) whether a death was a suicide. We might, for instance, begin with a dictionary definition or something we will call a theoretical definition of suicide as "the act of killing oneself intentionally." Then, we might decide to take any clear evidence of a deceased person's intention to take his life (for example, a note or a tape recording) as an indicator of suicide. We might look for more than one such indicator and would need to figure out what to do if some of the evidence was contradictory. Whatever decisions we make along the way would be part of our measurement strategy—one that usually starts with the broader issues of how to conceptualize something and then goes to the more specific issue of how to operationalize it.

Conceptualization

Researchers who start by creating a research question the way Michele Hoffnung did in the study discussed in the focal research section of Chapter 4 are working with the building blocks of all scientific, indeed of all thinking, concepts—words or signs that refer to phenomena that share common characteristics. In her work, Hoffnung asked questions about what would happen to young women who had graduated from college in 1993: Would they establish long-term relationships, have children, share family responsibilities, and settle into careers as valued as those of their spouses or life partners?

Whenever we refer to ideas such as "sharing family responsibilities," "long-term relationships," or "careers," we are using concepts. But let's notice something important: We each develop our basic understandings of concepts through our own idiosyncratic experiences. You initially might

BOX 6.1

What Do You Think About Life?

Take a moment to answer the following questions suggested by Diener, Emmons, Larsen, and Griffin (1985). Later in the chapter, we'll talk about what these questions are intended to measure.

Following are five statements with which you may agree or disagree. Using the 1–7 scale shown, indicate your agreement with each item by placing the appropriate number on the line preceding that item. Please be open and honest in responding.

The 7-point scale is as follows:

7—Strongly agree

6—Agree

5—Slightly agree

4—Neither agree nor disagree

3—Slightly disagree

2—Disagree

1—Strongly disagree

_____ In most ways my life is close to my ideal.

_____ The conditions of my life are excellent.

_____ I am satisfied with my life.

_____ So far I have gotten the important things I want in life.

_____ If I could live my life over, I would change almost nothing.

have learned what an "accident" is when, as a child, you spilled some milk, and a parent assured you that the spilling was an accident. Or you might have heard of a friend's bicycle "accident." Someone else might have had a baby brother who had an "accident" in his diaper. We might all have reasonably similar ideas of what an accident is, but we come to these ideas in ways that permit a fair amount of variation in just what we mean by the term.

So beware: Concepts are *not* phenomena; they are only symbols we use to refer to (sometimes quite different) phenomena. Concepts are abstractions that permit us to communicate with one another, but they do not, in and of themselves, guarantee precise and lucid communication—a goal of science. In fact, in nonscientific settings, concepts are frequently used to evoke multiple meanings and allusions, as when poets write or politicians speak. If, however, concepts are to be useful for social science purposes, we must use them with clarity and precision. The two related processes that help us with clarity and precision are called conceptualization and operationalization.

conceptualization, the
process of clarifying just
what we mean by a
concept.

conceptual definition, a
definition of a concept
through other concepts.
Also a theoretical
definition.

dimensions, aspects or
parts of a larger concept.

multidimensionality, the
degree to which a
concept has more than
one discernible aspect.

Conceptualization is the process of clarifying just what we mean by a concept. This process usually involves providing a theoretical or **conceptual definition** of the concept. Such a definition describes the concept through other concepts.

What concepts are part of our definition of suicide?

If we use Webster's dictionary definition (or conceptualization) of "suicide" as "the act of killing oneself intentionally," this definition employs the concepts of "killing" and "intention." Or, if we define "satisfaction with life" using Ed Diener et al.'s (1985) definition as "a cognitive judgment of one's life as a whole," we'd have to understand what "cognitive judgment" and "life as a whole" mean. We could define "power," using Max Weber's classic formulation, as the capacity to make others do what we want them to do despite their opposition—a definition that refers, among other things, to the concepts of "making others do" and "opposition." In each case, we would be defining a concept through other concepts: ideas (like "intention," "opposition," and "cognitive judgment") that themselves might deserve further clarification, but which nonetheless substantially narrow the possible range of meanings a concept might entail. Thus, for instance, in Diener's conceptual definition, he is implicitly suggesting that life satisfaction involves the conscious overall evaluative judgment of life using the person's own criteria. He implies that the concept doesn't involve the degree of frequency that the person feels joy, sadness, or anxiety—things that others might associate with life satisfaction. In so doing, he narrows and clarifies the meaning of his concept. Being specific and clear about a concept are the goals of conceptualization.

As part of creating and considering a conceptual definition, we should ask if the concept is so broad and inclusive that it could profitably be divided up into aspects or **dimensions** to further clarify it. A concept that refers to a phenomenon that is expected to have different ways of showing up or manifesting itself is said to be **multidimensional.** For example, we often hear about the "fear of crime," which can be conceptually defined as the fear of victimization of criminal activity. However, we could follow Mark Warr and Christopher Ellison's (2000) lead and divide it up into two dimensions or two kinds of fear of crime: a "personal fear" of crime, which is when people worry about their own personal safety, and an "altruistic fear," when they worry about the safety of others, such as spouses, children, and friends.

Ronda Priest (2005, personal communication) argues that, at the conceptualization stage, researchers might find it useful to think in terms of the "3-A's of Conceptualization": 1) a general concept may involve **A**ction. Thus, for instance, when we think of "religion" it may be thought of as involving a performance dimension, "like going to church," 2) a general concept may involve an **A**ttribute. When we think of religion, we might be interested in a characteristic of a person, like whether she or he belongs to a particular religious denomination, and 3) a general concept may involve an **A**ttitude. It is, of course, not necessary to conceptualize all concepts in terms of actions, attributes, and attitudes. But Priest's schema enables us to consider which of these dimensions we're most interested in. Thus, when we think of religion, we may be interested in how much faith a person possesses or what his or her belief system entails. Researchers often find themselves trying to conceptualize more general concepts as either an action, or an attribute, or an attitude, and then to clarify to others which of these they are focused on. Dividing a concept into dimensions as part of the process of conceptualization can help to specify what is being measured.

STOP AND THINK *See if you can apply Priest's "3-A's of Conceptualization" to the concept of "sense of personal power."*

Operationalization

As we've said, we often can't perceive the real-world phenomena to which our concepts refer, even after we've provided relatively unambiguous conceptual definitions of what we mean by the concepts. "Satisfaction with life" and "fear of crime" and "religion" are very hard to observe directly, and even such a "concrete" concept as "suicide" can elude our direct perception. Someone found dead in a body of water might have been drowned (died by homicide), or accidentally drowned (died by accident), or died as result of his or her own effort (died by suicide)—or even died of natural causes. When the phenomena to which concepts refer can't be observed directly, we might make inferences about their occurrence (or lack or degree thereof).

operationalization, the process of defining specific ways to infer the absence, presence, or degree of presence of a phenomenon.

The process of defining specific ways to infer the occurrence of specific phenomena is called **operationalization.** Operationalization involves providing **operational definitions,** or declarations of the specific ways in which the existence or the degree of existence of phenomena described by concepts are to be determined in a specific instance. For an example, let's take a concept that we're all familiar with—age. Sounds simple enough, right? However, reading the gerontology literature (see Atchley, 2003: 6–8), we could notice that there are several ways to conceptualize age: as chronological age (time since birth), functional age (the way people look or the things they can do), and life stages (adolescence, young adulthood, and so on).

operational definition, declarations of the specific ways in which the absence, presence, or the degree of presence of a phenomenon will be determined in a specific instance.

STOP AND THINK *Atchley's conceptualizations of age seem to focus on attributes (chronological age and life stages) and action (functional age). Use Priest's 3A's to see if you can come up with another way to conceptualize age.*

In any case, how we conceptualize age will make a huge difference in what we do when we operationalize the concept. Let's say we conceptualize age in a particular way—as chronological age; we'll still have to decide on an operational definition. We could, for example, operationalize chronological age by answers to any of the following questions:

1. Are you old?

2. How old are you? Circle the letter of your age category:

   a. 25 or younger   b. 26 to 64   c. 65 or older

3. How old are you?

And many others. Nevertheless, some of these might be much more useful than others are.

indicators, observations that we think reflect the presence or absence of the phenomenon to which a concept refers.

An operational definition includes one or more indicators. **Indicators** are observations that we think reflect the presence or absence of the phenomenon to which a concept refers. Indicators are answers to questions or comments, things we observe, or material from documents and other existing sources. In the previous discussion, the answer to each question is a possible indicator of age.

An Example of Conceptualization and Operationalization

Errol Hamarat and associates (2001: 182) were interested in quality of life and how stress and coping resources affect life satisfaction, with a particular interest in comparing younger, middle-aged, and older adults. To compare the three age groups, Errol Hamarat and associates could not have used question 1 (Are you old?) as their indicator of age because it wouldn't have yielded useful data.

STOP AND THINK   *Why not?*

So the way a concept is operationalized needs to be relevant to research concerns. Questions 2 and 3 would permit the operationalization of age in ways relevant to the research question, but they are not equivalent. One way in which they differ is in number of categories they permit. Question 2 creates three categories; Question 3 allows as many categories as the respondents provide.

In their study, Hamarat et al. (2001) asked the participants to identify their ages and, using the information, created three distinct categories: 18 to 40, 41 to 65, and 66 and over. The researchers also needed to measure each of their other concepts. Let's look at one of their other concepts—life satisfaction—to examine a slightly more complicated example of operationalization. Operationalizing life satisfaction is more complicated than operationalizing age, in part because conceptualizing life satisfaction is difficult. Hamarat et al. (2001) selected the conceptualization advanced by Diener et al. (1985)—that is, satisfaction with life as a whole or a global cognitive judgment of one's life. But there are other ways to define life satisfaction. Among other things, life satisfaction could be a judgment about the relative amounts of pleasure versus pain in one's life, the degree to which one feels (or lacks) anxiety on a daily basis, or a composite of feeling based on the satisfactions in all the various areas of life—work, friendship, love, and so on. Let's assume we can agree that "satisfaction with life as a whole" is an appropriate conceptualization of life satisfaction. What do we do now? How should we operationalize such a definition? Again, the social science literature provides quite a few options, but few are as simple as asking the respondent "Are you satisfied with life as a whole?" Such a simple question would miss the mark on at least two major counts: (1) Its very simplicity makes it too vague to justify its use, and (2) it fails to tap some of the ways in which one might be satisfied or dissatisfied with life.

index, a composite measure that is constructed by adding scores from several indicators.

In their study, Hamarat et al. (2001) used Diener's operational definitions for life satisfaction. Its indicators are the five questions with a seven-point answer scale that we asked you to answer at the beginning of this chapter (in Box 6.1). The **index**—or composite measure that is constructed by adding scores of individual items—for life satisfaction is created by adding the scores of the five items so that the totals range from 5 (low satisfaction) to 35 (high satisfaction).

STOP AND THINK   *Just for fun, why don't you use your answers to the five statements to calculate a "life satisfaction" score for yourself.*

In addition to studying people of different ages, researchers have used this life satisfaction measure in a variety of countries, including the United States, Japan, Australia, Northern Ireland, Portugal, Cape Verde, Korea, and England (Kim and Hatfield, 2004; Lewis, Dorahy, and Schumaker, 1999; Neto and Barros, 2000; Schumaker and Shea, 1993; Smith, Sim, Scharf, and Phillipson, 2004), among army spouses (2005) and among healthy people and those with severe injuries (Corrigan, Smith-Knapp, and Granger, 1998). Hamarat et al. (2001: 189) found that in their sample of 192 people, older adults experienced the highest satisfaction and younger adults the lowest. The average scores were 21.9 for those 18 to 40, 24.2 for those 41 to 65, and 26.4 for those 66 and over.

STOP AND THINK   *How did your score compare?*

Composite Measures

composite measure, a measure with more than one indicator.

Indexes like one of life satisfaction, and other **composite measures,** abound in the social sciences. All composite measures, or measures composed of more than one indicator, are designed to solve the problem of ambiguity that is sometimes associated with single indicators by including several indicators of a variable in one measure. The effort to confirm the presence (or absence or the degree of presence) of a phenomenon by using multiple indicators occurs in both qualitative and quantitative research.

In qualitative research (see especially Chapters 10 and 11 on Qualitative Interviewing and Observation Techniques), a researcher might or might not need to prepare measurement strategies in advance of, say, an interview or observation (see, for example, Miles and Huberman, 1994: 34–39). The careful qualitative researcher is likely to use multiple indicators of key concepts. A researcher can judge a study participant's "emotional tone" and body language and pay attention to the actual answer to a question like "How are you today?" Also, a question can be followed up with other questions. In Box 6.2, Hoffnung's coding of interviewees' answers to questions for two of her variables is described.

In quantitative research, there has been much sophisticated thinking about the preparation of composite measures in advance of data collection. A full discussion of this thinking is beyond the scope of this chapter, but we would like to introduce you to it briefly. All strategies for making composite measures of concepts using questions involve "pooling" the information

Coding Qualitative Data[1]

To illustrate how social researchers code qualitative data, let's return to Michele Hoffnung's work on women's career and family choices.

In 2000, seven years after they graduated, Michele Hoffnung re-interviewed the young women in her study, asking them a series of questions about their current lives and future plans. (See Chapter 4 for Hoffnung's focal research article.) After the taped interviews were transcribed, Hoffnung and an associate read each interview, coded the data, and then identified one category of each variable in which to locate each participant. In most cases, the two coders agreed on the category each participant's answers seemed to best fit. (Comparing coders' decisions is called checking intercoder reliability, a topic we'll cover later in the chapter.) In the section that follows, Hoffnung generously contributed her questions, coding decisions, and examples from interviews about the variable: Number of Children Wanted.

Variable: Number of Children Wanted

The indicators were different for women who were mothers and those who were not. Women without children were asked, "Would you like to have children? Why or why not?" If they said they wanted children, they were later asked, "How many children would you like to have?" Women with children were asked, "Do you plan to have any more children? Under what circumstances?"

The categories were the actual numbers of children that the women specified, including none. If the woman listed a range of numbers, then the midpoint of the range was used as the category.

THREE EXAMPLES:

Q. Would you like to have children?
A. Yes.

Q. Why?
A. In my mind they've always been a part of a family. Um, you know I grew up in a family with six kids. And have always been used to having, you know, kids around at least when I was growing up. I see myself having between two and four kids.

Q. How many children would you like to have?
A. We've said four, but we are going to go one at a time.

Q. How many children would you like to have?
A. Three.

[1] The data were collected by Michele Hoffnung, Professor of Psychology, Quinnipiac University, as part of her study of women's lives, and is published with permission.

gleaned from more than one item or question. Two such strategies, the index and the scale, can be distinguished by the rigor of their construction. An index is constructed by adding scores given to individual items, as occurs in the computation of the life satisfaction score. The items don't have to "go together" in any systematic way because indexes (sometimes called indices) are usually constructed with the explicit understanding that the given phenomenon, say "life satisfaction," can be indicated in different ways in different elements of the population. (Incidentally, we define a **scale** as an index in which some items can be, or are, weighed more than others and will illustrate scales a bit later in the chapter.)

scales, indexes in which some items are given more weight than others in the determination of the final measure of a concept.

If a concept is multidimensional—that is, has different ways of showing up or manifesting itself—then an index can sum up all the dimensions. The Consumer Price Index, for instance, summarizes the price of a variety of goods and services (for example, food, housing, clothing, transportation, recreation, and so on) in the United States as a whole in different times and for different cities at a given time. It is an index (of a multidimensional concept) because one wouldn't necessarily expect the price of individual items (say, the price of food and the price of recreation) to go up and down together. The Consumer Price Index is, however, a good way of gauging changes in the cost of living over time or space.

Other examples of indexes are evident in this chapter's focal research on impostor tendencies and academic dishonesty, by Joseph Ferrari. Ferrari became interested in the concept of impostor tendencies when a colleague in clinical psychology introduced him to the Impostor Phenomenon and Ferrari began to wonder whether many students might "shortchange" themselves by thinking they can't be as bright or capable as their accomplishments make them appear to be. He hoped that by examining the impostor phenomenon, he might shed light on it and help students see their own merits. While you read Ferrari's essay, see if you can answer the questions: How does Ferrari conceptualize and operationalize the concepts of "impostor tendencies" and "academic dishonesty?"

FOCAL RESEARCH

Impostor Tendencies and Academic Dishonesty

By Joseph R. Ferrari[2]

The impostor phenomenon refers to *feelings of intellectual phoniness experienced by high achieving individuals* (Chance & Imes, 1978), despite objective evidence to the contrary. Impostors believe others will discover that they are not truly intelligent, but are in fact "impostors" (Clance, 1985). Repeat successes fail to weaken impostors' feelings of fraudulence or to strengthen

[2] This is an abridged version of an article that appeared in the *Social Behavior and Personality*. It is reprinted by permission from the *Social Behavior and Personality*, 2005, Vol. 33, No. 1, 11–18 published by the Society for Personality Research (Inc.).

a belief in their ability. Impostors react either by extreme overpreparation, or by initial procrastination followed by frenzied preparations (Chrisman, Pieper, Clance, Holland, & Glickauf-Hughes, 1995).

Cowman and Ferrari (2002) found that shame-proneness and self-handicapping best predicted impostor fears among students. Consistent with this outcome, Ferrari and Thompson (2004) found that impostor scores by students correlated positively with favorable impression management strategies, and that impostors were more likely than were nonimpostors to engage in self-handicapping acts (like not practicing before an upcoming task) in order to reduce the likelihood of unexpected performance success. Together, these studies suggest that impostors may sabotage their task performances, perhaps to demonstrate that they do not deserve their obtained successes—to show they are, in fact, frauds. Within a college setting, the opportunity to engage in destructive behaviors may include cheating on assignments and examinations, or plagiarizing written works. Because impostors state that they self-handicap, and that they believe they do not deserve their successes (Cowman & Ferrari), it is possible that impostors would engage in academic dishonesty in order to "prove" to others that they do not deserve success.

On the other hand, Whitley and Keith-Spiegel (2002) report that students who engage in academic dishonesty have moderate expectations for success, anticipate high rewards for their success, and are competitive about obtaining good grades. Such a profile seems opposite to the characteristics of an impostor, who does not believe success is warranted. Consequently, and as an alternative to the previous hypothesis, one might expect impostors would not cheat as an academic self-handicapping strategy: that, in fact, impostors, compared to nonimpostors, would not use cheating and plagiarism to obtain academic success. The present study explores whether increased or decreased reports of academic dishonesty were more likely among impostors or nonimpostors.

Method

Participants

A total of 124 students (92 women, 32 men; mean age = 20.9 years old) enrolled in a medium-size, private, urban Midwestern teaching university participated in the present study as part of course credit for an introductory psychology class.

Psychometric Measures

Participants completed Clance's (1985) 20-item self-reported unidimensional Impostor Phenomenon Scale (CIPS) to assess individuals' experience of impostor fears. Sample items are "I can give the impression that I'm more competent than I really am" and "At times I feel my success is due to some kind of luck." Respondents indicate their endorsement of items on five-point scales with end-point designations *not at all true* (1) and *very true* (5). With the present sample, the coefficient alpha was 0.91 (mean CIPS score = 54.24).

All participants also completed Roig and DeTommaso's (1995) Academic Practices Survey, a 24-item scale that assesses students' understanding and practices associated with (1) *plagiarism* when asked to perform a written assignment (16 items: "I've added sources not read to the reference section of a paper." "I've taken one or two sentences from someone else's written work, changed them moderately, and inserted this information into my paper."); and, (2) *cheating* while completing an examination (eight items: "Copied answers from another student during an exam." "Used hidden notes, books, or calculators during an exam even though such use was prohibited."). All items are rated on a five-point scale (1 = *never*; 5 = *very frequently*). Within the present sample, coefficient alpha for the *plagiarism* subscale was 0.89 and 0.76 for the *cheating* subscale. In addition, all participants completed both sections of Paulhus' (1991) Balanced Inventory of Desirable Responding (BIDR), a well-known, reliable and valid 40-item measure of social desirability responding (see Endler & Parker, 1998, for details). This measure permits a control for the degree to which respondents engage in *self-deception enhancement* (that is, the degree to which they respond honestly but with an inflated sense of confidence, overclaiming their successes, and self-inflation of their skills) and in *impression management* (or with "faking" or "lying" on self-report measures).

Procedure

During a large testing session, a research assistant distributed to students the psychometric scales, along with forms to collect background information, after each student had returned a signed, confidential consent form. After scales had been collected, the assistant explained the purpose of this study.

Results and Discussion

Scores on the impostor scale were divided into extreme low or extreme high categories, to ensure the creation of two independent samples based on the personality style, using the bottom 25 percent and top 25 percent of scores, respectively. Persons who scored lower than or equal to 45 were labeled *nonimpostors* (n = 31: 20 women, 11 men: M score 37.58), whereas persons who scored higher than, or equal to, 63 were labeled *impostors* (n = 32: 22 women, 10 men: M score 71.00). There were no significant differences between men and women in their tendency to be impostors or nonimpostors. There were also no significant differences between impostors and nonimpostors in their likelihood of engaging in *self-deception enhancement* or *impression management*, as measured by the BIDR indexes.

Nonimpostors were significantly more likely than impostors to cheat on examinations (mean cheating score for impostors = 11.38; mean score for nonimpostors = 14.04) and to plagiarize in written assignments (mean plagiarism score for impostors = 24.38; mean score for nonimpostors = 31.38), and they were likely to do so, even when self-deception enhancement and impression management were controlled using the BIDR measures. These results indicate that perhaps among college students, nonimpostors engage in dishonest behaviors to succeed more often than do impostors. Alternatively, it is possible that impostors simply do not report academic dishonesty.

Still, the present study suggests that the academic success of impostors is unlikely to be due to unusually high levels of academic dishonesty. Some scholars (e.g., Cozzarelli and Major, 1990) have claimed that impostors have strong cognitive intelligence despite their beliefs to the contrary. The present study suggests that impostors do not handicap their academic performance through dishonest practices, even if they do engage in self-sabotaging behaviors at other times and in other ways (Cowman & Ferrari, 2002 Ferrari and Thompson, 2004). It should be noted that psychometric self-report scales were administered together in the same testing session to convenience samples and that the results might not reflect a causal relationship between the impostor phenomenon and academic honesty. The results, however, merit further investigation in other contexts. In summary, though, it seems that impostors are persons who are not likely to "cheat their way to the top." Nonimpostors are actually more likely to practice academic dishonesty to succeed.

REFERENCES

Chrisman, S. M., Pieper, W. A., Clance, P. R., Holland, C. L., & Glickauf-Hughes, C. 1995. Validation of the Clance Impostor Phenomenon scale. *Journal of Personality Assessment*, 65, 456–467.

Clance, P. R. 1985. *The impostor phenomenon: Overcoming the fear that haunts your success*. Atlanta: Peachtree Publishers.

Clance, P. R., & Imes, S. A. 1978. The impostor phenomenon in high achieving women: Dynamics and therapeutic intervention. *Psychotherapy: Therapy, Research and Practice*, 15, 241–247.

Cowman, S., & Ferrari, J. R. 2002. "Am I for real?" Predicting impostor tendencies from self-handicapping and affective components. *Social Behavior and Personality*, 30, 119–126.

Cozzarelli, C., & Major, B. 1990. Exploring the validity of the impostor phenomenon. *Journal of Social and Clinical Psychology*, 9, 410–417.

Endler, N. S., & Parker, J. D. A. 1998. *Paulhus deceptions scales*. (Manual available from Mental Health Services, Inc.)

Ferrari, J. R., & Thompson, T. 2005 or 2006. Impostor fears: Links with self-perfection concerns and self-handicapping. Forthcoming in *Personality and Individual Differences*.

Paulhus, D. L. 1991. Measurement and control of response bias. In J. P. Robinson, P. R. Shaver, & L. S. Wrightman (Eds.), *Measures of personality and social psychological attitudes* (pp. 17–59). New York: Academic Press.

Whitley, B. E., & Keith-Spiegel, P. 2002. *Academic dishonesty: An educator's guide*. Mahwah, NJ: Lawrence Erlbaum, Publishers.

STOP AND THINK *How is the impostor phenomenon conceptualized by Ferrari and by other social psychologists? Are you surprised by this way of thinking about "impostors"?*

Ferrari conceptualizes the impostor phenomenon, as other social psychologists, and particularly Clance, have as "feelings of intellectual phoniness experienced by high achieving individuals." Like all scientific conceptualizations, this one winnows away many possible connotations of the word "impostor" in favor of one with relative precision. If you think about Priest's 3-A's of conceptualization, you'll note that this conceptualization stresses an attitude some people might have (that their successes are undeserved), rather than an action (like misrepresenting themselves to others) or an attribute (like actually being a fraud). Ferrari further clarifies what he means by the "impostor phenomenon" in the way he operationalizes the concept, in terms of the series of questions called the Clance Impostor Phenomenon Scale (CIPS) that he asks his student respondents to answer. The 20 questions that constitute CIPS are presented in Box 6.3.

STOP AND THINK *Why don't you answer the CIPS questions and calculate your own score on it.*

Ferrari's students, we learn from his article, scored, on average, 54.24 on CIPS, and that the top 25 percent, whom Ferrari calls "impostors," scored over 63 and that the bottom 25 percent, whom he call "nonimpostors," scored under 45.

STOP AND THINK *To calculate your own score on CIPS, add the scores you gave to each item in Box 6.3. How does your score compare to the mean score among Ferrari's students? Would you have qualified as an impostor or a nonimpostor?*

Ferrari uses two other sets of indexes: the Academic Practices Survey (APS) and the Balanced Inventory of Desirable Responding (BIDR). The APS is used to find out which of two alternative hypotheses, that impostors are unusually likely to plagiarize and cheat, on the one hand, and that they're unusually likely not to plagiarize and cheat, on the other, better describes the world. He finds, of course, that they are actually less likely to plagiarize and cheat than nonimpostors.

But Ferrari wondered whether part of this difference might be due to the fact that impostors, more than nonimpostors, might be interested in giving a good impression of themselves. It is possible, for instance, that, when asked whether they'd ever "copied answers from another student during an exam," impostors might be more likely than nonimpostors to lie. Ferrari has, in personal communication (2005), asked us to remind you that sometimes people falsely state information in surveys to look good. Weisberg (2005, 86–89) suggests that this is a major source of **measurement error** in all surveys. Measurement error, unlike the kinds of error we discussed in Chapter 5—coverage error, nonresponse error, and sampling error—refers to the kind of error that occurs when the measurement we obtain is not an accurate portrayal of what we tried to measure. We'll have more to say about measurement error associated with surveys in Chapter 9. To deal with the

measurement error, the kind of error that occurs when the measurement we obtain is not an accurate portrayal of what we tried to measure.

BOX 6.3

The Clance Impostor Phenomenon Scale

Instrument scoring: 1 = not at all true, 2 = rarely, 3 = sometimes, 4 = often, 5 = very true

Question

- I have often succeeded on a test or task even though I was afraid I would not do well before I undertook the task.

- I tend to remember the incidents in which I have not done my best more than those times I have done my best.

- If I'm going to receive a promotion or gain recognition of some kind, I hesitate to tell others until it is an accomplished fact.

- I can give the impression that I'm more competent than I really am.

- I often compare my ability to those around me and think they may be more intelligent than I am.

- If I receive a great deal of praise and recognition for something I've accomplished, I tend to discount the importance of what I have done.

- I often worry about not succeeding with a project or on an examination, even though others around me have considerable confidence that I will do well.

- It's hard for me to accept compliments or praise about my intelligence or accomplishments.

- I feel bad and discouraged if I'm not "the best" or at least "very special" in situations that involve achievement.

- I'm often afraid I may fail at a new assignment or undertaking, even though I generally do well at what I attempt.

---

measurement problem associated with people trying to give a good impression of themselves on surveys, Ferrari employed the BIDR, which not only measures how much respondents are likely to deceive themselves (by asking 20 questions like "I always know why I do things" [Bonnano et al., 2005]) but also measures how much they are likely to try to give a good impression (by asking 20 questions about, for instance, whether they sometimes tell lies and have some pretty awful habits [von Hippel et al., 2005]). Using a statistical procedure (analysis of covariance), he is able to determine that impostors are less likely to plagiarize and cheat than nonimpostors, even when you hold constant any differences that might exist between impostors and nonimpostors in their chances of lying to give a good impression.

Indexes can be used to measure multidimensional concepts, then, and we often compute an index score, as we did for CIPS in the previous Stop and Think by adding the scores on individual items. Methodologists often

BOX 6.3 (*continued*)

- I rarely do a project or task as well as I'd like to do it.
- Sometimes I'm afraid others will discover how much knowledge or ability I really lack.
- When people praise me for something I've accomplished, I'm afraid I won't be able to live up to their expectations of me in the future.
- I avoid evaluations if possible and have a dread of others evaluating me.
- I'm disappointed at times in my present accomplishments and think I should have accomplished much more.
- I'm afraid people important to me may find out that I'm not as capable as they think I am.
- When I've succeeded at something and received recognition for my accomplishments, I have doubts that I can keep repeating that success.
- I sometimes think I obtained my present position or gained my present success because I happened to be in the right place at the right time.
- At times, I feel my success has been due to some kind of luck.
- Sometimes I feel or believe that my success in life or in my job has been the result of some kind of error.

Source: Oriel, Plane, and Mundt, 2004.

call indexes in which some items can be, or are, weighed more than others as **scales.** One kind of scale, the Bogardus social distance scale (designed by Emory Bogardus), was created to ascertain how much people were willing to interact with other kinds of people. Suppose you wanted to find out how accepting people are of Martians. You could ask them questions like:

1. Would you be inclined to let Martians live in the world?
2. Would you be inclined to let Martians live in your state?
3. Would you be inclined to let Martians live in your neighborhood?
4. Would you be inclined to let a Martian be your next door neighbor?
5. Would you be inclined to let a Martian marry your child?

Because you might expect that a person who answered yes to question 5 would also answer yes to all the other questions, you could reasonably give a higher score to a yes for question 5 than the others. Similarly, you could reasonably give a higher score to a yes for question 4 than for yeses to questions 3, 2, and 1.

STOP AND THINK   *Sometimes indexes that are referred to as scales don't use items that might be sensibly weighted the way our Bogardus scale does . . . but sometimes they do. Have a look at the items in the Clance Impostor Phenomenon Scale (in Box 6.3). When used by Ferrari, all items were given equal weight. But do you think some items could be given greater weight as indicators of the impostor phenomenon? Is it a genuine scale in this sense?*

Indexes and scales are not hard to make up, though assuring yourself of their validity and reliability can be challenging. (See the discussion of these topics that follows.) You're frequently well advised to find an index whose validity and reliability has already been tested if one exists for the variable you're trying to measure. (See Miller, 1991, or Touliatos, Perlmutter, and Straus, 2001, and see also the journal, *Social Indicators Research,* for a list of indexes that are often used in the social sciences.) Keep your eyes peeled during your literature review and you might see ways (or references to ways) that others have operationalized the concepts you're interested in. If you think the same approach might work for you, try it out (and remember to cite your source when you write up your results). There's no sense in reinventing the wheel. If you'd like advice about creating indexes or scales of your own, we recommend either Earl Babbie (2005) or Chara Frankfort-Nachmias and David Nachmias (2000) as a source.

Exhaustive and Mutually Exclusive Categories

Measuring a concept (for example, life satisfaction or age) means providing categories for classifying units of analysis as well as identifying indicators. For "life satisfaction," for example, Diener (2001) suggests creating seven categories by taking the answers to the five questions, adding the individual scores together (for a total of between 5 and 35), and classifying the score as follows:

31 to 35	extremely satisfied
26 to 30	satisfied
21 to 25	slightly satisfied
20	neutral
15 to 19	slightly dissatisfied
10 to 14	dissatisfied
5 to 9	extremely dissatisfied

But what of the nature of these categories? In this section, we'd like to emphasize two rules that should stand as self-checks any time you engage in measuring a variable concept.

The first rule for constructing variable categories is that they should be **exhaustive.** That is, the categories should enable you to classify every subject in your study. The aim of creating exhaustive categories doesn't require any particular number of categories, although the minimum number of categories for a variable concept is two. As an example, consider Hoffnung's classification of whether women's first birth will be (was) planned or not, for women who are trying to conceive or who already are mothers. She used

exhaustiveness, the capacity of a variable's categories to permit the classification of every unit of analysis.

three categories: first birth not planned, first birth planned/cl..
first birth planned and trying to conceive.

STOP AND THINK

Can you think of any likely instance that couldn't be "captured" by the net offered by Hoffnung? If you can't, the categories she's using are exhaustive.

mutual exclusiveness,
the capacity of a variable's
categories to permit the
classification of each unit
of analysis into one and
only one category.

The second rule for variable construction is that the categories should be **mutually exclusive.** This means you should be able to classify every unit of analysis into one and only one category. A woman can give birth as the result of a planned pregnancy, but then it won't be unplanned. A man can fall into the category "65 or older," but if he does, he can't also fall into the "26 to 64" category.

STOP AND THINK

Suppose you asked the question "How old are you?" and then told the person to circle the letter associated with his or her age category, giving them the answer categories of (a) 25 or less, (b) 25 to 65, and (c) 65 and older. Which of the rules for variable construction would you be offending? Why?

STOP AND THINK AGAIN

Based on your life satisfaction score, you can place yourself into one of the seven exhaustive and mutually exclusive categories that Diener suggests. What do you think of your placement in this category?

Quality of Measurement

reliability, the degree to
which a measure yields
consistent results.

validity, the degree to
which a measure taps
what we think it's
measuring.

We'd now like to introduce two other important qualities of measurement: **reliability** and **validity.** A measurement strategy is said to be reliable if its application yields consistent results time after time. If you were to use a scale to measure how much a rock weighs, the scale would provide you with reasonably reliable measurements of its weight if it tells you that the rock weighs the same today, tomorrow, and next week. A measurement strategy is valid if it measures what you think it is measuring. But, however reliable (and valid) your scale is when measuring the rock's weight, it cannot validly measure its shape or texture—it's just not supposed to.

STOP AND THINK

Suppose you attempted to measure how prejudiced people are with the question "What is your sex?" Which quality of measurement—reliability or validity—is most apt to be offended by this measurement strategy? Why?

Checking Reliability

The reliability of a measurement strategy is its capacity to produce consistent results when applied time after time. A yardstick can consistently tell you that your desk is 30 inches wide, so it's pretty reliable (even if it's possibly inaccurate—we'd have to check whether the yardstick's first inch or so has been sawed off). There are many ways to check the reliability of a

test-retest method, a method of checking the reliability of a test that involves comparing its results at one time with results, using the same subjects, at a later time.

measurement strategy. One is to measure the same object(s)—like your desk—two or more times in close succession. This **test-retest method** has been used by the researchers who created the life satisfaction measure. Diener et al. (1985) gave the life satisfaction questions to 76 students twice, two months apart. They found that students who scored high on the index the first time around were much more likely to do so the second time than were students who had not scored high the first time.

The major problems with test-retest methods of reliability assessment are twofold. First, the phenomenon under investigation might actually change between the test and the retest. Your desk won't grow, but a person's life satisfaction might actually change in two months. So if people's categories change BECAUSE there really have been changes in what's being measured, this does not mean that the measurement strategy is unreliable. Second, earlier test results can influence the results of the second. Do you think that if you completed a life satisfaction questionnaire today, it might influence your answers on the same questionnaire next month?

split-half method, a method of checking the reliability of several measures by dividing the measures into two sets of measures and determining whether the two sets are associated with each other.

The **split-half method** is one way to deal with the problems of the test-retest method. It relies on making more than one measurement of a phenomenon at essentially the same time. This method is grounded in the same logic as the complex measures (indexes and scales) mentioned earlier: We frequently want to take more than one measure of a phenomenon. The Clance Impostor Phenomenon Scale used by Ferrari consisted of the 20 items shown in Box 6.3. The split-half method of checking the reliability of this scale would entail randomly assigning those twenty items to two sets of ten and using statistics to see if the two sets corresponded in the way they classify respondents to the survey. Ferrari reports, for instance, that the statistic he used, Cronbach's alpha, is .91. Cronbach's alpha is actually a way of estimating the average correspondence of *all* split-half combinations of items. An alpha of .70 or greater suggests that items in a scale "go together" very well, that is that they are internally consistent, so an alpha of .91 indicates high split-half reliability.

STOP AND THINK   *What were the alphas for the "plagiarism" and "cheating" subscales of the Academic Practices Survey among Ferrari's respondents?*

interobserver (or inter-rater) reliability method, a way of checking the reliability of a measurement strategy by comparing results obtained by one observer with results obtained by another using exactly the same method.

Another common way to check the reliability of a measurement strategy is to compare results obtained by one observer with results obtained by another using exactly the same method. We'll see this **interobserver reliability** or **interrater reliability method** figure extensively in the focal research of later chapters. In Chapter 13, on content analysis, Clark, Guilmain, Saucier, and Tavarez report that, when trying to assess the traits of individual characters in children's books using a pre-defined measurement scheme, they agreed 88 percent of the time on average.

STOP AND THINK   *Suppose that you found a way of measuring homophobia with 10 questionnaire items and you wanted to use those questions in a questionnaire to see if there's a connection between homophobia and knowledge of AIDS. Describe a way in which you might check the reliability of the homophobia measure you were using.*

Checking Validity

Validity refers to how well a measurement strategy taps what it intends to measure. You might wonder how to tell whether a strategy is valid. After all, most measurement in the social sciences is indirect. Asking people whether, generally speaking, they are "very," "fairly," or "not too" happy, isn't exactly the most direct way of determining their personal well-being, even though this question figures into a frequently used measure of such feelings, as in a Gallup (1998) poll on well-being.

Ultimately the validity of a measure cannot be proven, but we can lend credence to it. Perhaps the most intuitively satisfying (and best) check on validity is this question: Does this measurement strategy *feel* as if it's getting at what it's supposed to? If a measure seems as if it measures what you and I (and maybe a whole lot of others) believe it's supposed to be measuring, we say it has **face validity,** and we can feel pretty comfortable with it. The answer to the question, "How old are you?" seems like a pretty good measure of age and, therefore, has high face validity. On the other hand, the question, "What, in your opinion, is the proper role of the military in foreign affairs?" would yield a poor measure of age and, therefore, has low face validity. And, generally, if most people who know something about a concept (say, life satisfaction) feel that a given set of questions (say, those used by Diener and others) "gets at" that concept in an important way, the measure is said to have face validity. (In fact, if they also agreed that the set of five questions asked by Diener and others covers the usual range of "content" implied by the concept of "life satisfaction" [for example, a judgment of one's life as close to ideal, lived under excellent conditions, etc.], they would also have affirmed the set's **content validity.**)

The obvious problem with face validity is that it is so subjective. Its virtue is that it is simple to use and simple to understand and that, once you "look behind" more objective procedures, they can look a bit subjective, too. Establishing **predictive validity,** for instance, involves comparing the results of your measurement scheme with a criterion variable that it should predict (Zeller, 1997: 824). If, for instance, you compared the SAT scores of high school seniors with their performance in college (that is, something those scores should "predict to") and found that students with higher test scores did better in college, then you could say the SAT test had predictive validity.

face validity, the degree to which a measure seems to be measuring what it's supposed to be measuring.

content validity, how well a measure covers the range of meanings associated with a concept.

predictive validity, how well a measure is associated with future behaviors you'd expect it to be associated with.

STOP AND THINK *Can you see why checking the results of an SAT test against a criterion such as performance in college still involves an element of subjectivity? (Hint: What does it mean to do "well" in college?; how does one measure such performance?; doesn't judging the validity of SAT scores by this criterion, in turn, depend on the criterion's face validity?)*

A major difficulty in validating social science measures using behavioral criteria, however, is the problem of finding behaviors that they obviously "predict to" (Carmines and Zeller, 1979). Many of our indicators are meant to measure abstract concepts (for example, life satisfaction or self-esteem) that are valued primarily because of their expected association with other

construct validity, how well a measure of a concept is associated with a measure of another concept that some theory says the first concept should be associated with.

abstract concepts (for example, happiness or sense of control). As a result, social scientists tend to pay more attention to the **construct validity** of their measures than to their predictive validity. Construct validity refers to how much a measure of one concept is associated with a measure of another concept that some theory says it should be associated with. Desrochers, Hilton, and Larwood (2005), for instance, developed an index to measure the degree to which parents blurred the distinction between their work and family roles, or the sense that work and family are essentially indistinguishable (and that seems especially likely to occur in highly integrated work-family arrangements). They then checked the construct validity of this index by comparing it with measures their theories led them to expect to be associated with such blurring: like the number of hours respondents worked at home and away from home, the presence of distractions when working, and the presence of work-family conflict.

STOP AND THINK   *What are two ways by which you might check the validity of the measure of life satisfaction that we've been discussing?*

There are, of course, many sources of measurement error associated with the wording of questions (Weisberg, 2005) other than that they might engender self-enhancing responses or might not have mutually exclusive and exhaustive answer categories. Some other considerations are that questions ask respondents about topics that are relevant to them or about which they've given some thought and that they not be misleading. We'll have more to say about such sources of measurement error in Chapters 9 and 10, but, for now, consider the different kinds of information that can come when questions are asked differently by having a look at the poll results on stem cell research in Box 6.4.

STOP AND THINK
ABOUT VALIDITY   *Comparing the actual questions asked by the three polls in Box 6.4, what do you think of their validity? Which questions are the best indicators of the attitudes under consideration? Why?*

STOP AND THINK
ABOUT RELIABILITY   *Gallup did another poll on the same topic on August 3–5 and 55 percent of Americans said they have followed the issue "very" or "somewhat" closely, compared with 38 percent who said that in the July poll. Among people who were following the issue closely, support and opposition for this research was higher than in July. In August, by a margin of 59 percent to 37 percent, the relatively attentive public supported government funding for this type of embryonic stem cell research, whereas in July the percentages were 55 percent versus 24 percent. Based on the differences in results, do you think the indicators are unreliable? Why or why not?*

One last quality of measurement is its precision or exactness. Some measurement strategies allow greater precision than others do. If we ask people how old they were at their last birthday, we can be more precise in our categorization than if we ask them to tell us their ages using broad categories such as "26 to 64." If we use the seven categories of life satisfaction (from extremely satisfied to extremely dissatisfied) that Diener (2001) suggests, we can be more precise than if we use only two categories (satisfied and dissatisfied), though less precise than using the 5-to-35 point scale on which the categories are based.

BOX 6.4

Different Indicators, Different Results

Excerpts from studies of attitudes toward government funding of stem cell research

In creating or selecting indicators, a researcher has a wide variety of choices to make. The specific choice of indicators will affect the study's results. This was clear during the summer of 2001, when the topic of human embryonic stem cell research became a topic of conversation among Americans from President George W. Bush to the average citizen. As people were hearing about stem cell research—many for the first time—they were asked by a variety of polling organizations what they thought about it.

With an important political decision about federal funding for medical research using cells obtained from human embryos on the horizon, pollsters conducted surveys on attitudes toward stem cell research. Each study focused on "attitudes toward government funding of human stem cell research," but they obtained very different results. An ABC poll reported that 58 percent of Americans supported federal funding of such research; the Conference of Catholic Bishops reported a poll indicating that 70 percent opposed it; and Gallup presented data that the majority of Americans didn't know enough to declare an opinion. With results like this, you may wonder about the validity and reliability of polling data. Before you reach a conclusion, let's examine some of the indicators and the findings.

* * * * * * * * * * * *

An ABC News/Beliefnet Poll was conducted on June 20–24, 2001, by telephone with a national random sample of 1,022 adults.[3]

The indicators:

1. Sometimes fertility clinics produce extra fertilized eggs, also called embryos, that are not implanted in a woman's womb. These extra embryos either are discarded, or couples can donate them for use in medical research, called stem-cell research. Some people support stem-cell research, saying it's an important way to find treatments for many diseases. Other people oppose stem-cell research, saying it's wrong to use any human embryos for research purposes. What about you—do you support or oppose stem-cell research?

2. The federal government provides funding to support a variety of medical research. Do you think federal funding for medical research should or should not include funding for stem-cell research?

The findings:

In this poll, 58 percent supported stem-cell research, 30 percent opposed it, and 12 percent did not answer the question. Sixty percent supported

[3] The information about the ABC News/Beliefnet Poll was retrieved from the Beliefnet website on August 11, 2001, at http://www.beliefnet.com/features/stemcell.html.

(continued)

BOX 6.4 (*continued*)

federal funding for it, 31 percent opposed such funding, and 9 percent did not answer the question.

* * * * * * * * * * * *

A CNN/USA Today/Gallup Poll was conducted on July 10–11, 2001, by telephone with a randomly selected national sample of 998 adults, 18 years and older.[4]

The indicators:

1. As you may know, the federal government is considering whether to fund certain kinds of medical research known as "stem cell research." How closely have you followed the debate about government funding of stem cell research? Very closely, somewhat closely, not too closely, or not closely at all?

2. Do you think the federal government should or should not fund this type of research or don't you know enough to say? Should, should not, don't know enough to say (depends can be volunteered)

3. How important are the following to you personally?

   ■ Finding cures or medical treatments for diseases such as Alzheimer's, diabetes, heart disease and spinal cord injury? Very important, somewhat important, not at all important.

   ■ Preventing human embryos from being used in medical research? Very important, somewhat important, not at all important.

4. Which position comes closest to your own? Embryonic stem cell research is morally wrong and unnecessary, it is morally wrong but may be necessary, it may be necessary and is not morally wrong, or it is unnecessary but not morally wrong.

The findings:

Only 38 percent of Americans said they were following the stem cell research issue very or somewhat closely.

A majority of Americans, 57 percent, say they "don't know enough to say" when asked whether the federal government should fund the particular kind of stem cell research being debated. Of the remainder, 30 percent say the government should fund it while 13 percent disagree. In June, the subset of Americans who were closely following the issue supported it by a wide margin: 55 percent of the currently attentive group favor it and just 24 percent are opposed.

The overwhelming majority of Americans say that finding cures or medical treatments for diseases and spinal cord injury is very important: 82 percent feel this way, and another 16 percent say it is somewhat

[4] The information about the CNN/USA Today/Gallup Poll was retrieved from Gallup's website on August 11, 2001, at http://www.gallup.com/poll/releases/pr010720.asp.

BOX 6.4 (*continued*)

important. Only 30 percent of Americans consider it very important to prevent human embryos from being used in medical research. Another 29 percent say this is somewhat important, but 36 percent say this concern is not important to them.

In response to the question about which of four positions comes closest to their own, only 20 percent of Americans report that the research is morally wrong and unnecessary, 34 percent believe the research is morally wrong, but say it may be necessary, 35 percent say it may be necessary and is not morally wrong, while 4 percent hold the view that the research is neither necessary nor morally wrong.

* * * * * * * * * * * *

An International Communications Research Poll, commissioned by the National Conference of Catholic Bishops, was conducted on June 1 and June 5, 2001, by telephone with a national sample of 1,013 American adults.[5]

The Indicators:

1. Stem cells are the basic cells from which all of a person's tissues and organs develop. Congress is considering whether to provide federal funding for experiments using stem cells from human embryos. The live embryos would be destroyed in their first week of development to obtain these cells. Do you support or oppose using your federal tax dollars for such experiments?

2. Stem cells for research can be obtained by destroying human embryos. They can also be obtained from adults, from placentas left over from live births, and in other ways that do no harm to the donor. Scientists disagree on which source may end up being most successful in treating diseases. How would you prefer your tax dollars to be used this year for stem cell research?

The findings:

The results for the first question were that 24 percent said they support using federal tax dollars for experiments using stem cells from human embryos in which the live embryos would be destroyed in their first week of development. 70 percent opposed such use of tax dollars, while 5 percent didn't know and 1 percent refused to answer.

The results for the other question were that 67 percent supported research using adult stem cells and other alternatives, while 18 percent supported other methods, including those that require destroying human embryos, 9 percent said they support neither, 6 percent didn't know, and 1 percent refused to answer.

[5] The information about the ICR poll was retrieved from the National Conference of Catholic Bishops' website on August 11, 2001, at http://www.nccbuscc.org/comm/archives/2001/01-101.htm.

Level of Measurement

Consideration of precision leads us to consider another way that variable categories can be related to one another—through their level of measurement or their type of categorization. Let's look at four levels of measurement: nominal, ordinal, interval, and ratio, and the implications of these levels.

Nominal Level Variables

nominal measure, a level of measurement that describes a variable whose categories have names.

Nominal (from the Latin, *nomen,* or name) level variables are variables whose categories have names or labels. Any variable worth the name is nominal. If your variable is "cause of death" and your categories are "suicide," "homicide," "natural causes," "accident," and "undetermined," your variable is a nominal level variable. If your variable is "sex" and your categories are "male" and "female," your variable is nominally scaled. Nominal level variables permit us to sort our data into categories that are mutually exclusive, as the variable "religion" (with categories of "Christian," "Jewish," "Muslim," "none," and "other") would. Another example of a nominal level variable is marital status, if its categories are "never married," "married," "divorced," "widowed," and "other."

A nominal level of measurement is said to be the lowest level of measurement, not because nominal level variables are themselves low or base, but because it's virtually impossible to create a variable that isn't nominally scaled.

Ordinal Level Variables

ordinal measure, a level of measurement that describes a variable whose categories have names and whose categories can be rank-ordered in some sensible way.

Ordinal (from the Latin, *ordinalis,* or order or place in a series) level variables are variables whose categories have names or labels and whose categories can be rank-ordered in some sensible fashion. The different categories of an ordinal level variable represent more or less of a variable. If racism were measured using categories like "very racist," "moderately racist," and "not racist," the "very racist" category could be said to indicate more racism than either of the other two. Similarly, if one measured social class using categories like "upper class," "middle class," and "lower class," one category (say "upper class") could be seen to be "higher" than the others.

All ordinal level variables can be treated as nominal level variables, but not all nominal level variables are ordinal. If your variable is "sex" and your categories are "male" and "female," you'd have to be pretty sexist to say that one category could be ranked ahead of the other. Sex is a nominal level variable, but not an ordinal level variable. On the other hand, if your categories of age are "less than 25 years old," "between 25 and 64," and "65 or older," you can say that one category is higher or lower than another. Age, in this case, is ordinal, though it can also be treated as nominal. Ordinal level variables guarantee that categories are mutually exclusive and that the categories can be ranked, as the variable "team standing" (with the categories of "1st place," "2nd place," and so on) does in many sports.

STOP AND THINK *What, do you think, is the level of measurement of the life satisfaction measure that has seven categories, ranging from extremely satisfied to extremely dissatisfied? Why?*

Interval Level Variables

interval measure, a level of measurement that describes a variable whose categories have names, whose categories can be rank-ordered in some sensible way, and whose adjacent categories are a standard distance from one another.

Interval level variables are those whose categories have names, whose categories can be rank-ordered in some sensible way, and whose adjacent categories are a standard distance from one another. Because of the constant distance criterion, the categories of interval scale variables can be meaningfully added and subtracted. SAT scores constitute an interval scale variable, because the difference between 550 and 600 on the math aptitude test can be seen as the same as the difference between 500 and 550. As a result, it is meaningful to add a score of 600 to a score of 500 and say that the sum of the scores is 1100. Similarly, the Fahrenheit temperature scale is an interval level variable because the difference between 45 and 46 degrees can be seen to be the same as the difference between 46 and 47 degrees. All interval level variables can be treated as ordinal and nominal level variables, even though not all ordinal and nominal level variables are interval level.

Ratio Level Variables

ratio measure, a level of measurement that describes a variable whose categories have names, whose categories may be rank-ordered in some sensible way, whose adjacent categories are a standard distance from one another, and one of whose categories is an absolute zero point—a point at which there is a complete absence of the phenomenon in question.

Ratio level variables are those whose categories have names, whose categories can be rank-ordered in some sensible way, whose adjacent categories are a standard distance from one another, and one of whose categories is an absolute zero point—a point at which there is a complete absence of the phenomenon in question. Age, in years since birth, is often measured as a ratio level variable (with categories like 1, 2, 3, 4 years, and so on) because one of its possible categories is zero. Other variables such as income, weight, length, and area can also have an absolute zero point and are therefore ratio level variables. They are also interval, ordinal, and nominal level variables. For practical purposes, however, interval and ratio level variables are similar. This raises the question of what practical purposes there are for learning about levels of measurement in the first place, however, and that's the topic of our next section.

STOP AND THINK *When Ferrari compares impostors with nonimpostors, what level of measurement is he using? What makes you think so? When he reports that the mean score of his student respondents on the Clance Impostor Phenomenon Scale was 54.24, what level of measurement is he ascribing to CIPS? What makes you think so?*

The Practical Significance of Level of Measurement

Perhaps the most practical reason for being concerned with levels of measurement is that frequently researchers find themselves in the position of wanting to summarize the information they've collected about their subjects, and to do that they have to use statistics (see Chapter 15). An interesting thing about statistics, however, is that they've all been designed with a particular level of measurement in mind. Thus, for instance, statisticians have given us three ways of describing the average or central tendency of a variable: the mean, the median, and the mode. The mean is defined as the sum of a set of values divided by the number of values. The mean of the ages 3 years, 4 years, and 5 years is $(3 + 4 + 5)/3 = 4$ years. Thus, the mean has been designed to summarize interval-level variables. If age were measured only by categories "under 25 years old," "25 to 64," and "65 and older," you couldn't hope to calculate a mean age for your subjects. The median, however, has been designed to describe ordinal level variables. It's defined as the middle value, when all values are arranged in order. You could calculate a median age if your (five) respondents reported ages of, say, 3, 4, 5, 4, and 4 years. (It would be 4, because the middle value would be 4, when these values were arranged in order—3, 4, 4, 4, 5). You could calculate a median because all interval level variables (as age would be in this case) are ordinal level variables. But you could also calculate a median age even if your categories were "under 25 years old," "25 to 64," and "65 and older."

STOP AND THINK   *Suppose you had five respondents and three reported their ages to be "under 25 years old," one reported it to be "25 to 64," and one reported it to be "65 and older." What would the median age for this group be?*

A third measure of central tendency or average, the mode, has been designed for nominal-level variables. The mode is simply the category that occurs most frequently. The modal age in the previous Stop and Think problem would be "under 25 years old" because more of the five respondents fell into this category (three) than into either of the other categories. The modal age for a group with ages 3, 4, 5, 4, 4, would be 4, for the same reason. You can also calculate the modal sex, if of five participants, three reported they were female and two reported they were male. (It would be "female.") And, you wouldn't have been able to calculate either a median or mean sex.

STOP AND THINK   *Suppose three students in your class were classified as "extremely satisfied" with life, five as "satisfied," four as "slightly satisfied," two as "neutral," three as "slightly dissatisfied," one as "dissatisfied," and none as "extremely dissatisfied." Which of the measures of central tendency could you use to describe the level of satisfaction of the class? Why?*

The upshot is that the scale of measurement used to collect information can have important implications for what we can do with the data once they've been collected. In general, the higher the level of measurement, the greater the variety of statistical procedures that can be performed. As it turns out, all statistical procedures that scientists use are applicable to interval (and, of course, ratio) data; a smaller subset of such procedures is applicable to ordinal data; a smaller subset still to nominal data.

What's the practical implication of all this? Perhaps the most important implication is that, given a choice and other things being equal, you'd probably want to measure a variable in a higher level of measurement than in a lower one. You can always, for instance, take information collected at the interval level (say, age in years) and reduce it to an ordinal level (say, age in three categories, "under 25 years," "25 to 64," and "65 and above") and use statistics that are appropriate for the ordinal level. The flexibility of future data analyses is maximized when variables are measured at higher levels.

Summary

Measurement means classifying units of analysis by categories to represent a variable concept. Measuring abstract concepts involves conceptualizing, or clarifying, what we mean by them, and then operationalizing them, or defining the specific ways we will make observations about the concepts in the real world.

The first step of measurement is to construct a conceptual definition of a concept. The second step is to operationalize the concept by identifying indicators with which to classify units of analysis into two or more categories. The categories should be exhaustive, permitting you to classify every subject in the study, and mutually exclusive, permitting you to classify every subject into one and only one category. Depending on the complexity of the concept involved, you can use simple or complex measures of it and be more or less precise in your measurement. Indexes, like those used by Ferrari in his study of impostors, frequently involve the summation of scores on a variety of items used to measure a concept.

The best measurement strategies are reliable and valid. Reliability refers to whether a measurement strategy yields consistent results. We can examine the reliability of a measurement strategy with a test-retest, split-half, or interrater reliability check. The calculation of Cronbach's alphas, like those used by Ferrari in his study, involve a kind of split-half check in which all items in an index are used as checks for consistency against all other items. Validity refers to whether a measurement strategy measures what you think you are measuring. We can examine the validity of a measurement by considering its face validity—by seeing how we and others feel about its validity—or by examining its predictive criterion-related validity—by seeing how well results we obtain with it correlate with some behavior we think it should predict to—or by investigating its construct validity—by seeing how well it correlates with some other measure that a theory leads us to believe it should correlate with.

There are three significant levels of measurement (nominal, ordinal, interval-ratio) in the social sciences. In general, the data analyses that can be used with a given set of variables depend on the level of measurement of the variables.

Putting Measurement to Work

This exercise gives you the opportunity to develop a measurement strategy of your own.

1. Identify an area of the social sciences you know something about (aging, marriage and the family, crime and criminal justice, social problems, social class, and so on).

2. Identify a concept (for example, attitudes toward the elderly, marital satisfaction, delinquency, social class) that receives attention in this area and that interests you.

3. Describe how you might conceptualize this concept.

4. Describe one or more measurement strategies you might use to operationalize this concept. List at least one or more indicators and two or more categories.

5. Identify the level of measurement of this concept given the operationalization you suggested. Give support for your answer.

6. Discuss the reliability of your measurement strategy. How might you check the reliability of this strategy?

7. Discuss the validity of your measurement strategy. How might you check the validity of this strategy?

Comparing Indicators of Life Satisfaction: Global Evaluation versus Areas of Life

Some researchers have asked people to evaluate how satisfied they are in various domains of their lives rather than asking for an overall evaluation. The Gallup Poll, for example, asks a series of questions about different areas of life as one of the ways it measures life satisfaction. In this exercise, you're asked to compare two alternative sets of indicators for life satisfaction—the one suggested by Diener and his associates and one used by the Gallup Poll.

1. Make two copies of the five questions on life satisfaction suggested by Diener et al. at the beginning of the chapter and find two people to answer the questionnaires.

2. After each of the people completes the questionnaire, interview them and ask the following questions adapted from a Gallup Poll. Read the

questions to them and write in the answers they give. You could also answer the questions yourself to see how you feel about the issues.

Using the answers "very satisfied," "somewhat satisfied," "somewhat dissatisfied," or "very dissatisfied," how satisfied are you with each of the following aspects of your life?

a. Your community as a place to live in _____

b. Your current housing _____

c. Your education _____

d. Your family life _____

e. Your financial situation _____

f. Your personal health _____

g. Your safety from physical harm or violence _____

h. The opportunities you have had to succeed in life _____

i. Your job, or the work you do _____

3. Americans responded to these questions in a survey conducted on June 11–17, 2001. Review Table 6.1 from Gallup.

TABLE 6.1   **Percentage Answering "Very Satisfied" to Questions**

Adults	*Men*	*Women*	*Non-Hispanic*	*Whites*	*Blacks*	*Hispanics*
Your family life	69%	67%	72%	72%	52%	73%
Your current housing	63	63	63	66	41	45
Your community as a place to live in	58	59	57	61	42	53
Your education	45	48	44	48	38	38
Your financial situation	26	28	25	29	15	23
Your personal health	45	57	52	54	54	57
Your safety from physical harm or violence	55	61	50	59	33	47
The opportunities you have had to succeed in life	48	53	43	49	40	44
Your job, or the work you do	50	50	50	51	36	51

Source: Saad, L. "Blacks Less Satisfied Than Hispanics with Their Quality of Life," retrieved from http://www.gallup.com/poll/releases/pr010622.asp

4. Discuss the measurement of "life satisfaction" by answering the following questions:

   a. Which set of questions—Diener's or Gallup's—do you think results in a more valid measure of life satisfaction? Why?

   b. Which set of questions do you think results in a more reliable measure of life satisfaction? Why?

   c. What would be an advantage of using both the Gallup and the Diener measures of satisfaction in a study?

   d. Can you suggest an index that could be constructed based on the nine questions asked by the Gallup organization?

   e. Suggest one or more ways of measuring "life satisfaction" in addition to the two sets of questions presented here. Describe one or more additional questions that could be asked, one or more observations that could be made, or one or more pieces of information that could be obtained from records or documents.

7

Cross-Sectional, Longitudinal, and Case Study Designs

Introduction

Think about your life so far and how it has changed since you were an adolescent. What changes have occurred in your values and your attitudes? What changes have occurred in your family, your community, and the larger social world? What things have stayed the same? If you were to create a hypothesis based on your own personal experience, what would it be? And, what kind of a study could you do to test it?

Let's say you created a hypothesis about your family's social class background and your current career aspirations. At this point, you could plan most aspects of a study to test such a hypothesis, including making some connections between theory and research, planning to be ethical in doing your study, figuring out a methodology that is feasible in terms of time, money, and access, and considering validity and reliability when deciding how to measure the variables of interest. We now turn to the remaining research decisions you'd need to make—those related to selecting a study design and to choosing one or more methods of data collection.

Study Design

study design, a research strategy specifying the number of cases to be studied, the number of times data will be collected, the number of samples that will be used, and whether or not the researcher will try to control or manipulate the independent variable in some way.

causal hypothesis, a testable expectation about an independent variable's affect on a dependent variable.

A **study design** results from several interconnected decisions about research strategies. Obviously data must be collected at least once from at least one unit of analysis. But, because data *can* be collected two, three, or more times, an important decision is the *number of times* to collect data. Other decisions focus on the *size of the sample* and *the number of samples* needed. Samples can be small or large, and one or more than one sample must be selected. (See Box 7.1.)

For a study with an explanatory purpose that has a causal hypothesis to test, there is an additional decision. You might recall that in testing a **causal hypothesis,** a researcher speculates that one variable, the independent variable, is the *cause* of another variable, the dependent variable. When a researcher has a causal hypothesis, it's necessary to decide if it is appropriate to *try to control or manipulate the independent variable* in some way. In the study we will be describing in Chapter 8, for instance, Emily and a colleague decided it would be useful for one group of students to have the experience of reading and discussing stories involving "caring" behavior while a second group of students had a different experience. The independent variable was whether or not students were exposed to stories focusing on caring, and the researchers thought it best to try to "control" it. The study design in this case is an experiment—a design in which the "control" of an independent variable is integral.

Researchers have to balance the methodological concerns against practical matters and ethical responsibilities when making study design choices. Researchers must always consider if a study is practical—that is, "doable" with the available resources. Because it is almost always easier, less time consuming, and less expensive to collect data once, from one sample, and to

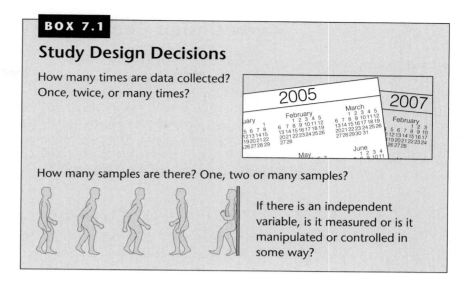

measure rather than control or manipulate independent variables, making decisions that will cost more in time or money must be justified. In some cases, it is just not possible to control the independent variable; in other cases, it might be unethical to do so. Researchers know that controlling or manipulating something is an important responsibility because of the potential effect on people's lives. In addition, if a researcher wants to study a sample more than once, the practical issue of keeping the names and addresses of study participants raises the ethical concern of the risk of confidentiality being violated.

Connections Between the Uses of Research, Theory, and Study Design

Decisions about study design are based partly on the uses or purposes of the research, the researcher's interest in testing causal hypotheses, and the use of theory. You might remember that in earlier chapters we talked about using research for purposes of exploration (breaking new ground), description (describing groups, situations, activities, events, and the like), and explanation (explaining why things are the way they are). We described theory as an explanation about how and why something is as it is and compared the deductive and inductive approaches to theory, to see that theory can be used to deduce a testable hypothesis or can be generated after data is collected and analyzed. Our point in recalling the purposes of research and the deductive and inductive approaches here is that these purposes and approaches can have important implications for research design decisions.

Study Design Choices

Let's say you were interested in studying middle school students, to learn something about their behavior, such as how they get along with their parents, how frequently they use drugs, or how well they do in school. How might you go about planning this study? Would you decide to talk to students? Observe them? Ask for information from their parents or their schools? How might you pick your population and sample of students? Would you need information about many of them or just a small group? What about if you wanted to find out about changes in their behavior as they move from adolescence into adulthood? Would you need to collect data more than once?

Cross-Sectional Study Design

Araxi Macaulay and a team of researchers were interested in finding out about adolescent drug use and the role that parenting practices play in preventing it (Macaulay, et al., 2005). Reviewing the research literature, they located quite a few studies that indicated that one or more parenting behaviors were connected to lower drug use rates among children. However, they found that the existing work in the field did not focus on the mechanisms by which parenting behaviors was protective (Macaulay, et al., 2005: 69). The researchers set out to do a study that would help them *describe* the frequency of behaviors and the relationships among the variables and help them *explain* why some adolescents are more likely to use drugs than others. In their study, the researchers were not trying to control or manipulate parental behavior, but to measure it. They recruited a sample of more than 2,000 middle school students from 34 middle and junior high schools in rural and suburban New York, each of whom filled out a questionnaire one time. The survey included questions on the students' perceptions of parental disciplinary efforts and the monitoring of children's behavior, and questions about the students' drug-related knowledge, attitudes, and behavior.

cross-sectional study, a study design in which data are collected for all the variables of interest using one sample at one time.

The research described above is an example of the **cross-sectional study,** the most frequently used design in social science research. In the cross-sectional design, data are collected about one sample at one point in time, even if that "one time" lasts for hours, days, months, or years. It might have taken the research team weeks or even months to obtain all of the completed questionnaires, but each student answered the questions only once. Another example of the cross-sectional design is Joseph Ferrari's study of the impostor phenomenon—the focal research from Chapter 6. In Ferrari's study, a sample of college students completed a questionnaire that included the "impostor scale" and indicators of other variables only one time.

Cross-sectional designs are widely used because they enable the description of samples or populations on several variables, and because it is least expensive to collect data once from one sample. Cross-sectional studies often have large samples and usually collect data that lend themselves to statistical analyses. In looking for patterns of relationships among variables, the sample can be divided into two or more categories of an independent variable to identify differences and similarities among the groups on some

FIGURE 7.1

The Cross-Sectional Study Design

one sample (divided up into categories of the independent variable during analysis)		time 1
	category 1 of an independent variable	measure of a dependent variable
	category 2 of an independent variable	measure of a dependent variable

dependent variable. Macaulay and associates (2005) used statistical analyses to find associations between the independent variables of parenting practices and the dependent variables of children's drug use. More specifically, they were able to describe the relationships as follows: Effective parenting practices, such as greater monitoring and enforcement of parental rules, were associated with lower rates of adolescent smoking, alcohol, and marijuana use (Macaulay et al., 2005: 75).

Studies like Macaulay et al.'s that have at least one independent and one dependent variable and that use a cross-sectional design can be diagrammed[1] as shown in Figure 7.1.

STOP AND THINK   *Here are four variables that might "go together" in one or more patterns:*
- *student's gender*
- *parental annual income*
- *student's perceptions of peers' alcohol use*
- *student's alcohol use*

Try to use at least two of these variables and construct one or more hypotheses. Think about the time order and speculate about which of these variables might "go before" others. State your hypothesis and think about how to do a cross-sectional study to test your hypothesis.

Cross-Sectional Studies and Causal Relationships

Cross-sectional designs like the one used by Macaulay and associates are sometimes used to examine causal hypotheses. You might remember from our discussion in Chapter 2 that just because two variables "go together" doesn't mean that one variable *causes* change in another. Let's take a simple example of a causal hypothesis and hypothesize that the sight of a bone makes your dog salivate.

Our example would need to meet three conditions before we could say a **causal relationship** exists. First, in a causal relationship, there must be an *empirical association* between the independent and dependent variables. In

causal relationship, a nonspurious relationship between an independent and dependent variable with the independent variable occurring before the dependent variable.

---

[1] This diagram and the others used in this chapter and Chapter 8 have been influenced by those presented by Samuel Stouffer (1950) in his classic article "Some Observations on Study Design."

the case of the dog and the bone, we'd need to see the dog generally salivating in the presence of a bone and generally failing to salivate in its absence. Second, we'd need to make sure that the *temporal precedence* or *time order* was correct—that is, we'd need to make sure that the bone appeared before, rather than after, salivation. (This is to make sure that the bone causes salivation, rather than salivation "causing," in some sense, the bone.)

spurious relationship, a noncausal relationship between two variables.

Finally, we'd want to make sure that the relationship is not **spurious,** or caused by the action of some third variable, sometimes called *antecedent variable,* that comes before, and is actually responsible for, making the independent variable and the dependent variable vary together. We'll illustrate spuriousness with a different example. Consider, for instance, that in some regions there is a relationship between two autumnal events: leaves falling from deciduous trees and the shortening of days. One might be tempted, after years of observing leaves falling while days shorten (or, put another way, days shortening while leaves fall) to hypothesize that one of these changes causes the other. But, the relationship between the two events is really spurious, or noncausal. Both are caused by an antecedent factor: the change of seasons from summer to winter. It is perfectly true, that in some regions, leaves fall while days shorten, but it is also true that both of these events are due to the changing of the seasons, and so neither event causes the other. Therefore, the two events are spuriously related.

requirements for supporting causality, the requirements needed to support a causal relationship include a pattern or relationship between the independent and dependent variables, determination that the independent variable occurs first, and support for the conclusion that the apparent relationship is not caused by the effect of one or more third variables.

The reason we mention the criteria for determining causation in the context of research designs is simple: Some research designs are better than are others for examining the **requirements for supporting causality.** Thus, for instance, although it is possible to use a cross-sectional design to test causal hypotheses, this design might be less than ideal for such tests in some circumstances. One data collection can tell us about the current state of two or more variables. Sometimes the time order is clear in cross-sectional studies. Vivian Tseng (2004) did a cross-sectional analysis of family interdependence and academic adjustment. Comparing young adults of different ethnicities and foreign-born to non-immigrants, Tseng found that Asian youth and those who were foreign-born put greater emphasis on family obligations than did those of European backgrounds and those born in the United States. While the effects of third variables still need to be considered, the time order is clear: ethnicity and place of birth come before attitudes toward family obligations. However, in other studies, it can be difficult to disentangle the time order. It is always possible to ask questions about past behavior, events, or attitudes, but such retrospective data might involve distortions and inaccuracies of memory.

STOP AND THINK

*Can you think of a problem with concluding that parental practices such as monitoring children's behavior **cause** decreased drug use in middle-school children?*

In their study of parental practices and student drug use, Macauley et al. (2005: 75) acknowledge the concern about using a cross-sectional study for explanatory purposes by noting that their ability to infer causality is limited by the cross-sectional nature of their study. In their data analysis, they find evidence of a relationship between the variables, but they don't know with certainty whether student drug use or parental behavior came first. Time order would have been better established by using an alternative study design, one that enabled the collection of data more than once. That way the researchers could determine if parental behavior at one time had an impact on children's drug use at a later time.

So, be cautious in interpreting the results of studies that use a cross-sectional design. This study design is very appropriate for describing a sample on one or more variables and for seeing connections between the variables. A cross-sectional design can also be useful for studying causal hypotheses when the time order between the variables is easy to determine, and when sophisticated statistical analyses can be used to control for possible antecedent variables.

Longitudinal Designs: Panel, Trend, and Cohort Designs

In the mid 1990s, a group of British researchers observed that policy and curriculum decisions were being based on the assumption that the youth in Britain were experiencing ethical decline and needed to be taught morality. These researchers decided to study the values that young people already had, and to consider the changes in their moral landscapes over time. The goals of the research were to understand many aspects of young people's lives, to learn how social divisions such as class, locality, gender, ethnicity, disability, and sexuality play out over time in the making of inequality, and to test theoretical ideas such as the view that the self-identity is created rather than inherited or static (Thomson, personal communication). In the focal research that follows, we'll read about the study that this team of researchers designed and some of their findings.

FOCAL RESEARCH

Inventing Adulthoods: A Qualitative Longitudinal Study of Youth Transitions

By Rachel Thomson, Sheila Henderson, Janet Holland, Sheena McGrellis, and Sue Sharpe[2]

Background

This study of young people's accounts of adulthood is situated within a historical and theoretical account of social change. In the UK young people are faced with problems of *social mobility* (defending against a loss in social

[2] Several pages of this article have been adapted from "Hindsight, foresight and insight: the challenges of longitudinal qualitative research" by Rachel Thomson and Janet Holland, 2003 *International Journal of Social Research Methodology* 6 (2): 233–244 and are reprinted with permission. Information about the journal is available at http://www.tandf.co.uk/journals.

status relative to parents, as well as seeking to increase status relative to parents), and problems of *social (re)production* (creating forms of adulthood that are similar or different from those of their parents and significant others).

These problems are structured by a range of factors that include locality, gender, sexuality, ethnicity and disability, each providing young people with access to particular resources and identities and constraining access to others. The rapidity and particular character of social changes mean that young people are forced into projects of *reinvention* as well as reproduction, which may demand new kinds of resources and skills as well as the transmission of more traditional forms of social, cultural, symbolic, and material capital (Bourdieu 1986). So although there may be less room for social mobility than in a previous generation, it is possible that there may also be a greater role for individual agency and entrepreneurialism in the formation of new social positions.

We have adopted a longitudinal, qualitative approach, following young people through the transition to adulthood—collecting, comparing, and interpreting their changing accounts of adulthood over time. The aims and objectives of the research include documenting young people's accounts of their own transitions, identifying "critical moments" in the construction of adult identities, and analyzing the relationships between socially structured opportunities and individual biographies.

Sample and Data Collection

Data collection had three phases. In the first phase (Youth Values), we selected a very large sample of young people from five contrasting locations of the UK, including inner city sites, suburbs, and a rural village. Using multiple methods, including questionnaires and interviews, we collected data from 1996 to 1999 by asking questions focusing on the moral frameworks of the youth. When we received additional funding, we began the next phase by selecting 120 research participants (aged 14 to 23) from the phase one sample. In this second phase (Inventing Adulthood) we followed the participants from 1999 to 2001, interviewing each of them three times at nine-month intervals and collecting other data as well (see Thomson et al., 2004 for more details). In the third stage (Youth Transitions), we continued to follow the sample from 2002 to 2006, conducting two additional interviews.

Remarkably, we have been able to secure continuous funding from the UK Economic and Social Research Council, enabling us to hold together a core research team and maintain our original sample. By establishing consistent researcher/participant relationships over time we believe we have been able to minimise attrition and facilitate local understanding. And while researchers worked at some geographical distance, they also met regularly to compare emergent findings, share what they were learning, and encourage consistency in approach.

Each fieldworker took responsibility for a particular research location and group of young people and maintained telephone contact with participants between interviews. Regular mailings were sent to all research participants by the central research administrator. These included information about the study, Christmas cards, regular newsletters, and competitions. We also developed an interactive website.

Methods of Analysis and Interpretation

In our approach we look both cross-sectionally to identify discourses through which identities are constructed as well as longitudinally at the development of a particular narrative over time. The complexity of the interview data set has demanded a number of complementary strategies for analysis.

Narrative Analysis of Individual Cases Over Time and Within Localities

In order to capture the narrative character of individuals and changes over time, the interviewers conducted a "narrative analysis," which included substantive content and the researchers' personal reflections on the interview and their hopes, fears, and predictions for each young person at each round. At the end of each round of interviewing a "summary narrative analysis" was written for each location, identifying local themes in young people's accounts. After the third interview, researchers drew together the three narrative analyses for each young person to produce a "case profile" tracing changes and continuities in their narrative over time.

Longitudinal qualitative data is intimidating in that there is no closure of analysis, and the next round of data can challenge interpretations. When is it appropriate to start making interpretations of the data? When do you start writing up? Analysis and data collection are never finished, interpretation is always provisional. Recording researchers' observations enabled us to observe shifts in our provisional interpretations, and to trace the process of interpretative revision in an explicit way.

This recognition of the contingency of individual accounts was in part balanced by our attempts to accumulate and compare accounts, both of individuals over time and at any one time within a single locality. In moving from the narratives of individuals to those of a community we were able to identify the local stock of narrative resources that may be available to individual young people (Finnegan 1997) while also gaining a sense of the diversity that exists within particular localities (Plummer 2001). Successive interviews gave us a better understanding of the individual, if not the "truth" of that person. Rather than moving towards "saturation" as suggested by some, we felt that the case profiles captured the "kaleidoscope approach" in which "each time you look you see something rather different, composed mainly of the same elements but in a new configuration" (Stanley 1992:158).

Analysis of Narratives Across the Sample

In our view it was not sufficient to rely on the narrative analyses and case profiles as our main source of data. Each round of interview data was transcribed and coded descriptively and conceptually, using qualitative data analysis software. We have endeavoured to undertake cross-sectional analyses for each round of data collection in order to capture a particular moment in time, often highlighting biographically structured temporal themes such as waiting for examination results and leaving school as well as wider cultural and political contexts. These analyses highlight differences and similarities within the sample, and by accumulating further rounds of analysis, begin to identify the relationship between individual narratives and wider social processes.

Being Involved in a Longitudinal Study

We neither sought to make an intervention into young people's lives, nor denied that we might well be doing so. Our decision to ensure a continuity of interviewer over time was both pragmatic and guided by a concern with the quality of the research relationship. We recognised that it is not 'normal' for young people to be invited to participate in regular in-depth interviews by researchers from a university, and that impact of the research process would have to be addressed in the process of data collection, analysis, and interpretation. Throughout the research process we have attempted to make space for young people to talk about the impact of the research on them.

At the end of their first interview for this stage of the research, young people were invited to comment on the research process. Most were positive, commenting that it was unusual and enjoyable. A number said that the interview had a cathartic effect, helping them to express and resolve emotions as well to clarify their feelings and ideas about the future. Some young people actively *used* the space provided by the confidential interviews in a project of self understanding.

> Allan[3]: It's made me feel more stronger, with my feelings about things. The last one, the last interview that we had, I wasn't as open about stuff [. . .] I'm starting to speak out more. [. . .] Like if I have any problems you know I would speak them out to somebody that I could trust.

By the second round of interviews young people were much more forthcoming in their comments, which reflected their increasing confidence in the method. They knew what was going to happen and were able to plan what they wanted to do in the interview. Young people were also more forthcoming as to the status that being involved in the research had conferred on them in peer groups, commenting that being selected and listened to in this way made them feel special.

> Judy: It's something no one else is doing. Well some other people are but most people aren't which is nice. [. . .] you feel like the chosen you [laughs] I don't know why you chose me [laughs].

[3] To protect confidentiality, all names of study participants are pseudonyms.

The character of the research relationship was also mentioned, and young people compared it to the ways in which they were able to talk to parents, friends, and careers advisers. Of particular interest here are their comments about the impact of the research on their way of thinking.

It should be noted that not all young people reported this; some denied that the research had any impact at all and cited pragmatic reasons for continuing (for example, in the early interviews, missing lessons). The young people commented on not "having" to talk about something that they did not wish to, and in some cases they resisted the demands that the research placed on them. The interviewers were aware that the structure of the research encouraged young people to present themselves as being involved in a progressive and developmental process of change. Some of those young people who did drop out of the study were certainly those experiencing difficult circumstances.

Reasons given for withdrawing from the study included being too busy and lack of interest. The factors contributing to attrition included age (when they left school young people were more difficult to contact and much more busy), mobility (moving house/country) and marginality. Some who withdrew from the research subsequently returned.

Over the course of the study most of the young people became increasingly confident in interviews, more able to direct the course of the conversation, and more instrumental in how they used the interview space for themselves. This could give rise to tensions between the interview schedule and the story that the young person wanted to tell.

The longitudinal research process also affects the researchers involved, generating genuine familiarity with research participants. As a team we attempted to be reflexive about this relationship, making space for the interviewers to record and discuss their feelings on the basis that such emotions were an important part of understanding the research process. While we might expect a longitudinal method to complicate the already complex boundaries of the research relationship, the researchers generally felt comfortable in their role. Yet researchers' feelings about the research process also gave us important insight such as the normative effects of the repeat interview structure as noted by Sheena McGrellis:

> I am increasingly conscious of the fact that we appear in young people's lives on a semi-regular basis, and ask about their thoughts and aspirations for the future, among many other things. We hear about relationships and their investment in new relationships and friendships, about their experiences of education and plans for further education, for employment and travel. We allow them to talk and record their thoughts and feelings. And then we meet them again and ask for an update. They know we have a recording of what they said last time. For those whose lives remain on course this, I suspect, is a very gratifying and confirming experience; for those whose lives don't turn out quite as planned, this may not be such a great experience. I'm not sure how we measure this and whether or not this is a factor that will be linked to attrition. [fieldnote]

Looking Across the Sample

By comparing accounts of adulthood across the sample, we have been struck by the centrality of locality in shaping young people's orientation to adulthood as well as the variations in the "embeddedness" of young people within their localities. The model of extended dependency demanded by participation in further and higher education was often in tension with local models of adulthood, particularly in working-class communities, even where the conditions for these adult identities have largely disappeared (Thomson and Holland 2002). The following extract from a case profile provides a strong sense of the way in which locality frames the meanings of adulthood for young people, and the contingent nature of this embeddedness.

> Cheryl framed her life, initially, in the same picture as the adult world around her. She subscribed to the political and social mores available, and promoted in her community. She had no reason to question or confront these. She accepted and conformed to the rules of her tight-knit community. She toyed with the idea of moving out through education but was never sure where she wanted to go or what she wanted to do—there were few local alternative adulthoods to model on. Instead she worked on what she saw and knew—the part-time job, house, home and kids, husband out working. She gave up on education as a route out and opted for a full-time local job. By her third interview she was drawing on other accounts of adulthood. She had been exposed to alternative paths, or to people on other pathways, from other backgrounds and was using this as a way out. She had moved into a more flexible, broader world, where she might train and work, go to university at some time in the future, travel at some point, and generally build up life experiences. How long she will be able to sustain this life or idea is questionable, depends very much on how things work out for her in the immediate future. [case profile]

By looking across the sample we found that although many of the markers of adulthood are fragmented and contested, parenthood and an independent "home" appear to be at the centre of most young people's understandings of adulthood. We found little evidence of reinvention or re-sequencing, or that detraditionalised models of adulthood have gained moral legitimacy (Thomson, Henderson, and Holland 2002). But we did find that some of the markers traditionally associated with adulthood (such as sexual activity, drinking, drug taking, mobility, and consumption) were being reworked as markers of youth (Henderson 2001). Where young people were caught in an extended economic dependence on parents they were more likely to invest in consumption and lifestyle-based "youth" identities. Where they had embarked directly into the world of work from school they tended to invest in more traditional aspects of adult identity. We illustrate this with extracts from case profiles of two young people, Jimmy and Una.

> Jimmy's version of adulthood revolves around being in a particular clique of grunge rock lovers—it's all about trusting close friends, a "them and

us" mentality, being different, and takes place in houses, rooms, garages—playing music, pool, smoking a bit of dope. Education and employment careers seem to take a back seat compared to friendships and social life—that's what really seems to be driving him over these three interviews. No real sense of longer term planning, apart from what comes out in lifeline, but he's fairly consistent that he wants to get away from the area. [case profile]

Una's idea and experience of adulthood has been one of increasing pressure, stress and responsibility over time. She entered the part-time job market at 17, the permanent market at 18, and continues, despite herself, to look to education for opportunities to further or define her adult pathway. Her position as eldest daughter still at home is central to how she has taken on adulthood. She has set herself up as a replacement breadwinner and partner for her mother. In her social life she has met a steady boy friend and is looking towards a traditional relationship, living with mum until marriage. She is saving now (although not sure what for) but is buying "mad adult things". The accumulation of serious commodities also represents a move towards a certain kind of traditional adulthood—one she sees (but is strongly resisting) as fully fledged in her work. [case profile]

We found tensions between an individualised model of adulthood in which young people stress feelings of maturity and autonomy and a socialised (relational) model of adulthood in which young people stress responsibilities of care for others. Our data suggest that adulthood can be both gendered and individualised with young people (especially young working-class women) assuming responsibility for economic independence and caring. Not surprisingly, young people expressed ambivalent orientations towards adulthood. A significant proportion sought to delay adulthood, associating it with onerous responsibility. Others saw themselves as having accelerated passages to adulthood, and this could be a source of pride, grievance or both. The extent of parental support as well as the impact of experiences such as bereavement, caring responsibilities, parental conflict, etc., was significant in shaping these orientations.

Looking at Individuals Across Time

Young people are likely to have different levels of responsibility and autonomy in different areas of their lives (education, work, domestic, leisure). The extent to which they invest in different aspects of adult identity is both a response to the recognition of their competence in these arenas and a response to a lack of recognition in other areas. For example, young people who experience themselves as incompetent in education are more likely to invest in an adult identity within a romantic relationship, work, consumption, or in alternative leisure or criminal careers (Thomson et al., 2004). This process of *"competence—recognition—investment"* can make sense of the choices that young people make, without reducing these either to individual psychology or to the determination of social structures.

Social structures manifest themselves within individual biographies in term of events, experiences, and responses. We approached our data with the express intention of identifying "critical moments" in the construction of adulthood. Such moments may be institutionally defined—for example, leaving school can be understood as a critical moment of transition that is experienced collectively or individually defined (for example accounts of emotional crises) and may be more or less in the control of the individual concerned. We have plotted such moments on a choice/fate continuum, suggesting that the distribution of critical moments may reflect the uneven distribution of risk in the lives of young people. In analysing our data we have come to recognise that "critical moments" themselves are important narrative devices, and in comparing accounts over time we have been able to distinguish between these and accounts of events that endure over time as having biographical significance. Again this has drawn our attention to the centrality of both timing and resources, finding that what seem to be very similar events in young people's lives may have very different consequences. For example, two young people in our sample lost one of their parents during the course of the study. While this was the dominant critical moment of both of their lives, the consequences were quite different. One young man went into a downward spiral of family conflict, educational failure, drinking, and fighting while the other was drawn into more equal relationships with adults through the bereavement process and was able to maintain his educational progress.

Future Work

The data analysis and interpretation are ongoing and we are increasingly focusing on issues of leaving home, the establishment of couple relationships, the transition to full-time employment and parenthood (Henderson et al., 2007). We have secured funding to explore the feasibility of archiving the data set and opening it up to secondary analysis. For more information see www.lsbu.ac.uk/inventingadulthoods

REFERENCES

Bourdieu, P. 1986. The forms of capital. In *Handbook of Theory of Research for the Sociology of Education*, edited by J. E. Richardson. New York: Greenwood Press.

Finnegan, R. 1997. Storying the self: Personal narrative and identity. In H. MacKay (ed.) *Consumption and Everyday Life.* London: Sage.

Henderson, S. 2001. "Me Dad says I can if I want": The place and meaning of drugs in young people's strategies for transitions. Presented at the annual meetings of the European Sociological Association, Helsinki, Finland, September 2001.

Henderson, S., J. Holland, S. McGrellis, S. Sharpe, & R. Thomson, R. 2007. *Inventing adulthoods: A biographical approach to youth transitions.* London: Sage.

Plummer, K., 2001. *Documents of life 2: An invitation to critical humanism.* London: Sage.

Stanley, L. 1992. *The auto/biographical I: Theory and practice of feminist auto/biography.* Manchester: Manchester University Press.

Thomson, R. & J. Holland. 2002. Young people, social change and the negotiation of moral authority. *Children and Society* 16 (2): 103–115.

Thomson, R., Henderson, S. & Holland, J. 2003. Making the most of what you've got: Resources, values and inequalities in young people's transitions to adulthood. *Educational Review* 55 (1): 33–46.

Thomson, R., J. Holland, S. McGrellis, R. Bell, S. Henderson & S. Sharpe, S. 2004. Inventing adulthoods: A biographical approach to understanding youth citizenship. *The Sociological Review* 52 (2): 218–239.

Thomson and her colleagues were interested in understanding the variety of transitions to adulthood experienced by a group of young people in the UK. From previous research they knew that these transitions are differentiated and structured by background characteristics, including gender and parental social class. Their research was intended to describe young people's portrayals of their transitions and the "critical moments" in the process of moving into adulthood, to examine theoretical perspectives on the nature of identity, and to explain some of the cause and effect relationships between aspects of economic position, social structures, and personal biographies on adult identities.

STOP AND THINK   *These researchers collected data several times from their sample and, in doing so, went beyond the cross-sectional design. Why do you think Thomson and her colleagues chose not to do a cross-sectional study?*

internal validity, agreement between a study's conclusions about causal connections and what is actually true.

longitudinal research, a research design in which data are collected at least two different times, such as a panel, trends, or cohort study.

panel study, a study design in which data are collected about one sample at least two times where the independent variable is not controlled by the researcher.

Thomson and her colleagues could have waited until the young people in their study were all in their twenties and then, using a cross-sectional design, asked questions about their childhoods and adolescent years. However, using a cross-sectional design with retrospective questions would have been problematic for the **internal validity** or accuracy of the study's conclusions. Validity concerns would result, in part, because of reliance on people's answers about what they felt or did years before. Memories fade, and current realities affect recollections and interpretations of past experiences and attitudes. The researchers instead chose an over-time or **longitudinal approach;** between 1999 and 2006, during the second and third phases of the study, they conducted five in-depth interviews with the members of the sample. However, for their data analysis not only can they examine changes over time, they can also analyze the data cross-sectionally by comparing and contrasting the study participants at particular moments in time. An advantage of longitudinal studies is that it is possible to do over-time analyses and to analyze each set of data as if it were cross-sectional.

The Panel Study

The study design used by Thomson and her colleagues is a **panel study,** a choice that requires following one sample over time. A panel study needs

FIGURE 7.2

The Minimum Panel Study
Design

one sample	time 1	time 2
	independent and dependent variables measured	independent and dependent variables measured

only two data collections, separated by months or years—although Thomson and her colleagues followed their sample for many years and collected data many times. (See Figure 7.2 for a representation of the minimum panel design.)

Thomson describes the advantages of the panel study:

> The duration of the research is crucial in a repeat interview study such as ours. So much research simply captures a snapshot, which prevents us from exploring processes that unfold over time. Not all processes take years to unfold; some play out over the course of a day. Our study has convinced us of both the value of and the challenges posed by a processual perspective on social phenomena. (Thomson, personal communication)

The longitudinal or over-time nature of a panel study allows researchers to document patterns of change and establish time order sequences. Of course, as the research findings in Box 7.2 demonstrate, selecting the time frame for one or more follow-ups is an important decision. In the study done by Thomson and her colleagues, the sample was followed for 10 years, allowing the researchers to see many changes in identity, understand the various paths to adulthood, and examine the relationships between variables. One strength of the panel design for this research is that it allowed the researchers to maintain the integrity of individual narratives and did not force them to cut up the data into small chunks of text (Thomson, personal communication). Among the study's findings are that youths invest in and create identities based on their competences and the recognition they receive in the various spheres of their lives, such as education, work, and domestic relationships.

STOP AND THINK *Although contacting each person in the sample two or more times has advantages, it also has disadvantages. What do you think are some of the difficulties facing a researcher who collects data more than once from the same sample?*

The ability to see change over time is very important, but there are offsetting costs and difficulties in doing a panel study. The most obvious concerns are time and money. Assuming identical samples and no inflation, a study with three data collections takes much longer and costs three times as much as a one-time study. For three data collections, with the same resources, a researcher could select a sample only one-third as large as for a study using a cross-sectional design. Therefore, it's important to consider the

---

BOX 7.2

A Question of Timing

There have been many longitudinal studies done of children's development. One that has raised questions while answering them is the Study of Early Child Care and Youth Development, a longitudinal study of a large group of children from birth through adolescence. One area of interest in this research is the connection between child care experiences and characteristics and the children's developmental outcomes. Over the course of the study, as the researchers have released results, one of the study's unintended messages is that **timing matters.** One of the first findings presented was that children's attachments to their mothers were not connected to kind of care: At 15 months, children in day care were as strongly attached to their mothers as children reared at home. But the findings at age two were different and showed some negative effects: Those spending many hours in day care showed less social competence and more behavior problems. At age three, the news was different: Those in day care at least 10 hours a week had better cognitive and language skills. But a year later, more findings were announced: Children who spent more than 30 hours a week in day care displayed higher incidence of disobedience, aggression, and conflict with caregivers (Barry, 2002). Whereas the actual findings of the study should be of interest to parents, educators, and policy makers, methodologists will be especially interested in what the results say about the importance of timing in doing follow-up studies. For more information about this study, see http://secc.rti.org/summary.cfm

---

need for a longitudinal approach to establish time order or improve data validity before incurring extra expenses and time or decreasing sample size.

Thomson and her colleagues have worked on their project for many years. During this time, they have sometimes encountered unexpected events and developments:

> Some of the unexpected events were external like the advent of mobile phone culture and the impact that that made on young people's lives. Some events were internal to the research process. For example we had to learn how to manage the boundaries of these ongoing research relationships over time. Things also happened to us as a research team: we had babies, changed jobs, moved home, lost loved ones, and had crises of various kinds. For us all these events are written into the research process and often into the data in subtle ways. Perhaps the most surprising thing to us was our ability to keep the study going. I doubt that we would have had the ambition to design a 10-year study from the outset. Instead we had the intention to be longitudinal and worked in an incremental way, paying attention to the quality of the relationships and attempting to learn from our experience as we went. (Thomson, personal communication)

Funding is always a concern for research and in a longitudinal study, this is especially true. However, this research team felt that, in retrospect, getting the study off the ground was the hardest part. Feeling very fortunate in getting funding for all three phases of the project, they note that, "As the study matured, it became easier to demonstrate its value for answering research questions. We are delighted that the Economic and Social Research Council has now recognized the value of qualitative longitudinal research and is proposing to fund a new prospective study" (Thomson, personal communication).

Keeping track of study participants, and encouraging them to continue participation, are very important for panel studies. Data should be kept confidential, but anonymity cannot be promised to participants because identifying information, including names and addresses, is necessary to re-contact them.[4] In our highly mobile society, keeping track of respondents is difficult, even when the participants are highly motivated. **Panel attrition,** the loss of subjects from the study because of disinterest, death, illness, or moving without leaving a forwarding address, can become a significant issue. In the Inventing Adulthood study, researchers maintained telephone contact with participants between interviews and sent out regular mailings, which helped to keep track of participants and encouraged continued participation. Despite these efforts, there was panel attrition. In phase 2 of the study, from 1999 to 2001, the participants were interviewed three times: 120 during the first round of interviews, 98 during the second round and 83 during the third round. During phase 3, beginning in 2002, the numbers dropped to 70 and 67 for the next two sets of interviews (Thomson, personal communication). As noted in the focal research, participants dropped out when they moved, were experiencing difficult circumstances, or were very busy; sometimes participants returned to the study for subsequent interviews. If some kinds of participants drop out of panel studies more than others, attrition may lead to bias in the study's findings.

Another concern for panel studies is **panel conditioning,** the effect of repeated measurement of variables. Participants tend to become more conscious of their attitudes, emotions, and behavior with repeated data collections. This awareness can be at least partially responsible for differences in reports of attitudes, emotions, or behavior over time (Menard, 1991: 39). The British researchers are aware of panel conditioning. They note that being invited to participate in regular in-depth interviews by researchers from a university is not a "normal" part of the lives of adolescents. Although they found that participants had a positive view of the interview experience, they do believe that being part of a research project had an impact on the participants' lives.

panel attrition, the loss of subjects from a study because of disinterest, death, illness, or inability to locate them.

panel conditioning, the effect of repeatedly measuring variables on members of a panel study.

---

[4] The ethical issues concerning anonymity and confidentiality were covered in Chapter 3. We will be discussing anonymity and confidentiality as they relate to practicality and validity of data in our discussion of questionnaires and interviews in Chapters 9 and 10.

FIGURE 7.3

The Trend Study

	time 1	time 2
first sample	measure independent and dependent variables	
second sample		measure independent and dependent variables

Finally, the issue of changes due to time period need to be considered. "The transition to adulthood is old as humankind and as new as today . . . But while each generation's path shares many milestones and processes in common, each generation's journey is unique" (Smith, 2005: 177). The historical events of each time period are intertwined with the experiences of individual transition, and unless multiple samples of the individuals who are the same age in different decades are followed over time, a panel study like the one done by Thomson et al. can not sort out the effects of the historical era from other variables.

STOP AND THINK *Can you think of any situations where you'd want to see changes over time and would want to do a longitudinal study, but would prefer to select a new sample each time rather than contact the original sample multiple times?*

The Trend Study

Selecting new samples for longitudinal studies can be very useful when researchers are interested in identifying changes over time in a large population, such as registered voters in one country. It can be very time consuming and expensive to re-locate the same individuals to track changes in attitudes, opinions, and behaviors of a very large sample over a number of months or years. It is much more practical to select a new random sample each time. In addition, selecting new samples allows for anonymous data collection, which can be a more valid way to find out about embarrassing, illegal, or deviant behavior.

The longitudinal design that calls for the selection of a new random sample from a population for at least two data collections is called a **trend study,** the design diagrammed in Figure 7.3.

trend study, a study design in which data are collected at least two times with a new sample selected from a population each time.

STOP AND THINK *As an example, let's think about investigating changes in altruism and empathy in a society. Suppose we did a trend study by studying the attitudes and values of a population two or more times, selecting a new sample each time. What would be the advantages and disadvantages of this strategy?*

Trend studies avoid panel attrition and panel conditioning, save the expense of finding the original participants, and enable the researcher to collect data anonymously. Furthermore, trend studies are useful for describing changes in a population over time. For these reasons, studies of changes in public opinion and election behavior often use the trend study design. Such studies can tell us when social phenomena emerge in a society. The American National Election Studies (ANES) provide interesting data in this regard. Each data collection in this series is based on 1,000 to 2,000 interviews with a new, multistage representative sample of citizens of voting age that live in private households (Miller and Traugott, 1989). Jeff Manza and Clem Brooks (1998) analyzed the ANES results from 1952 to 1992 to see if the "gender gap" in voting began with the 1980 presidential election. They found the difference between men's and women's voting patterns had emerged much earlier, with women disproportionately supporting the Democratic candidates since 1952. They conclude that women's rising labor force participation rates over time best explains the gender gap (Manza and Brooks, 1998: 1261).

Another widely used source of information is the General Social Survey (GSS) which is conducted periodically in a wide variety of countries, including Japan, German, Great Britain, and Australia. In the American GSSs, probability samples of more than 1,000 adults living in households in the U.S. are interviewed with some topics rotating in and out and others continuing (Smith et al., 2005). Recent uses of the GSS as trend data include studies of attitudes toward the importance and timing of transitions to adulthood (Smith, 2004), support for funding for public schools (Plutzer and Berkman, 2005), and levels of altruism and empathy (Smith, 2006). See Box 7.3 for a brief discussion of trends in altruism and empathy using the General Social Surveys from 2002 and 2004.

An advantage of the trend study is its use in studying the impact of time periods. Using a trend design, Celia Lo (2000) compared data collected annually from high school students from 1977 to 1997. Using a multistage sampling procedure, each year approximately 130 schools were selected, and all students who attended classes on the day of data collection were asked to complete a questionnaire that included questions on drinking and drug use. A one-time, cross-sectional study would be appropriate for an analysis of the connection between the onset age of drinking and other drug-using behaviors, but by using a trend study, Lo (2000) was able to determine if the variables and their relationships had been consistent and stable over two decades. Using calendar year as a unit of analysis, she found that delaying the onset of drinking reduced illicit drug-using behaviors and that, starting in 1993, an increase in the onset drinking age was associated with drops in alcohol-use measures.

Trend studies have limitations. Because a trend study does not compare a specific sample's responses at two or more times, the trend study can only identify *aggregated* changes, not changes in individuals. People move in and out of populations, so identified changes in attitudes, opinions, or characteristics over time could reflect changes in the composition of the population or in the sample rather than changes in the individuals. If a trend study finds a change over time, such as in empathy, we don't know if there would have been no change, the same change, or a more extreme change if the *same* people had been used at two or more times (that is, if a

Trends in Altruism and Empathy

How would you answer these questions?

During the last few months how often have you done each of the following things?

- Donated blood
- Given food or money to a homeless person
- Allowed a stranger to go ahead of you in line
- Done volunteer work for a charity

How well do these statements describe you?

- I sometimes have tender, concerned feelings for people less fortunate than me.
- Sometimes I don't feel very sorry for other people when they are having problems.
- I would describe myself as a pretty soft-hearted person.
- Other people's misfortunes do not usually disturb me a great deal.

These questions are from the General Social Survey. The first set is used as indicators of altruism and the second is used to measure empathy. The GSS surveys done in 2002 and 2004 show some changes in empathy, with greater percentages of the sample in 2004 as compared to the 2002 sample being more likely to describe themselves as having tender, more concerned feelings toward the less fortunate and as being more soft-hearted. In addition, there was a slight increase in altruism scores over the two years (Smith, 2006).

If you want to use data from the General Social Surveys go to http://sda.berkeley.edu/D3/GSS04/Doc/gs04.htm and check out the list of variables from 1972 to 2004 to select variables to compare over time.

---

cohort study, a study that follows a cohort over time.

cohort, a group of people born within a given time frame or experiencing a life event, such as marriage or graduation from high school, in the same time period.

panel study had been done). In addition, it is possible for "no change" to result for the aggregate even if individual members of the population change in opposite directions between data collections. True to its name, the trend study design is best for describing trends in a population; it cannot identify how individuals have changed, nor can it pinpoint the causes of change.

The Cohort Study

A **cohort study** is a longitudinal study that follows a cohort over time. In social science research, a **cohort** is a group of people who have experienced the same significant life event at the same time. The most frequently selected cohorts are birth cohorts, that is, people born in the same year. Cohorts can also be groups that experience a life event at the same time, such as graduating from college or getting divorced. Although people can "exit" from a cohort (e.g., by dying), no one can join after its formation.

In a cohort study, the same population is followed at least two times using either a panel or trend study design. In the panel study approach, a sample that is a cohort is followed with at least two data collections. A cohort study can examine one birth cohort or an entire generation, such as the "Baby Boomers" or "Generation X." A project of the Department of Labor is the National Longitudinal Survey of Mature Women Cohort, a project that started with a nationally representative sample of more than 5,000 women in 1967 when the sample was between 30 and 44 years old, and followed this sample until 2003 (BLS, 2001). Periodic data collection originally focused on labor force experiences, marital and fertility histories, and then on retirement and pensions. Elizabeth Hill (1995), for example, analyzed the women's experiences between 1967 and 1984. One of her questions was whether there were any advantages in receiving at least one form of training—formal schooling, on-the-job training, or some other kind of training—after the usual schooling age. Hill found that women who had obtained training any time between 1967 and 1984 had received larger wage increases during those years and were more likely to be working in 1984 than were women who had not received it. More recently, over-time data about the women in this cohort were analyzed to determine the ages at which significant physical functional limitations arose for the members of the cohort (Long and Pavalko, 2004).

The focal research study on transitions to adulthoods is a cohort study and also a panel study, because it followed one British cohort of young adults (ages 14 to 23) over time. Similarly, Michele Hoffnung's study of 200 women who graduated from college in 1992 which was discussed in Chapter 4 is also both a panel and a cohort study.

However, it is possible to use a trend study approach in cohort studies. In a regular trend study, if we study the same population over time (for example, all married men in Rhode Island, all university students in France, all registered voters in Canada), the population itself changes over time as people move in and out of the population. But in a cohort study using the trend design, the population remains the same. If, for example, we were studying the birth cohort of Americans born in 1989 over time, we could ask random samples of this cohort questions about their attitudes toward cohabitation, marriage, and divorce every other year. We would then have the ability to describe changes in attitudes in this cohort as they aged. We might find the cohort becoming more liberal, more conservative or remaining constant in their attitudes as they moved through their teens and into their twenties.

Norval Glenn (1998) did a cohort study using a trend study design to investigate a hypothesis derived from cross-sectional research that marital satisfaction is "U shaped," that is higher in the beginning and later stages of marriage and lower in the middle. Glenn used data from the American General Social Survey (GSS) from 1973–1994. Using the almost annual survey data from these different nationally representative samples of Americans 18 and older, Glenn computed the year of first marriage and the age at first marriage to compare marital satisfaction over time among five marriage cohorts. He concludes that, *in the aggregate,* among all cohorts, the average levels of marital satisfaction decline steeply in the first decade of marriage and continue to decline moderately over the next two decades (Glenn, 1998: 574).

The Case Study

case study, a research strategy that focuses on one case (an individual, a group, an organization, and so on) within its social context at one point in time, even if that one time spans months or years.

The other nonexperimental design is the **case study,** a design with a long and respected history. A case study purposely selects one or a few individuals, groups, organizations, or the like and analyzes the selected case(s) within their social context(s). Each case might be studied over a brief period of time or for months or years. The distinguishing feature of the case study is its *holistic* approach, which means looking at the case as a whole embedded in its natural context and using an open-ended rather than prestructured strategy for collecting data (Verschuren, 2003). Typically, the case study relies on several data sources, is conducted in great detail, and results in an in-depth, multifaceted investigation of a single social phenomenon (Orum, Feagin, and Sjoberg, 1991: 2). The case study is distinguished from a cross-sectional design with a small sample in that when using the case study approach, researchers do *not* analyze relationships between variables; instead, they try to make sense of the case or cases as a whole.

Examples of case studies include classic analyses like *The Urban Villagers* (Gans, 1982) and *Middletown* (Lynd and Lynd, 1956), and anthropological ethnographies (also called field studies), like *Occasions of Faith* (Taylor, 1995), an analysis of Donegal county, Ireland. Sociologist Terry Williams (1989) used a case study approach when he "hung out" with a group of teenage cocaine dealers for four years to understand their daily lives, perceptions of the future, and relationships with families and friends. During his research, Williams met and gained the confidence of the young members of a mostly Dominican "crew" or gang; Williams came to believe that in many ways the dealers were no different from those of other ambitious ethnic groups attracted to illegal businesses because of limited opportunities for legal enterprises.

Leigh Culver (2004) wanted to investigate the impact of new immigration patterns on police-Latino community relations in rural Midwestern communities. Culver selected three neighboring communities in central Missouri that had recently experienced substantial increases in the Latino population and collected data using a variety of strategies, including observations of uniformed patrol officers and formal and informal interviews with officers, community, and government leaders. Noting the language barriers and the amount of fear and distrust in interactions, Culver (2004) was

able to describe the difficulties in developing cooperative, working relationships between the Hispanic community and the police.

Robert Stake (2003: 136–138) identifies three types of case studies: the
intrinsic, undertaken because the particular case itself, with its particularity
and ordinariness is of interest, the *instrumental*, where a typical case is
selected to provide insight into and understanding of, or a generalization
about, an issue that it illustrates, and the *collective*, where the instrumental
approach is extended to several cases selected to lead to a better understanding or theory about a larger collection of cases. Jo Reger's (2004) study of the
New York City chapter of the National Organization for Women illustrates
the instrumental case study. Using documents and in-depth interviews with
chapter members, Reger (2004: 220) finds that the case illustrates the importance of an external-internal dynamic in social movements: Responses to
political events and social structures pull people into a social change organization and then internal organizational strategies channel the emotions into
activism.

Case study analysis is somewhat paradoxical. On the one hand, case
studies "frankly imply particularity—cases are situationally grounded, limited views of social life. On the other hand they are something more ...
they invest the study of a particular social setting with some sense of generality" (Walton, 1992: 121). The studies that have become social science
classics describe specific people and places, but also provide a sense of understanding about general categories of the social world. Using the case
study approach, "Goffman (1961) tells us what goes on in mental institutions, Sykes (1958) explains the operation of prisons, Whyte (1943) and
Leibow (1967) reveal the attractions of street corner gangs, and Thompson
(1971) makes food riots sensible" (Walton, 1992: 125).

A disadvantage of the case study approach is its limited generalizability.
With samples of one or a few, we cannot know if what is observed is unique
or typical. On the other hand, an advantage of the case study is its "close
reading" of social life and its attention to the broader social context
(Feagin, Orum, and Sjoberg, 1991: 274). The case study can be very important in generating new ideas and theories (Orum, Feagin, and Sjoberg,
1991: 13) and can be particularly appropriate for exploratory and descriptive purposes.

Summary

STOP AND THINK *Two researchers, Drs. Baker and Miller, are interested in studying changes in college
students' attitudes about attending graduate school. Dr. Baker decides to do a
cross-sectional study with a sample of freshmen and seniors at one university. Based
on his interviews, Baker notes that the seniors have more positive opinions toward
graduate school than the freshmen. In contrast, Dr. Miller selects a panel design.
She interviews a sample of freshmen at one university and re-interviews them three*

years later when most are seniors. Miller finds that the students' attitudes as seniors are more positive toward graduate school than they were as freshmen. Both Dr. Baker and Dr. Miller conclude that student attitudes toward graduate education change, becoming more positive as students progress through four years of college. In whose conclusions do you have more confidence, Dr. Baker's or Dr. Miller's? Why?

Researchers begin with specific questions and purposes. They might be interested in exploring new areas, generating theories, describing samples and populations, seeing patterns between variables, critically assessing some aspect of the social world, testing a causal hypothesis, or, as in the example of Drs. Baker and Miller, documenting changes over time.

The purposes or reasons for the research directly influence the choice of study design. If, for example, we are interested in describing student attitudes toward graduate school, selecting a cross-sectional design would be effective and sufficient. A cross-sectional design, like the study by Dr. Baker, allows us to describe the samples' attitudes toward graduate school and to *compare* the attitudes of freshmen and seniors. On the other hand, a panel design, although more expensive and time consuming, is better if we want to study *changes* in attitudes. A panel allows us to ask about current attitudes at two times rather than depend on retrospective data; it documents the time order of the variables; and lets us analyze changes in the attitudes of individuals not just aggregated changes.

Judging which design is most appropriate for a given project involves a series of decisions. The answers to the following questions provide a basis for selecting one of the five study designs covered in this chapter and the experimental designs that will be discussed in Chapter 8:

1. Is it useful to analyze only one case or a small number of cases, or does the sample need to be larger?

2. Is it useful and possible to collect data more than one time from the same sample?

3. Is there a cohort that can be studied to answer the research question?

4. Is it useful and possible to select a new sample each time if data will be collected more than once?

5. If there is a causal hypothesis, is it possible, useful, and ethical to try to control or manipulate the independent variable in some way?

The focus of this chapter has been on the strengths and weaknesses of the various non-experimental designs as related to internal validity and generalizability. When selecting a study design, it is important to balance methodological concerns with ethical considerations and practical needs. Although the summary shown in Figure 7.4 focuses on the methodological issues, ethical and practical criteria must also be considered when selecting an appropriate research design.

FIGURE 7.4

Summary of Cross-Sectional, Panel, and Trend Study Designs

Summary of Cross-Sectional, Panel, and Trend Study Designs

Study Design	Design Features	Design Uses
CROSS-SECTIONAL STUDY	One sample of cases studied one time with data about one or more variables	Useful for describing samples and the relationship between variables

One Sample
(divided up into categories of the independent variable during analysis)

	time 1
category 1 of an independent variable	measure of a dependent variable
category 2 of an independent variable	measure of a dependent variable

Useful for explanatory purposes especially if the time order of variables is known and sophisticated statistical analyses possible

PANEL STUDY	One sample of cases is studied at least two times	Useful for describing changes in individuals and groups over time (including developmental issues)

	time 1	time 2
one sample	variables measured	variables measured

Useful for explanatory purposes

TREND STUDIES	Different samples from the same population are selected	Useful for describing changes in populations and for analyzing the effect of time period

	time 1	time 2
first sample	measure several variables	
second sample		measure the same variables

EXERCISE 7.1

Identifying Study Designs

Find an article in a social science journal that reports the results of actual research. (Be sure it is not an article that only reviews and summarizes other research.) Answer the following questions about the article and the research.

1. List the author(s), title of article, journal, month and year of publication, and pages.

2. Did the researcher(s) begin with at least one hypothesis? If so, write one hypothesis and identify the independent and dependent variables. If there was no hypothesis, but there was a research question, write it down.

3. Identify the main purpose(s) of this study as exploratory, descriptive, explanatory, critical, or applied.

4. Describe the population and the sample in this study. Is the population a cohort?

5. If this study used only one or a few cases or analyzed the cases as a whole rather than by focusing on variables, then it might be a *case study*. If you believe it is a case study, give support for this conclusion and skip questions 6–9. If you think it is NOT a case study, skip this question and answer the rest.

6. How many times were data collected in this study?

7. If data were collected more than once, how many samples were selected?

8. Did the researchers try to control or manipulate the independent variable in some way?

9. Did the researchers use a case, cross-sectional, panel, trend, or cohort study design? If so, which one? Give support for your answer.

10. Was the design useful for the researcher's purposes and goals or would another design have been better? Give support for your answer.

EXERCISE 7.2

Planning a Case Study of Your Town or City

In his book *Code of the Street,* Elijah Andersen introduces his study by describing the 8.5-mile major artery in Philadelphia that links the northwest suburbs with the heart of the inner city. Anderson (1999: 15) says,

> Germantown Avenue provides an excellent cross section of the social ecology of a major American city. Along this artery live the well-to-do, the middle classes, the working poor, and the very poor—the diverse segments of urban society. The story of Germantown Avenue can therefore serve in many respects as a metaphor for the whole city.

In a case study of a geographic entity such as a city or a town, one or more streets, areas, organizations, groups, or individuals can be selected purposively. Think about the town or city in which you currently live or one in which you have lived in the recent past. Describe who or what your focus would be if you wanted to use a case study approach to study this town or city. Think of a research question you could answer with your case study.

1. Name the geographic area you want to focus on and what research question you would like to answer about it.

2. Describe who or what you would study. (You can identify one unit of analysis or you can propose to study more than one.)

3. Give support for your choices by discussing why they would be useful for doing a case study of your city or town that could answer your research question.

Evaluating Samples and Study Designs

Janice Jones, a graduate student, designs a research project to find out if women who are very career-oriented are more likely to be mothers than women who are less career-oriented. Jones selects a random sample of women who are corporate employees and hold jobs as president or vice-president. Interviewing the women she finds that their average age is 45 and that 35 percent are mothers, 55 percent are childless by choice, and 10 percent are childless involuntarily. She concludes that strong career orientations cause women to choose childlessness.

1. Discuss some of the challenges to the internal validity of Jones's conclusions.

2. Think of a different way this study could have been done. Specifically describe a sample and a study design that you think would be useful in doing research to answer Jones's question.

© Robert Finkin/Index Stock Imagery

8

Experimental Research

Introduction

explanatory research,
research that seeks to
explain the cause of a
phenomenon, and
typically asks "what
causes what?" or "why is
it this way?"

causal hypothesis, a
testable expectation
about an independent
variable's affect on a
dependent variable.

STOP AND THINK

Do you remember what your favorite book was when you were a child? Did you like stories with happy endings? Did you read adventure stories or books about families or friends? Sometimes children's books teach us things while they are entertaining us. In this chapter, we'll learn more about study designs and read about some research conducted by Emily and a colleague which tested a hypothesis about children's literature.

We'll first continue our discussion of study designs by considering those that are especially useful for research with an explanatory purpose—research that seeks to explain *why* cases vary on one or more variables. **Explanatory research** begins with a **causal hypothesis** or a statement that one variable, the independent variable, *causes* another, the dependent variable. Frequently the researcher has deduced the hypothesis from an existing theory.

Explanatory research can be conducted on any area of social life. If we were interested in studying some aspect of children's literature, there are many research questions or hypotheses we could consider. Think of one causal hypothesis that concerns the impact of children's literature on something else.

Causal Hypotheses and Experimental Designs

Thinking about existing theories of socialization and how children learn, we can construct any number of causal hypotheses about children's literature. You might have thought of others, but here are three:

1. Reading books in which the main character faces and solves a problem helps children become better problem solvers.

2. Reading books with characters from a variety of ethnic and racial backgrounds increases children's knowledge of the cultures of those racial and ethnic groups.

3. Reading books in which the main characters care for and about others increases a child's own caring behavior and attitudes.

Each of these hypotheses is a causal hypothesis. To see if there is support for any one of them, we'd need a research strategy designed to meet the requirements for showing causality. Causality cannot be directly observed but, instead, must be inferred from other criteria. As we discussed in Chapters 2 and 7, factors that support the existence of a causal relationship include the following:

1. An independent and dependent variable that are associated, or vary in some way, with each other. This is the condition of *empirical association*.

2. An independent variable that occurs *before* the dependent variable. This is the condition of *temporal precedence* or *time order*.

3. The relationship between the independent and dependent variable is not explained by the action of a third variable (sometimes called an antecedent variable) which makes the first two vary together. That is, the relationship between the independent and dependent variables is not *spurious* or non-causal.

As in any study, it is necessary to make the series of choices we've discussed in the previous chapters. We'd need to find a way to measure the variables, select a population and sampling strategy, determine the number of times to collect data, and establish the number of samples to use. We might decide that it is appropriate to use one of the study designs that we presented in Chapter 7 and design a panel, trend, cohort, case, or cross-sectional study. Before making a final decision, however, we'd want to consider the possibility of using an **experimental design.** Although the study designs we considered in Chapter 7 typically allow the researcher to determine if there is a pattern or relationship between independent and dependent variables, experimental designs are especially helpful for determining the time order of variables and for minimizing the third-variable effects.

experimental design, a study design in which the independent variable is controlled, manipulated, or introduced in some way by the researcher.

The Classic Experiment: Data Collection Technique or Study Design?

As we discussed in Chapter 6, a critical set of decisions focuses on how to measure variables. That is, for each variable, the researcher has to figure out how to classify each unit of analysis into one of two or more categories. The uniqueness of the experimental design becomes apparent when we think about the issue of measurement because, in a traditional experiment, the independent variable is not measured at all! Instead, in an experiment, the independent variable is "introduced," "manipulated," or "controlled" by the researcher. That is, the independent variable does not occur naturally, but is the result of an action taken by the researcher.

In each of the three hypotheses we listed earlier, the independent variable is "type of books read by children." If we design a panel or cross-sectional study for any of these hypotheses, we will need to measure this variable using one or more indicators. We could measure the type of books read by children by asking them to list every book they'd read in the past six months. Alternatively, we could ask libraries and booksellers for lists of their most popular children's or young adult books.

On the other hand, if we decide to do an experiment, we need to do something different. Rather than measuring the kinds of books students read, we'd have to find a way to have *some* types of books read by some students and *other* types of books read by others. We're saying that we'd need to *control* the placement of sample members into the categories of the independent variable.

Controlling the placement of sample members into two or more categories of the independent variable is the unique feature of the classic experimental design. If we control placement and measure the dependent variable at least two times, the experimental strategy allows us to see the effect of the independent variable by comparing the measurements of the dependent variable among those in the various categories.

In the next few chapters, we will be describing methods of data collection, such as asking questions, observing, and using available data. Because of the unique way that independent variables are handled in an experiment,

> ### BOX 8.1
>
> ## What Is Character Education and Why Should We Care About Its Effects?
>
> Values are staging a comeback in the classroom these days. Eager to deal with a host of ills, such as bullying and school violence, school districts all over the United States are trying a variety of programs known as "character education," with 40 states requiring or encouraging this type of curriculum (Lord, 2001: 50). Each fiscal year from 2002 to 2007, President Bush's education budget has included at least $24 million annually to allow states and school districts to compete for funds for classroom-based character education (http://www.ed.gov/about/overview/budget/budget07/summary/edlite-section2a.html#character). Some states, including California and Kentucky, have designated a month as "Character Education Month."
>
> Others are not big fans of character education. Sheldon Berman, the superintendent of schools in Hudson, Massachusetts, has noted, "There's a lot of fluff out there and it's really disappointing . . . it doesn't have much impact on behavior," and author Alfie Kohn argues that character education is "geared toward creating obedient sheep" (cited by Lord, 2001: 51). With all the time and energy that our schools are committing to these efforts and the taxes that we're paying for this type of curriculum, shouldn't we examine the outcomes of such programs?

some textbooks define the experiment as a method of data collection or as a mode of scientific observation. Although we recognize the usefulness of this approach, we want to highlight the experiment's special treatment of the independent variable as a unique *design* element. For this reason, we include experiments in our discussion of designs and see this chapter as a natural transition between the consideration of study designs and the presentations on data collection techniques that follow.

The History of One Experiment

In a joint research project, Emily and Roger studied how the transition from childhood to adulthood was portrayed in children's literature by analyzing the way that adolescence was depicted in literature for children and teens (Adler and Clark, 1991). In our review of the literature, we examined research on children's books in a wide variety of disciplines and were intrigued to discover an area of study previously unfamiliar to us—bibliotherapy, a multidisciplinary field at the intersection of education, social work, counseling, and social science.

Advocates of bibliotherapy encouraged the use of literature-based approaches as part of therapy and to help children and adults deal with normal developmental transitions. Many of its supporters claimed bibliotherapy to be a wonderfully useful technique, so you can imagine our surprise at finding little to support this assertion. We found almost no research that tested the effects of literature on behavior, attitudes, or social adjustment.

After the original project was completed, Emily had a conversation about the use of bibliotherapy in school settings with Paula Foster, a social worker employed in a local community mental health agency. They noted that "character education" was on an increasing number of political and social agendas (see Box 8.1) and decided to collaborate on a research project that connected bibliotherapy and character education. Further discussion led them to develop a specific hypothesis and plans for a research project to test it. This chapter's focal research is the result of that conversation.

FOCAL RESEARCH

A Literature-Based Approach to Teaching Values to Adolescents: Does It Work?

by Emily Stier Adler and Paula J. Foster[1]

Introduction

The intentional teaching of morals and values was common in American public education for almost three hundred years and had widespread support until the 1930s. After decades of a "hands off" approach to the teaching of values, there have been calls for schools to return to explicit moral or character education in recent years (Kilpatrick, 1992; Wynne and Ryan, 1993). One proposed curriculum revision suggests the use of literature to help youngsters develop their capacity for courage, charity, justice, and other virtues (Kilpatrick, 1992: 268). At the same time, advocacy of literature-based approaches has increased in educational psychology and social work practice. The literature-based approach, sometimes called bibliotherapy, has been championed for numerous uses—a way to teach children to relate moral principles to real life (Dana and Lynch-Brown, 1991), an effective deterrent to substance abuse (Bump, 1990; Pardeck, 1991), a guide to self-understanding and improved self-concept (Calhoun, 1987; Hebert, 1991; Lenkowsky and Lenkowsky, 1978; Miller, 1993), and as a way to help children deal with family problems, such as parental divorce (Early, 1993). The advocacy has occurred despite the caution that the approach has been awarded a "scientific respectability it does not have" and that the application of the technique "far outstrips the tight validating studies supporting its use" (Riordan, 1991: 306). In fact, our review of the literature found no studies of bibliotherapy or other

---

[1] This article was adapted from Adler and Foster's article of the same name published in *Adolescence* 32 (Summer, 1997): 275–286, with permission.

literature-based approaches to teaching attitudes, values, or character traits that used an experimental design to evaluate the effectiveness of the method. It is this evaluation which our study seeks to provide.

The Hypothesis

We were interested in determining the impact of a literature-based curriculum on student values. Constructing a causal hypothesis, we designed a specific curriculum for use as our independent variable and identified a specific value, the value of caring for others, as the dependent variable. We hypothesized that the use of a literature-based curriculum would increase students' belief in the value of caring for others. We went into the field—the real world of students, classrooms, and teachers—to test our hypothesis.

The Sample

A middle school in an urban community with a fairly homogeneous population of approximately 40,000 (94 percent were non-Hispanic and white, 3 percent Hispanic, 2 percent Asian, and 1 percent African American) was selected because administrators were willing to allow a new curricular approach to be implemented on a small scale. The school principal and the teachers of one large seventh-grade class agreed to work with us. We did not select the community, school, or students randomly and we recognize that our results cannot be generalized beyond the students in our sample.

The class was team-taught, with 2 teachers, 57 students, and a room which could be divided into two rooms with a movable partition. The class structure varied between the teachers teaching all 57 students in one space, the teachers working with many small groups, and two separate classrooms, each with a teacher. The students were an average of 13.1 years old when the study began. There were 31 girls and 26 boys; 50 students were white and 7 were Asian.

Using the characteristics of gender, race, age, and academic ability, the teachers matched the students into pairs and created two approximately equal groups. One of the groups was randomly selected to be the experimental group while the other became the control group.

The Independent Variable: The Reading Project

The independent variable was the 10-week reading project developed for this study. In consultation with the teachers, we selected three books with appropriate themes and designed class exercises and discussions to accompany them.[2] The theme of each book is the importance of caring for others.

[2] Activities for students in the experimental group were based on methods presented in *Webbing With Literature* (Bromley, 1991). In "webbing," a core concept is selected and emphasized by using "webs" or lines connecting the concept to information. In the experiment, the concept of caring was highlighted during the discussion of each book. Class exercises included writing favorite quotes and feelings in a journal, making a group collage expressing the theme and the feelings of the novel, and webbing how the characters in the books were connected to each other (by marriage, birth, friendship, etc.).

The main characters make choices and ultimately decide to care for others, even at the expense of achieving personal goals. In all three books, the main characters are adolescents; two are female and one is male, two are white and one is Asian.[3] During the weeks that the experimental group read these books, several times each week the students participated in classroom discussions and exercises designed to reinforce the theme of caring for others. During the same time, the students in the control group read and discussed regular seventh-grade literature (books with animal main characters, such as *Call of the Wild*) and worked with the other teacher in the other half of the divided classroom space.

The Dependent Variable: Caring for Others

The dependent variable, support for the value of caring for others, is defined conceptually as the extent to which students "see and respond to need" and advocate "taking care of the world by sustaining the web of relationship so that no one is left alone" (Gilligan, 1982: 62). The extent to which the students supported this value was measured both before and after the reading project by their answers using an indirect measure—a set of three essays on caring for others. Students in both groups completed the essays during the two weeks before and the two weeks after the 10-week reading project. The teachers used class time and asked the students to write the essays during the language arts class. The students were told that there were no "right" answers and were asked to express their opinions or make up a story without worrying about what they were "supposed to say." Student essays were identified by number. Due to absenteeism, some students did not complete all six essays.

After considering several additional essay questions, we used three to measure caring attitudes:

1. John and Bill both have jobs in a movie theater. The manager is deciding on the work schedule for the weekend. Both John and Bill want Friday night off. John wants to spend the evening visiting his grandmother who is sick and in the hospital, and Bill has plans to go out with his friends.

[3] *Friends Are Like That* by Patricia Hermes (New York: Scholastic Inc., 1984) is the story of Tracy, an eighth grader who weighs popularity against supporting her best friend, Kelly, a nonconformist and a social pariah. Tempted to become part of the school's "popular" crowd, Tracy instead chooses friendship. *Red Cap* by G. Clifton Wisler (New York: Lodestar Books, 1991) is the true story of Ransom Powell, who in 1862, at age thirteen, joined the Union Army, was captured along with his company and sent to Camp Sumter, the Confederate prison at Andersonville. Under terrible conditions, the prisoners in Ransom's company helped and supported each other in ways that a family might, which allowed Ransom to survive imprisonment. *The Clay Marble* by Minfong Ho (New York: Farrar Straus Giroux, 1991) is set in Cambodia in the early 1980s. Twelve-year-old Dara and her family are among the thousands who flee from their villages to a refugee camp on the border of Cambodia and Thailand. Her family meets and becomes emotionally attached to another family. Even though some family members die, the two units become one at the story's end.

The manager tells them that he needs one of them to work on Friday night, but that they can either decide between themselves who will work or he'll flip a coin to decide. What do you think should happen and why? (GRANDMOTHER)

2. Susan reads in the local newspaper about a family that lost all their belongings when there was a fire in their home. She reads that they didn't have much insurance to replace their belongings. She decides to help out by sending them her whole allowance of $5.00 a week for the next three weeks. Describe how you feel about Susan's behavior. (ALLOWANCE)

3. Sometimes people have small families. Do you think that friends can be like families? In what ways? In what ways are they different from families? Write about your opinions. (FRIENDS)

Each essay was read for content by both researchers and coded into one of three categories. For GRANDMOTHER and ALLOWANCE, each was coded from most supportive for caring for others to least supportive of caring for others. For GRANDMOTHER, answers that defined John's and Bill's needs as equivalent were defined as least supportive of caring (for example, "Bill and John should split the hours in half that Friday," or "The manager should flip a coin since it's fair and quicker."). Answers that gave some priority to John's visiting his grandmother were coded as moderately supportive of caring (for example, "I think Bill should work because he should understand that John really needs to see his grandmother this week. Next week John will work and let Bill have the time off"). Essays that identified John's desire to visit his grandmother as having more priority than Bill's spending time with friends were classified as most supportive of the value of caring. For example, "If Bill has any decency, then he should work instead of John. John seeing his sick grandmother is more important than going out with friends. Friends last forever, but grandparents don't."

Intercoder reliability was 95 percent for GRANDMOTHER. On this essay, 63 percent of the sample (57 percent of the experimental group and 68 percent of the control group) was classified as most supportive of caring for others at the pretest.

Essays in response to the ALLOWANCE question were judged to be least supportive of the value of caring if the student disagreed with Susan's behavior. Examples include the student who said, "I wouldn't do that. Susan's five dollars wouldn't help very much anyway"; the one who said, "I don't care. I think she was nice, but stupid to do that"; and the response, "Susan's behavior is not logical. If she read the newspaper I read, then she would be broke. I doubt $15.00 will help the poor family anyway." Essays that were partially enthusiastic about Susan's behavior or felt she would get some sort of reward for it were coded as being somewhat supportive of caring. The student who said, "She is very kind, but I don't think she should give all her money away. She should get a group of people and start a collection for the family so Susan won't have to give all her money" was placed in this category. If the student felt that Susan had done the right thing and expected nothing in return, the answer was classified as most supportive of caring. Illustrative of this response was the student who said, "I think Susan is doing a great thing because most people wouldn't care. I also think she has a big heart to give up her allowance for three weeks. That is a really great thing to do."

Intercoder reliability was 95 percent for this essay. Seventy-five percent of the sample (78 percent of the experimental group and 71 percent of the control group) was classified as most supportive of caring before the introduction of the literature project.

For the essay FRIENDS, student answers were coded into three categories: friends and families seen as very different from each other, friends and families seen as somewhat similar, and friends seen as serving the same emotional and support functions as families do, even if the people live in different households. An example of the first category is the essay that said, "I don't think friends are like family. Friends are people to hang around with and have fun with. You can have fun with family but it's different. I wouldn't consider my friends my family." An example of the second category, seeing some similarities between the two units, is the student who wrote, "In some ways friends can be like family, sometimes you can trust them and they can care for you. In other ways, they aren't like family because they can't know you as well as family members do." Illustrative of the essays that were coded as seeing the units as serving the same emotional and support functions are these two: "I think friends can be like family. You can like them as well as family and they can like you in return. You can trust them even if you haven't grown up with them" and "I do think friends can be like families. They can do special things for you, or just show you they care. The only way they are different is that they are not blood-related. There could be other differences, but it doesn't matter as long as they care for you."

The intercoder reliability for this essay was 94 percent. At the pretest, 59 percent of the sample (50 percent of the experimental group and 71 percent of the control group) thought that friends can provide the same sorts of emotional and social supports as families.

Findings

To determine the effect of the reading project, the essays each student wrote before and after completing one of the 10-week curriculums were compared. Based on the two essays, each student was classified as becoming more supportive of caring for others, less supportive, or showing no change in values. Most students in the experimental and control groups were consistent in their values at both time 1 and time 2 and showed no change in support for the value of caring (see Table 8.1). For the students who did change, the results for two of the essays were in the hypothesized direction.

There was a statistically significant difference between the two groups for the FRIENDS essay. Thirty percent of the experimental group and none of the control group became more supportive of the belief that friends can be like family in providing emotional and social support to people. For GRANDMOTHER, the results approached statistical significance, with 20 percent of the control group and none of the experimental group becoming

TABLE 8.1   Comparison of Experimental and Control Groups in Changes in Support for the Value of Caring for Others

		Experimental Group	Control Group
FRIENDS	Friends Seen Less Like Family for Emotional and Social Support	0 (0%)	2 (10.5%)
	No Change	16 (69.6%)	17 (89.5%)
	Friends Seen More Like Family for Emotional and Social Support	7 (30.4%)	0 (0%)
	$(N = 42, \chi^2 = 9.7, \text{d.f.} = 2, p = .013)$		
GRANDMOTHER	Less Support for Caring	0 (0%)	4 (20%)
	No Change	18 (81.8%)	13 (65%)
	More Support for Caring	4 (18.2%)	3 (15%)
	$(N = 42, \chi^2 = 4.8, \text{d.f.} = 2, p = .09)$		
ALLOWANCE	Less Support for Caring	1 (3.8%)	2 (9.1%)
	No Change	23 (88.5%)	15 (68.2%)
	More Support for Caring	2 (7.7%)	5 (22.7%)
	$(N = 48, \chi^2 = 2.99, \text{d.f.} = 2, p = .22)$		

less supportive of caring for others after the reading project, and slightly more (3 percent) of the experimental group becoming more caring. For ALLOWANCE, the results, which were not statistically significant, were nevertheless not supportive of the hypotheses. Twenty-three percent of the control group and only 8 percent of the experimental group became more supportive of caring values after the completion of the reading project. Analyses controlling for gender are very similar to the results with no control variable, and also demonstrate some support for the hypothesis that a literature-based curriculum can have an impact on caring values.

Discussion and Conclusion

The strongest support for the hypothesis that a literature-based curriculum can influence values is the finding that the experimental group became significantly more committed to the view that friends can fulfill many of the same emotional and social support functions as families. The data for one of the other two essays also showed changes in the hypothesized direction. The experimental group remained more supportive of the caring value for the GRANDMOTHER essay, providing weak support for the hypothesis. However, for the ALLOWANCE essay, the control group became more supportive of caring than the experimental group, a finding that does not support the effectiveness of the reading project.

Several issues, including methodological ones, are important for interpreting the results of the research. The specific indicators used to measure the value of caring can be questioned. The large percentage of the student essays at time 1 which advocated caring values could be a source of concern. It might be that students wrote the essays that reflected the values they thought their teachers wanted to see, rather than sharing their own beliefs. Another concern is variability. With such a large proportion of students being categorized as "most caring" at time 1, only a relatively small increase in caring was possible. A final methodological concern is the amount of missing data. With a large number of students not completing the essays both times due to absences, as many as sixteen cases were omitted from each analysis.

Even with the methodological weaknesses, one interpretation of the findings is that there were some changes in caring values *as a result* of reading and discussing selected literature. However, the changes were small, perhaps because values are not quickly or easily affected by a modest change in the curriculum. We can assume that these students, like other American children and teens, watch an average of three hours of television per day (*Time,* 1995: 74), engage in many out-of-school activities, and are exposed to other curricula materials that present values explicitly and implicitly. Therefore, it should not be surprising that the 10-week reading project in this study had, at best, a modest effect. Teaching values and developing character is a complex and time-consuming task. If the value system presented in popular culture and the rest of the school curriculum does not include a great deal of support for positive values and pro-social behavior, substantial character or moral education with small, discrete projects might not be feasible. Schools interested in character education should consider a multiyear curriculum starting in the early grades with frequent reinforcement. The curriculum could include the reading and discussion of books with specific values in addition to community service projects, the development of rules within the school, the inclusion of families in curriculum development, and other techniques. Although small reading projects may be able to have some effect on some values, only when they are part of a larger framework will they be able to provide the kind of socialization that the proponents of moral and character education advocate.

REFERENCES

Bromley, K. 1991. *Webbing with literature: A practical guide.* Boston: Allyn & Bacon.

Bump, J. 1990. Innovative bibliotherapy approaches to substance abuse education. *Arts in Psychotherapy* 17: 355–362.

Calhoun, G. 1987. Enhancing self-perception through bibliotherapy. *Adolescence* 22 (Winter): 939–943.

Dana, N. Fichtman, and C. Lynch-Brown. 1991. Moral development of the gifted: Making a case for children's literature. *Roeper Review* 14 (Sept.): 13–16.

Early, B. P. 1993. The healing magic of myth: Allegorical tales and the treatment of children of divorce. *Child and Adolescent Social Work Journal* 10 (April): 97–106.

Gilligan, C. 1982. *In a different voice.* Cambridge, MA: Harvard University Press.

Hebert, T. P. 1991. Meeting the affective needs of bright boys through bibliotherapy. *Roeper Review* 13 (June): 207–212.

Kilpatrick, W. 1992. *Why Johnny can't tell right from wrong.* New York: Simon & Schuster.

Lenkowsky, B. E., and R. S. Lenkowsky. 1978. Bibliotherapy for the LD adolescent. *Academic Therapy* 14 (1978): 179–185.

Miller, D. 1993. The Literature Project: Using literature to improve the self-concept of at-risk adolescent females. *Journal of Reading* 36 (March): 442–448.

Pardeck, J. 1991. Using books to prevent and treat adolescent chemical dependency. *Adolescence* 26 (Spring): 201–208.

Riordan, R. J. 1991. Bibliotherapy revisited. *Psychological Reports* 68: 306.

Time. 1995. Special Report: the State of the Union. Jan. 30: 74.

Wynne, E. A., and K. Ryan. 1993. *Reclaiming our schools.* New York: Macmillan.

The Reading Project as an Experiment

The reading project focused on whether an independent variable had *caused* a change in or *had an effect* on the dependent variable. As such, it was an explanatory study with a causal hypothesis. Having an explanatory purpose influenced the methodology, including the choices for ways to measure variables and the number of times data were collected. Looking at this reading project, we'll see some of the benefits and disadvantages inherent in the controlled experiment, a study design that is very useful for testing causal hypotheses.

STOP AND THINK

Review the focal research article and think about the hypothesis and design strategy. The dependent variable, caring, was measured twice. Do you think this was a useful choice? The independent variable was the reading and discussion of specially selected literature. Do you think the independent variable was "measured" or "introduced" in this study?

To test the hypothesis that participating in the reading project would affect "caring," it was important to determine changes in the dependent variable. For this reason, student support for caring was measured two times—first before and then after the implementation of the reading project. In experiments, the before and after measurements of the dependent variable are called the **pretest** and the **posttest.**

In this study, the independent variable—whether a student participated in the literature-based curriculum focused on caring or the regular curriculum—was not "measured." Instead, the literature-based curriculum was "introduced" to the experimental group while the control group received the regular seventh-grade reading curriculum. Determining who

pretest, the measurement of the dependent variable that occurs before the introduction of the stimulus or independent variable.

posttest, the measurement of the dependent variable that occurs after the introduction of the stimulus or the independent variable.

receives which condition of the independent variable, and controlling its timing are central elements in the experimental design.

In experiments, the control group is exposed to *all* the influences that the experimental group is exposed to *except* for the **stimulus** (the experimental condition of the independent variable). The researcher tries to have the two groups treated exactly alike, except that instead of a stimulus, the control group receives no treatment, an alternative treatment, or a **placebo** (a simulated treatment that is supposed to appear authentic). For maximum internal validity, beyond introducing the independent variable, the experimental design tries to eliminate systematic differences between the experimental and control group. Remember that **internal validity** is concerned with factors that affect the internal links between the independent and dependent variables, specifically factors that support alternative explanations for variations in the dependent variable. In experiments, internal validity focuses on whether the independent variable or other factors, such as prior differences between the experimental and control groups, are explanations for the observed differences in the dependent variable.

If, as in the reading project, the data analysis reveals at least some association between the independent and dependent variables, then the advantages of the experimental design become evident. An experiment offers evidence of time-order because the dependent variable is measured both *before* and *after* the independent variable is introduced, and *only one group* is exposed to the stimulus in the interval between the measurements. In addition, the method of selecting the experimental and control groups can minimize the chances of preexisting systematic differences between the two groups either by matching their members or assigning members randomly.

In the reading project study, two of the three indicators of the dependent variable showed small changes in the hypothesized direction. In addition, the time-order of the variables is known. However, demonstrating change in the dependent variable and a time-order in which the independent variable precedes changes in the dependent variable offer *necessary,* but not *sufficient* support for a cause and effect relationship. Another, unobserved factor could have caused the change in the dependent variable. In the reading project, for example, perhaps one or more significant events occurred in the community, school, or classroom during the weeks that students were reading and discussing the books. It could be that new staff members or students arrived at the school in those weeks. Or, maybe a new television show, game, or movie became very popular during the experimental period. We might be concerned that these or other factors influenced student caring.

Concern about the effects of other factors is the rationale for one essential design element in the controlled experiment—the use of two groups, an experimental and a control group. In the focal research, although all the students in the class might have been exposed to new students, staff, games, movies, or television programs, we know that only those in the experimental group were directly exposed to the stimulus. With two groups and the ability to control who is exposed to a stimulus, researchers using an

stimulus, the experimental condition of the independent variable that is controlled or "introduced" by the researcher in an experiment.

placebo, a simulated treatment of the control group that is designed to appear authentic.

internal validity, agreement between a study's conclusions about causal connections and what is actually true.

experimental design can be fairly confident of the separating effects of the independent variable from many possible influences.

Can you think of any challenges to the assumption that there were no systematic differences between the two groups except for their participation in the reading project?

There are several possible challenges to the assumption that there were no systematic differences between the two groups. It is possible, for example, that the two teachers were different in a way that could have affected the results—perhaps one was a better teacher, more caring, or more enthusiastic. Another possibility is that the kinds of exercises used by the reading project encouraged closer student and teacher interaction than did the traditional class work. It could also be that the two groups were unequal in some important way to begin with, even though the teachers matched them on age, race, gender, and academic ability. For instance, we don't know if there were more students in the experimental group who were avid readers or if there were more in the control group who had recently moved to the community. Using matching to select experimental and control groups in this and other studies has clear limitations. In hindsight, both Emily and her co-author believe that an alternative method of selecting the two groups might have been better for internal validity. In the next section, we'll explore several options for selecting groups in experiments and then consider design strategies that can enhance internal validity.

Experimental Designs

Pretest-Posttest Control Group Experiment

pretest-posttest control group experiment, an experimental design with two or more randomly selected groups (an experimental and control group) in which the researcher controls or "introduces" the independent variable and measures the dependent variable at least two times (pretest and posttest measurement).

probability sample, a sample that gives every member of the population a known (nonzero) chance of inclusion.

The Adler and Foster research is an example of the kind of experiment called the classic controlled experiment or the **pretest-posttest control group experiment.** This kind of experiment is the most highly recommended of all experimental designs (Campbell and Stanley, 1963). It uses two or more samples or groups—the experimental and the control group—that can be selected in one of several ways. The preferred method of picking the two groups is to identify an appropriate population and, using **probability sampling,** to select two random samples from the population. Figure 8.1 illustrates the use of probability sampling to pick experimental and control groups. This method gives the greatest generalizability of results because selecting the samples randomly from a larger population increases the chances that the samples represent the population.

However, participants in experiments are rarely selected from a larger population usually because it's typically not practical and might not be ethical. Instead, participants are more typically a group of people who have volunteered or, as in this chapter's focal research, have been selected by someone in the setting. In such cases, the sample is really a convenience sample and might not be representative of a larger population.

There are two ways to sort a sample into experimental and control groups to minimize preexisting, systematic differences between the groups.

FIGURE 8.1

Use of Probability Sampling to Select Groups

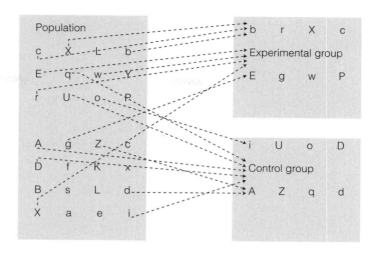

By minimizing such differences, experiments can offer support for the conclusion that no third variable is making the independent and dependent variables appear to "go together"—the third requirement of causality.

The first way to select groups is to assign members of the sample to the experimental and control groups using **random assignment,** the process of randomly selecting experimental and control groups from the study participants. This can be done by flipping a coin to determine which subject is assigned to which group or by assigning each subject a number and using either a random number table or computer-driven random number generator to select members of each group (experimental or control). Figure 8.2 illustrates the use of random assignment to create two groups.

Another way to assign members of the sample to experimental and control groups is through **matching.** Using this method, the sample's members are matched on as many characteristics as possible so that the whole group is "paired up." The pairs are then split, forming two matched groups. (See Figure 8.3.) A concern is that typically several characteristics in the sample are matched while others are not. After the sample members are matched into groups, one group is randomly selected to become the experimental group and the other becomes the control group.

The pretest and posttest experiment has several elements that distinguish it from other study designs:

1. The study uses at least one experimental and one control group, selected using a strategy to make the groups as similar as possible (probability sampling, random assignment, or matching). Experimental groups are sometimes called treatment groups because an experiment will have at least one and possibly more experimental treatments.

2. The dependent variable is measured at least two times for the experimental and control groups. The first measurement is before and the second is after the independent variable is introduced. (These measurements are called the pretest and the posttest.)

random assignment, a technique for assigning members of the sample to experimental and control groups by chance to maximize the likelihood that the groups are similar at the beginning of the experiment.

matching, assigning members of the sample to groups by matching members of the sample on one or more characteristics and separating the pairs into two groups with one group randomly selected to become the experimental group.

FIGURE 8.2

Use of Random Assignment to Select Groups

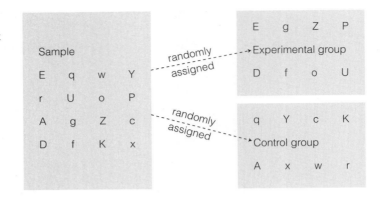

FIGURE 8.3

Use of Matching to Select Groups

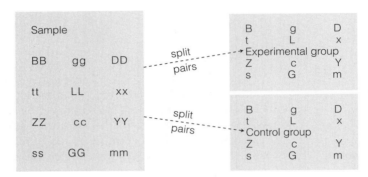

independent variable, a variable that is seen as affecting or influencing another variable.

dependent variable, a variable that is seen as being affected or influenced by another variable.

3. The **independent variable** is introduced, manipulated, or controlled by the researcher between the two measurements of the dependent variable. While the experimental group receives the stimulus, the control group gets nothing, an alternative treatment, or a placebo.

4. The differences in the **dependent variable** between the pretest and posttest are calculated for the experimental group(s) and for the control group. The differences in the dependent variable for the experimental and control groups are compared.

Figure 8.4 shows a diagram of the pretest-posttest control group experiment.

Internal Validity and Experiments

We've already noted that having a pretest and posttest and an experimental and control group can handle many concerns about internal validity by regulating and identifying systematic differences between the groups. The following are some other advantages of this kind of experiment that Donald Campbell and Julian Stanley (1963) identified:

maturation, the biological and psychological processes that cause people to change over time.

1. The experiment is useful for separating the effects of the independent variable from those of **maturation,** the biological and psychological processes that cause people to change over time, because, in experiments, both the experimental and control groups are maturing for the same amount of time.

FIGURE 8.4

Pretest-Posttest Control Group Experiment

Groups	Time		
	1 (pretest)		*2 (posttest)*
Experimental	Measure dependent variable	*Stimulus*	Measure dependent variable again
Control	Measure dependent variable	*	Measure dependent variable again

• The control group receives no stimulus, receives an alternative treatment, or receives a placebo.

testing effect, the sensitizing effect on subjects of the pretest.

history, the effects of general historical events on study participants.

selection bias, a bias in the way the experimental and control or comparison groups are selected that is responsible for preexisting differences between the groups.

2. The experiment can control for some aspects of the **testing effect,** the separate effects on subjects that result from taking the pretest, because both groups take the pretest.

3. The experiment can control for some aspects of **history,** the effects of general historical events not connected to the independent variable that influence the dependent variable, because both groups are influenced by the same historical events.

4. The methods of selecting the control and experimental groups are designed to minimize **selection bias,** or preexisting differences between the groups.

STOP AND THINK

Walking out of the shopping mall, you notice that someone has left a handbill advertising a new store under your car's windshield wiper. What would you do? Leave it there and hope it blows away as you drive? Take it and dispose of it in the trash can at the end of the row of cars? Put it in your car? Do you think your behavior would be different if you saw a big "DON'T LITTER" sign while walking to your car? How about if you saw someone else picking up litter? Now turn your imagined experience into the basis of a research study. Construct a hypothesis to test about littering. Can you think of a way to test it?

Posttest-Only Control Group Experiment

posttest-only control group experiment, an experimental design with no pretest.

Another experimental design, the **posttest-only control group experiment,** has no time 1 measurement (no pretest), either because it is not possible to do a pretest or because of a concern that using a pretest would sensitize the experimental group to the stimulus. Otherwise, it has the same design elements as other experiments: control or manipulation of the stimulus and two or more groups using random selection or assignment. This design is illustrated in Figure 8.5.

Experiments using posttest-only designs have been conducted to study the effects of norms on behavior. Kallgren, Reno, and Cialdini (2000), using a theory that social norms direct behavior when they are focal,

FIGURE 8.5

Posttest-Only Control
Groups Experiment

Groups		Posttest
Experimental	Stimulus	Measure dependent variable
Control		Measure dependent variable

hypothesized that situational social norms against littering will have an impact on actual littering. They conducted several experiments on this topic, one that might be quite like what you thought of when answering the last "Stop and Think" question. They did an experiment in a public urban hospital garage, which was a fairly littered environment. The sample consisted of 149 people who were returning to their cars, where each encountered one or two large handbills tucked under the windshield wiper so that vision was obscured. The independent variable was introduced by having an experimental confederate of college age walk toward each member of the sample. In the half of the sample that became the experimental group, the stimulus of "high norm and high focus" was introduced by having a male experimental confederate pick up a crumpled fast food bag approximately four meters from the participant. For the other half of the sample, the control group, he just walked by. From a hidden vantage point, a researcher noticed what the driver did. As there were no trash barrels in the garage, if a handbill was deposited outside of the vehicle, the participant was defined as littering. The hypothesis received support because only 9 percent of the drivers in the experimental group littered, whereas 34 percent of the control group did (Kallgren, Reno, and Cialdini, 2000).

A major challenge to the internal validity of this experimental design is the unverifiable assumption that a difference between the experimental and control groups on the dependent variable represents a *change* in the experimental group caused by the independent variable. Without a pretest, it is difficult to determine if the difference between the groups on the dependent variable represents a *change* in the experimental group or a continuation of a preexisting difference between the groups. After all, though unlikely, it is possible that the drivers in the experimental group were usually less likely to litter than the control group drivers. The use of the posttest-only design calls for caution in reaching the conclusion that a causal hypothesis has been supported. However, as in the littering experiment, when a pretest is not an option, this design provides many of the advantages of the pretest and posttest experiment.

**Solomon four-group
design,** a controlled
experiment with an
additional experimental
and control group with
each receiving a posttest
only.

Extended Experimental Design

We will present one other experimental design that is useful in increasing internal validity. The **Solomon four-group design,** which is illustrated in

FIGURE 8.6

Solomon Four-Group
Experiment

Groups	Time		
	1 (pretest)		2 (posttest)
Experimental	Measure dependent variable	Stimulus	Measure dependent variable again
Control	Measure dependent variable		Measure dependent variable again
Experimental with no pretest		Stimulus	Measure dependent variable
Control with no pretest			Measure dependent variable

Figure 8.6, has four groups—two experimental and two control groups, with only one set of groups having a pretest. It is most useful when there is a concern about a possible interaction effect between the first measurement of the dependent variable and the introduction of the stimulus. The compensation for the extra effort, time, and expense of this design is the ability to determine the main effects of testing, the interaction effects of testing, and the combined effect of maturation and history (Campbell and Stanley, 1963: 25). A Solomon four-group experiment was used by Divett, Critteden, and Henderson (2003) in a study of consumer loyalty and behavior among subscription patrons of a regional theater in Australia. They selected this design because of the possibility that the pretest would sensitize the respondents to the stimulus. In this study, four groups of subscribers were selected and the first experimental and control group received a pretest (a questionnaire focusing on the views of the theater's approachability and responsiveness and the respondent's loyalty and likelihood of expressing concerns directly to the theater staff). Then both experimental groups received the stimulus, which was a brochure outlining the way the theater could be contacted and examples of subscribers' experiences, while the two control groups received nothing. One year later, all four groups received the posttest (the same questionnaire as the pretest). In this experiment, the researchers found no pretest effect but found that the brochure did have an impact on subscriber attitudes.

STOP AND THINK   *What additions to the study would Adler and Foster have had to make if they had used the Solomon four-group experimental design instead of the pretest-posttest control group experiment?*

Quasi-Experimental Design

STOP AND THINK

Let's say you want to evaluate a six-month "quit smoking" program. You know that you could evaluate its effectiveness by taking a sample of smokers who volunteer, divide them into two groups at random, and enroll only the experimental group in the program. At the end of six months, you could see how many in each group have quit smoking and what health changes, if any, have occurred in each group. But, what about the ethical concern of depriving potential "quitters" of the chance to stop smoking and the practical problem that members of the control group might get help elsewhere if asked to wait six months for help? What other study design could you use to study the impact of the "quit smoking" program?

quasi-experiment, an experimental design that is missing one or more aspects of a true experiment, most frequently random assignment into experimental and control groups.

In situations where it is not ethical or practical to do a true controlled experiment, the researcher can use a **quasi-experimental design.** In this design, the researcher does not have full control over the all aspects of the experiment; control over timing, the selection of subjects, or the ability to randomize exposure is lacking (Campbell and Stanley, 1963: 34). The element that is most frequently missing is random assignment into experimental and control groups. In the comparison group quasi-experiment, for example, there are two groups, both of which are given a pretest and posttest, but the groups are not selected by random selection, randomization, or matching to have pre-experimental equivalence. In this design, there is an experimental group and a comparison group in place of the control group (see Figure 8.7).

So, for example, if we were interested in the causal hypothesis that an anti-smoking program is effective in reducing smoking and improving health, we could do a comparison group quasi-experiment. We'd compare a group of volunteers who, on their own, decided to enroll in our program and find other similar people—let's say through a health clinic—who weren't taking the program, and study them over the same time period. Because we didn't pick or assign the groups randomly, they'd serve as our experimental and comparison groups.

A study of the impact of a police program on gang violence used a quasi-experiment for practical reasons. Fritsch, Caeti, and Taylor (1999) examined the impact of an innovative community policing strategy. The experimental group was composed of five geographic areas that the Dallas Police targeted. In these areas, police officers were freed from calls for service and spent their time aggressively enforcing truancy laws, conducting high-visibility patrols, stopping and frisking suspected gang members, and making arrests when appropriate. The police department picked the experimental group, so the researchers selected four similar areas as the comparison group. For both sets of areas, the average number of violent gang-related offenses (as determined by a follow-up investigation by the city's Gang Unit detectives) per month was compared for the year before the program started and the year of the program. The researchers found a decrease in gang violence in both groups from one year to the next with a larger decrease in the experimental group. Overall, the decrease in gang-related offenses was 57 percent in the experimental areas and 37 percent in the comparison areas (Fritsch, Caeti, and Taylor, 1999). Although the two sets of areas were similar, the use of a comparison

FIGURE 8.7

Comparison Group Quasi-
Experiment

Groups	Time		
	1 (pretest)		2 (posttest)
Experimental	Measure dependent variable	Stimulus	Measure dependent variable again
Comparison	Measure dependent variable	*	Measure dependent variable again

• The control group receives no stimulus, receives an alternative treatment, or receives a placebo.

rather than a control group raises some challenges to the study's internal validity: Were there important preexisting differences between the two sets of areas? Could such differences interact in some way with the independent variable to create the outcome differences in the dependent variable?

Although using a true control group typically gives a study better internal validity, the researchers did not have the option of creating one. Quasi-experiments like the one on gang violence can provide important information. They can provide an understanding of causal relationships and handle some challenges to internal validity, such as maturation, the testing effect, and history.

The Experimental Setting: Field and Laboratory Experiments

All research is conducted within social settings, but there is variability between settings and the degree to which the researcher has control over them. A **field experiment** is an experiment that takes place in "real life" settings where people congregate naturally—schools, supermarkets, hospitals, prisons, workplaces, summer camps, parking garages, and the like. The focal research project, conducted in a classroom, is an example of a field experiment where the researchers had the ability to select participants, had some control over what happened in the setting, and could decide which subjects would constitute the experimental and control groups, but could not influence other aspects of the school setting.

Joanne Miller and Jon Krosnick (2004) designed a field experiment in a real-world setting of political action and fund raising to measure the effectiveness of political appeals. In a study designed to explore the effect of threat and opportunity on support, the researchers designed a project where different versions of a letter soliciting contributions to the National Abortion and Reproductive Rights Action League (NARAL) of Ohio were sent to 9,000 women who had been randomly assigned to one of three groups. The responses of the two experimental groups and one control group helped

field experiment, an experiment done in the "real world" of classrooms, offices, factories, homes, playgrounds, and the like.

the researchers determine that a "policy change threat" (i.e., abortions might become more difficult to obtain) was more effective than a "policy change opportunity" (abortions might become easier to obtain) in obtaining contributions.

The "realness" of field experiments like these is why they typically have better **generalizability** (also called external validity) than other kinds of experiments. In a natural setting, the introduction of the independent variable (such as a curriculum, treatment, political strategy, administrative decision, and so on) is more the way things really are, and the researchers can use samples that are more like (or actually may be) the actual students, patients, workers, customers, citizens, and other groups that they want to generalize their results to.

Although working in real settings has advantages, it also can present difficulties. In the focal research project, for example, the researchers could not match the two groups of students by social class even though they wanted to, as the teachers could not identify parental occupations, incomes, or levels of education, and the researchers did not have access to students' files. Researchers doing field experiments are typically guests of the institutions or community settings and might not be able to control as much of the experiment as would be ideal. Looking at school-based evaluations of problem behavior (such as smoking) prevention interventions, Flay and Collins (2005) find that although study designs have improved over the past twenty years, often there are difficulties in doing experiments: school personnel resist randomization, appropriate control schools might not be used, the intervention (the stimulus) might not be fully implemented in some of the experimental group schools, permission for student participation is difficult to obtain, and biases can be introduced by data collection when the same people deliver the intervention and measure its outcome.

generalizability, the ability to apply the results of a study to groups or situations beyond those actually studied.

STOP AND THINK

In the Adler and Foster field experiment, the teachers did not tell the students that they were part of an experiment. Instead, the teachers presented each set of reading materials as the curriculums they would be covering for ten weeks. What were the advantages of not telling the students that they were part of an experiment? What were the disadvantages?

Another, perhaps more common setting for experiments is the "laboratory," which is a setting controlled by the researcher. The literature of psychology, social psychology, education, medical sociology, delinquency and corrections, political sociology, and organizational sociology is replete with reports of **laboratory research,** much of it conducted in universities using students as participants. In an experiment conducted at Princeton University, for example, a sample of 55 undergraduates was divided into experimental and control groups; the two groups received different descriptions of the competency of an elderly person, specifically in terms of their memory (the independent variable) (Cuddy, Norton, and Fiske, 2005). To measure the dependent variable, the views of the elderly, the students were asked to rate the hypothetical person in terms of warmth. The researchers found that in the "high-incompetence condition," the elderly person was rated as warmer.

laboratory research, research done in settings that allows the researcher control over the conditions, such as in a university or medical setting.

In another laboratory experiment with undergraduates, Arthur Brief and his associates (1995) conducted an experiment to test a hypothesis based on Milgram's work on authority (see Chapter 3 for a discussion of Milgram's work). The specific hypothesis tested was that when organizational members are instructed to do so, they will use race as a selection criterion in hiring decisions. A sample of 76 undergraduates enrolled in business courses was randomly assigned to one of three groups. In each group, the students were asked to play the role of chief financial officer for a hypothetical Mexican restaurant chain and to make a variety of managerial decisions, including evaluating candidates for jobs and making decisions about salary offers. In two groups the students were given instructions from the president of the company to consider each applicant's race, education, and experience when making hiring recommendations. One group got a "pro-white" message, the second got the alternate treatment of a "pro-black" message, and the third became the control group and was given instructions to consider the candidates' education and experience only. The dependent variable was measured by both the students' rating of the quality of eight job candidates (three of whom were black) on a scale from 1 to 5 and the three candidates they selected for a final interview. Based on comparisons of the groups' ratings of the black candidates and the number of black candidates selected for a final interview, the researchers concluded that there was support for Milgram's ideas about the influence of legitimate authority (Brief, Buttram, Elliot, Reizenstein, and McCline, 1995).

In these two examples of lab experiments the researchers had control over the setting and could easily randomly assign sample members to groups. On the other hand, although realistic situations can be created in labs and subjects can be kept in the dark about the real purpose of the study,[4] the participants in these experiments typically know that they were involved in research. If a setting's artificiality and the staging of an experiment contribute to changes in participants' attitudes and behaviors, the study's internal validity will be hampered.

It's also important to evaluate a study's external validity or the ability to generalize the results from the lab to the "real world." The following issues must be considered when you evaluate the generalizability of the results of a lab experiment:

1. Was the situation very artificial, or did it approximate "real life?"

2. How different were study participants from other populations?

3. To what extent did the participants believe that they were up for inspection, serving as guinea pigs or play-acting, or have other feelings that would affect responses to the stimulus (Campbell and Stanley, 1963: 21)?

4. To what extent did the researcher communicate his or her expectations for results to the subjects with verbal or nonverbal cues?

One way to handle the issue of **experimenter expectations** is to use a design called the **double-blind experiment,** which involves keeping

experimenter expectations, when expected behaviors or outcomes are communicated to subjects by the researcher.

double-blind experiment, an experiment in which neither the subjects nor research staff who interact with them knows the memberships of the experimental or control groups.

[4] See Chapter 3 for a discussion of the ethical issues involved in not informing study participants about a study's topic.

both the subjects and the research staff that directly interacts with them from knowing which members of the sample are in the experimental or the control group. In this way, the researcher's expectations about the experimental group are less likely to affect the outcome.

Natural Experiments

STOP AND THINK *What if you were interested in studying how work affects marriage? For example, if you wanted to know if the amount of work-related travel that a spouse does has an impact on his or her chances of getting divorced, could you design an experiment to answer your research question?*

There are many situations where it's just not possible to control the independent variable for practical and/or ethical reasons. So, even if we had an explanatory purpose, such as trying to find out if work schedules involving extended travel put stress on marriages and families, we couldn't do a traditional experiment because it's not possible for a researcher to control people's work schedules. Some researchers, however, do what they call **natural experiments** by finding real-world phenomena that can approximate the experimental design even though the independent variable is not controlled, manipulated, or introduced by the researcher. Wars, hurricanes, political events, and so on, might affect some groups and not others, and these can be used as pseudo-experimental and control groups. For example, Joshua Angrist and John Johnson (2000) used data from the armed services to look at the effects of deployment away from home during the Gulf War on the marriages and families of armed services personnel. Although the *researchers* could not control deployment, in 1991, during the Gulf War, the military assigned some personnel to overseas duty, either in the Gulf area or elsewhere, while others remained at home. In addition, in 1992, the military conducted a survey of officers and enlisted personnel that included questions on marriage and family that can be used as a sort of "posttest." Using information obtained from the military, the researchers did a statistical analysis of soldiers by time away from home. They found that time away does have an impact—especially for women serving in the military, with each month away raising divorce probabilities by more than one percent. However, because this is not a true experiment, there are questions about internal validity. The researchers caution that, although there is a temptation to interpret the results as causal, the differences in divorce could be the result of other factors (Angrist and Johnson, 2000). In other words, as in all explanatory studies, we need to be cautious about reaching conclusions about causality.

natural experiment, a study using real-world phenomena that approximates an experimental design even though the independent variable is not controlled, manipulated, or introduced by the researcher.

Comparing Experiments to Other Designs

Although the experimental study design is best suited for explanatory research and is a useful choice when you are considering causal hypotheses, many research questions cannot and should not be studied experimentally.

For descriptive and exploratory research purposes, when large samples are needed, and in situations where it is not practical, possible, or ethical to control or manipulate the independent variable, researchers are wise to consider other study designs. As we discussed in Chapter 7, the panel, trend, cross-sectional, and case study designs are more appropriate for many research purposes.

STOP AND THINK   *Can you think of hypotheses that do not lend themselves to being studied with an experiment? What is one such hypothesis? What design could be used to test your hypothesis?*

Social scientists often construct hypotheses using background variables like gender, race, age, or level of education as independent variables. Let's say that we hypothesize that level of education affects the extent to which young adults are involved in community service and political activity; more specifically, that increases in education, the independent variable, cause increases in the dependent variables (amounts of community service and political activity). We can't control how much education an individual receives, so we'd need to use a nonexperimental design to examine these relationships. For example, we could do a panel study and observe changes in the variables over time by starting with a sample of 18-year-olds. Following the sample for a number of years, we could measure the independent variable (years of schooling completed) and the dependent variables (involvement in community service and political activities) at least twice. We'd be able to see the pattern or relationship between the variables and their time-order. We could use statistical analyses as a way to disentangle possible "third variables" and spuriousness.

Panel studies have been widely used to study the cause and effect relationship between behaviors such as drinking coffee, smoking tobacco, and eating a diet rich in red meat, and the onset and progression of illnesses such as heart disease and cancer. Health researchers can't and wouldn't randomly select groups of people and force them to drink coffee, eat red meat, or smoke cigarettes for a number of years while prohibiting the members of other randomly selected groups from doing so, but they can follow people who make these choices themselves. Panel studies like the Nurses' Health Study have followed people since 1976. The youngest women were 25 when they joined the studies and the oldest are now in their eighties (NHS News, 2005). Every two years, 122,000 nurses enrolled in the first study, and 116,000 nurses enrolled in the second study are asked for medical histories, asked questions about daily diet and major life events, and periodically provide blood, urine, toenail, and cell samples to the researchers. The response rate for the questionnaires is typically 90 percent for each two-year cycle. The study is ethical because the researchers do not attempt to intervene in subjects' behavior, and because they keep all data confidential. Although some are critical of the lack of an experimental design and of the study's expense, other public health officials argue that it is better than an experiment in identifying long-term or delayed effects (Brophy, 1999). This panel study has contributed many important insights about the connections between behavior, life style, and health, including support for explanatory hypotheses.

Let's consider another nonexperimental design, the cross-sectional study, for its use in explanation. Assume we hypothesize that gender affects the degree of social isolation among senior citizens. While we can't "control" or randomly assign gender, we could measure this independent variable and any number of dependent variables, including social isolation in a sample of senior citizens. Because the time-order of these independent and dependent variables is clear (the gender of a senior citizen obviously occurs before his or her social isolation), if we found an association between the variables, we could then look for other support for causality. We could, for example, use statistical analysis techniques to eliminate the effect of other variables that might make the original relationship spurious.

Summary

This chapter adds to the researcher's toolbox by describing the study design options of controlled experiments and extended and quasi-experimental designs. Because experiments allow researchers to control the time-order of the independent and dependent variables and to identify changes in the dependent variable, they are very useful for testing causal hypotheses.

Experiments can be done as field experiments in the "real world," or they can be conducted in laboratories. Although the field experiment might be less artificial, the laboratory experiment can provide a greater ability to regulate the situation. Another option is the natural experiment, which is really not an experiment in the traditional sense, but which attempts to approximate an experimental design.

Experimental designs can be extended by adding extra experimental and control groups, or can be truncated into a quasi-experiment by eliminating one or more aspects of the classic design. Some design choices are more practical than others, whereas others have greater internal validity. Experiments typically use small, non-random samples and therefore usually have limited generalizability.

Ethical considerations and the practical concerns of time, cost, and the feasibility of controlling the independent variable are important issues in designing a study. Cross-sectional studies are typically the most practical, and when a large and representative sample is selected, the study will have good generalizability. However, the tradeoff for selecting a nonexperimental design may be questions about internal validity if the time-order is not clear, or if there are pre-existing differences between groups.

Each study design has strengths and weaknesses. As always, it's important to consider a project's goals against each design's costs and benefits and to balance methodological, ethical, and practical considerations when selecting a research strategy. Both experimental and nonexperimental designs should be considered when evaluating options. (See Table 8.2)

TABLE 8.2 Summary of Design Studies

Study Design	Design Features	Design Uses
Pretest-Posttest Control Group Experiment	This has at least one experimental and one control group; the independent variable is manipulated, controlled, or introduced by the researcher; the dependent variable is measured at least two times in both groups (before and after the independent variable is introduced).	Useful for explanatory research testing causal hypotheses.
Posttest-Only Control Group Experiment	Same as above except that the dependent variable is measured only after the independent variable is introduced.	Same as above.
Extended Experimental Design	Same as the pretest-posttest control group experiment with an additional experimental and control group measured only at the posttest.	Same as above but can also sort out the effects of pretesting.
Quasi-Experiment	The researcher does not have complete control of experimental stimuli. In the comparison group quasi-experiment, an experimental and a comparison group are given a pretest and posttest.	Same as experiment but with increased concerns about internal validity. May be more ethical or practical than a control-group experiment.
Cross-Sectional Study	One sample of cases studied one time with data about one or more variables. Typically, the sample is divided into categories of the independent variable and compared on the dependent variable during analysis.	Useful for describing samples and the relationship between variables; useful for explanatory purposes especially if the time-order of variables is known and sophisticated statistical analysis is possible.
Panel Study	One sample of cases is studied at least two times.	Useful for describing changes in individuals and groups over time (including developmental issues); useful for explanatory purposes.
Trend or Cohort Study	Two or more different samples are selected from the same population at least two times.	Useful for describing changes in populations.
Case Study	In-depth analysis of one or a few cases (including individuals, groups and organizations) within their social context.	Useful for exploratory, descriptive and, with caution, explanatory purposes; useful for generating theory and developing concepts; very limited generalizability.

Reading About Experiments

Find a research article in a psychology or social science journal that reports the results of an experiment. Answer the following questions about the study.

1. List the author(s), title of article, journal, month and year of publication, and pages.

2. What is one hypothesis that the research was designed to test? Identify the independent and dependent variables.

3. Describe the sample.

4. How many groups were used and how were they selected? Is there a control group?

5. Was there a pretest? If so, describe it.

6. How was the independent variable (or the stimulus) manipulated, controlled, or introduced?

7. Was there a posttest? If so, describe it.

8. What kind of experimental design was used—pretest and posttest experiment, a posttest-only experiment, a Solomon four-group, a comparison group quasi-experiment, or some other kind? Give support for your answer.

9. Comment on the issues of internal validity and generalizability as they apply to this study.

Conducting a Quasi-Experiment

What to do:

First, check with your professor about the need to seek approval from the Human Subjects Committee, the Institutional Review Board, or the committee at your college or university concerned with evaluating ethical issues in research. If necessary, obtain approval before completing this exercise.

In this quasi-experiment, you will be testing the hypothesis that when there is a "conversational icebreaker," people are more likely to make conversation with strangers than when there is no icebreaker.

Working in "the field," select a situation with which you are familiar from your daily life where strangers typically come into contact with each other—such as in an elevator, in a doctor's waiting room, waiting for a bus, standing in the check-out line in a supermarket, and so on.

Select one item to use as a conversational icebreaker. (Make sure it is something that could be commented on but is not something that could be perceived as threatening, hostile, or dangerous.) Some examples of icebreakers that you can use are carrying a plant or a bouquet of flowers, holding a cake with birthday candles on it, pushing an infant in a stroller, carrying a fish-bowl with water and tropical fish in it, or wearing a very unusual item of clothing or jewelry.

Over the course of several hours or days, put yourself in the same situation six times: three times with your icebreaker and three times without it. To the extent that you can, make sure that as much as possible, you keep everything else under your control the same each time (such as how you dress, the extent to which you make eye contact with strangers, whether you are alone or with friends, and so on). But, recognize that the groups of people you interact with, with and without the icebreaker, are two convenience samples—with one an experimental and the other a comparison group.

Immediately after leaving each situation, record field notes, which are detailed descriptions of what happened. In your field notes, record information about the number of strangers you interacted with, what they were like, and the way they interacted with you, specifically recording any comments that were made.

What to write:

After completing your experiment, answer the following questions:

1. What situation did you use for your experiment?

2. What icebreaker did you use?

3. Describe the dependent variable in the experimental group—that is, the kind of interactions that you had and any comments that strangers made in the three situations where you had an icebreaker.

4. Describe the dependent variable in the comparison group—that is, the kind of interactions that you had and any comments that strangers made in the three situations where you did not have an icebreaker.

5. Does your data support the hypothesis that an icebreaker is more likely to foster conversation and interaction with strangers?

6. Comment on the internal validity and generalizability of your study.

7. Comment on the experience of doing this exercise.

EXERCISE 8.3

Considering Nonexperimental Study Designs

In some situations, it won't be practical, ethical, or feasible to do a controlled experiment. Think about what you would do if, like Adler and Foster, you had a hypothesis about the affect of reading some sort of literature on students' attitudes or values, but could not use an experimental design. Construct a specific hypothesis that has reading a specific kind of literature as the independent variable and design a study using a nonexperimental design (see Chapter 7, Figure 7.4, for these designs).

1. State your hypothesis, identifying the specific independent and dependent variables you have selected.

2. Identify, describe, and diagram a study design you could use to examine your hypothesis.

3. Describe how you would measure your independent and dependent variables in this study.

4. Comment on the issues of internal validity and generalizability for the study you propose.

5. Compare the internal validity, generalizability, and practicality of your proposed study design with the kind of study conducted by Adler and Foster.

9

Questionnaires and Structured Interviews

© Emily Stier Adler

Introduction

To reduce the threat of terrorism, would you be willing or not willing to allow government agencies to monitor the telephone calls and e-mails of Americans? How concerned are you about losing some of your civil liberties as a result of the measures enacted by the administration to fight terrorism? In a 2006 CNN/USA Today/Gallup poll, 50 percent of Americans said it was okay to forgo warrants when ordering electronic surveillance of people suspected of having ties to terrorists abroad. That same year, responding to a poll done by the Pew Research Center for People and the Press, only 33 percent of respondents felt the government had gone too far in restricting civil liberties. What are your opinions on these issues?

questionnaire, a data collection instrument with questions and statements that are designed to solicit information from respondents.

structured interview, a data collection method in which an interviewer reads a standardized list of questions to the respondent and records the respondent's answers.

survey, a study in which the same data, usually in the form of answers to questions, are collected from all members of the sample.

respondent, the participant in a survey who completes a questionnaire or interview.

self-report method, another name for questionnaires and interviews because respondents are most often asked to report their own characteristics, behaviors, and attitudes.

Have you ever received a phone call asking you to participate in a survey? Do you remember someone in your household completing a questionnaire from the U.S. Bureau of the Census? Have you heard or read about the results of recent polls that asked questions about government policies? Most of us are familiar with interviews and questionnaires; it's hard to avoid these methods of data collection because so many of them are conducted annually. **Questionnaires** and their first cousins, **structured interviews,** are the most widely used methods of data collection in the social sciences. Research projects that use structured interviews and questionnaires are sometimes referred to as surveys because, although these methods of data collection can be used in experiments or in case studies, they are most commonly used in surveys.

A **survey** is a study in which the same data are collected (most often a specified set of questions, but possibly observations or available data) from one or more samples (usually individuals, but possibly other units of analysis) and analyzed statistically. Surveys often use large probability samples and the cross-sectional study design, which is when data are collected about many variables at one time from one sample. However, surveys can also employ nonprobability samples and trend and panel study designs. See Box 9.1 for two examples of recent surveys. Although these examples are from the United States, survey research is now found around the world.

The various modes of data collection that use communication between respondents and researchers can be seen as falling along a continuum from the least interactive to the most interactive—that is, from self-administered questionnaires to in-person interviews (Singleton, Jr. and Straits, 2001). In all of the modes of data collection covered in this chapter, data are collected by asking study participants, typically called **respondents,** a set of predetermined questions. This technique is also called the **self-report method** because respondents are usually asked to provide information about themselves. The researcher can take the answers as facts or interpret them in some way, but the information provided by respondents is the basic data in questionnaires and interviews.

Answering questions in a survey is a social act in the same way that attending a class or browsing in a shopping mall is. The social context of the data collection and the views of those we question are influenced by the surrounding culture. When we meet new acquaintances in our everyday lives,

BOX 9.1

Surveys are all around us. If we want to know what Americans think about an issue, we often can find a survey that's asked questions on the topic. Two sources of interesting and methodologically careful surveys are Public Agenda (www.publicagenda.org) and the Pew Research Center for the People and the Press (http://people-press.org). Both of these nonprofit organizations conduct many studies annually. The work of Public Agenda has included studies on families, children, crime, corrections, education, health care, and the economy, The Pew Center focuses on attitudes toward the press, politics, and public policy.

STOP AND THINK *What's your opinion about getting a college degree? Did you have someone who encouraged you do go to college? Do you go online to access the Internet or World Wide Web or to send or receive e-mail? Do people ever complain that you spend too much time on the Internet? Do you think your feelings or behaviors in these areas are unusual or typical?*

A national sample of 1,000 young adults, aged 18 through 25, was interviewed by phone in 2004 for *Life After High School,* a study by Public Agenda. The study found most young Americans feel attending college makes a real difference in how they'll fare in the world and that their parents and teachers encouraged them to aim for college. See more findings of this survey at www.publicagenda.org/research/research_reports_details.cfm?list=31.

A national sample of 2,200 adults was interviewed by phone in 2004 for the Pew Internet and American Life Project. In this survey, 69 percent of the respondents reported going online and 87 percent said that other people did not complain that they spent too much time on the Internet. For more of this study's findings see http://www.pewinternet.org/pdfs/PIP_Internet_ties.pdf

we often ask questions. In the same way, researchers use questions to elicit information about the backgrounds, behavior, and opinions of respondents. And, just as in our personal conversations, researchers try to figure out if the answers they've received are honest and accurate.

STOP AND THINK *Suggest a list of topics that you think you could ask questions about using a questionnaire. Would you be concerned about the accuracy of the answers on any of these topics?*

In the research described in the focal research that follows, Gray, Palileo, and Johnson wanted to find out some things about their sample of college students. Specifically, they were interested in knowing to what extent their respondents accepted common myths about rape, if there were any differences of opinion between women and men, how many women had been sexually victimized, and how many men were sexual assailants. In addition, they wanted to explore possible causes of some of the attitudes. As you will see, these researchers were interested in doing a study that had both descriptive and explanatory purposes.

In reviewing the literature about attributing blame for criminal acts, Gray, Palileo, and Johnson discovered two very different theories or models about the way that blame is assigned. To discover patterns in the acceptance of rape myths and to see which of the competing theories about the causes of these patterns received support, they collected data by asking questions.

FOCAL **RESEARCH**

Explaining Rape Victim Blame: A Test of Attribution Theory

by Norma B. Gray, Gloria J. Palileo, and G. David Johnson[1]

Introduction

"Rape culture" refers to the set of beliefs that members of a social collectivity hold about rape, its causes, and the relative responsibility attributed to assailants and victims. It often has been argued that the prevailing rape culture in the United States contributes to its high incidence of rape. The victims of rape are frequently blamed for their own victimization, perhaps more frequently than the victims of any other crime. Such beliefs inhibit the reporting of rape, limit the provision of social support to victims, and thus facilitate the commission of the crime by assailants.

In this paper we examine victim-blaming attitudes with data from a probability sample of students at a southern university. Our objectives are to measure the pervasiveness of rape myths and examine sources of variation in their acceptance.

We test hypotheses derived from attribution theory (Heider, 1958; Kelley, 1971) in order to explain variations in rape myth acceptance. A basic assumption of attribution theory is that individuals will attribute causality for observed behavior, i.e., they will make decisions about who or what is responsible for the behavior they observe. Individuals will attribute causality and responsibility either to the actor's "personal disposition" or to external (e.g., other persons, the environment, chance) causes (Thorton, Robbins, and Johnson, 1981). The attributions an individual makes about an actor's responsibility for his or her own behavior are influenced by (1) the social

[1] Adapted from an article in *Sociological Spectrum,* 1993, Vol. 13(4), pp. 377–392, by N. B. Gray, G. J. Palileo, and G. D. Johnson. Reproduced by permission of Taylor & Francis, Inc., www.routledge-ny.com.

location of the observer and the observed (e.g., their class position, gender, and race); (2) the observer's previous experience (particularly in related activities); (3) the setting; and (4) the potential psychological benefit to be derived from alternate attributions (Shaver, 1975).

Previous Research

Much of the research on rape victim blaming has focused on characteristics of victims. Victims dressed in sexy attire are more likely to be blamed than those more modestly dressed (Lewis and Johnson, 1989; Cahoon and Edmonds, 1989). Those with histories of sexual activity are more likely to be held responsible (Cann, Calhoun, and Selby, 1979). The setting of the rape is also associated with the attribution process. Terry and Doerge (1979), for example, found that victims were more likely to be blamed when the rape occurred outside the home.

We examine the association between the social characteristics of the observers and the attribution of responsibility from the opposing perspectives of the "defensive attribution" (Shaver, 1975) and "need for control" (Walster, 1966) models. The defensive attribution model states that the greater the similarity in social characteristics or experiences between observer and observed (e.g., if they are the same gender and/or the same race), the greater the likelihood that the observer will attribute responsibility to someone or something other than the observed victim. This attribution is motivated by the need for the observer to protect his or her self-esteem and to avoid self-blame if he or she should be similarly victimized in the future.

The need for control model, in contrast, assumes that, typically, individuals will seek to maintain a more predictable, and therefore more controllable, environment by discovering underlying dispositions in an actor's behavior (Heider, cited in Shaver, 1975). Consequently, observers attribute responsibility to the observed (including victims of crimes) in order to maintain a psychological sense that they, the observers, have control over their own lives. According to this model, if the observers attributed causality to external events, they would have to confront their lack of control over their own fates.

Previous sexual victimization has also been found to be modestly correlated with rape myth acceptance (Burt, 1980), suggesting that individuals who have experienced sexual violence are more likely to identify with other victims and, therefore, are more likely to avoid victim blaming.

There are a number of variables, not addressed in the current literature, that we also predict will be associated with victim blaming. We predict that victims of nonsexual violent crimes will also be more likely to identify with rape victims and, therefore, more likely to avoid victim blaming. Conversely, we predict that individuals who have committed acts of aggression, particularly sexual aggression, will be more likely to attribute blame to justify their own aggression.

Another characteristic we examine is risk-taking behavior. The defensive attribution model predicts risk takers are more likely to identify with rape victims. Individuals who take chances they regard as reasonable (such as walking across campus alone at night) will not hold others to a behavioral standard to which they, themselves, do not adhere. On the other hand, the control model makes the opposite prediction: Risk takers will be more likely to blame the victim. According to this theory, individuals need to maintain a sense of control over their lives and, unless they hold victims responsible, they will need to admit their vulnerability they have taken similar chances.

Hypotheses

We test four hypotheses from the defensive attribution model:

1. Women will attribute less rape victim responsibility than will men.
2. Women who have experienced sexual victimization will attribute less victim responsibility than women who have not.
3. Men who have committed acts of sexual aggression will attribute greater victim responsibility than men who have not.
4. Men and women who have been victims of nonsexual crimes will attribute less responsibility to rape victims than those who have not.

We also construct a hypothesis from the perspective of each of the two opposing theories and use the data to choose between them.

5a. From the defensive attribution model: High risk takers will attribute less responsibility to rape victims than will more cautious individuals.

5b. From the need for control attribution model: High risk takers will attribute more responsibility to rape victims than will more cautious individuals.

Methodology

The data were collected in January 1991 from 1,177 students at a southern university with an enrollment of approximately 11,000. The sampling frame consisted of Monday 10:00 A.M. and 6:00 P.M. classes, which were the university's peak scheduling hours. Instructors from 58 out of the 63 classes selected by simple one-stage cluster probability sampling allowed our proctors to administer the survey questionnaires, yielding a total of 511 men and 666 women (more than 99 percent completion rate).

Rape victim blame was measured with an eight-item scale of commonly held rape myths, developed from the much larger Burt (1980) scale. A factor analysis revealed that the eight items had a single underlying dimension. Consequently, an eight-item additive index was created with an acceptable level of inter-item reliability. See the appendix for the specific indicators of this and the other variables.

Previous sexual victimization (asked only of women) was measured by a series of "yes/no" questions on various forced sexual acts, modified with permission from Koss, Gidycz, and Wisniewski (1987). A respondent who said "yes" to any of the items was classified as having been sexually victimized.

Previous sexual aggression (asked only of men) was measured by a similar series of "yes/no" questions on various forced sexual acts committed by the respondent within the past year. A man who said "yes" to any of the items was classified as having been a sexual victimizer.

Nonsexual victimization was measured by two "yes/no" questions concerning assault and mugging. These two crimes were selected because they most closely resembled the experience of rape victimization.

A risk-taking behavior scale included seven questions that formed an additive index with acceptable inter-item reliability. In addition, two questions were asked about alcohol consumption. Frequency of drink was measured by an ordinal scale from 0 (none in the past year) to 4 (more than twice a week). Quantity of drinking was measured on a scale from 0 (none) to 4 (more than 6 cans of beer or 5 glasses of wine or spirits).

Additional variables included age, gender, race, marital status, class standing (freshman through graduate school), current residence (on-campus, off-campus), knowledge of rape prevention, and national residence (citizen or resident versus noncitizen and nonresident).

Results

The sample's mean age was 24.7, with the students fairly evenly split between all four undergraduate years and 13 percent in graduate school. The sample was 84 percent white, 8 percent black, 5 percent Asian, 1 percent Hispanic, and 2 percent American Indian or other. Ninety-four percent were U.S. citizens or residents, 76 percent lived off-campus, 76 percent were unmarried, and 57 percent were women.

The possible range of scores for the rape myths acceptance scale was from 8 (total rejection of myths) to 40 (total acceptance of myths). The mean score for the women in the sample was 13.0 (s.d. $= .17$) and for men it was 17.5 (s.d. $= .27$), indicating that overall the sample tended not to blame the victim.

We first analyzed the strength of relationships between two variables for the total sample and for the men and women separately. As predicted by Hypothesis 1, we found gender to be strongly associated with rape victim blaming, with men more likely than women to hold rape victims responsible ($r = .42$, $p < .05$). The findings also confirm the predicted relationship between previous sexual victimization and victim blaming. Women who had experienced sexual victimization are less likely to accept rape myths ($r = .09$, $p < .05$). Also as predicted, men who have committed sexual aggression are more likely to blame rape victims ($r = .21$, $p < .01$). Nonsexual victimization, however, is not related to rape myth acceptance for either gender.

There is significant interaction with gender in the results for risk taking. Among women, there is a negative and significant association between risk taking and victim blaming ($r = -.10$, $p < .05$), a finding more consistent with a defensive attribution model. For men, the relationship is positive ($r = .07$; increased risk taking is associated with increased myth acceptance), but is not statistically significant.

An interaction exists between gender and alcohol consumption. For men, increased alcohol consumption is associated with increased myth acceptance, but for women, increased alcohol consumption is associated with reduced victim blaming.

The other variables related to myth acceptance in bivariate analyses are national residence (i.e., noncitizens and nonresidents are more likely to accept myths), rape knowledge (i.e., the more knowledgeable are less likely to accept myths), and class standing (i.e., senior students are less accepting of myths).

However, after conducting a stepwise multiple regression analysis (a method of statistical analysis designed to control for the other variables under consideration when considering a relationship between two variables), only a few of the original relationships remain significant. As expected, gender remains the most important predictor of rape myth acceptance. Nationality, rape prevention knowledge, and class standing are also significantly related to the dependent variable in the analysis of the whole sample.

For the subsample of women, only two predictors of myth acceptance remain after multivariate analysis: frequency of drinking and degree of risk taking. Each relationship is negative, suggesting that the defensive attribution model best fits the data from the women.

For the men, several independent variables continue to be related to myth acceptance. As predicted, previous sexual aggression is associated with higher rape myth acceptance while previous nonsexual victimization ("ever assaulted") is associated negatively with victim blaming. Quantity of alcohol consumption is positively associated with the dependent variable. In addition, foreign students and nonwhites are more likely to accept rape myths. Class standing and degree of rape prevention knowledge are negatively related to myth acceptance among men.

Conclusions

Most of the hypotheses are at least partially confirmed. As anticipated, gender is the most powerful predictor of rape victim blaming, with women substantially less likely to accept rape myths. This finding is fully consistent with a defensive attribution model, supporting the interpretation that women are better able to identify with victims.

Many of the relationships that applied to the female sample in the initial analysis became statistically insignificant when controlling for the other variables. In the multivariate analysis, only risk-taking behavior (including frequency of alcohol consumption) remains a significant predictor for women, with women who take risks more likely to empathize with victims, presumably because they are sensitive to their own vulnerability.

For men, a greater amount of the variance in victim blame attitudes is explained by the independent variables of past sexual aggression, previous nonsexual victimization, and alcohol consumption. Men, particularly those who have been sexual aggressors, are more likely to identify with the assailant, as indicated by their greater acceptance of rape myths. Men who take risks are less likely to empathize with rape victims, thus protecting their own subjective sense of invulnerability.

Rape prevention knowledge and class standing help explain variation among the men, suggesting that education influences attitudes about rape. Men who report knowledge of rape prevention are less likely to blame rape victims. In addition, even controlling for age, men who have been in college longer are less likely to embrace rape myths. Perhaps there is reason to hope that rape culture can be altered by education and knowledge about rape and its prevention.

REFERENCES

Burt, M. R. 1980. Culture myths and supports for rape. *Journal of Personality and Social Psychology* 38: 217–230.

Cahoon, D., and E. Edmonds. 1989. Male and female estimates of opposite sex first impressions concerning females' clothing styles. *Bulletin of the Psychonomic Society* 27: 280–281.

Cann, A., L. Calhoun, and J. Selby. 1979. Attributing responsibility to the victim of rape: Influence of information regarding past sexual experience. *Human Relations* 32: 57–67.

Heider, F. 1958. *The psychology of interpersonal relations.* New York: Wiley.

Kelley, H. H. 1971. *Attribution in social interaction.* Morristown, NJ: General Learning.

Koss, M. P., C. Gidycz, and N. Wisniewski, 1987. The scope of rape: Incidence and prevalence of sexual aggression and victimization in a national sample of higher education students. *Journal of Consulting and Clinical Psychology* 55: 162–171.

Lewis, L., and K. Johnson. 1989. Effects of dress, cosmetics, sex of subject and causal influence on attribution of victim responsibility. *Clothing and Textile Research Journal* 8: 22–27.

Shaver, K. 1975. *An introduction to attribution processes.* Cambridge, MA: Winthrop.

Terry, R., and S. Doerge. 1979. Dress, posture and setting as additive factors in subjective probabilities of rape. *Perceptual and Motor Skills* 48: 903–906.

Thorton, B., M. Robbins, and J. Johnson. 1981. Social perceptions of the rape victim's culpability: The influence of respondent person-environmental causal attribution tendencies. *Human Relations* 39: 225–237.

Walster, E. 1966. Assignment of responsibility for an accident. *Journal of Personality and Social Psychology* 3: 73–79.

Appendix A: Group-Administered Questionnaire

The participants in this study each received a questionnaire with approximately 50 closed-ended questions. The respondent was asked to check the box of the appropriate answer category for each question. The men's and women's questionnaires were identical except for one set of questions as specified below. Each questionnaire included the following introductory statement:

This is a survey of the opinions and experiences of students at the University of South Alabama. The survey is being conducted by the Department of Sociology and Anthropology. The questions are of a personal nature and your participation is voluntary. We would like for you to answer the following questions as truthfully as you can. To insure your anonymity, please do not write your name anywhere on the survey. Your cooperation is very much appreciated. Please drop the completed form in a box provided by the monitor. If you do not wish to participate, please return the questionnaire to the monitor and remain in your seat.

The indicators for the variables described in the focal research are as follows:

A. Risk-taking behavior *[The answer categories provided for each of these 7 statements were "never," "seldom," "sometimes" and "often."]*

1. How frequently would you say you do the following?
   a. Walk alone on campus at night.
   b. Walk alone off campus at night.
   c. Leave dorm/apartment door open.
   d. Open door/let in strangers.
   e. Leave car unlocked.
   f. Go out on blind dates.
   g. Enter or work alone in an isolated building late at night or weekends.

2. How often do you drink alcohol? (Check one)

   None in the past year

   Less than once a month but at least once in the past year

   One to three times a month

   One to two times a week

   More than twice a week

3. On a typical drinking occasion, how much do you usually drink? (Check one)

   Never

   No more than 3 cans of beer (or 2 glasses of wine or 2 distilled spirits)

   No more than 4 cans of beer (or 3 glasses of wine or 3 drinks of distilled spirits)

   No more than 5 or 6 cans of beer (or 4 glasses of wine or 4 drinks of distilled spirits)

   More than 6 cans of beer (or 5 or more glasses of wine or distilled spirits)

B. Rape victim blame *[The answer categories provided for each of these statements was strongly agree, agree, undecided, disagree and strongly disagree]*

Now we want to ask you about some of your opinions. Please tell us if you strongly agree, agree, are undecided, disagree, or strongly disagree with each of the following statements.

1. A female who goes to the home or apartment of a man on their first date implies that she is willing to have sex.

2. A man has a right to have sex with his wife even if it is sometimes against her will.

3. Any healthy female can successfully resist a rapist if she really wants to.

4. When females go around braless or wearing short skirts and tight tops, they are asking for trouble.

5. If a female engages in necking or petting and she lets things get out of hand, it is her own fault if her partner uses some force to get her to have sex with him.

6. Many females who get raped have done something to provoke it.

7. Most women, although they might not at first admit it, would enjoy being forced to have intercourse with a man.

8. If a female gets drunk at a party and has intercourse with a man she's just met there, she can be considered available to other males at the party who want to have sex with her too, whether she wants to or not.

C. *Nonsexual victimization*

1. Have you ever been a victim of a (non-sexual) crime? If yes, what kind of crime were you a victim of? (Check all that happened to you).

   No (Skip to Question xx)

2. Physical assault   No   Yes

3. Mugging   No   Yes

   [questions about other crimes were included but not used as indicators of nonsexual victimization]

D. *Previous sexual victimization* (on the women's questionnaire only)
   [The answer categories provided were "no" and "yes"]

The following questions are about your sexual experience from age 14 on.

1. Have you given in to sex play (fondling, kissing, or petting, but not intercourse) when you didn't want to because you were overwhelmed by a man's continual arguments and pressure? (From age 14 on)

2. Have you had sex play (fondling, kissing, or petting, but not intercourse) when you didn't want to because a man used his position of authority (boss, teacher, camp counselor, supervisor) to make you? (From age 14 on)

3. Have you had sex play (fondling, kissing, or petting) when you didn't want to because a man threatened or used some degree of physical force (twisting your arm, holding you down, etc.)? (From age 14 on)

The following are questions about sexual intercourse. By sexual intercourse we mean penetration of a woman's vagina, no matter how slight, by a man's penis. Ejaculation is not required. Whenever you see the words sexual intercourse, please use this definition.

4. Have you had a man attempt sexual intercourse (attempt to insert his penis) when you didn't want to by threatening or using some degree of force (twisting your arm, holding you down, etc.) but intercourse did not occur? (From age 14 on)

5. Have you had a man attempt sexual intercourse (attempt to insert penis) with you by giving you alcohol or drugs, but intercourse did not occur? (From age 14 on)

6. Have you given in to sexual intercourse when you didn't want to because you were overwhelmed by a man's continual arguments and pressure? (From age 14 on)

7. Have you had sexual intercourse when you didn't want to because a man used his position of authority (boss, teacher, camp counselor, supervisor)? (From age 14 on)

8. Have you had sexual intercourse when you didn't want to because a man gave you alcohol or drugs? (From age 14 on)

9. Have you had sexual intercourse when you didn't want to because a man threatened or used some degree of physical force (twisting your arm, holding you down, etc.) to make you? (From age 14 on)

10. Have you had sex acts (anal or oral intercourse or penetration by objects other than the penis) when you didn't want to because a man threatened or used some degree of physical force (twisting your arm, holding you down, etc.) to make you? (From age 14 on)

Previous sexual aggression *(on the men's questionnaire only) [The answer categories provided were "no" and "yes"]*

1. Have you engaged in sex play (fondling, kissing, or petting, but not intercourse) with a woman when she didn't want to by overwhelming her with continual arguments and pressure? (From age 14 on)

2. Have you engaged in sex play (fondling, kissing, or petting, but not intercourse) with a woman when she didn't want to by using your position of authority (boss, teacher, camp counselor, supervisor)? (From age 14 on)

3. Have you engaged in sex play (fondling, kissing, or petting, but not intercourse) with a woman when she didn't want to by threatening or using some degree of physical force (twisting her arm, holding her down, etc.)? (From age 14 on)

The following are questions about sexual intercourse. By sexual intercourse we mean penetration of a woman's vagina, no matter how slight, by a man's penis. Ejaculation is not required. Whenever you see the words sexual intercourse, please use this definition.

4. Have you attempted sexual intercourse (attempted to insert penis) with a woman when she didn't want it by threatening or using some degree of physical force (twisting her arm, holding her down, etc.) but intercourse did not occur? (From age 14 on)

5. Have you attempted sexual intercourse (attempted to insert penis) with a woman when she didn't want it by giving her alcohol or drugs, but intercourse did not occur? (From age 14 on)

6. Have you engaged in sexual intercourse with a woman when she didn't want to by overwhelming her with continual arguments and pressure? (From age 14 on)

7. Have you engaged in sexual intercourse with a woman when she didn't want to by using your position of authority (boss, teacher, camp counselor, supervisor)? (From age 14 on)

8. Have you engaged in sexual intercourse with a woman when she didn't want to by giving her alcohol or drugs? (From age 14 on)

9. Have you engaged in sexual intercourse with a woman when she didn't want to by threatening or using some degree of physical force (twisting her arm, holding her down, etc.)? (From age 14 on)

10. Have you engaged in sex acts (anal or oral intercourse by objects other than the penis) with a woman when she didn't want to by threatening or using some degree of physical force (twisting her arm, holding her down, etc.)? (From age 14 on)

The Uses of Questionnaires and Interviews

Once researchers decide on the kind of information required, they'll typically consider if asking questions is an appropriate way to collect this information and then think about ways of administering those questions. Looking at the examples in Box 9.2 you'll see that questions can be asked about a variety of topics. It is very common to use a self-report method to collect information about *attitudes, beliefs, values, goals,* and *expectations,* although data on these topics can sometimes be inferred from observed behavior. Similarly, although data about *social characteristics* and *past experiences of individuals* and *information about organizations or institutions* might be accessible through records and documents, they are suitable topics for questions.

Questions can be used to measure a person's level of *knowledge* about a topic. They are also frequently employed to obtain information about *behavior,* even though observational methods will often yield more valid data. If, like Gray and her colleagues in the focal research, you are interested in finding out about behavior that typically occurs without observers (such as sexual aggression or victimization), or if you want to find out about behavior that would be unethical, impractical, or illegal to observe (such as voting behavior or getting tested for HIV), then a self-report method might be the only data collection option.

STOP AND THINK   *In addition to the one presented in this chapter, several other studies in the focal research sections of the book have collected data by asking questions. Can you remember any of the studies that have used this method? Do you have any concerns about the validity of the self-report data collected in any of the focal research studies?*

The self-report method has a wide variety of applications, including public opinion polls, election forecasts, market research, and publicly funded

BOX 9.2

Topics and Examples of Questions

Questions can be asked about attitudes.
An example: Employers should be allowed to require drug testing of employees.

☐ agree strongly
☐ agree somewhat
☐ undecided
☐ disagree somewhat
☐ disagree strongly

Questions can be asked about organizations.
An example: Does your current place of employment have a policy concerning sexual harassment?

☐ no
☐ yes

Questions can be asked about an expectation.
An example: What is your best guess that you will graduate with honors from college?

☐ I have no chance
☐ I have very little chance
☐ I have some chance
☐ I have a very good chance

Questions can be asked about behavior.
An example: During a typical week, how much time do you spend exercising or playing sports?

☐ none
☐ less than an hour
☐ one to two hours
☐ three to five hours
☐ six or more hours

Questions can be asked about a social characteristic or attribute.
An example: How old will you be on December 31 of this year?

☐ under 18
☐ 18–20
☐ 21–23
☐ 24–26
☐ 27–30
☐ 31 and over

surveys. Questions can be asked in less structured ways, options which we will consider in the next chapter, or they can be asked using questionnaires and structured interviews, the formats we'll consider in this chapter. The more structured ways of asking questions can facilitate the collection of the same information from large samples, especially when the samples have been selected from regional, national, or multinational populations. The resulting analyses are often used by local, state, federal, and international organizations and are frequently the focus of articles in the mass media. For example, twice a year, the Bureau of Justice Statistics collects interview data about people's experiences with criminal victimization from 42,000 American households for the National Crime Victimization Survey (NCVS). Researchers and policymakers benefit from the National Longitudinal Surveys of Youth, sponsored by the U.S. Department of Labor, which provide information about children's abilities, attitudes, and family relationships and the transition from school to work. The General Social Survey (GSS) and the International Social Survey Programs (ISSP) are two surveys that allow the documentation and comparison of attitudes and behavior over time. The GSS does so with samples of residents of the United States; the ISSP uses international ones. The numerous surveys make it possible to gather large amounts of information from mass publics which can be used descriptively and to inform public policy.

Although surveys are very widely used, there are important concerns about the *validity* of the information that they generate. The use of all self-report methods is based on the implicit *assumption* that people have the information being asked about and that they will answer based on their core values and beliefs. However, even assuming that questions are understood, responses could represent casual thoughts rather than deep feelings and might reflect what the respondent heard on the news most recently rather than personally salient answers (Geer, 1991; Kane and Schuman, 1991: 82). In addition, people in more positive moods have been found to give more favorable answers to questions than do those who are feeling more negative (Martin, Abend, Sedikides, and Green, 1997).

It is fairly simple to design questions that ask respondents to describe their attitudes toward abortion, to share their views on laws that ban assault weapons, or to recall the number of sex partners they had in the past 12 months. However, as we noted in Chapter 6, we must consider **measurement error,** the giving of inaccurate answers to questions. In addition to the issue of question wording, which we will discuss later in this chapter, measurement error is affected by the respondents' level of knowledge, whether or not they have an opinion about the topic, their being able to recall events accurately from months or years before, and having the intent to answer honestly. It is problematic when the respondent wants to present a good image or tries to conform to what are perceived to be the researcher's expectations.

Recall of prior events by respondents has been found to be influenced by the time that has elapsed between the event and the data collection, the importance of the event, the amount of detail required to be reported, and the social psychological motivation of the respondents (Groves, 1989: 422).

measurement error, the kind of error that occurs when the measurement we obtain is not an accurate portrayal of what we tried to measure.

For example, it is not uncommon to find a 20 percent difference between the official turnout in presidential elections and the self-reports of voting, with the most common explanation being that respondents are motivated to give socially desirable responses so as to present themselves in a good light (Karp and Brockington, 2005).

In one study, patients of a health maintenance organization were asked if they had specific procedures, such as a flu shot, in the recent past. When the self-reports were compared with medical records, the median overreporting of procedures was 13 percent when asked about the past six months and 4 percent for two months (Loftus, Klinger, Smith, and Fielder, 1990: 343). In another study, researchers counted attendees at worship, obtained Sunday school records, and then asked members of a large church if they had attended that past Sunday. They found more than a 20 percentage-point rate of overreporting church attendance (Hadaway and Marler, 1993: 127).

Other factors, such as gender role stereotypes, can also influence responses. When studying the play activities of fifth grade students, Janet Lever (1981) compared answers to the question "Where do you usually play after school, indoors or outdoors?" with activity logs that the children kept, and found that boys overestimated and girls underestimated their outdoor play. Although studies that provide data by asking questions can be valuable, it's important to examine the specific questions carefully before accepting the results of analyses.

Participant Involvement

response rate, the percentage of the sample contacted that actually participates in a study.

One critical issue for all studies is the **response rate,** the percentage of the sample that is contacted and, once contacted, agrees to participate. The response rate is affected by the number of people who cannot be reached (noncontacts), the number who choose not to participate (refusals), and the number who are incapable of performing the tasks required of them (due to illness or language barriers, for example).

The response rates of studies using self-report methods vary greatly, with the in-person interview tending to have the highest rate. For example, the in-person interview for the National Health Interview Survey, conducted by the Census Bureau since 1960, continues to get at least a 90 percent response rate (National Center for Health Statistics, 2005), and 80 percent of those approached for a national survey of adult sexual behavior agreed to be interviewed (Michael, Gagnon, Laumann and Kolata, 1994: 33). On the other hand, we've seen published studies with quite low response rates.

Refusal rates have grown in the past few decades. Yehuda Baruch (1999) examined 175 published studies over three decades and found that the average participation rate of studies published in 1975 was 64 percent but only 48 percent in 1995. Curtin, Presser, and Singer (2005) note that the University of Michigan's Survey of Consumer Attitudes had experienced declines in response rates over the years it was an in-person interview (1954–1976) but, when it became a phone interview, obtaining interviews became even harder. For example, they found that the number of calls to complete an interview doubled from approximately four to eight between 1976 and 1996 with the response rate dropping from 72 percent to 60 percent. Since 1996, the deterioration has become much greater, so that by 2003, it was 48 percent (Curtin, Presser, and Singer, 2005: 90). The greatest difference in recent years is the rise in the noncontact rate, attributable to caller ID and growth in nonvoice and unassigned numbers.

nonresponse errors,
errors that result from
differences between
nonresponders and
responders to a survey.

The major reason for concern about declines in response rates is **nonresponse error,** which is the error that results from the differences between those who participate and those who do not. In comparing response rates, some studies have found higher refusal rates among older persons, nonblacks, and those with less than a high school education (Groves, 1989; Krysan, Schuman, Scott, and Beatty, 1994). When participants and nonparticipants differ in social characteristics, opinions, values, attitudes, or behavior, then generalization from the sample to the larger population becomes problematic. Recently, however, some have questioned whether higher rates of refusal automatically mean more bias (Curtin, Presser and Singer, 2005; Groves, Presser, and Dipko, 2004). While greater participation is found among those interested in the topic of a survey, some researchers note that there are many other influences on participation and that only factors linked to survey statistics should be causes of concern (Groves, Presser, and Dipko, 2004).

STOP AND THINK   *Have you ever been asked to complete a questionnaire or an interview? What factors did you consider when deciding whether or not to participate? Think about what you would do if you were on the other side. What would you do to try to convince someone to participate in a survey?*

Participation in studies can best be understood within a social exchange context. That is, once potential sample members are contacted, they must decide about cooperation after thinking about the costs and benefits. Respondents are asked to give up their time, engage in interactions controlled by the interviewer, think about issues or topics that might cause discomfort, and take the risk of being asked to reveal embarrassing information. Even though they are usually assured that the information they provide is anonymous or will be confidential, potential participants might worry about privacy and lack of control over the information. On the other hand, potential respondents might want to participate because of an interest in the topic, a desire to share their views, or knowledge that their information will be useful to science and society.

Advance mailings are increasingly being used to combat the decline in response rates (Hembroff et al., 2005). However, because they tend to disproportionately raise participation rates among some segments of the population, potential bias needs to be considered (Link and Mokdad, 2005). Other useful approaches to increasing participation include giving small gifts or token cash payments, focusing on the interesting aspects of participation, re-contacting participants to encourage participation, making it clear that

> **BOX 9.3**
>
> ## Recruiting a Sample
>
> Australian sociologist Jo Monash found that it wasn't easy to find a sample of nonprofessional workers for a survey on sexual health, alcohol, and drug consumption. Once she gained access to workplaces, she put up fliers about the project, but didn't find it a successful strategy. Instead, she did better when she made in-person contact with potential participants at lunchtime. She says, "In the pitch I made to the young workers, I presented myself as their advocate. I told them it was their right to have good health information and that I would give them a voice in the health issues affecting them. I emphasized the low impact of the research again. I talked about the limited time the survey would take and the confidentiality of the research" (Monash, 2005: 124).

the research is legitimate with a bona fide sponsor, and minimizing the costs of participation (such as time and possible embarrassment). See Box 9.3 for a discussion of one researcher's efforts to recruit a sample for her study.

Gray, Palileo, and Johnson had a very high response rate in the focal research—99 percent of those present and approximately 85 percent of those enrolled in the classes selected for the sample completed the questionnaire. After receiving permission from faculty members, Gray and associates hired students to make unannounced visits to each of 58 classes to describe the study and request participation. The classes were told that the survey was anonymous, sponsored by the dean of students, and would take approximately 20 to 30 minutes to complete. Students were asked to give the questionnaire serious consideration and, if willing, to complete it without discussion. The monitors answered questions about the survey, told participants how to obtain the results, and provided referral information for those who wanted help in dealing with any victimization experiences. The study's sponsorship by the university, the fact that the students were a "captive audience," the college's intention to use the results for campus crime prevention programs, the relative brevity of the questionnaire, and the anonymity of the responses were all likely to have contributed to the high response rate.

Choices of Method

Self-Administered Questionnaires

self-administered questionnaire, a questionnaire that the respondent completes by him or herself.

There are different methods or modes of data collection based on questioning respondents. Recently a variety of new technologies have resulted in more ways of asking and answering surveys. **Self-administered questionnaires** can be offered in paper-and-pencil versions or in computer-assisted formats (CASI) such as web-based designs, versions that use touch screen responses, or ones that have respondents answer computer-controlled questions by pressing the keypad on a touch-tone telephone. An **interview** can be done by phone or in person, with the interviewer recording the responses for later data entry and transcription, or by using computer-assisted technology to enter data as the interview proceeds. One analysis of 48 studies that compared respondents' answers on various methods of administering questions found that, overall, computer administration did not have an overall affect in how respondents answer questions (Richman, Kiesler, Weisband, and Drasgow, 1999), but other studies have found differences due to method. One study which focused on risk-taking behavior among adolescents found more reports of health-risk behaviors in self-administered computer questionnaires than in either face-to-face interviews or written questionnaires (Rew et al., 2004).

interview, a data collection method in which respondents answer questions asked by an interviewer.

Gray, Palileo, and Johnson used a paper questionnaire that they administered in a group setting by distributing it to all students attending participating classes on a specific day. If you wanted to conduct a survey using a questionnaire and if, like Gray, Palileo, and Johnson, you could locate a group setting for an appropriate sample (such as church-goers, club members, students, and the like) and obtain permission to recruit participants, then you might want to use the **group-administered questionnaire** as they did. Administering a questionnaire to a group of people who have congregated gives the researcher the opportunity to explain the study and answer respondents' questions, provides the researcher with some control over the setting in which the questionnaire is completed, allows the respondents to participate anonymously (which can aid in getting honest answers to "sensitive" questions like those Gray and her associates asked), and usually results in a good response rate.

group-administered questionnaire, questionnaire administered to respondents in a group setting.

Group-administered questionnaires are typically inexpensive. With Gray, Palileo, and Johnson volunteering their time to develop the instrument and analyze the data, the study expenses, covered by university funds, were $200 to print the questionnaire and $1,000 to pay the proctors. The cost of about $1 per completed questionnaire demonstrates the bargain that this method of data collection can be.

On the other hand, group-administered questionnaires have drawbacks, beyond the fact that there might be no group setting for the population the researcher wants to study. The extra pressure to participate that people might feel when in a group setting raises an ethical concern about voluntary participation, and the limit to the length of time that groups will allot to data collection mandates using a relatively short questionnaire.

mailed questionnaire, questionnaire mailed to the respondent's residence or workplace.

If mailing addresses are available, another way to administer questionnaires is to send them to respondents' homes or, less commonly, to their workplaces. The **mailed questionnaire** remains a commonly used method, with advantages to recommend it. Michele Hoffnung, in the panel study of college graduates described in the focal research in Chapter 4, uses a brief mailed questionnaire for some data collections.

Mailed questionnaires are fairly inexpensive and can be reasonably effective. Recent studies like one on women's health (Whiteman et al., 2003) and one on the drinking behavior of undergraduates (Delva et al., 2004) had response rates to mailed questionnaires of 38 percent and 31 percent respectively, but others, like one that questioned a sample of college administrators about university drinking policies, had a response rate of 68 percent (Wechsler et al., 2004). If enough respondents can be encouraged to complete and return the questionnaire, the mailed questionnaire can be useful, as the costs of paper and postage aren't excessive, and filling it out without time deadlines in their homes or workplaces permits respondents to check their records and not feel rushed.

In addition, respondents answering questions in private face fewer social pressures to conform to the expectations of others and don't need to try to please an interviewer. This might explain some of the differences that compare answers to the same question using different modes of data collection. Susan Gano-Phillips and Frank Fincham (1992), for example, found that married couples expressed greater satisfaction with marital quality over the phone than in their written responses; Krysan, Schuman, Scott, and Beatty (1994) found mail respondents were more likely than were phone respondents to indicate that they were having problems in their neighborhoods and less likely to express support for affirmative action (both for women and blacks) or for open housing laws. A study of attitudes in Los Angeles found that phone respondents were more likely than those who completed a mailed questionnaire to voice approval for the police and express fear of crime, but were less likely to perceive social cohesion in their neighborhoods (Hennigan et al., 2002).

individually administered questionnaire, questionnaire that is hand delivered to a respondent and picked up after completion.

Internet questionnaire, questionnaire that is sent by e-mail or posted on a website.

A questionnaire can be **individually administered** or administered as an **Internet questionnaire.** Individual administration means that a research associate delivers and picks up the instrument. Approaching respondents in person is similar to using a mailed questionnaire, although it is more expensive and often has a better response rate.

The rapid rise in the use of computers has created an explosion in the surveys using the Internet. With 55 percent of Americans having Internet access in their households in 2003—up 5 percent since 2001—and another 5 percent having access to the Internet elsewhere (U.S. Census, 2005), the population that can be reached electronically is rapidly growing. Questionnaires can be e-mailed or put on a website for respondents to complete. Sending a questionnaire by e-mail is inexpensive, and although there are software development costs for website questionnaires, researchers have found using the Internet to be cost effective when collecting data on large samples and when samples are geographically dispersed (Schoniau, Fricker, and Elliot, 2001; Fox, Murray, and Warm, 2003). In addition, the Web-based survey can be interactive and can be programmed to vary the questions and keep respondents from skipping questions. (See Box 9.4 for an internet survey.)

Time, money, population, kind of sample, and response rate are issues to consider when selecting a questionnaire mode. With 40 percent of the U.S. population not being online, an issue for Internet surveys is coverage

> **BOX 9.4**
>
> Do you want to participate in an online survey? Check out the website www.survey.net. If you'd like, answer the questions for one of the online questionnaires and, afterward, check out the compiled results for the thousands of respondents who have also filled it out. Before you take the results too seriously, remember that some people might have answered multiple times and that many people didn't have the opportunity to fill it out or thought about it and decided not to respond.

error, the kind of sampling error that arises when the sampling frame is different from the intended population. However, if a convenience sample of self-selected Internet users is fine for your purposes, you can recruit respondents by posting invitations to participate on websites or sending them to Internet discussion groups, chat rooms, e-mail lists, and other venues. On the other hand, if a list of e-mail addresses of a population is available and accessible, such as at a workplace, a school, or from an organization, then a random sample of those with access can be invited to participate.

One concern is that Internet surveys tend to have low response rates. In a comparison to mailed questionnaires, a RAND analysis of surveys done between 1992 and 2001 found that the response rates of the Internet surveys ranged from 7 percent to 44 percent and the rates for the mailed surveys were between 6 percent and 68 percent (Schoniau, Fricker, and Elliot, 2001: 20). In addition, researchers have found relatively high rates of abandonment— that is, when the participants begin but do not complete the study (Crawford, Couper, and Lamias, 2001). However, studies indicate that response rates to Internet surveys can be comparable to mailed questionnaires when they are preceded by advance notification (Kaplowitz, Hadlock, and Levine, 2004), when phone follow-ups are used to encourage participation, and when mail and Internet modes are used together (Schoniau, Fricker, and Elliot, 2001: 20).

cover letter, the letter accompanying a questionnaire that explains the research and invites participation.

Whether distributed in person, by mail or electronically, self-administered questionnaires require reading and language skills, and are less likely to be completed and returned than interviews. Without an interviewer, a researcher using a questionnaire depends on an introductory or **cover letter** to persuade respondents to participate. The cover letter provides a context for the research and information about the legitimacy of the project, the researcher, and the sponsoring organization (see Box 9.5). The physical appearance of the questionnaire and the difficulty of completing it can affect the response rate. The format should be neat and easy to follow, with clear instructions, including statements about whether one or more than one answer is appropriate, and sufficient room to answer each question. The question order should be easy to follow, and should begin with interesting and nonthreatening questions, leaving difficult or "sensitive" questions for the middle or end.

BOX 9.5

Desirable Qualities of a Cover Letter

A good cover letter should

- Look professional and be brief
- State the purpose of the study
- Tell how the respondent was selected
- Describe the benefits of participation
- Appeal to the potential participant to respond
- Explain that participation is voluntary
- Describe the confidentiality or anonymity of responses
- Include information about a contact person in case of questions or concerns
- Explain how to return the questionnaire

Multiple contacts, both before and after the questionnaire has been distributed, and for mailed questionnaires, including a return envelope with first class postage, can improve response rates. Response rates tend to increase by 20 to 25 percent after the first follow-up with smaller returns after additional efforts (Dillman, Sinclair, and Clark, 1993). When Hoffnung uses annual mailed questionnaires in her longitudinal study described in Chapter 4, she finds that about 50 percent are returned within a month, another 35 percent come in after follow-up letters, e-mails, or phone calls, and a few additional responses are returned months later.

Unless everyone in the sample will be getting a follow-up letter, the researcher needs to know who has not returned their questionnaires. This can be accomplished by using some sort of identification on the instrument. This means, however, that replies will not be anonymous. Although respondents can be assured of confidentiality, some respondents might be less willing to participate when the survey is not anonymous.

In-Person and Phone Interviews

An alternate way of getting answers to questions is by having an interviewer ask them. In a structured interview, an interviewer reads a standardized set of questions and the response options for closed-ended questions. The interview has some similarities to a conversation, except that the interviewer controls the topic, asks the questions, and does not share opinions or experiences. In addition, the respondent is often given relatively few options for responses. The set of instructions to the interviewer, the list of questions, and the answer categories make up the **interview schedule,** but it is basically a questionnaire that is read to the respondent.

interview schedule, the list of questions read to a respondent in a structured or semi-structured interview.

Interviewers are supposed to be neutral and use their interpersonal skills to encourage the expression of attitudes and information but not to help *construct* them (Gubrium and Holstein, 2001: 14). The goal is to make the questioning of all respondents as similar as possible. The assumption is that differences in answers are the result of real differences among respondents rather than because of differences in the instrument (Denzin, 1989: 104).

The use of a structured interview allows for some flexibility in administration, clarification of questions, and the use of follow-up questions. Interviews can be useful for respondents with limited literacy and education. Although rates have dropped for all surveys in recent years, interviews, especially in-person interviews, tend to have good response rates.

On the other hand, interviews are more expensive than questionnaires because interviewers need to be hired and trained. In addition, using an interviewer adds another factor to the data collection process—the **interviewer effect,** or the changes in respondents' behaviors or answers that result from some aspect of the interview situation. As we noted previously, some studies have found that answers to the same question will vary depending on how the question is administered. The interviewer effect can be the result of the interviewers' personal and social characteristics, and of the specific way each interviewer presents questions. Moreover, the same interviewer might present questions differently on different days or in different settings.

The majority of paid interviewers in many large surveys are women, and interviewers and respondents frequently differ from each other on important background traits including age, ethnicity, gender, and/or race. Characteristics such as race and gender have been found in some studies to affect responses for at least some topics (Davis, 1997; Finkel, Gutterbok, and Borg, 1991; Kane and Macaulay, 1993). For example, in a study of attitudes on race and race-related issues, using both African American and white respondents and interviewers, the results varied by question topic. The researchers found no race-of-interviewer effects for some questions, but for other questions, African Americans provided less liberal racial attitudes to white interviewers, while white interviewers seemed to reduce the negative attitudes of white respondents (Krysan and Couper, 2003).

The **in-person interview** was the dominant mode of survey data collection in this country from 1940 to 1970 (Groves, 1989: 501). Depending on the sampling procedure—as, for example, when the researcher has a list of home addresses—and when there is a need for privacy in conducting the interview, interviews can be conducted in respondents' homes. Interviews can also be conducted in public places, such as an employee cafeteria or an office. The in-person interview is a good choice for questions involving complex reports of behavior, for groups difficult to reach by phone, for respondents who need to see materials or to consult records as part of the data collection (Bradburn and Sudman, 1988: 102), when the interview is long, and when high response rates are essential. The General Social Survey (GSS), for example, collects data by means of an in-person interview of approximately 3,000 people every other year. The GSS response rate in 1975 was 76 percent and has stayed fairly constant, although it did drop to 70 percent during data collections from 2000 to 2004 (NORC, 2006).

interviewer effect, the change in a respondent's behavior or answers that is the result of being interviewed by a specific interviewer.

in-person interview, an interview conducted face-to-face.

rapport, a sense of interpersonal harmony, connection, or compatibility between an interviewer and a respondent.

Another benefit of the in-person interview is **rapport,** or sense of interpersonal harmony, connection, or compatibility between the interviewer and respondent. Increased rapport might be why people are more willing to provide information on such topics as income in face-to-face interviews than on the phone (Weisberg, Krosnick, and Bowen, 1989: 100). Furthermore, the in-person interviewer can obtain visual information about respondents and their surroundings and a sense of the honesty of the answers.

On the other hand, beyond the interviewer effect, there are other disadvantages of in-person interviewing, including the inability of interviewers to make unscheduled contact with respondents in some buildings, such as in apartment buildings with security personnel at the entrances, and the reluctance of interviewers to go into rough neighborhoods (Bradburn and Sudman, 1988: 102). In addition, although it is standard practice to instruct interviewers to do interviews in private, some researchers report the presence of a third party during interviews, most typically a spouse, which has been found to affect some responses (Zipp and Toth, 2002).

In-person interviews are the most expensive of the self-report methods because they involve time to locate and contact respondents, and travel time to the interview. Some interview studies now use computer-assisted personal interviewing methods (CAPI) where the interviewer uses a handheld computer to collect, store, and transmit interview data. As this method can be cost effective, it is likely to increase in the future. However, research indicates that the addition of a computer into the interview affects the interaction in the way the interviewer asks questions, the way the interviewer enters responses, and the interviewer-respondent interaction when the computer demands take the interviewer's attention from the respondent (Couper and Hansen, 2001).

Before the 1960s, the proportion of households with telephones in the United States was too small to do interviews by phone for national probability samples. By the 1970s, 90 percent of U.S. households had telephones, and the phone interview was commonplace (Lavrakas, 1987: 10). Today, the **telephone interview** is the dominant mode of survey data collection in the United Sates (Holbrook, Green, and Krosnick, 2003). It became the preferred approach because it can yield substantially the same information as an interview conducted face-to-face at about half the cost (Groves, 1989; Herzog and Rogers, 1988) and, using random-digit dialing, can be conducted without names and addresses. Perhaps the greater feeling of anonymity is why phone interviews generated higher estimates of self-reported harm caused by alcohol use than did in-person interviews (Midanik, Greenfield, and Rogers, 2001). Phone interviews are especially useful for groups that feel they are too busy for in-person interviews, such as physicians, managers, or sales personnel (Sudman, 1967: 65). The interviewers do not have to travel, can do more interviews in the same amount of time, and require fewer supervisors (Bradburn and Sudman, 1988: 99). In addition, newer technology makes it cost effective to do computer-assisted telephone interviews (CATI), in which data are collected, stored, and transmitted during the interview.

telephone interview, an interview conducted over the telephone.

On the other hand, disadvantages of phone interviews include the previously mentioned lower response rates when compared to in-person interviews and the elimination of those without landline phones from the sample. Approximately 6 percent of Americans have cell phones only and another 7 percent have neither household landlines nor cell phones (Belinfante, 2005: 2). Other disadvantages include a limit on the length of the interview and the kinds of questions that can be asked. Phone interviews tend to have a quicker pace and elicit shorter answers than the in-person version (Groves, 1989: 551). It is unlikely for respondents to stay on the phone for more than 20 or 30 minutes, whereas in-person interviews often last 40 or more minutes. Complicated questions are more difficult to explain over the phone than in person. The phone interviewer can't show something to the respondent and may not be able to judge the respondent's comprehension. Phone interviews are more likely to generate higher proportions of "no opinion" answers (Holbrook, Green, and Krosnick, 2003). "Sensitive" questions are answered less frequently and perhaps less accurately on the phone (Aquilino, 1994; Frey, 1983: 71; Johnson, Hougland, and Clayton, 1989). Finally, the increasing use of technology designed to screen calls means a greater inability to contact potential respondents.

Because the request for a phone interview is frequently the first contact with a respondent, and requests for interviews are often mistaken for solicitations or telemarketing, the interviewer's introductory statement is crucial. It's up to interviewers to convince potential respondents that they aren't soliciting contributions or selling anything. The interviewer should make it clear that the research is legitimate and be open about the time the interview will take, the topic, and the sponsor of the study.

In interviews where an appointment is scheduled, the rest of life is "put on hold" to some extent while the interview takes place. But, in most surveys, interviewers call or ring doorbells while respondents are in the midst of daily life. Interruptions, like a baby crying or a neighbor stopping over, are often part of the interview process (Converse and Schuman, 1974).

For phone interviews, the interviewer must often call back to contact the household or find the person specified by sampling decisions. From 1979 to 1996, for example, the interviewers for the Survey of Consumer attitudes had to double the number of call backs to obtain the same response rate (Curtin, Presser and Singer, 2000). Trying to reach potential respondents over a period of weeks rather than over a few days can result in a much higher response rate (Keeter, Miller, Kohut, Groves, and Presser, 2000).

Whether they are volunteers or paid staff, interviewers must be trained so that they understand the general purpose of the study, are familiar with the introductory statement, the questions, and the possible answers. The interviewer's manner should be polite, open, and neutral, neither critical nor overly friendly. In other words, the interviewer should try to establish rapport or a sense of connection with the respondent, although this is more critical for qualitative interviewing, the topic of Chapter 10. In keeping with ethical standards of research, all interviewers should enter into an agreement not to violate the confidentiality of the respondents.

> ### BOX 9.6
>
> ## One Question—Two Formats
>
> **Open-Ended Version:**
> Thinking about your job and your spouse's, whose would you say is considered more important in your family?
>
> **Probe Questions for Follow-Up:**
> Could you tell me a little more about what you mean?
> Could you give me an example?
>
> **Closed-Ended Version:**
> Whose job is considered more important in your family?
> Would you say your spouse's job is considered more important, both of your jobs are considered equally important, or your job is considered more important?

Constructing Questions

Researchers make a series of decisions based on assumptions about how respondents read or hear the questions that are asked. Theories of answering questions assume that the stages are understanding and interpreting the question, retrieving or constructing an answer, and then reporting the answer using the appropriate format (Schaeffer and Presser, 2003: 67).

Types of Questions

open-ended question, a question that allows respondents to answer in their own words.

closed-ended question, a question that includes a list of predetermined answers.

Questions can be asked with or without answer categories, as in the examples in Box 9.6. **Open-ended questions** allow the respondents to answer in their own words. On a questionnaire, one or more blank lines can be used to indicate an approximate length for the answer. In an interview, the interviewer waits for the response and, if appropriate, can ask for an expanded answer or specific details.

    Closed-ended questions use a multiple-choice format and ask the respondent to pick from a list of answer categories. One common practice is to use open-ended questions in a preliminary draft and then develop closed-ended choices from the responses. Answer categories for closed-ended questions must be exhaustive and, if only one answer is wanted, mutually exclusive. Exhaustive answers should have sufficient answer choices so that every respondent can find one that represents or is similar to what he or she would have said had the question been open-ended. (Sometimes "other" is included as one of the answers to make the list exhaustive.) Answer categories should be mutually exclusive so that each respondent can find *only* *one* appropriate answer. For example, Gray, Palileo, and Johnson used the less useful choice of "single" rather than "never married" when asking about marital status on their questionnaire. Some divorced and widowed respondents might define themselves as single in addition to divorced or widowed, and be tempted to check two answers. Another example is the common error made by beginning researchers to list overlapping categories, such as when age groups are used as closed-ended answers to a question about age. If ages are listed as 15 to 20, 20 to 25, and 25 to 30, the category set will not be mutually exclusive.

STOP AND THINK   *Gray and her associates used only closed-ended questions in their study of rape victim blame. They included five answer categories from "strongly agree" to "strongly disagree" as responses to each of the rape myth statements; the questions on risk taking had four possible answers, ranging from "often" to "never." Why do you think they used closed-ended rather than open-ended questions?*

Selecting whether to use open- or closed-ended questions involves several issues. Answer choices can provide a context for the question, and they can make the completion and coding of questionnaires and interviews easier. On the other hand, respondents might not find the response that best fits what they want to say, and possible answer categories (such as "often" or "few") can be interpreted differently by different respondents. In addition, respondents' "presentation of self" might be influenced by using the response alternatives to infer which behavior is "usual," if they assume that the "average" person is represented by the middle category of a response scale (Schwarz and Hippler, 1987: 167).

Although a small percentage of respondents will not answer open-ended questions (Geer, 1988), and other respondents give incomplete or vague answers, leaving the answers open-ended does permit respondents to use their own words and frames of reference. As a result, open-ended questions typically encourage fuller and more complex answers that closed-ended categories miss. One cost-effective alternative is to use mostly closed-ended questions and include only a few open-ended ones to obtain greater detail because open-ended questions are more expensive to record or transcribe. In questionnaires and interviews, all open-ended responses must be categorized before the researcher does statistical analyses. A limited number of answer categories must first be created for each question so that the data can be coded. **Coding,** the process of assigning data to categories, is an expensive and time-consuming task that can result in the loss of some of the data's richness.

coding, the process of assigning data to categories.

In the focal research, Gray, Palileo, and Johnson restricted their instrument to closed-ended questions. They wanted to minimize the time necessary to complete the instrument to encourage faculty to allow the survey to be conducted. As Dave Johnson told us,

> The most precious commodity in this study was classroom time—time the students had paid for, and time the professors were jealous of. We feared that open-ended questions could drag the time out on at least

---

BOX 9.7

Using a Contingency Question

Have you applied to a graduate or professional program?

☐ no

☐ yes → If yes, please list the program(s)

_____

_____

_____

---

screening question, a question that asks for information before asking the question of interest.

contingency question, a question that depends on the answers to previous questions.

some occasions. In hindsight, I think I could have asked at least one or perhaps two questions asking for explication of the date rape incidents (personal communication).

A useful type of question is the **screening question,** a question that asks for information before asking the question of interest. It's important to ask if someone voted in November, for example, before asking for whom they voted. Screening questions can help reduce the problem of "nonattitudes" (when people with no genuine attitudes respond to questions anyway). By asking "Do you have an opinion on this or not?" or "Have you thought much about this issue?" the interviewer can make it socially acceptable for respondents to say they are unfamiliar with a topic; when this type of question is used, some studies have found a sizable number saying they have no opinion (Asher, 1992). A screening question is typically followed by one or more **contingency questions,** which are the questions that are based on (or are contingent upon) the answer to the previous question (see Box 9.7). For example, in an interview, those who answer "yes" to the question, "Do you work for pay outside the home?" can then be asked the contingency question, "How many hours per week are you employed?" On a questionnaire, the respondent can be directed to applicable contingency questions with instructions or arrows. Gray and associates could have used gender as a screening question and followed it with contingency questions on aggression for the men and victimization for the women. Instead, they chose to give the men and women different questionnaires, rather than asking respondents to identify the appropriate questions.

Most self-report methods use direct questions to ask for the information desired, such as "What is your opinion about whether people convicted of murder should be subject to the death penalty?" One way to find out how people spend their time is to ask them to keep a time dairy of daily events or create an event history calendar by listing information reconstructing their daily life. Box 9.8 describes one study that uses this technique. Occasionally researchers use indirect questions, where the link between the

BOX 9.8

- How do you spend your days?
- Are you happy?

Daniel Kahneman and colleagues (2004) wanted to find out if happiness and daily activities were connected. They used the time diary technique called the Day Reconstruction Method, which first asks people to reconstruct how they spent their time and then to describe their feelings about each activity and setting. With data collected from 909 employed women in Texas, the researchers concluded that life enjoyment is strongly influenced by things like quality of sleep and the level of time pressure in the work situation, rather than general circumstances like income and education.

The questionnaire included the following questions:

About what time did you wake up yesterday? _____

And when did you go to sleep? _____

On the next three pages, please describe your day. Think of your day as a continuous series of scenes or episodes in a film. Give each episode a brief name that will help you remember it (for example, "commuting to work", or "at lunch with B") . . . Write down the approximate times at which each episode began and ended. The episodes people identify usually last between 15 minutes and two hours.

Kahneman et al. (2004) also asked respondents to rate each episode on a seven-point scale (from not at all to very much) using each of 12 qualities for each episode (happy, angry/hostile, enjoying myself, frustrated/annoyed, etc.)

What do you think of this method of data collection? Do you think it asks the right things? Do you think most people remember their days and their feelings accurately?

---

vignettes, scenarios about people or situations that the researcher creates to use as part of the data collection method.

information desired and the question is not as obvious. They can do this by asking what others think as an indicator of the person's own opinion, as in the question, "How do you think your co-workers would feel if a woman manager were appointed to your group?" Using **vignettes,** scenarios or stories about people or situations that the researcher creates, is another approach. Holden, Miller, and Harris (1999) used this method in a study of parents' beliefs about discipline. Mothers and fathers of 3-year-olds were asked a series of questions, such as how likely they'd be to spank the children, after reading six vignettes of child misbehavior, including one about a child interrupting an important phone conversation and another concerning a child hitting a peer. Adler and Foster, in the focal research piece in Chapter 8, use vignettes and ask for opinions of "Susan" and others to measure caring attitudes.

How to Ask Questions

To keep the questionnaire or interview as short as possible, only the questions that are necessary for the planned analysis should be asked. It's useful to start by making a list of the variables that are of interest and then do a literature search to see if another study's measurement strategy can be used or adapted. As we mentioned in Chapter 6, you can adapt questions that other studies have used. Check out sources like *Handbook of Research Design and Social Measurement* (Miller and Salkind, 2002) for ways to operationalize variables. A preliminary draft or **pilot test** should be used with a small sample of respondents similar to those who will be selected for the actual study. Giving the interview or questionnaire in this kind of pre-test will help determine if the questions and instructions are clear, if the open-ended questions generate the kinds of information wanted, and if the closed-ended questions have sufficient choices and elicit useful responses.

pilot test, a preliminary draft of a set of questions that is tested before the actual data collection.

STOP AND THINK

An interviewer was thinking of asking respondents, "Do you think that men's and women's attitudes toward marriage have changed in recent years?" and "Do you feel that women police officers are unfeminine?" Try to answer these questions. Do you see anything wrong with the wording?

Guidelines for Question Wording

Interviews and questionnaires are based on language, so it's important to remember that words can have multiple meanings and can be interpreted differently by different respondents. Keeping questions concise and using carefully selected words is important, as the specific choice of words can affect the results. A study of affirmative action, for example, found 35 percent agreeing that where there has been job discrimination against blacks in the past, "preference in hiring and promotion" should be given, while 55 percent said they favored "special efforts" to help minorities get ahead (Verhovek, 1997: 32). Studies with questions about public spending priorities found much more support for "providing assistance to the poor" than "spending for welfare" (Smith, 1989). One study found 45 percent saying they favored physician-assisted suicide when the question was "Do you favor or oppose physician-assisted suicide," while another found 61 percent saying "yes" to the question "When a person has a disease that cannot be cured and is living in severe pain, do you think doctors should be allowed by law to assist the patient to commit suicide if the patient requests it, or not?" (Public Agenda, 2002). In cross-national studies and within countries with large immigrant populations, translation becomes important. The choice of words is particularly significant because there might be no functional equivalents of some words in other languages. In one study, for example, Japanese researchers concluded that no appropriate word or phrase in Japanese approximated the Judeo-Christian-Islamic concept of God (Jowell, 1998).

Our suggestions for question construction include:

- Avoid *loaded words,* words that trigger an emotional response or strong association by their use.

- Avoid *ambiguous* words, words that can be interpreted in more than one way.

- Don't use *double negative* questions, questions that ask people to disagree with a negative statement.

- Don't use *leading questions,* questions that encourage the respondent to answer in a certain way, typically by indicating which is the "right" or "correct" answer.

- Avoid *"threatening"* questions, that is, questions that make respondents afraid or embarrassed to give an honest answer.

- Don't use *double-barreled* or *compound* questions—questions that ask two or more questions in one.

- Ask questions in the language of your respondents, using the idioms and vernacular appropriate to the sample's level of education, vocabulary of the region, and so on.

Now review the poorly worded questions and the suggested revisions in Box 9.9 to see how these general guidelines can be applied to specific question wordings.

STOP AND THINK   *Did you identify what was wrong with the questions in the previous Stop and Think? Did you see that the first question is double-barreled in that it asks two questions about marriage, one about men's attitudes, and one about women's? Did you notice that the word "unfeminine" in the question about women police officers is really a loaded word?*

Response Categories Guidelines

When response categories are used, a large enough range of answers is needed. Some common closed-ended responses are listings of answers from strongly agree to strongly disagree, excellent to poor, very satisfied to very dissatisfied, or a numerical rating scale with only the end numbers (such as 1 and 10) given as descriptors. For questions with more than one possible answer, a list can be used and the respondent can be asked to identify all that apply, by asking, for example, "Which of the following television shows have you watched in the past week? Check all that apply." Arguing against a "mark all that apply" approach, Rasinki, Mingay, and Bradburn (1994) assert that more accurate responses are elicited when the question is broken into separate questions, each followed by "yes" or "no" answers. Another debate in the research literature is the question of whether a "no opinion," "don't know," or "middle category" should be offered. (For example, "Should we spend less money for defense, more money for defense, or continue at the present level?") Some argue that it is better to offer just two polar alternatives, whereas others claim it is more valid to offer a middle choice. Studies have found people to be much more likely to select a middle response if it is offered to them (Asher, 1992; Schuman and Presser, 1981) and that listing the "middle" option last rather than in the middle makes it more likely to be selected (Bishop, 1987).

BOX 9.9

Examples of Poorly Worded Questions and Suggested Revisions

Ambiguous question (adapted from Fowler, 1988):

How many times in the past year have you seen or talked with a doctor about your health?

Revised:

We are going to ask about visits to doctors and getting medical advice from them. How many office visits have you had this year with all professional personnel who have M.D. degrees or those who work directly for an M.D. in the office, such as a nurse or medical assistant?

Double negative question:

Do you disagree with the view that the President has done a poor job in dealing with foreign policy issues this past year?

Revised:

What do you think of the job that the President has done in dealing with foreign policy issues this past year?

Double-barreled question:

What is your opinion of the state's current economic situation and the measures the governor has taken recently?

Revised:

What is your opinion of the economic situation in the state at present?

What do you think of the measures that the governor has taken in the past month?

Leading question:

Like most former presidents, the President faces opposition as well as support for his economic policies. How good a job do you think that the President is doing in leading the country out of the economic crisis created by the national debt?

Revised:

Are you aware of the President's policies to reduce the national debt?

If so, what is your opinion of them?

Threatening question (adapted from Bradburn and Sudman, 1979):

Have you ever smoked marijuana?

Revised:

Different people use different terms for marijuana. What do you think we should call it so you understand us? *(If no response or awkward phrase, use "marijuana" in the following questions. Otherwise use respondent's word(s).)* _____ is commonly used. People smoke _____ in private to relax, with friends at parties, with friends to relax, and in other situations. Have you, yourself, at any time in your life smoked _____?

Question Order and Context

Responses to questions can be affected by the question order as earlier questions provide a context for later ones and people may try to be consistent in their answers to questions on the same topic. For example, one study found that reports of alcohol consumption were higher when questions were included on the Semi-Quantitative Food Frequency Questionnaire than when they were used on a questionnaire exclusively targeting alcohol use (King, 1994). Another study found that answers to questions about people's faith in elections and government were affected when these questions were preceded by questions on government waste and crooked politicians (Bartels, 2000). Some research on question order in surveys of attitudes on privacy issues indicates that when order effects are found they are not especially large and may be relatively local, that is, restricted to a few nearby items that are related conceptually (Tourangeau el al., 2003).

There are no easy answers to the issue of question order because each choice involves trade-offs. Consider a logical order that makes participation easier for respondents. To encourage respondent participation, start with interesting, nonthreatening questions, and save questions about sensitive topics for the middle or near the end. If questions about emotionally difficult topics are included, it's preferable to have some "cool down" questions follow them so as to minimize psychological discomfort.

Gray, Palileo, and Johnson began their questionnaire with nonthreatening questions on age and class standing before going on to those on risk taking and sexual victimization or sexual aggression and going finally to questions on rape victim blame. Sociologist Dave Johnson (personal communication) had this to say about the existing question order:

> We decided to ask nonthreatening questions first to ease the respondents into the questionnaire. We asked the sexual victimization questions fairly soon because it was the most important information we wanted, and we did not want to reveal our "agenda" prior to asking the victimization questions. We felt that the rape myth questions might have done this, so we held those until the end.

Thus, in this questionnaire, the questions on victimization or aggression provided a context for answering questions about blaming victims. Answers to the victim-blaming questions might have been different had they been asked first.

Summary

Self-report methods are widely used in social research. They can be used profitably to collect information about many different kinds of variables, including attitudes, opinions, levels of knowledge, social characteristics, and some kinds of behavior. Clear introductory statements and instructions and the judicious use of carefully worded open- and closed-ended questions are critical if questionnaires and interviews are to provide useful results. As we saw in the focal research article on rape victim blame, the researchers were able to collect a great deal of data with clearly worded questions and answer choices and a sample of respondents who were willing to participate.

Questionnaires and structured interviews are less useful when the topic under study is difficult to talk about or when asking questions about a topic is unlikely to generate honest or accurate responses.

In choosing between the various kinds of questionnaires and interviews, factors like cost, response rate, anonymity, and interviewer effect must be considered. Questionnaires are typically less expensive than interviews. Group-administered questionnaires, like the one used by Gray, Palileo, and Johnson are among the least expensive and typically have good response rates. In addition, questionnaires can usually be completed anonymously and privately, which makes them a very good choice for sensitive topics. If the questionnaire is delivered on the Internet or needs to be mailed back, the researcher will need to think of ways to encourage participation, as these methods tend to have lower response rates.

Interviews, especially in-person ones, usually have good response rates and little missing data. They can be especially helpful with populations that have difficulty reading or writing, and when topics are complex. Phone interviews allow researchers to collect self-reports from national samples in a less costly way than in-person interviews, but the latter allow for better rapport. For both kinds of interviews, interviewer effect should be considered.

EXERCISE 9.1

Open- and Closed-Ended Questions

Check with your instructor about the need to seek approval from the Human Subjects Committee, the Institutional Review Board, or the committee at your college or university concerned with evaluating ethical issues in research. If necessary, obtain approval before completing the rest of the exercise.

1. Make five copies of the list of questions that follows. Find five people who are working for pay and are willing to be interviewed. Introduce yourself and your project, assure potential participants of the confidentiality of their answers, and ask them if they are willing to participate in a brief study on occupations. Write down their answers to these open-ended questions as completely as possible. At the end of each interview, thank the respondent for his or her time.

    a. Are you currently employed? (If yes, continue. If no, end the interview politely.)

b. What is your occupation?

c. On average, how many hours per week do you work?

d. Overall, how satisfied are you with your current job?

e. In general, how does your current job compare to the job you expected when you took this job?

f. [*If the job is different from what was expected*]
   In what ways is the job different from what you expected?

g. If it were up to you, what parts of your job would you change?

h. What parts of your job do you find most satisfying?

2. Before data can be analyzed statistically, it must be coded. Coding involves creating a limited number of categories for each question so that each of the responses can be put into one of the categories. From your interview data, select the answers to three questions from d to h and for each respondent, **code** the data using *no more than five* response categories. We've started a table below that you can use as a model for your coding.

Respondent	Answers to question d coded:	Answers to question __ coded	Answers to question __ coded
1	She is very extremely satisfied with her job		
2	He thinks that his job is okay		
3	She thinks her job is awful and is very dissatisfied		
4			
5			

3. Based on your respondents' answers, your coding schemes, and other possible answers to the questions, rewrite the three interview questions you selected for question 2 and make them closed-ended.

4. Describe your response rate by noting the number of people you asked to participate to get five respondents. Based on your experience, comment on the reasons you think some people agree to participate in a short interview and others decline.

5. Attach the five completed interviews to your exercise.

EXERCISE 9.2

Asking Questions (or Have I Got a Survey for You!)

Check with your instructor about the need to seek approval from the Human Subjects Committee, the Institutional Review Board, or the committee at your college or university concerned with evaluating ethical issues in research. If necessary, obtain approval before completing the rest of the exercise.

1. Select between two and four concepts or variables to study. Do not choose "sensitive" or threatening topics. (Appropriate topics include opinions about candidates running for office, career aspirations, involvement in community service, marriage and family expectations, social characteristics, and so forth.) List the variables you have selected.

2. Construct between one and five questions to measure each variable. (Feel free to use or adapt questions from articles you've read or those listed in sources like *Handbook of Research design and Social Measurement* (Miller and Salkind, 2002). Develop a list of no more than 20 questions.

3. Make six copies of the interview questions, leaving enough space to write down respondents' answers. Contact potential respondents to find six people who are willing to be interviewed. Introduce yourself and your project, assure potential participants of the confidentiality of their answers, and ask them to participate. Do three of the interviews over the phone and three in person.

4. Turn the list of interview questions into a questionnaire, complete with instructions and a cover sheet explaining the research and its purpose. (Introducing yourself as a student and the survey as a part of the requirements of your course would be appropriate.) Give the questionnaire to three new people to complete.

5. Compare the information you obtained using the three methods of data collection. Comment on the validity and reliability of the information you received using each method. Which did you feel most comfortable using? Which was the most time-consuming method? If you were planning an actual study using these questions, which of the three methods would you select? Why?

6. Based on your experience, select one of the methods, and make any changes to the questionnaire or list of questions that you think would make the answers more valid indicators of your variables in an actual study.

7. Attach the six completed interviews and the three completed questionnaires to this exercise as an appendix.

EXERCISE 9.3

Direct and Indirect Questions

Check with your instructor about the need to seek approval from the Human Subjects Committee, the Institutional Review Board, or the committee at your college or university concerned with evaluating ethical issues in research. If necessary, obtain approval before completing the rest of the exercise. In this exercise, you'll be comparing asking direct questions versus asking indirect questions using a vignette.

1. Make five copies of the questionnaire that follows, leaving enough space after each of the first four questions for the respondent to write a few sentences. Get five envelopes to give with the questionnaires.

2. Using a convenience sample, find five people who are willing to answer this questionnaire anonymously. Have them complete the questionnaire while you wait, and ask them to seal it in the envelope before returning it to you.

3. Analyze the information on the completed questionnaires. Specifically, compare the information from the first four direct questions to the information from question 5, an indirect question. Comment on the validity and reliability of the information you received using each way of asking questions. Which way of asking questions do you think is better? Why?

4. Attach the five completed questionnaires to this exercise as an appendix.

Questionnaire on Family Roles

Thank you for agreeing to participate in a study on family roles for a course on research methods. Please don't put your name on this questionnaire so that your responses will be anonymous. Answer the first four questions with a few sentences. For the last set of questions, place an X on the spot on the scale that best reflects your point of view.

When you have completed your responses, please put this questionnaire in the envelope provided, seal it and return it to me.

1. All couples in long-term relationships need to make decisions about family roles. How do you think married couples should divide household and economic responsibilities?

2. If there are children in a two-parent household, do you think that the mother's and father's roles should be different in some ways when it comes to caring for the children?

3. Do you think husbands should be expected to make career sacrifices to accommodate their families? If so, what kinds of sacrifices?

4. Do you think wives should be expected to make career sacrifices to accommodate their families? If so, what kinds of sacrifices?

5. Susan and Steve have been married for five years and both have worked full-time during this time. Susan makes a slightly higher salary than Steve and is expecting a promotion soon. Steve's company has been bought by another company, and he has been told that his job will disappear after the two companies merge. Susan is pregnant with their first child, and both she and Steve agree that it will be better for their child if one parent stays home for the first few years of the child's life. After discussing their situations, Steve and Susan agree that she will continue to work full-time while he will stay home with the baby.

What's your impression of Susan and Steve? Put an X on each of the following scales.

Susan is:

1	2	3	4	5
dedicated to family				not dedicated to family

Susan is:

1	2	3	4	5
a caring person				not a caring person

Susan is:

1	2	3	4	5
ambitious				not ambitious

Steve is:

1	2	3	4	5
dedicated to family				not dedicated to family

Steve is:

1	2	3	4	5
a caring person				not a caring person

Steve is:

1	2	3	4	5
ambitious				not ambitious

EXERCISE 9.4

Using Structured Interview Questions from the General Social Survey

The General Social Survey (GSS), conducted by the National Opinion Research Center (NORC), uses a structured interview schedule. The GSS was administered annually since 1972 and then biennially from 1994 to a national probability sample. Some questions are asked on every survey, other questions have rotated, and some have been used only in a single survey.

The GSS measures a great many variables using answers to questions as indicators. The variables included on past surveys cover a wide range of behaviors, among them drinking behavior, marijuana use, membership in voluntary associations and the practice of religion, and numerous attitudes, such as opinions about abortion, affirmative action, capital punishment, family roles, and confidence in the military.

Browse through the GSS codebook of questions asked between 1972 and 2004—available online at http://sda.berkeley.edu/D3/GSS04/Doc/gs04.htm—and select a few variables of interest.

1. List the names of the variable and the questions on the GSS that are the indicator(s) of those variables.

2. Comment on the specific questions that were used. Do you think they are valid indicators for the variables? Are the questions clear and easy to understand? What do you think of the answer categories?

EXERCISE 9.5

Doing a Web Survey

There are many websites that allow users to create and distribute their own Web-based surveys. Some of them are free of charge. Check out this process by using a website to construct and implement a questionnaire. Afterwards, evaluate the data you obtained.

As of 2006, websites that allowed for a free trial for doing surveys included:

http://www.inquisite.com/default.aspx?campaignid=701300000003MkX

http://www.zapsurvey.com/

www.zoomerang.com/index.zgi

http://www.websurveyor.com/easily-create-surveys.asp#

www.intercom.virginia.edu/cgi-bin/cgiwrap/intercom/SurveySuite/ss_index.pl

www.formsite.com

When you have completed the survey, write a brief report about this experience including the following:

1. What was the topic you were interested in?

2. Print a copy of your questionnaire.

3. Evaluate the quantity and quality of the data you obtained using the Internet survey method.

4. Based on your experience, what you think of Web-based surveys?

© Pictor International/Pictor International, Ltd./PictureQuest

10

Qualitative Interviewing

Introduction

Suppose a researcher called you and asked you to spend an hour or two talking about your college experiences and your plans for the future. Would you agree to be interviewed? Would you feel flattered that someone wanted to hear some of your life story or annoyed that you were being asked to spend the hour or two? Would you feel intimidated by having your views recorded? On the other hand, consider being the interviewer. How would you feel about conducting a series of interviews with your peers?

In this chapter, we'll learn about qualitative interviewing, a social science tool for more than a century—and a method whose uses have been debated for almost as long (Lazarsfeld, 1944; Thurlow, 1935). The less structured interview, which has been called "the art of sociological sociability" (Benney and Hughes, 1956: 137), is an important method of data collection. These more intensive interviews are opportunities to learn a wide variety of things, among them people's backgrounds and experiences, their attitudes, expectations, and perceptions of themselves, their life histories, their sense of the meaning of events, and their views about groups of which they are a part and organizations with which they interact.

STOP AND THINK *We discussed the structured interview in the last chapter. How do you think it differs from the qualitative interview that we'll be discussing in this chapter?*

qualitative interview, a data collection method in which an interviewer adapts and modifies the interview for each interviewee.

structured interview, a data collection method in which an interviewer reads a standardized interview schedule to the respondent and records the respondent's answers.

The **qualitative interview,** also called the in-depth or intensive interview, has much in common with the **structured interview** discussed in the last chapter. Both kinds of interviews rely on self-reports and anticipate that the interviewer will do more of the asking and the respondent will do more of the answering. Both kinds of interviews, "far from being a kind of snapshot or tape-recording . . . in which the interviewer is a neutral agent who simply trips the shutter or triggers the response" (Kuhn, 1962: 194), are opportunities for data to emerge *from the social interaction* between the interviewer and interviewee.

Despite these commonalties, the two kinds of interviews are distinct. In the structured interviews that we discussed in Chapter 9, the interviewer uses a standardized list of questions, heavily weighted toward closed-ended questions. The questions and closed-ended answer choices are delivered to each respondent with as little variation as possible to achieve uniformity of questioning. In addition, some kinds of structured interviews, such as phone interviews using random-digit dialing, can be done anonymously without identifying the respondent.

In contrast, the qualitative interview is designed to allow the study's participants to structure and control much more of the interaction. Some researchers use the interview to obtain information about the interviewees' lives and social worlds; others describe it as an opportunity for participants to tell their stories or construct narratives. In the qualitative interview, as Box 10.1 illustrates, the interviewer asks open-ended questions (either preformulated or constructed during the interview), frequently modifies the order and the wording of the questions, and typically asks respondents to elaborate or clarify their answers.

> **BOX 10.1**
>
> ## What Would You Say?
>
> You're seated across from the interviewer and she continues the interview by asking you, "So, how has your family been of help as you've pursued your undergraduate education?"
>
> Perhaps while you think about what to say, you wonder exactly who you should talk about in your answer. You might think she means for you to talk about your mother or father, who have saved over the years to help pay your tuition. But you wonder if she also means for you to talk about your great-aunt, whom you rarely see but who sometimes sends you little notes of encouragement. Then you consider if she can possibly mean for you to tell about your "cousin," who is not a relative at all, but the family friend who has helped you find a summer job each year. You might even consider mentioning your roommate, the person who gives you emotional support when you most need it. If you're like most participants in an interview, you have your own unique understanding of what is meant by each word and question. Fortunately, the in-depth interview can provide an opportunity to foster the kind of give-and-take where the participants can ask for clarifications about the questions and discuss their meanings.

Before the interview, the interviewer typically knows the topics the interview is intended to cover and the kind of information that is desired, but has the opportunity to tailor each interview to fit the participant and situation. In the in-depth interview, the answers are usually longer; the interviewees can impose their own structure, ask for clarification about questions and answer in their own terms; and interviewers can follow-up by asking additional questions. Because these interviews are almost always scheduled in advance and usually occur face-to-face, the interviewer typically knows the identity of the interviewee and, ethically, must keep confidential all information obtained.

Qualitative interviewing has been used as the sole data collection technique in a wide variety of studies, including Ruth Frankenberg's (1993) analysis of white women's social constructions of race, Lillian Rubin's (1976; 1994) studies of working-class families, research by Ibañez et al. (2003) on conflict and support among Mexican and Mexican-American disaster survivors, and Kaufman and Feldman's (2004) study of the formation of college students' subjective identities. Informal interviewing has also been

observational techniques, methods of collecting data by observing people, most typically in their natural settings.

used in conjunction with **observational techniques** (collecting data by observing people in their natural settings, which we will discuss in detail in Chapter 11), as in, for example, research on resources and the organization of daily routines among low-income families (Roy, Tubbs, and Barton, 2004) and in a study of urban merchant-customer interactions (Lee, 2002).

This chapter's focal research is by Sandra Enos, a sociologist whose research interests included both families and corrections. She combined these interests by asking how the social processes of mothering are worked out when mothers are in prison. She first collected data while working as a volunteer in the prison parenting program, observing and talking informally to the participants. In the excerpt that follows, she draws on her qualitative interviews for her analyses of "mothering" while in prison.

FOCAL **RESEARCH**

Managing Motherhood in Prison

by Sandra Enos[1]

Introduction

The United States has experienced an incarceration boom. By mid-year 2000, the population behind bars had grown from 732,000 in 1985 to just under 2 million prisoners (Beck and Karberg, 2001). The U.S. has one of the highest incarceration rates in the world, which means that proportionately we send more individuals to prison than does almost any other country in the world (Mauer, 1997). Although males are more likely to be imprisoned, a growing number of women are finding themselves serving terms behind bars. While the "imprisonment boom" increased the male population behind bars by 165 percent, the number of sentenced women and those awaiting trial increased by over 300 percent during the same time (Beck and Karberg, 2001).

The children of these inmates are also affected by imprisonment. It is estimated that there were 1.5 million children with parents in prison in 1999 (Mumola, 2000). Research has shown that children of inmates are likely to follow their parents' criminal careers; their grades are likely to suffer in school; they display behavioral problems; and they become early users of drugs and alcohol (Johnston, 1995). The impact of having either parent in prison is likely to be detrimental, but because mothers are more often the primary caretakers of children, the incarceration of a female inmate is especially burdensome for children and for families. Most women in prison are mothers, with approximately 65 to 80 percent of them having children under the age of 18 (Baunach, 1985). Because arrest, pretrial holding, and court processing are usually unplanned, the placement of children may be haphazard; imprisonment means that existing family and extended family units must reorganize to adjust to life without the offender. Nationally, 90 percent of children who have fathers in prison live with their mothers, while only 25 percent of those with mothers in prison reside with their fathers (Snell, 1994). This means that the children of inmate mothers are

---

[1] This article has been revised by Sandra Enos from her paper of the same name that appeared in the first edition of this book and is published with permission.

more likely to experience changes in their living arrangements when their mothers come to prison and may, in fact, be significantly more affected by the incarceration of their mothers than by that of their fathers.

Inmates as Mothers

Most people agree that mothering under normal conditions is challenging and often stressful. A variety of myths and ideologies about the "family" and about mothering carry enormous weight in influencing our values, norms, and behavior patterns. These ideas about what it means to be a family and what it means to be a mother have also affected the research that social scientists have done (Baca Zinn and Eitzen, 1993). Traditional sociological work has proposed one normative family form and defined other forms as deviant or at least underdeveloped (O'Barr, Pope, and Wyer, 1990).

Some of the most important myths that underlie contemporary American views of mothering include the following: mothers are totally absorbed in mothering, the mother is the only figure who can give the child what s/he needs, the mother has resources to give the child what s/he needs, and the mother should be able to manipulate the child's behavior toward successful adulthood (Caplan and Hall-McCorquodale, 1985; Chodorow and Contratto, 1982). These assumptions do not make room for mothering styles outside the middle class (Dill, 1988; Maher, 1992) or consider that motherhood is experienced differently by women in different race/ethnic categories or life cycle positions. There is only a small body of work (Collins, 1990; Rubin, 1994; Stack, 1974; Young and Wilmott, 1957) that has investigated how mothers under strain and economic stress "do mothering." There is little room for models of mothering where caretakers share mothering responsibilities, such as "other mothers" identified by Collins (1990). Few have studied mothering in a way that makes room for interpretations of the mothers themselves. It is important to analyze mothering, including race and class differences, in a way that goes beyond overly idealistic images or negative categorizations of mothers as evil or neglectful caretakers. In other words, it is important to "unpack" motherhood if we wish to learn what is really involved in this work.

Because images of mothers are idealized, we are likely to look harshly upon women who violate our expectations. Women's involvement in crime and drug use is viewed through a gendered lens. Women are strongly condemned for these behaviors because they suggest ignoring the welfare of their children. These expectations seldom come into play for male offenders. When women who are mothers are sent to prison, this not only brings the case to the attention of criminal justice agencies but also to the attention of child welfare authorities. In some instances, a sentence of longer than one year can be grounds to move to terminate a mother's right to a child. So, the importance of motherhood carries more weight for a female inmate than does fatherhood for a male inmate.

How mothers remain connected to their children, how they maintain a place and position in their families, how they arrange and evaluate child-care options and the impact on their long-term careers as mothers are elements of the social process of "managing motherhood" in prison. They are the research interests which guided this project.

The Sample and Method

A women's prison in the Northeast served as the study site. The facility is small in size, holding a maximum of 250 women awaiting trial or serving sentences. Approximately half of the mothers in the prison were involved with the state's child welfare agency as a result of complaints of child abuse, neglect, or the presence of drugs in infants at birth. In other cases, the women sought help from child welfare to place children when they felt they had no other viable option.

The parenting program in this prison is operated off-site and allows considerable freedom from usual security constraints. The program facility is filled with toys and more closely resembles a school than a correctional site. Child welfare authorities view participation in the program as an indication of inmate mothers' interest in rehabilitation. Participation in the program is a way for inmate mothers to demonstrate their "fitness" as parents. This may help them to regain custody and resume care of their children upon release. For these reasons, participation in the program is a welcome opportunity.

To participate in the parenting program, the women had to be mothers or guardians of minor children, have either minimum or work release security status, and be free from work obligations at the time of the program. The participants in the parenting program were serving sentences similar to those of the general population—ranging from six months to a few years for nonviolent offenses such as embezzlement, fraud, soliciting for prostitution, drug offenses, breaking and entering.

I was granted access to the parenting program by the Warden and the Director of the parenting program in 1993. I assumed two roles—observer and volunteer/program aide. The mothers gave me permission to sit in on the parenting group (an hour or more discussion group designed to help mothers work out issues related to reunification with children) and to observe the children's visits. Attendance at the program varied from week to week, attracting from 7 to 12 women and between 12 and 25 children. There was turnover in the population; approximately 20 women participated during the four months I was an observer.

I attended 16 sessions and held informal conversations with both the women and their children during the parent-child visits. The conversations were typically about the child and the women's path to prison. Although I was able to interview some women fairly completely during this time, the informal conversations with the women and children were typically limited to a few minutes at each session so that my research interests did not detract from the real purposes of the parenting program—to allow contact and enhance communication between parents and their children. During this time, I took extensive field notes, and generated methodological and analytical memos, a method that supports the grounded theory approach (Charmaz, 1983).

I observed two differences during this preliminary work: Both the placement of children and their mothers' paths to prison was different for African Americans and whites. An interest in the patterns and possible connections among them led to the interview stage of data collection. I conducted 25 additional interviews, most with women who had participated in the parenting program. I gathered a purposive sample consisting of women serving short sentences and those with longer terms; those with involvement with child welfare and those without; and with women who were recidivists and those serving their first terms in prison. Because I was interested in the effect race and ethnicity played on the management of motherhood in prison, the sample reflected the racial/ethnic composition of the prison population. Finally, because I was investigating where children were placed, I located mothers whose children were living in a variety of placements: with grandmothers, fathers, in foster care, and with other relatives. The 25 women had given birth to 77 children, 18 of whom had been adopted by relatives or other individuals.

I identified myself as a researcher and told the mothers that the interview was about the challenges female inmates faced in maintaining relationships with their children while they were in prison. I made it clear that unlike other conversations about their children with representatives from child welfare, counseling agencies, drug treatment facilities, school departments, and so on, their conversations with me would not directly affect their status at the prison. All signed informed consent forms and were assured that our conversations were confidential. Most were curious about my study and wanted to share their stories, hoping it might help others or lead to changes in the parenting program.

In the lower security levels, the interviews were held in a hallway or program area. In the higher security level, a correctional official stood outside the room where I was allowed to conduct interviews. Some of the participants allowed me to tape record the interview and all but one of the rest agreed to my taking notes. One woman insisted on neither notes nor tape, which required reconstructing the interview from memory immediately after leaving the site. The interviews ranged from one to two and one-half hours.

The Emotional Content of Interviewing

I was concerned about how forthcoming the women would be. Given the dynamics related to the presentation of self and to the pressures for women to conform to idealized conceptions of mothering, I was concerned about the fruitfulness of certain lines of questioning. For the most part, I directed conversations along paths initially suggested by women themselves. In later interviews, I occasionally paraphrased the responses from previous interviews as a way of giving the women something to respond to. In very few instances did I ask questions that would result in a yes or no answer.

My aims were to strike a balance between protecting each woman's privacy, getting an accurate picture of how she viewed her role and place as a mother, and hearing her story in her own voice. To establish this balance, I asked very broad questions in a semi-structured interview format, such as

"What brought you to prison?" (See the Appendix to this essay for the entire list of questions.) The answers often required additional clarification. Some women had been to prison several times and were confused about whether I was referring to their first incarceration or their most recent term. The question about where their children were living was too vague, as some women had been caring for children who were not their biological children and others had already lost custody of their children to the state or other caretakers before imprisonment. Many issues were brought into the conversation by the women themselves. For instance, several women, more typically the middle-class women or those in prison for the first time, brought up problems with the conditions of confinement, such as the lack of meaningful programming and the arbitrary nature of the rules, although I did not ask about them.

In the interviews, the women were direct and forthcoming. Some interviewees stated very simply that they were not emotionally bonded to certain of their children; others expressed overly optimistic views of the possibility of reestablishing good and easy relations with their children; others were not in touch with their children because current caretakers prevented contact; some discussed in detail strategies for maintaining a drug habit while care taking their children. In other words, while the emotional context of the interviews was rich and the conversation occasionally difficult, the mothers rarely presented an idealized view of themselves as mothers or a view of mothering as nonproblematic.

The women expressed an appreciation for the time spent speaking about their children and their hopes for the future. Another rewarding aspect for the mothers was that I offered to photograph their children, giving them both the photographs and negatives. Since many of the women didn't have recent pictures of themselves or their children, this opportunity was appreciated.

I transcribed the interviews and coded them using a software program called HyperRESEARCH. The social situations that affect inmate mothers' careers as mothers that emerged from the analysis include arranging care and managing caretakers, demonstrating fitness, negotiating ownership of children, constructing and managing motherhood, and balancing crime, drugs, and motherhood. Because of space limitations, my discussion of results will be brief.[2]

Findings

Arranging Care and Managing Caretakers

A recent survey has reported that 35 percent of state and 15 percent of federal women inmates have not lived with their children prior to their incarceration (Mumola, 2000). For those who co-reside with their children, child placement becomes a challenge upon imprisonment. As in previous

---

[2] For the complete findings, see Enos, S. 2001. *Mothering from the Inside: Parenting in a Women's Prison.* Albany: State University of New York Press.

TABLE 10.1   **Race/Ethnicity of the Sample of Female Inmates (N = 25) and Current Residences (N = 29) of Their Children***

	Grandparent	Foster	Husband	Other Relative	Total
White	4	1	5	-	10
African American	7	2	-	5	14
Hispanic	2	-	2	-	4
Native American	-	-	1	-	1
Total	13	3	8	5	29

*Several siblings do not reside in the same residence.

research (Snell, 1994), there were differences in white, African American, and Hispanic child placements (see Table 10.1). While children most frequently reside with grandparents, white and Hispanic mothers are much more likely to leave their children with husbands; African American women are more likely to place their children with their mothers or other relatives. Making a decision about placing a child is not easy. Stacey,[3] a white woman in her twenties, wanted to protect her child but also wished to minimize her involvement with child welfare. Her four-year-old son was living with his father, and she had heard from friends that the father had resumed a drug habit and the selling of narcotics. Stacey knew that contacting the child welfare authorities would prompt an investigation of the father. However, this could also mean having an agency remove the child from the home and place him in foster care. As she noted, "Most of the time, trying to get help means getting into trouble."

Constructing and Managing Motherhood

Maintaining presence as a mother is difficult when a woman is absent from the home and is even more problematic when she has few resources. Bernice, a pregnant African American mother of a five-year-old son who had been in many foster homes, echoed Stacey's concern. She said, "I feel like I'm having babies for the state. Once they [the child welfare agency] get in your life, they're always there. They stick with you forever." This mother remarked that foster care was the worst place for an African American child to be. Her child was placed with a foster family only because her family and friends wouldn't (or couldn't) respond to her need for a place for him to live during her incarceration.

Other women were more positive about putting their children in the care of foster parents. Some white women characterized it as a "life saver." Louise talked about her satisfaction with foster care, and she was careful to

[3] To protect confidentiality, Stacey and all the other names of interviewees are pseudonyms.

make a distinction between being a child's mother and the care taking of the child. Her comments indicate that she was able to separate the biological identity of mother and the social and psychological work of "mothering."

Enos: So, you had a baby just before you came to jail. Where is he now?

Louise: He's with foster parents who really love him a lot. They are just great with him.

Enos: Does he visit you here?

Louise: Sure he does, a little, the foster parents bring him to me.

Enos: And does he know you?

Louise: Well, he knows who I am. But, he doesn't know me know me.

Enos: Can you explain that?

Louise: Well, he knows I'm his mother, but he doesn't know me that way.

The distinction Louise makes may imply special challenges when the mother resumes care taking of the child or when she reflects upon how she may handle the tasks associated with care taking. Pam, a white middle-class woman, noted that her husband had taken up much of the responsibility for their developmentally delayed son. She felt that some activities can be delegated, while others are better for mothers to do. As an example, she described what happened when the father took the son shopping for school shoes. She was dismayed at his choice, saying "I looked at what they picked out and thought to myself, 'My God, I would have never bought such a thing for him. No mother would have.' There is a difference in what a mother would have done."

Unlike other women, Pam had no difficulty maintaining her position in her family. She was still defined as the family's major caretaker. She reported that her husband visited often and that they maintained the close working relationship that parenting a special child required.

Negotiating Ownership

Family members who assume caretaking responsibilities immediately after a child's birth were viewed differently than those who became involved later in the child's life. Vanessa, an African American woman, said, "Just after I had the baby, my mom took him. Then, I got a short bid. But ever since he was born, he was her baby. I never had a chance." Vanessa is suggesting she has a marginal role in the child's life as his mother—either because others have asserted ownership in her absence or because of her lack of effort.

Other women noted the same development, but suggested that this may have turned out for the best. Kate, a white woman, said, "I don't know what was wrong with me then. I just never had any feeling for my first. I just had it and then my mom took over. We used to do drugs together and party all the time and then she went straight and was preaching to me all the time. Then, the baby came and she took over." Kate conveyed a simpler delineation of responsibility by putting her child in her mother's hands with little regret.

Significant differences were also revealed as women described their paths to prison. White women were much more likely to attribute blame to their families of origin for their problems, suggesting that they were pushed out of their homes or subjected to early sexual and physical abuse and neglect. Few white women relied on their families to take care of their children, and most were wary of incurring "unpayable debts" if they utilized family resources to help them with child care.

On the other hand, African American women tended to trace their paths to prison to their attraction to the "fast life," arguing that their families had tried unsuccessfully to divert them from a life of crime. Placing children with family and friends was expected and acceptable. In many cases, the mothers had previously prevailed upon relatives to care for their children when they were engaged in criminal activities. These children were taken care of by a variety of adults before the women's imprisonment, and most African American women expected to continue using these shared child caring arrangements after release.

There were exceptions to these patterns. Not all children of white women were in the care of husbands, the state, or other non-kin, and not all children of African American mothers lived with relatives or other kin. White women with access to deviant life styles had placements similar to the African American inmates. That is, white women with parents or relatives involved in law-breaking behavior, or who lived in areas where entry into this lifestyle was easy, did place children with their own mothers. On the other hand, African American women who traced their imprisonment to deficient families of origin and were estranged from them placed their children in foster care.

Family "expectations of trouble" occur in white and African American families. Histories of conflict with the law and with other institutions create capacities within families to extend resources when needed. Some women suggested that racial and ethnic differences characterized families of white women and those of color. They stated that expectations for girls in white families were more rigid and punitive, with young women expected to "stay out of trouble," and to follow gendered prescriptions for behavior. Most of the white women did report that their families had expected them to follow the rules, and handled deviations by constricting the family's resources to protect the other members of the family and to keep the offender isolated.

Balancing Crime, Drugs, and Motherhood

Relationships with family before and during incarceration, the quality of those relationships, whether they were supportive or undermined the relationship between mothers and their children, the extent of the involvement of women in crime and drugs, all affect the long-term career of inmates as mothers. Research indicates that about one-third of these mothers will not maintain custody of their children (Martin, 1997). Some mothers maintained that their involvement in criminal behavior (such as larceny, selling drugs, or assault) supported the work they did as mothers by supplementing their meager incomes or by protecting children from physical danger. Others stated that they managed to keep their long-term involvement in drugs and crime from their partners and from their small children. There was disagreement among the women, however, on the issue of balancing drug use and motherhood. Some inmate mothers claimed that one could separate what one did as a drug user or an offender and what one did as a mother by confining each activity to a separate part of the day and making certain that the drug use did not grow out of control. Other women challenged this perception, as demonstrated by the comments of Tee and Margaret.

> You think you are doing right, but you are really not. You are doing so many things, you really don't know what you're doing really. You think you're being good, but you're not. You're neglecting them because you're taking a chance by going stealing and risking yourself. You might not come back to them knowing that they love you.

> Can't nobody say they didn't do it [use drugs] in front of their kids, because they're lying. That addiction is always there.

There were a variety of mother-careers among incarcerated mothers. Some were primary caretakers of children before incarceration; some shared care with others; some did not live with children or assume major responsibilities for their care. Upon release, some inmate mothers will resume the major responsibility of caretaker. Others will assume this role for the first time as other caretakers terminate their roles as the child's primary care giver. Still others will lose their rights to children in formal judicial hearings. Finally, some will continue to share child care with others.

Summary and Conclusion

Imprisonment provides an important vantage point for sociologists to examine family responses to women in trouble and how these responses are affected by larger structural forces. The availability of families to assist women in prison is very much influenced by normative expectations. The financial resources and capabilities of families to lend a hand—here evidenced by their taking care of children—appear to be less important than cultural beliefs about what members do for each other in crisis. Race and ethnicity have a powerful impact on the resources women inmates have in arranging care for their children. Few child welfare organizations recognize these differences and few provide support that matches the needs of families from different racial and ethnic groups.

This examination has also focused on the complex nature and meanings of motherhood. Motherhood might mean a biological connection to child, taking care of a child, maintaining overall responsibility for a child's welfare (even if not directly engaged in the care), or sustaining a unique and irreplaceable relationship with a child. We gain insight into the management of motherhood by seeing how women in prison understand motherhood and meet its challenges. While parenting programs may provide opportunities to visit with children, the variety of mother careers supports the need for a range of programs to help inmate mothers with the challenges they'll face upon release from prison.

REFERENCES

Baca Zinn, M. and D. S. Eitzen. 1993. *Diversity in families,* 3rd ed. New York: Harper-Collins College.

Baunach, P. J. 1985. *Mothers in prison.* New Brunswick, NJ: Transaction.

Beck, A. J. and J. Karberg. 2001. *Prison and jail inmates at midyear 2000.* Bureau of Justice Statistics: Washington, D.C.: U.S. Department of Justice.

Caplan, P. J. and I. Hall-McCorquodale. 1985. Mother-blaming in major clinical journals. *American Journal of Orthopsychiatry* 55: 345–353.

Charmaz, J. 1983. The grounded theory method: An explication and interpretation. In *Contemporary field research: A collection of readings,* edited by R.M. Emerson, pp. 109–126. Boston: Little, Brown.

Chodorow, N. and S. Contratto. 1982. The fantasy of the perfect mother. In *Rethinking the family,* edited by B. Thorne and M. Yalom. New York: Longman.

Collins, P. Hill. 1990. *Black feminist thought: Knowledge, consciousness, and the politics of empowerment.* New York: Routledge.

Dill, B. T. 1988. Our mothers' grief: Racial ethnic women and the maintenance of families. In *Race, class and gender: An anthology,* edited by M. L. Anderson and P. H. Collins, pp. 215–237. Newbury Park, CA: Sage.

Johnston, D. 1995. The effects of parental incarceration. In *Children of incarcerated parents,* edited by K. Gabel and D. Johnston, pp. 59–88. New York: Lexington.

Maher, L. 1992. Punishment and welfare: Crack cocaine and the regulation of mothering. In *The criminalization of a woman's body,* edited by C. Feinman, pp. 157–192. New York: Harrington Park.

Martin, M. 1997. Connected mothers: A follow-up study of incarcerated mothers and their children. *Women and Criminal Justice* 8(4): 1–23.

Mauer, M. 1997. *Americans behind bars: U.S. and the international use of incarceration.* The Sentencing Project. Washington D.C.

Mumola, C. J. 2000. *Incarcerated parents and their children.* Bureau of Justice Statistics: Washington, D.C.: U.S. Department of Justice.

O'Barr, J., D. Pope and M. Wyer. 1990. Introduction. In *Ties that bind: Essays on mothering and patriarchy,* edited by J. O'Barr, D. Pope, and M. Wyer, pp. 1–14. Chicago: University of Illinois Press.

Rubin, L. B. 1994. *Families on the fault line.* New York: HarperCollins.

Snell, T. L. 1994. *Women in prison: Survey of state inmates, 1991.* Bureau of Justice Statistics. Washington, D.C.: U.S. Department of Justice.

Stack, C., 1974. *All our kin: Strategies for survival in the Black community.* New York: Harper & Row.

Young, M. and P. Wilmott. 1957. *Family and kinship in East London.* London: Routledge & Kegan Paul.

Appendix: Interview Questions

Information was first collected from the correctional staff and official records and checked in the course of the interview, as appropriate. This information included the woman's race and ethnicity, living arrangement of child(ren), length of sentence, recidivism status, and involvement of child welfare.

Paths to Prison

Can you tell me what brought you to prison?

How did you get involved in drugs/crime?

Did your family and friends provide an entry? Did a boyfriend?

How would you describe your family when you were growing up?

What has been the hardest thing about being in prison?

Children

Where are your children living right now?

Were you living with your children before you came to prison? If not, who was taking care of them?

How old are your children?

Is DCF (child welfare) involved? If so, in what way?

What are your plans after you are released? Will you live with your children? Immediately? Eventually?

Have any of your children been adopted?

People say that crime and getting into trouble with the law might be passed down through generations. Are you concerned about your children getting into trouble? What can be done, if anything, to prevent this?

Caretaker Characteristics

How did you decide where to place your children?

Did you feel you had some good options?

What are the pros and cons of placing your children with your parents?/ foster care?/your husband?/relatives?

Do you think your child's caretaker is doing a good job? A better job than you can just because of your situation?

Are you involved in making decisions about your children, like where they will go to school? If they need to go to the doctor?

What kind of burden do you think it is for other people to take care of your children? Do you think it is hard for them?

What kinds of obligations do you think families have for each other?

What are family members supposed to do if somebody in the family needs help?

Are you comfortable asking your family for help?

Management of Motherhood

Women who are serving a long sentence must have to make lots of arrangements for their children. What are some of these and how is that different from women who are serving a short sentence?

Are there things that only mothers can give and do for their children? If so, what are these?

Are there things that you think your children are missing because someone else is taking care of them?

Do your children understand that you are in prison? What do you think they think about this?

How often do you see your children? How often are you in touch with them?

Do your children understand that you are their mother even though someone else is taking care of them?

Have you heard of instances where children call women other than their mother "Mom?"

What are some things mothers are supposed to do for their children?

How do you think women in prison try to make up for the fact that they are in prison?

What are some of the things that make it hard to be a good mother to your children?

In terms of your family and friends, do you need to prove to them that you are a good mother? How do you do that?

In terms of child welfare, do you need to prove to them that you are a good mother? How do you do that?

Women's Understanding of Other Mothers

There are some racial differences in where children live when their mothers come to prison. African American and Hispanic children seem to be more likely to live with relatives while white kids seem more likely to go into foster care or live with husbands. Can you explain why you think that happens?

Can you tell when women in prison are ready to make a change in their lives and go straight? What are some of the signs?

There has been a lot of talk about making it easier for the state child welfare agency to terminate parental rights if there has been a child death in the family or if the children have been exposed to drugs. What do you think about these new laws?

Do you think it is possible to tell if someone is a good mother by seeing how she acts with her children in prison?

Do you think the courts are easier on women who have children or harder? Why does this happen?

STOP AND THINK *Enos used an in-person semi-structured interview as one of her methods of data collection. Do you think a more structured interview method, such as an individually administered questionnaire, would have been as effective?*

Qualitative versus Structured Interviews

The techniques of the structured interview are quite appropriate when studying homogeneous populations not too different from the investigator (Benney and Hughes, 1956: 137), when the focus is on topics that lend themselves to standardized questions, and when large samples and cost containment are necessary. As we noted in the last chapter, structured interviews can be useful when the researcher wants to collect data on many topics, especially when surveying representative samples.

On the other hand, structure in an interview can limit the researcher's ability to obtain in-depth information on any given issue. Furthermore, using a standardized format implicitly assumes that all respondents understand and interpret questions in the same way. Another concern is that the interaction in such interviews is very asymmetric and hierarchical, as interviewers control the interaction and direct the talk while keeping their own views private (Mishler, 1986). Some critics have gone so far as to argue that the approach used by the structured interview "breaks the living connection with subjects so that it is forever engaged in the dissection of corpses" (Mies, 1991: 66).

For researchers less interested in measuring variables and more interested in understanding how individuals subjectively see the world and make sense of their lives, less structured approaches, like the one used by Enos, are quite useful. In qualitative interviewing, the interviewer can "break the frame of the interview script" and shift her or his perspective to that of the interviewee and adapt the questions for the person being interviewed rather than using standardized wordings.

Encouragement to elaborate can increase the interviewee's personal investment in the interview and decrease the sense of being a machine producing acceptable answers to questions (Suchman and Jordan, 1992: 251). The women with whom Enos talked were able to tell their stories in ways that were meaningful to them and to use their own words rather than selecting from a set of responses. Qualitative interviews allow the researcher an insight into the meanings of their participants' everyday lives by exploring with them their practices, roles, and attitudes. Participants in qualitative interviews can talk about how they felt, believed, or acted, and they can describe *why* by focusing on what influenced their feelings, beliefs, or actions (Massey, Cameron, Ouellette, and Fine, 1998). Although the interviewer and interviewee are usually strangers, the qualitative interview can provide vibrant data. Such interviews can facilitate research with greater depth and breadth and allow us to understand expectations, experiences, and worldviews of interviewees without imposing the external structure of standardized questions.

Perhaps because of its engaging nature, the qualitative interview typically has a high response rate. Most of those asked agree to participate and complete an interview. For example, in a study of heroin and methadone users, 90 percent of those approached in two methadone clinics agreed to complete a one- to two-hour interview (Friedman and Alicea, 1995: 435).

Variations in Qualitative Interviews

Number and Length of Interviews

Qualitative interviews can vary in several ways: the *number of times* each member of the sample is interviewed, the *length of each interview,* the *degree of structure* in the interview, and the *number of interviewees and interviewers* participating in the session. Interviews can be used in longitudinal research, with multiple interviews as in Hoffnung's long-term study of college graduates, the focal research in Chapter 4, and in the 10-year study of the transition to adulthood by Thomson and her colleagues described in Chapter 7. However, in studies that do not follow participants over time, each member of the sample is typically interviewed only once. Despite the additional cost, however, some researchers suggest dividing the material to be covered into sequential interviews. Irving Seidman (1998: 11–12), for example, suggests a three-session series, with the first focused on the past and background to provide a context for the participant's experience, the second to cover the concrete details of the participant's present experience in the topic area, and the third directed to the participant's reflections on the meaning of the experience. Other researchers advocate multiple interviews because they feel that the additional contact will give interviewees more confidence in the procedure and increase their willingness to report fully. In one study, Robert Weiss (1994) found that only in a fourth interview did one man talk about his wife's alcoholism. In another study, women who were single parents were interviewed every two weeks for about five months but typically did not talk about the emotional ups and downs in relationships with boyfriends until the fifth or sixth interview (Weiss, 1994: 57).

In any given study, there is typically variation, sometimes quite a lot, in how long each interview takes. The interviews Enos conducted lasted between 1 and 2½ hours. In Pauline Bart and Patricia O'Brien's (1985) study of rape victims and avoiders, women were encouraged to talk as long as they wanted to and the interviews lasted from 1½ to 6 hours.

STOP AND THINK

Sometimes an interviewer doing a structured interview feels foolish reading the questions almost rigidly, not being able to modify, add, or delete questions. How do you think you'd feel in the opposite situation—without a "script," and needing to spontaneously construct questions as the interview proceeded?

semi-structured interview, interview with an interview guide containing primarily open-ended questions that can be modified for each interview.

Degree of Structure

Qualitative interviews can vary from unstructured to semi-structured interactions. **Semi-structured interviews** are designed ahead of time but are

interview guide, the list of topics to cover and the order in which to cover them that can be used to guide less structured interviews.

modified as appropriate for each participant. They begin either with a list of interview questions that can be modified, or with an **interview guide,** which is a list of topics to cover in a suggested order. Although some semi-structured interviews use at least a few closed-ended questions, more typically there is a core set of open-ended questions, supplemented liberally with probes and questions tailored to the specific person. Such an approach will generate some quantifiable data and some material that allows for in-depth analysis. Constructing questions ahead of time makes the interviewer's job easier because there is a "script" to ensure coverage of all the topics in each interview. The interviewer must judge whether the questions are appropriate or not, re-order and re-word them if necessary, and use follow-up questions and encouragement to help the interviewee answer fully.

In her study of college graduates, described in Chapter 4, Hoffnung chose the semi-structured interview for two of her data collections to make the interview more like a conversation, to be able to identify new issues as they came up and, given her heterogeneous sample, to have the ability to re-word questions for each individual respondent. Similarly, Rachel Thomson and her colleagues in the study on the transition to adulthood featured in Chapter 7 decided that this method was best suited to their goal of obtaining "narratives of self" from the participants. In her work, Sandra Enos chose the semi-structured interview because using pre-formulated questions helped her organize her thoughts and paths of investigation, but left things open enough to pursue "surprises"—unexpected but interesting avenues of investigation (personal communication).

The semi-structured approach is most useful if you know in advance the kinds of questions to ask, feel fairly sure that you and the interviewees "speak the same language," and plan an analysis that requires the same information from each participant. The questions for a semi-structured interview can be made available to other researchers, allowing them the opportunity to evaluate the questions and to replicate the interview in other settings.

unstructured interview, a data collection method in which the interviewer starts with only a general sense of the topics to be discussed and creates questions as the interaction proceeds.

Another approach is an **unstructured interview.** Researchers doing unstructured interviews start with a sense of what information is needed and formulate questions as the interview unfolds. Marjorie DeVault, for example, interviewing those working in the area of dietetics and nutritional counseling, did not plan specific questions, but typically began with the statement, "I usually start by asking if you'll tell me the story of your career" (1999: 144). She used the word "story" to be informal, to keep from imposing structure, and to signal her openness to hearing complex narratives. In their study of disaster survivors, Ibañez et al. (2003: 7), began each interview with the broad and non-threatening question, "It [the disaster] must have been awful, how did the disaster affect you and others who lived here?" and followed up with "example" and "experience" questions, such as "Can you give me an example of how others helped you?" and "Tell me about your experiences in the shelter." Although some who use unstructured interviews develop an interview guide with a list of topics before interviewing, the list may not be used. In all unstructured interviews, interviewee "digression" tends to be valued as much as core information. Flexibility in questioning can provide insight into the participant's viewpoint and the meaning behind statements.

The unstructured interview lends itself to situations where the researcher wants people to tell about their lives as a whole from their own perspectives, as in "life story interviews" (Atkinson, 2001), when the researcher wants to understand the social context of the participant and when the researcher is developing hypotheses during data collection. As understandings or hypotheses are formulated, additional questions can be added to the interview. In their study of women attempting to move into the labor market, Acker, Barry, and Esseveld conducted unstructured interviews, trying not to anticipate what would be important before the interviews but rather to let the concepts, explanations, and interpretations of the participants become the data. Only as the interview process proceeded, did they bring up questions if they did not emerge in the interviews (1991: 138).

Shulamit Reinharz (1992) labels the unstructured approach an "interviewee-guided" interview because the focus is more on understanding the interviewee than on getting specific questions answered. She notes that this kind of interview "requires great attention on the part of the interviewer and a kind of trust that the interviewee will lead the interviewer in fruitful directions" (Reinharz, 1992: 24). This method is most useful when the researcher does not know in advance which questions are important to ask, when interviewees are expected to find different meanings in questions or possess different vocabularies (Berg, 1989: 16), or when the researcher wants to work inductively (as discussed in Chapter 2). The less structured interview can also be used *first* to develop themes and topics which are then covered in more structured ways, as Garza and Landeck (2004) did in their study exploring risk factors for dropping out of college, or can be used *after* a survey to obtain more complex narratives as Fine et al. (2003) did in their study of urban youth and their experiences with and attitudes toward adult surveillance.

Because unstructured interviewing is very interactive, some see it as a data production technique in addition to a data collection technique. In this view of interviewing, interviewees are seen as engaging in a "meaning-making" and "reality-constructing" process as they tell their stories in collaboration with interviewers (Jarvinen, 2000), with the result being a negotiated text (Fontana and Frey, 2000).

STOP AND THINK   *Although being able to modify each interview to fit each specific situation has advantages, what do you think are some disadvantages?*

Conducting and transcribing interviews is a time- and labor-intensive task. As a result, qualitative interviewing is an expensive method that can sometimes mean a small sample with limited generalizability. As we saw in the focal research for this chapter, Enos conducted semi-structured interviews with 25 women, but other researchers have been able to use larger samples. Lillian Rubin (1994), for example, conducted 388 separate in-depth interviews in her study of 162 working-class and lower-middle-class families. Because of the benefits and the costs, some argue that the qualitative interview is most appropriate for exploratory research, for research that seeks an understanding of interviewees' worlds and meanings, and when the researcher is interested in generating "grounded theory"—that is, theory developed in intimate relationship with data, without preconceived notions, and with researchers aware of themselves as instruments for developing theory (Strauss, 1987: 6).

STOP AND THINK *So far, all the interviewing we've discussed has been one-to-one. Are there any situations you can think of when it might be better to have more than one interviewer or more than one interviewee?*

Joint Interviewers

The use of more than one interviewer is fairly uncommon. When used, it works best when the person being interviewed is not easily intimidated, and the perspective of two or more interviewers is important. For example, in the study of the women who served in the Rhode Island legislature (Adler and Lemons, 1990), one of us, Emily, interviewed several of the more than 50 women jointly with her co-researcher, historian J. Stanley Lemons. This enabled the researchers to develop similar interviewing styles and to elicit information that was of interest both sociologically and historically. After the joint interviews, we moved to one-to-one interviewing to be able to complete the interviews (each lasting two to three hours) in a timely fashion.

Group and Focus Group Interviews

group interview, a data collection method with one interviewer and two or more interviewees.

It is more usual to have more than one interviewee as in the **group interview,** where one interviewer or moderator directs the inquiry or interaction in an interview with at least two respondents. The individuals in the group are selected because they have something in common. They might know each other (such as a married couple or members of the same church) or be strangers (such as teachers from schools in different towns or patients in a given hospital). Such interviews can be based on a predetermined set of questions or can use an unstructured format.

Group interviews are useful when the group is the unit of analysis. Racher, Kaufert, and Havens (2000: 367) interviewed frail, rural, elderly couples jointly and, to maximize their understanding of the couple as a unit, used the interview itself as an opportunity to observe the couple's verbal and nonverbal interaction while focusing on the content of their answers. When the unit of interest is not the group itself, the group interview can have the advantage of releasing the inhibitions of individuals who are otherwise reluctant to disclose what for them are private matters. Because some "are more willing than the others to speak of personal experiences and responses, they tend to be among the first to take active part in the discussion. As one ventilates his experiences, this can encourage others to ventilate theirs . . . and establishes a standard for the rest who progressively report more personalized responses" (Merton, Fiske, and Kendall, 1956: 142–143).

In addition, the interaction in the group might remind each individual of details that would otherwise not be recalled (Merton, Fiske, and Kendall, 1956: 146). Researchers studying couples compared the responses of spouses given in separate, individual interviews to the information shared in a joint, couple interview and found that interviewees who provided limited answers in the individual session became more forthcoming and involved in the couple interview (Bennett and McAvity, 1994: 95).

Group sessions are less time consuming than interviewing the same number of participants individually, and have "the advantages of being inexpensive, data rich, flexible, stimulating to respondents, recall aiding, and cumulative and elaborative over and above individual responses" (Fontana and Frey, 1994: 365). Group interviews can be used as a way to obtain shared perceptions of family life or varying perceptions of family history, and the naturally unfolding dialog can be highly relevant (Bennett and McAvity, 1994). In addition to using the joint interview to provide either a shared or disparate view of their experiences, the researcher can watch interaction between participants while the interview is in progress.

STOP AND THINK   *Although interviewing several people at once clearly has advantages, there are drawbacks as well. Can you think of some of them?*

One concern about group interviews is the possible suppression of negative attitudes. William Aquilino (1993: 372) found that even though the absolute magnitude of effects was small, the presence of a spouse influenced subjective assessment of marriage in the direction of more positive assessment. Some argue that it is better to interview husbands and wives separately to enable them to talk more freely about their feelings and views of each other (Adler, 1981; Rubin, 1983). In addition, even among those who are strangers, participants who become group leaders can structure the situation for others, monopolize the discussion, or inhibit the comments of others (Merton, Fiske, and Kendall, 1956: 149).

focus group interview, a type of group interview where participants converse with each other and have minimal interaction with a moderator.

A special kind of group interview is the **focus group interview,** a research tool that uses group interaction on a topic to obtain data. Instead of asking questions of each participant in turn, the focus group has a moderator or facilitator who encourages the group's participants to talk to and ask questions of each other. Focus groups were used in social sciences in the 1940s and 1950s (Merton, Fiske, and Kendall, 1956), but they became more of a market research tool recently, when they were rediscovered by social scientists. By the end of the 1990s, more than 200 articles a year using focus groups appeared in academic journals (Morgan, 2001). Focus groups can be used alone or in combination with other methods; they can provide useful qualitative data, or can precede or supplement a questionnaire or structured interview. Examples of focus group research include an exploration of the peer groups of 11- and 12-year-olds (Michell, 1998) and a study of black college women's views of interracial dating (Chito Childs, 2005).

Focus groups are designed to have between 3 and 12 participants, selected because they are homogeneous on the characteristic for which the researcher recruited them, such as people who have been recently widowed or have specific health concerns. The participants usually don't know each

BOX 10.2

Focus Group Questions

Erica Chito Childs (2005) used focus groups as one source of data in her very interesting study of black college students' attitudes toward interracial dating. The participants in the focus groups were women between the ages of 18 to 23 who were active in black/African American student organizations on three college campuses. Chito Childs has very generously provided us with a description of the focus group questions she used. Notice that she began with general questions before asking more specific ones. She used follow-up questions to encourage participants to expand their answers.

> I began by asking about their general views on race relations with the question "How would you characterize race relations between blacks and whites?" I started with very general questions about societal views to encourage the discussion to flow, and then would follow up with more specific questions about their own views/experiences, encouraging them to talk. I next asked "Tell me about interracial dating. Are there certain ideas about these relationships?" I followed this up with questions about their own views. I also asked them about the idea that black women as a group are opposed to interracial dating and if they agreed with that view. After the women talked about their own feelings, some discussed their views of how other black women felt. I followed up with questions, such as "Where do these views come from?" "Explain that a little more"; "Why do you think that is?" "Could you give me some examples of a situation like that?" Since much of the discussion was on black men dating white women, I also asked "What about black women dating white men?" and followed up with questions to have them explain in more depth. Finally, I asked "How do you think whites feel about interracial dating?" I found using informal, conversational questions that did not steer the discussion too much worked best for me (Chito Childs, personal communication).

How would respond to questions like the ones Chito Childs asked? Do you think your answers would be different depending on whether the questions were asked in a focus group or in a one-to-one interview?

other before the group interaction. Projects vary in the number of focus groups that are used, ranging from just a few to dozens of groups.

Focus groups are usually conducted with participants seated around a table with a moderator who starts the interaction with one or more questions, such as those used by Chito Childs (2005), which are presented in Box 10.2. The participants' interaction within the group leads to a greater emphasis on their points of view and on hearing "plural voices" (Madriz, 2000). Some groups are more structured, with the moderator being more involved in controlling the group dynamics, keeping it on topic and focused

---

BOX 10.3

Constructing Boyhood in Focus Groups and Individual Interviews

In a study of boyhood in London, a group of researchers did individual and focus group interviews with 14-year-old boys. The data from the two methods were quite different. In the focus groups, the comments demonstrated hierarchies between the boys, included a great deal of ridicule of girls, and demonstrated the boys' humor and naughtiness. In the individual interviews, the boys were quieter and more serious, and talked about their close relationships with parents, grandparents, pets, children, and girls. Rather than privileging one account over the other, the researchers, Pittman, Frosh, and Phoenix (2005: 560) conclude that "What the very different and contradictory accounts of boys in the different modes of research suggested was that boys were *both* powerful and vulnerable."

---

on the researcher's interests while others are less structured, much more conversational and more self-managed (Morgan, 2001: 147). Some argue that the group interaction highlights issues and concerns that would have been neglected by questionnaires (Powell, Single, and Lloyd, 1996) and works well for participants who would be more likely to decline one-to-one interviews (Madriz, 2000).

While the face-to-face interaction is an important feature of the focus group, some researchers have begun to use "virtual focus groups." Turney and Pocknee (2004), for example, argue that new communications technology provides a secure and anonymous environment for participants and makes it possible to accurately record discursive data in text format. In their study of attitudes toward new technology, including DNA testing, they used a discussion forum, with the moderator posting a series of questions and probes, and gave participants one week to post responses, read other participants' and the moderators' probes, and post further comments.

The use of focus groups is not without problems. Some voices might not be heard. Michell (1998: 36) found, when comparing focus groups to interviews, that the lowest-status teen girls in the study were silent and withdrawn in the focus groups, but willing to reveal feelings and personal information in individual interviews. One-to-one interviews can also be more useful in eliciting answers about specific behaviors and experiences and can have less of a "polarization" effect, which is what happens when attitudes become more extreme after discussion (Morgan, 1996). In a focus group, "the emerging group culture may interfere with individual expression, the group makes it difficult to research sensitive topics, 'group think' is a possible outcome, and the requirements for interviewer skill are greater because of group dynamics" (Fontana and Frey, 1994: 365). Sometimes, however, researchers find it interesting to compare the data from the group and individual interviews. See Box 10.3 for an example.

Two final issues include sample size and ethical concerns. Like other studies that use small and nonrandom samples, focus group studies will have limited generalizability. In all group interviews, including the focus group, it is very important ethically for participants to keep confidential the information provided by others. An agreement to maintain confidentiality can be included on an informed consent form, but it is hard for the researcher to monitor compliance.

Locating Respondents and Presenting the Project

In all qualitative interviews, the researcher must decide on the population and the kind of sample before locating potential interviewees and contacting them. If the researcher has names and addresses or phone numbers, the approach can be a letter or a phone call. Hoffnung, whose panel study of women college graduates was discussed in Chapter 4, began with a list of seniors who had been selected at random from each of five colleges. She called each woman, introduced herself, and asked her to participate in a study of women's lives, starting with a one-hour in-person interview. In her study of working parents and family life, Hochschild (1989) contacted a random sample of employees at a large corporation and then asked that sample for the names of friends and neighbors. In her study of widowhood, Helena Lopata (1980) discovered that finding widows or a sample of them in a geographic area was not easy, even though at the time of her study, there were more than 10 million in the United States. She tried, with limited success, to locate widows using a modified area sample that assigned blocks to interviewers, but did better when she used lists of beneficiaries provided by the Social Security Administration.

STOP AND THINK *What about groups of people for whom there are no lists? For example, what would you do if you wanted to talk to homeless people, noncustodial fathers, or people planning to retire in the next few years? How could you locate samples like these?*

Qualitative researchers are frequently interested in studying groups of people for whom there are no lists—such as mothers in prison, fathers who do not live with their children (Hamer, 2001) and grandfathers raising grandchildren (Bullock, 2005)—so selecting a random sample might not be possible. Researchers can use friendship networks, newspaper ads, notices on bulletin boards, announcements at meeting, posts on websites, blogs, or discussion groups, or e-mails to recruit participants. Often, a snowball sample (discussed in Chapter 5) is useful. Another strategy is to contact participants through **gatekeepers,** that is, people who control access to others. Some gatekeepers are legitimate, such as the parents and guardians of children under 18 and, as in Enos's study, the heads of institutions, community organizations, agencies, or groups whose members you want to contact. However, you don't always need to gain access to adults through an authority. Irving Seidman (1998: 38) approached individual community college faculty

gatekeeper, someone who can get a researcher into a setting or facilitate access to participants.

at a number of schools directly, rather than through administrators of the colleges. Terje Gronning (1997), who wanted to contact union members employed by a corporation without going through management, used a company newspaper to identify employees' names and a local phone directory to locate them.

If you want to interview elites—those in positions of higher status or power—access can be difficult. Being able to interview members of Congress, for example, has become distinctly more difficult in recent decades, and researchers interested in conducting such interviews must be prepared to spend extended periods of time to schedule and complete them (Sinclair and Brady, 1987: 63).

STOP AND THINK   *What kind of respondents might be particularly reluctant to be interviewed? What do you think about offering monetary compensation to interviewees? Do you think payment encourages participation? Are there any drawbacks to offering payment?*

The more "political" or "deviant" one's topic is, the more difficult it is to get access to and participation of potential respondents. For example, when Frankenberg, a researcher interested in the social construction of race, told people that she was "doing research on white women and race," she was greeted with interest in some circles, but suspicion and hostility in others. She realized that her approach was "closing more doors—and mouths—than it was opening" (Frankenberg, 1993: 32). She was more successful when she told white women that she was interested in whether they interacted with people of different racial or cultural groups and whether they saw themselves as belonging to an ethnic or cultural group (Frankenberg, 1993: 35).

Other difficulties include the concern of potential interviewees that the information they provide will be used against them or that the researcher is not who she says she is. Enos presented herself as someone who was "outside the system." The women who Enos approached knew that she was not part of the prison administration or the child welfare system, and they believed her pledge of confidentiality of information, including her promise not to use their real names. Other researchers have faced suspicion. Eleanor Miller, for example, felt that some of the "street women" she approached thought she might be a vice officer or narcotics agent in disguise (1986: 186). Other researchers believed that some participants in their studies confused them with investigative reporters, union organizers, or industrial spies (Harkess and Warren, 1993: 324).

Some researchers provide interviewees with incentives to encourage participation. Ciambrone (Chapter 3) gave participants a small gift certificate to a local store. In their study of how Latino parents transmit their culture to their adolescent children, Umaña-Taylor and Bámaca did the same thing, noting that they felt that monetary compensation was useful for participants, many of whom were making a great effort to participate while working double shifts or more than eight hours a day (2004: 268). Similarly, in longitudinal study of drug users, the interviews were conducted in pubs or cafes so the interviewers could buy food and drinks for their respondents, most of whom were unemployed, and small payments were offered to make participation in the study more "professional" (Harocopos and Dennis, 2003).

Most researchers don't offer financial incentives, feeling that payment doesn't seem to have a big influence on the decision to participate. In most studies, the respondent agrees because he or she finds the interview experience interesting and believes it can make a contribution to social science or society. In her current study of retirement, for example, Emily finds that many of the interviewees comment on their interest in talking about their lives and plans, and how it helps them to clarify their thinking. The focal research by Thomson et al. presented in Chapter 7, discusses the participants' positive views of participating in the interview process, with most finding it enjoyable.

Even those talking about difficult subjects can find the experience valuable. Hamer interviewed fathers who lived apart from their children and found that they were grateful for the opportunity to talk about their children and the chance to reflect on both the good and bad aspects of their parenting (2001: 10). In Enos's study, the women appreciated her offer of photographs of their children, but participated because they enjoyed being listened to with interest and respect and because they thought the study could help others in their position.

Planning the Interview

Using Consent Forms

informed consent form,
a statement that describes
the study and the
researcher and formally
requests participation.

Anonymous surveys like those discussed in the previous chapter are often exempted from the requirements of using **informed consent forms.** In some studies using face-to-face interviews, researchers rely on verbal rather than written permission because it is possible in an interview to explain the purpose of a study, offer confidentiality of answers, and assume that the decision to participate implicitly means giving consent. However, as discussed in Chapter 3, institutional review boards or IRBs, the groups explicitly designed to consider ethical issues, usually require interviewers to use a written informed consent form. The researcher provides a document that informs potential participants about the purpose of the study, its sponsor, the benefits to participants and others, the identity and affiliation of the researcher, the nature and likelihood of risks for participants (such as the possibility of raising sensitive issues), and an account of who will have access to the study's records and for what purposes. With such information, potential participants can weigh the benefits and risks in deciding whether to participate. The informed consent form that Ciambrone used in her study of women's experiences with HIV/AIDS is in Chapter 3. The informed consent form used by Enos is in Box 10.4. In both studies, the form was approved by a university's IRB and researchers gave a signed copy of the form to each interviewee. Note that each form offers confidentiality to the extent provided by law, although some question the impact of this limitation on the validity and reliability of the data (Palys and Lowman, 2001).

BOX 10.4

Informed Consent Form Used by Sandra Enos

I have been asked to take part in a research project by Sandra Enos. The researcher will explain the project to me in detail. I should feel free to ask questions. If I have additional questions later, Sandra Enos, the person mainly responsible for the study, will come here to discuss them with me.

Description of the project: I have been asked to take part in a project that examines how women in prison manage motherhood while being incarcerated. The aim of the project is to learn about the challenges and obstacles that face women who are attempting to maintain relationships with their children while serving time in prison.

What will be done: If I decide to take part in the project, I will be asked about my children, their living arrangements, what brought me to prison and how imprisonment is affecting my relationship with children and other family members. I may also be asked about involvement with the child welfare agency. My part in the study will involve an interview which will last about 1–2 hours and which will be tape recorded. My name will not be on the tape and after the interview is typed, the tape recording will be destroyed.

Risks: The possible risks in the study are small. I may feel some discomfort as I talk about the past. There is a chance that some of the interview questions may result in my feeling uncomfortable or anxious. If that is the case, the interview may be suspended at that point if I wish. The researcher will ask several times during the interview if I want to stop. The decision to participate in the study is up to me. I may terminate the interview at any time. Whatever I decide will not be held against me. I understand that the researcher is not affiliated with the Department of Corrections and that my participation in the interview will not have an impact on my treatment, criminal processing or, any other matter.

Reportable child abuse: If, while talking with Sandra Enos, I tell her about some abuse or neglect of a child that I say has not been reported to DCYF, the researcher will inform me that (1) she is required by law to report the abuse/neglect to DCYF and (2) that we must terminate the interview. The tape recording of our interview will be immediately destroyed. I understand that the purpose of the study is not to track reportable incidents of abuse or neglect.

Benefits: There are no guarantees that my being in this research will provide any direct benefit to me. I understand that taking part in this research will have no effect on my parole, classification status and/or inmate record. My taking part will provide important information for people who are trying to understand the impact of prison on women and their families.

Confidentiality: My participation in the study is confidential to the extent permitted by law. None of the information collected will identify me by name. All information provided by me will be confidential. Within two

(continued)

> **BOX 10.4** (*continued*)
>
> weeks after the interview, the tape will be transcribed, and the recording will be destroyed. No information that is traceable to me will be on the transcript. Transcripts will be maintained in a locked cabinet in a secure location available only to the researcher. No information collected by the research that identifies me will be given to the Department of Corrections.
>
> **Decision to quit:** The decision whether or not to take part is up to me. I do not have to be in the study. If I decide to take part in the study, I can quit at any time. Whatever I decide is OK. If I want to quit, I simply tell the researcher.
>
> **Rights and complaints:** If I have any concerns about the research, I may contact Sandra Enos at 456-____ or ask the warden to contact her for me.
>
> I have read the consent form and understand what is stated. Any questions I have about the research have been answered. By signing the form, I am indicating my willingness to participate in the study. The consent form will be kept in a locked cabinet and will not be attached to any transcripts or other materials.
>
> _____   _____
> Researcher signature   Interviewee signature
>
> _____   _____
> Date   Date

Constructing an Interview Guide or Schedule

Once the study's objectives are set, the researcher determines the information that is essential to collect from participants. If the interview is to be unstructured, a list of topics, an introductory statement, and a general sense of the order of discussion is sufficient. The interviewer typically starts with general questions and follows up participants' comments. More information can be sought to provide the context, chronology, or additional meaning of the answer. Usually, it's better to wait until **rapport,** a sense of connection between the participants, is developed before asking questions about sensitive issues.

If you're doing semi-structured interviewing, you'll need to construct a list of questions, both basic and follow-up questions to gather information. Sometimes "filler questions" are needed to provide a transition from one topic to another. In less structured interviews, starting with broad, interesting questions is important. Box 10.5 includes examples of these kinds of questions. The guidelines for question construction provided in Chapter 9— such as staying away from double-barreled, double-negative, or threatening questions and not using wording that is ambiguous or leading—also apply

rapport, a sense of interpersonal harmony, connection, or compatibility between an interviewer and an interviewee.

> **BOX 10.5**
>
> ## Semi-Structured Interviews for Oral Histories
>
> - "What are some of your childhood memories?"
> - "How are holidays traditionally celebrated in your family?"
> - "Can you draw a map of your local community?"
>
> To help those interested in documenting the legacies of family folklore and community traditions, the Smithsonian Institution's oral history interviewing guide is available online. Providing questions like the ones above and a sample interview release form, the guide walks a new interviewer through the process of doing a semi-structured interview with kin, neighbors, or community members. Check it out at http://www.folklife.si.edu/resources/pdf/InterviewingGuide.pdf

to qualitative interviews. Doing practice interviews helps to prepare for the actual experience and allows the interviewer to gain experience covering issues and developing conversation generators. Questions that might elicit more information include "Can you tell me more about that?," "When did that happen?," "Who did you go with?," and "What happened next?" Noticing not only what is mentioned, but what is *not* mentioned, can lead to additional questions.

The choice of specific questions is affected by the intended analysis. If the idea is to describe experience, then the questions should help people tell their stories. In her study of female school superintendents, Susan Chase found that abstract, sociological questions (for example, asking the women if their experiences fit with sociologists' ideas about women in male-dominated professions) encouraged answers that had little to do with how the women lived their lives. On the other hand, questions about specific experiences—such as asking the subjects to describe their work histories—produced lively, lengthy, and engrossing stories that helped construct narratives about the educators (Chase, 1995: 8). Once an interview guide or schedule is constructed, it should be pilot tested with people similar to those who will be interviewed during the actual data collection. Trying out questions can be helpful in deciding which will be most useful for focusing the conversation on desired topics. The interviewees in these preliminary interviews can be asked about the interview experience—what was effective and what was difficult to understand. In a qualitative study, the list of questions can continue to evolve during the course of data collection, especially if the focus changes or becomes sharper.

STOP AND THINK   *As a college student, you use words and language in certain ways. Are there any groups of Americans that you think might use language in ways that are so different from your usage that you might initially have difficulty communicating with them?*

Speaking the Same Language

Obviously it is essential for the interviewer and interviewee to literally speak the same language, but for participants who speak more than one language, language of preference should be considered. It is necessary to be familiar with the cultural milieu of study participants. Given cultural and subcultural differences, it's important to use conceptually equivalent language rather than dictionary translations (Deutscher, 1978: 201). Umaña-Taylor and Bámaca (2004: 265), in their research with Latino parents, for example, note their participants sometimes switched between English and Spanish, perhaps because some words had no direct translation and others, like "abuelita," had more sentimental value than the English equivalent, "grandmother." They also point out that while Latinos share Spanish as a common language, there are expressions and words whose meanings are unique to specific nationalities. In studying behavior change among injection drug users, Booth and his associates (1993) found that the right word could be critical for getting accurate information. For example, drug users who had purchased syringes with others and then shared them typically would respond negatively to questions about how often they *lent or shared* someone else's syringe (perhaps because of the implied degree of intimacy), but would answer in the affirmative if asked if they'd injected with "works" that had been *used* by someone else (Booth, Koester, Reichardt, and Brewster, 1993: 179).

STOP AND THINK   *One woman reminisced about her life, telling Sandra Enos that "back in the day," things were great. Enos wondered about those words. Could it mean a long time ago? Or when things were different? Finally, she asked another inmate mother about the term and was told that it was when you were young and free or out on the streets—getting high, having fun or whatever. Did you know that? How would you go about finding out what your interviewee meant if you didn't understand the terminology?*

In *Street Woman,* Miller noted that many of the black women she interviewed accommodated her by using "White English" rather than the "Black English" they normally would have used (1986: 185). She commented on having problems with the meanings of particular words and phrases. "Although I had admitted my ignorance of their world, I didn't want to appear too square. As a result, when I couldn't determine the meaning of a word or phrase from the context in which it was used, I felt comfortable asking for one or two definitions. Beyond that, I would save my questions about terminology for another time or another informant" (Miller, 1986: 185).

Enos found that she'd picked up enough "prison lingo" during the observation part of her study that she knew what the interviewees meant when they reported "getting jammed up" (getting in trouble), "catching a bid" (getting sentenced), and the like. If she didn't understand something, she'd usually ask that it be repeated or explained. In the context of her study, it isn't surprising that she sometimes had to ask the women to explain family relationships because some of the children they talked about "mothering" were not their biological children, but those for whom they were informal guardians.

Conducting the Interview

Where and How to Interview

Interviews can be held in offices, in the interviewee's home, or elsewhere. Ching Yoon Louie preferred to conduct her interviews of Chinese, Korean, and Mexican sweatshop workers who had become leaders in the movement to improve work conditions wherever they were working: "camped out at movement offices, sandwiched between pickets and workshops, while driving across the state to demonstrations, leafleting at factory gates, and fighting with government officials" (2001: 8).

Interviews can be conducted with or without others present. In a study of married couples, when Emily was interviewing the wives (while her associate was interviewing the husbands), she found that, on occasion, children would be present for part of the interview—sometimes commenting on the woman's comments! If privacy is needed in an interview, it's important to consider that when scheduling the time and place.

Although the interview might not follow all the conventions of conversation, it *is* a social interaction. The interviewer should be aware of all that the interviewee is communicating: verbal and nonverbal messages and any inconsistencies in them. The interviewer should strive to be nonjudgmental in voice tone, inflection, phrasing of questions, and body language. A goal is to communicate openness to whatever the interviewee wants to share, rather than conveying expectations about his or her answers. Following up answers, the interviewer can ask for details, clarification, and additional information. One useful skill is in knowing when to keep quiet because interested and active listening can help in obtaining complete responses.

Recording the Interview

When an interview covers a large number of topics, a recording—either audio or video—is invaluable; the more unstructured the interview, the more necessary recording becomes. If there is no strict order of questioning, and probing is an important part of the process, the interviewer will not be able to attend adequately to what the interviewee is saying if trying to write everything down (Lofland, 1984: 60). But the impact of recording is hard to assess. On the one hand, it can help rapport by allowing the interviewer to make eye contact, nod, and show interest and help the interviewer concentrate on follow-up questions. On the other hand, being recorded can intimidate interviewees and inhibit frankness.

STOP AND THINK   *After describing his relationship with his father, the person you are interviewing asks you how you get along with your father. What would you do? Would you tell him that what you think really isn't important because he's the one being interviewed, answer that you'll tell about your father after the interview is over, describe your relationship with your father honestly but briefly, or launch into a detailed description of how you get along with your father?*

Being "Real" in the Interview

In more traditional qualitative interviewing, the interviewer maintains social distance from the interviewee. This means using a style that gives evidence of interest and understanding in what is being said (nodding, smiling, murmuring "uh-huh"), but that prohibits judgment, reciprocal sharing, or "real conversation." The interviewer is advised not to share opinions or any personal information with respondents because it can increase the chance of leading subjects to say what they think the interviewer wants to hear and shift attention from the interviewee.

Critics of the traditional method dispute the view that the traditional interviewer response is really neutral. Frankenberg (1993: 31) feels that "evasive or vague responses mark one as something specific by interviewees, be it 'close-mouthed,' 'scientific,' 'rude,' 'mainstream,' 'moderate,' or perhaps 'strange'—and many of those are negative characterizations in some or all of the communities in which I was interviewing." More generally, qualitative interviewers see the specific social context of each interview not as something to be controlled, but rather instead as data and an important part of the meaning making that the interview encourages (Warren, 2001: 91).

Some researchers argue that the interviewer and interviewee should treat each other as full human beings. Fontana and Frey (2000: 658) believe that the emphasis is shifting to allow the development of a closer relationship between the two interview participants as researchers attempt to minimize status differences and show their human side by answering questions and expressing feelings. Although Enos was only asked an occasional personal question about herself or her opinions, quite a few interviewers report interviewees asking numerous questions—about the study, about their personal lives, or for advice on the topic under study (Acker, Barry, and Esseveld, 1991; Oakley, 1981). Each researcher must decide how to respond. Joan Acker and her associates

> always responded as honestly as we could, talking about aspects of our lives that were similar to the things we had been discussing about the experience of the interviewee—our marriages, our children, our jobs, our parents. Often this meant also that our relationship was defined as something which existed beyond the limits of the interview situation. We formed friendships with many of the women in the study. We were offered hospitality and were asked to meet husbands, friends, and children. Sometimes we would provide help to one or another woman in the study. . . . However, we recognized a usually unarticulated tension between friendships and the goal of research. The researcher's goal is always to gather information; thus the danger always exists of manipulating friendships to that end. Given that the power difference between researcher and researched can not be completely eliminated, attempting to create a more equal relationship can paradoxically become exploitation and use. (Acker, Barry, and Esseveld, 1991: 141)

Listening to someone describe an intimate, personal story is a moving and powerful experience (Davison, 2004). Being "real" as an interviewer can mean acknowledging that at least some of what you hear is painful, difficult, or upsetting. The demanding nature of interviewing is not often discussed. Interviewers, even those with training, might not be prepared for or trained to handle the emotional impact of the interview process. In her work on juvenile prostitution, Melrose describes how she often felt distressed by the accounts she had heard: "[A]lone in a hotel in an unfamiliar place, struggling with my own feelings of anger and despair . . . I was forced to contain these feelings in order to carry on with the fieldwork the next day . . . I also imposed the expectation on myself that I should be able to 'manage' these feelings" (2002: 325). Eleanor Miller conducted her interviews of "street women" in a variety of settings, and at times felt afraid, intimidated, and uncomfortable. Hearing the details of the women's lives sometimes made her angry that their childhoods had been so awful, upset that people could be so brutal to one another, and depressed about the lack of realistic options for the women and the probable futures of their children. Sometimes, however, she found the fieldwork exhilarating and, in the long run, Miller (1986: 189) felt the study was worth the effort both personally and professionally.

Interviewing Across the Great Divides

STOP AND THINK

How effective do you think you would be as an interviewer if you were interviewing someone considerably older than you are? How about if the person were of the opposite gender, of a different race or from a different ethnic group? Do you think the topic of the interview would make a difference in interviewing someone whose background was very different from yours?

interviewer effect, the change in a respondent's behavior or answers that is the result of being interviewed by a specific interviewer.

As we discussed in Chapter 9, researchers using interviews need to think about **interviewer effect,** the change in a participant's behavior or the answers given to questions as the result of being interviewed by a specific interviewer. Now we'll add the issue that the researcher's identities, such as class, race, sexual orientation, etc., can affect all aspects of the research process, including data collection. Because each interviewer brings unique qualities and characteristics to the interview situation, an ongoing debate concerns the desirability of matching interviewers and interviewees on social characteristics, such as gender, race, ethnicity, age, and class.

The most common perspective is that it's advisable to match the participants in an interview because people of similar backgrounds are thought to develop better rapport. In addition, interviewers who differ from interviewees on class, ethnicity, gender, or race tend to get different responses than when the participants have the same backgrounds (Kane and Macaulay, 1993; Reisman, 1987). Enos believes that her gender was helpful in the interview process. "A man in that setting could have been problematic, especially since the women would have had some trouble understanding why a man would be interested in mothering. Also a number of the women expressed exasperation with relying on men and others were still looking for Prince Charming" (Enos, personal communication).

BOX 10.6

Reflections on Matching the Gender of the Interviewer and Interviewee

by Donald C. Naylor[4]

In planning a study in which I'd be interviewing men about gender, I knew that the men in my sample might not be completely honest with me. But, I didn't think I needed to be concerned about the interviewer effect or be worried about getting the polite answers that sometimes result when people are interviewed by someone who is perceived as different and possibly unreceptive to their views. In my study, the interviewees and I would be the same gender and would be similar in other ways—age, class, and sexual orientation. Of the many problems associated with interviewing, interviewer effect was *not* one I thought I would have to consider.

Upon reading transcripts of my interviews though, I noticed that what was said seemed to have been affected by something happening between myself and the men I was interviewing. In one instance, I noticed that the man's answer seemed like something he felt he had to say to appear masculine. I wondered if he believed what he'd said or if it was just what one guy thought he was "supposed" to say to another. While the literature on interviewing says how much better it is when the interviewer and interviewee are similar, I wondered if there were also some problems with similarity.

Reflecting on the interview process, I've had some thoughts about what was occurring. My first is that there is an effect even when the interviewer and interviewee are similar. It is a different dynamic than when the two are different, but it is still there. Similarity can affect the content of the interview, the types of questions asked, the answers given, and the way the interviewer and interviewee interact. In addition, it seems harder to observe critically a familiar interaction.

Second, I began to see that there was more than interviewer effect going on. The interviewee also had an active part in constructing the interview. This was especially true since I was doing an unstructured interview. When the men varied in what they said and how they said it

[4] Permission granted by Donald C. Naylor, who wrote this when he was a graduate student in the Department of Sociology, University of Southern California.

It is possible for interviewers and interviewees of different backgrounds to have a good research relationship and for those who are alike to have problems. The researchers in charge of the *Sex in America* study (Michael, Gagnon, Laumann, and Kolata, 1994: 32) asked people of different races in focus groups who they would feel most comfortable talking to. They were surprised to hear that almost everyone—men, African Americans, and Hispanics—preferred middle-aged white women as interviewers. Similarly, in her study of male clients of female prostitutes, Grenz (2005) asked

BOX 10.6 (*continued***)**

(amount of openness, attempts to control, emotions displayed), I would change my style and demeanor. They were affecting me as much as I was affecting them. I began to think not so much in terms of interviewer effect, but about interviewer-interviewee effect as we were jointly constructing the interview.

Third, it began to be apparent to me that the interview process is a gendered one. We were "doing" an interview not just as two people but as two men. We were "doing gender." The dialog being created was typical of how men often converse. Of course, variations resulted since there are a lot of masculinities (and femininities) and these men (and myself) were not identical. But the more I read the transcripts, the more apparent it became that we had been "doing gender." I noted that some men tried to control the situation (as did I) and most avoided topics with much emotional content (as did I some of the time). They all tried to present themselves as competent, and each wanted to avoid a discussion of problems, perhaps seeing these as failures rather than as normal difficulties (did I collude in this?). The ways I responded to these men and the ways they reacted to me seemed to be typically "male" ways of interacting.

As I thought about the basic structure of the kind of interview I used, it became obvious that this was a method that did not fit easily with the style of many men. Having a stranger walk into their living rooms and ask personal questions is not what men typically do. Many men are hesitant about disclosing personal information, especially to strangers in unfamiliar situations.

I found it most useful to try not to press the men for very personal information. I found that when I was accepting, uncritical, and nonthreatening, the men did not clam up or shut down. Oddly, at times, the less I asked, the more they told me. Perhaps unconsciously I was trying to conduct an interview in a way that was "guy like." This is "doing gender." But it also, I believe, allowed me to obtain more information than if I had tried a different style.

Interviewer-interviewee effect and "doing gender" cannot be avoided. The question for me was, given the inevitability of this, what do I consider the effects on the information obtained to be? We can acknowledge and understand these effects, and perhaps even profit from them.

prospective interviewees whether they preferred to be interviewed by a man or a woman and found that none wanted to be interviewed by a man and many indicated a preference to be interviewed by a woman. Perhaps, as Robb (2004: 402) found in his study of fathers, masculinity should be seen as something that is "worked on" in the interview process and that men being interviewed by men will be partly motivated by a desire to prove their masculinity. For an interesting commentary on this topic, see Don Naylor's discussion in Box 10.6.

Hall (2004), a white British woman, described both the positive and negative outcomes of being an outsider while interviewing women of South Asian heritage about their immigration experiences. As an outsider, she was sometimes viewed with suspicion, but also was judged to be a neutral researcher who was safe to talk to because she had few community links. In thinking about the issue, Enos feels that it might have been better to have worked with both African American and Hispanic co-researchers, but she didn't have the option of hiring interviewers. She believes that she was able to communicate effectively across race, class, and ethnic lines.

However, even when matching on key characteristics, there can still be differences. Interviewing Members of Parliament in Britain, Puwar (1997: 9.4) found that because of occupational differences, rapport did not necessarily follow when she interviewed MPs with whom she shared gender, ethnicity, parental background, and regional accent. Practically, it's typically not possible to match interviewer and interviewees on more than one or two key characteristics. Although an interviewer's multiple identities can not be ignored, they can benefit by seeing themselves as socially situated, and by taking seriously the concerns and challenges that subjects raise about these identities during the research process. (McCorkel and Myers, 2003)

Issues of Validity

As with the other self-report methods, there are questions about the validity of the data produced by qualitative interviews. Inaccurate memories, misunderstandings, and miscommunications must be considered in evaluating the information obtained. As early as 1935, researchers debated the "questionable value" of the interview for securing reliable data when people become defensive concerning their private and personal lives (Thurlow, 1935). Another concern is that in a qualitative interview, the interviewer is not a passive listener. The way an interviewer questions, responds, and acts can affect the way a participant responds. It's possible that a different interviewer or even the same interviewer using different wording would be told a different account. Norman Denzin (1970: 188) cautions that the tremendous range and variation in interviews in the same study can make comparability a fiction.

The counterpoint to these concerns is that standardizing the interview might be no better at producing comparability because different respondents can hear the same question in different ways. The ability to reword and restate questions can give the researcher a chance to present them in a way that's more understandable by each participant. In addition, a relatively natural interactional style allows the interviewer to better judge the information obtained. Qualitative researchers can try out what they think they have come to know as a result of interviewing others, so that interviewees can say "That's what I meant" or "No. What I meant was . . ." to validate and self-correct understandings (Gold, 1997).

Enos and many others who have used the qualitative interview believe that, overall, interviewees tell the truth as they understand it and rarely offer false information knowingly. Massey, Cameron, Ouellette, and Fine (1998) note that in their studies using the life story interview, analysis and data collection occur together. In this approach, interviewers make analytic decisions and ask follow-up questions during interviews to achieve greater understanding of the stories being told.

After the Interview's Over

STOP AND THINK

Imagine that you're almost at the end of an interview on college students' relationships with significant others. After describing how the most recent love relationship ended, the student you're interviewing looks up and says, "I'm so depressed, I feel like killing myself." What would you do?

At its end, the interviewer typically has the desired outcome—a completed interview—yet there may be many different responses from the interviewee. The respondent might rush off as soon as possible, "turn the tables" and interrogate the interviewer, bring up his or her own agenda or continue talking about the topic after the official end of the interview (Warren et al., 2003: 98). When covering emotionally difficult topics, researchers might need to prepare for an emotional aftermath or a request for help. Lopata, reporting on her study of widows, said, "over and over, we found the respondents expecting some sort of help as a result of the interview, a solution of problems, and even a complete change in life. They assumed that the interviewer, or at least the university staff, has the power to bring societal resources to them or to change the attitudes or behavior of significant others toward them. It is difficult to be faced with a respondent, caller, or letter writer who is obviously in pain or need whom we are not trained to help" (Lopata, 1980: 78).

At the very least, researchers usually include a series of "cool down" questions at the end of an interview so the interview doesn't end immediately after talking about sensitive subjects. Some researchers prepare something to leave with participants, most typically a list of local organizations that provide services in the area under discussion. Some researchers have offered to locate or provide counseling or therapy sessions for the interviewees after the interviewing process is over (Bart and O'Brien, 1985).

Analyzing Interview Data

STOP AND THINK

Imagine that you've conducted qualitative interviews with a small sample of nurses about their jobs. Let's say you've done 25 interviews, with each one lasting about 30 minutes and resulting in 20 pages of transcription. Now that you have approximately 500 pages of data, how would you go about analyzing the data?

If interviews have been recorded, they are usually transcribed. This is a time-consuming task. Enos found that, on average, it took her more than four hours to transcribe an hour interview (personal communication.)

BOX 10.7

Thinking About "You Know"

Sometimes researchers need to pay attention to the words between the words. Talking about her interviews on housework, Marjorie DeVault (1999: 69) says,

> I became aware that my transcripts were filled with notations of women saying to me, "you know." . . . This seemed like an incidental feature of their speech, but perhaps the phrase was not as empty as it seems. . . . The "you know" seems to mean something like, "OK, this next bit is going to be a little tricky. I can't say it quite right, but help me out for a little; meet me halfway and you'll understand what I mean." . . . If this is so, it provides a new way to think about these data.

Interviewing qualitatively generates a great deal of text and the process of analyzing such data is typically more inductive than deductive. That is, although the researcher might come to the data with some tentative hypotheses or ideas from another context, the most common approach is to read with an open mind, while looking for motifs. Careful attention to what participants say is essential, as Majorie DeVault points out in Box 10.7.

Here, we concentrate on some of the possible approaches to interview data and save the more technical aspects of data reduction and analysis for Chapter 15. One general approach to data analysis is to take the topics that the participants have talked about and to use them as a sorting scheme. For example, in a study of women's perceptions of abuse, Levendosky, Lynch, and Graham-Bermann (2000) asked "How do you think that the violence you have experienced from your partner has affected your parenting of your child?" After reading the responses, they developed 14 answer categories that four coders used with 90 percent interrater reliability. At that point, similar responses were combined to reduce the categories to seven, including "no impact on parenting" and "reducing the amount of emotional energy and time available for children." Each account was read to determine which category it best fit.

After categorizing material by topic, researchers can look for patterns in the accounts. In their study of rape victims, Bart and O'Brien saw a connection between the relationships of intended victims and attackers and the woman's likelihood of avoiding rape. For example, the women were more likely to avoid rape if attacked by a stranger and more likely to be raped if attacked by a man they knew (1985: 29). In her piece in this chapter, Enos discusses the different "expectations of trouble" that tended to distinguish the white and African American families.

A different approach is to construct life histories, profiles, or "types" that typify the patterns or "totalities" represented in the sample. Narratives such as life stories can be read as a whole (Atkinson, 2001) or general phenomena

---

BOX 10.8

Life Story Excerpts

If you want to read some excerpts from life stories, go to the archives at the University of Southern Maine's Center for the Study of Lives posted at http://www.usm.maine.edu/cehd/csl/excerpts.htm

---

can be described by focusing on their embodiment in specific life stories (Chase, 1995: 2). Some caution that a "battle of representation" can arise because the researchers ultimately write in their own voices, but also present long and edited narratives drawn from informants (Fine, Weis, Wessen, and Wong, 2000).

Cases can also be examined one at a time to see if most fit a hypothesis, and then the "deviant cases"—those that don't fit the hypothesis—can be examined more closely (Runcie, 1980: 186). Enos originally interviewed to hear people's stories, but as she heard them, she saw variables emerge. She was then able to see patterns in relationships between variables, and then look for exceptions (personal communication). By suspending judgment and focusing on the conditions under which the patterns did not hold, she was able to look for other explanations and consider additional independent variables.

If a description fits a series of respondents, the investigator can propose a more general statement as a theory. This is similar to what Billson (1991) calls progressive verification, and what Glaser and Strauss (1967) call theoretical saturation. In these approaches, the researcher determines that he or she is getting an accurate picture when successive interviewees repeat similar things.

Summary

The qualitative interview is an important and useful tool for researchers interested in understanding the world as others see it. These less structured interviews can be especially useful for exploratory and descriptive work. They allow the researcher to develop insights into other people's worlds and lend themselves to working inductively toward theoretical understandings.

The qualitative interviewer typically uses either a list of topics or questions. Using mostly open-ended questions, the interview is modified and adapted for each interview, and participants are encouraged to "tell their stories." In such interview settings, rapport may develop between the participants, and the flexibility of the qualitative approach lends itself to good response rates, complex topics, and interviewing for more lengthy sessions. However, because it is an expensive and time-consuming method of data collection, the qualitative interview is most frequently used with relatively small samples.

There are several choices to be made when using qualitative interviews. Among the most important are the amount of structure in the interview, whether they are done one-to-one or in groups, whether to interview or moderate, whether to "match" interviewer and interviewee, and the extent to which the interviewer shares opinions and information with interviewees during the interview.

When the interviewing process is completed, the creative process of looking for patterns and themes in the transcribed accounts begins. Data analysis can be a rewarding task, but is often very time consuming. Qualitative interviews can't provide comparable information about each member of a large sample as easily as more structured methods can, but they can be important sources of insight into specific realities and lives.

EXERCISE 10.1

Doing an Unstructured Interview

1. Pick an occupation about which you know something, but not a great deal (such as police officer, waiter, veterinarian, letter carrier, high school principal, and so on). Find someone who is currently employed in that occupation and is willing to be interviewed about her or his work.

2. Conduct an unstructured interview of about ½ hour in length, finding out how the person trained or prepared for her or his work, the kinds of activities the person does on the job, approximately how much time is spent on each activity, which activities are enjoyed and which are not, what the person's satisfactions and dissatisfactions with the job are, and whether or not he or she would like to continue doing this work for the next 10 years.

3. Transcribe at least 15 minutes of the interview, including your questions and comments, and the interviewee's replies. Include the transcription with your exercise.

4. Consider this a case study and write a brief account of the person and his or her occupation based on your interview. If you can, draw a tentative conclusion or construct a hypothesis that could be tested if you were to do additional interviewing. Include a paragraph describing your reactions to using this method of data collection.

EXERCISE 10.2

Constructing Questions

1. Construct a semi-structured interview guide for a research project on work in an occupation of your choice.

2. Following the ethical guidelines discussed in Chapter 3, write an introductory statement that describes the research project and write at least 12 open-ended interview questions. Include both basic and follow-up questions. Focus on the same kinds of information called for in exercise 10.1, including how the person trained or prepared for her or his work, the kinds of activities the person does on the job, the amount of time spent on each activity, which activities are enjoyed and which are not, what the person's satisfactions and dissatisfactions with the job are, and so on. Do a pilot test of your questions and then make modifications as appropriate. Turn in the introductory statement and the final version of the interview questions.

EXERCISE 10.3

Evaluating Methods

In her research with incarcerated mothers, Sandra Enos used an in-person semi-structured interview. Two other self-report methods are the in-person structured interview and the mailed questionnaire (both discussed in Chapter 9). Compare these other two methods to the method Enos used for her study.

1. Discuss the advantages and the disadvantages for Enos' study had she chosen to use a **mailed questionnaire** given her topic and the population she was interested in.

2. Discuss the advantages and the disadvantages for Enos' study had she chosen to use an **in-person structured interview** given her topic and the population she was interested in.

EXERCISE 10.4

Writing Part of Your Life Story

One kind of unstructured interview is the life story interview in which people tell about some aspect of their lives to an interviewer (Atkinson, 2001). The Center for the Study of Lives, a research, educational, and service unit which Atkinson founded at the University of Southern Maine, is involved in an ongoing effort to record, preserve, and disseminate the life stories of people of all ages and backgrounds. As the Center's website notes, "every person has an important story to tell about the life they have lived. Stories shared between multigenerational and multicultural groups become teaching tools by which we can gain a greater appreciation of both our commonalities and differences."[5]

[5] http://www.usm.maine.edu/cehd/csl/purpose.htm

In this exercise, you're asked to be both interviewer and interviewee and to write a part of your life story by answering these questions adapted from Atkinson's work.

1. What was going on in your family, your community, and the world at the time of your birth?

2. Are there any family stories told about you as a baby?

3. Are there any stories of family members or ancestors who immigrated to this country?

4. Was there a noticeable cultural flavor to the home you grew up in?

5. What was growing up in your house or neighborhood like?

6. What family or cultural celebrations, traditions, or rituals were important in your life?

7. Was your family different from other families in your neighborhood?

8. What cultural values were passed on to you, and by whom?

Observational Techniques

© David Young-Wolff/PhotoEdit, Inc.

Introduction

Did you participate in adult-organized afterschool activities when you were young? Maybe you played T-ball early or danced when you were in primary school. Emily and I didn't when we were little, back in the 1950s. These things weren't available when we were young. This relatively new social institution of adult-controlled afterschool activities, especially for young kids, merits research investigation. Unlike subjects such as attitudes toward rape victims (studied by Gray, Palileo, and Johnson in Chapter 9) and crime committed by immigrants (looked at by Martinez and Lee in Chapter 12), organized afterschool activities haven't received much public attention. Consequently, it's difficult to come up with many educated guesses (or hypotheses) about them. This was especially true when two parents and sociologists, Patricia A. Adler and Peter Adler, became so interested in their own children's afterschool activities that they produced the focal research of this chapter. It was also true for Kimberly Huisman and Pierrette Hondag-neu-Sotelo (2005) when they became interested in the meaning of dress practice, such as wearing headscarves, among Muslim refugee women in the United States; for Bruce Maycock and Peter Howat (2005) when they became curious about the barriers that men overcome prior to initiating anabolic steroid use; for Katherine Frank (2005) when she wanted to find out about the motivations of men who go to strip clubs; and for Lynn Harter and Charlene Berquist (2005) when they became interested in the ways in which the invisibility of the hidden homeless was achieved. All these authors practiced a brand of research that we broadly call **observational techniques,** and under which rubric we include methods that are sometimes called participant and nonparticipant observation. **Participant observation** is performed by observers who take part in the activities of the people they are studying; **nonparticipant observation** is conducted by those who remain as aloof as possible. Adler and Adler call their own research participant observation because each participated as "parent, friend, counselor, coach, volunteer, and carpooler" in the activities of the children they studied.

observational techniques, methods of collecting data by observing people, most typically in their natural settings.

participant observation, observation performed by observers who take part in the activities they observe.

nonparticipant observation, observation made by an observer who remains as aloof as possible from those observed.

FOCAL RESEARCH

The Institutionalization of Afterschool Activities

by Patricia A. Adler and Peter Adler[1]

Introduction

Since the early 1970s, we have witnessed the rise of a new phenomenon: the broad expansion and institutionalization of the adult-controlled

[1] This article is adapted from Patricia and Peter Adler's (1994b) "Social Reproduction and The Corporate Other: The Institutionalization of Afterschool Activities," *The Sociological Quarterly*, Vol. 35, No. 2, 309–328, copyrighted by the Midwest Sociology Society, by permission.

"afterschool" period (Berlage, 1982, Eitzen and Sage, 1989). Instead of merely coming home and playing in the house, neighborhood, or school-yard, elementary, middle school, and junior high youths are likely to be registered in some extracurricular activity organized through a local YMCA, recreation center, community center, or private association founded for this purpose. As a result, afterschool activities have become one of the most salient features of many families' childrearing experience and are instrumental in defining the developing identities of participating youth.

The rise of the institutionalized afterschool phenomenon may be rooted in two cultural conditions that have occurred over the last generation. First, with the massive entry of women (especially middle-class women) into the labor force, a need arose for childcare and/or child supervision following the school day. While we have seen the development of the "latchkey" kid (see Rodman, 1990), we have more often seen activities where children could be taken care of, entertained, and enriched. Afterschool activities thus derived reinforcement from the prevalence of dual-career families and a social consciousness that demanded the cultural edification of children apart from traditional instruction offered in the classroom. Second, there have been rising concerns about leaving children unsupervised in public, outdoor places (Cahill, 1990). Afterschool activities represented a safe place where children could spend recreational time.

Our work builds on previous research that has focused primarily on organized youth sport, some of which has extolled adult-run leisure (for example, Webb, 1969), while some has been less sanguine (for example, Devereaux, 1976). First, we describe more fully the characteristics of available afterschool activities. Second, we illustrate a model of developmental progression that young people frequently follow in advancing their extracurricular participation. Finally, we analyze the socializing effects of these experiences through the culture that is generated and the effect this has on the reproduction of social structure and the development of the self.

Methods

We draw on data gathered through participant observation with students at elementary and junior high schools. Over the course of six years (1987–1992), we observed and interacted with children both inside and outside of their schools. The children we studied attended public schools drawing predominantly on middle- and upper-middle-class neighborhoods in a large, mostly white university community. While doing our research, we occupied several roles: parent, friend, counselor, coach, volunteer, and carpooler (Fine and Sandstrom, 1988). We undertook these diverse roles both as they naturally presented themselves and as deliberate research strategies, sometimes combining them as opportunities for interacting with children became available through familial obligations or work/school requirements.

In interacting with children we varied our behavior; there were times when we acted naturally, expressing ourselves fully as responsible adults, yet there were times when we cast these attitudes and demeanors aside and tried to hang out with the children, getting into their gossip and (mis)adventures. Through these varied approaches, and through the often irrepressible candor of children, we were able to gain an insider's access to their thoughts, beliefs, and assessments.

Conducting most of our research outside of school settings, we tried to develop the parameters of the "parental" research roles by observing, casually conversing with, and interviewing children, children's friends, other parents, and teachers. We followed our daughter, son, and their neighbors and friends through their school experiences, gathering data on them as they developed. The children we befriended relished the role of research subjects because it raised their status in the eyes of adults to "experts," whose lives were important and who were seriously consulted about matters ranging from "chasing and kissing" games to the characteristics of "nerds."

In addition, we conducted informal interviews with a range of young people from a variety of ages and types of activities. We selected interview subjects on several bases: age, gender, broad interests, or degree of specialization in particular activities. In addition, new subjects were referred to us by people we had already talked to who had specific knowledge of or participation in something on a different level than our previous subjects. We continued this snowball referral (Biernaki and Waldorf, 1981) until we felt that we had adequately covered the range of available afterschool activities. The topics we focused on in discussing extracurricular participation with parents and children included the range and type they had experienced, perceptions and feelings about them, identification and commitment to them, organizations and individuals associated with them, and decisions to continue or disaffiliate with various activities.

The Extracurricular Career

A broad range of adult-organized afterschool activities are available to youth. Three categories emerge, varying in their degrees of organization, rationalization, competition, commitment, and professionalization: recreational, competitive, and elite. These categories represent a developmental model, where young people progress through various kinds of activities, increasing their depth, fervor, and skill of involvement as they escalate their participation. Not all young people follow the progression of the extracurricular career; some remain at more recreational levels or retreat from intense types of afterschool participation to more moderate and less demanding activities. However, these are exceptions to the norm, and the ensuing depiction represents the path followed by most young people.

Spontaneous Play

Traditionally, afterschool activities were those planned and directed by children themselves. Beginning in earliest childhood, children would come home after school, play in each other's houses, backyards, outside in the neighborhood, at the school playground or athletic field, and in child-organized games at any of these locations.

There are myriad organizational skills children have to master to accomplish spontaneous play. For instance, they have to plan what to do, decide on a location, establish rules and roles for participants, and set handicaps to ensure equitable and enjoyable play (see Coakley, 1990). Negotiation is another skill learned through spontaneous play. Children must routinely resolve competing desires, settle different interpretations of what has occurred and what it means, select among competing plans, and make adjustments when things are not going well. Finally, spontaneous play involves a considerable amount of problem solving. This involves significant power dynamics, compromise, and communication. For example, we observed one child who would often quit the game, without warning, when things were not going as he desired. The other children learned ways to deal with him, ranging from ignoring him to begging him to return, depending on their needs and his mood. Invariably, after this grandstanding behavior, he would reluctantly return to play.

Recreational

Recreationally organized afterschool activities vary widely in scope and character. They can be fundamentally centered on the social functions of companionship and play, such as those offered by established bureaucratic organizations (for example, YMCA, local Rec, or Community Centers). Second, these recreational activities can be centered around fitness. For instance, the Kidsport Fun and Fitness Club promotes their "Fun to be Fit" program for 8- to 12-year-olds, emphasizing the health, nutrition, grooming, and fitness advantages of Participation (Kidsport flyer). Third, recreational activities focus on extracurricular *learning*. Schools, seeking to capitalize on the afterschool market, offer a variety of (paid) programs designed to keep children on the grounds after the traditional day has concluded. These programs offer cooking, art, creative writing, and computer classes. Finally, the greatest majority of afterschool activities involve *skill development* in the arts (dance, acting, singing, music), crafts (cooking, sewing, needlework), and individual sports (horseback riding, skiing, martial arts, tennis, skating).

In recreationally organized team sports, scoring is de-emphasized. The YMCA organizes its youngest "T-ball" league so that players bat through the lineup once on each team per inning, until the designated time is reached. One eight-year-old boy expressed his feeling about this orientation:

> Yeah, it's fun. Me and my friends, we all play on the same team. Jack's dad is the coach and he's pretty nice. But sometimes it's not so fun, because everybody has to get an equal chance to play every position, no matter how good or bad they are. And some of those guys out there can't even throw or catch the ball. And he puts me in outfield and them at shortstop, and if I catch a fly ball, I can't throw it in to second base for the double play, because they can't catch it.

Past the youngest ages, recreational teams may keep score and acknowledge a winner and loser. However, there are no league standings or playoffs at the end of the season. Children, especially boys, who value displays of domination (Adler, Kless, and Adler, 1993; Best, 1983; Thorne, 1986), may still undercut this by infusing hierarchy. In a football league for nine-year-old boys, one team kept their own record of all the teams' victories so that they could track league standings, and taunted players on losing teams. Still, the emphasis at this level is on children forming broad interests in many fitness and extracurricular activities, without becoming particularly enmeshed in any specific activity.

These types of programs have several defining characteristics. First, they are adult-organized and supervised. This brings with them a set of rules and regulations for how, when, and where things should be done, introduces a teacher-student style relationship, and establishes a clear situation of authority and hierarchy (compare Coakley, 1990; Eitzen and Sage, 1989). Second, they are located outside of the home or neighborhood/yard, in institutional settings, where adults are responsible for decision-making and safety. Third, in contrast to child-directed play, where participation may be exclusive, these programs are geared toward democratic acceptance of interested parties. Finally, they are guided by a philosophy of noncompetitive structure (although participants may compare themselves informally to each other). Young participants are thus socialized to value acceptance and fairness, team spirit and camaraderie, knowledge acquisition and skill development, and submission to adult rules and authority.

Competitive

Succeeding this recreational base are afterschool activities (such as music, gymnastics, and team sports) organized on a more competitive plane. Although universal participation is initially held as an ideal, this progressively diminishes as competition escalates.

Along with the formalization of the activity's structure, children are ranked and placed in skill levels. A hierarchy is established, with children who do not measure up progressively excluded, ostracized, or denigrated. A continuum can be outlined, from an initial philosophy of democracy to one of increasing meritocracy, accompanying the increasingly competitive nature of children's play. All of the accompanying factors such as seriousness, commitment, bureaucratization, and professionalism shift as one moves from the lower end of this continuum toward the top.

Democratic At the democratic end of the competitive continuum, activities are structured hierarchically, yet the rhetoric of recreation is still espoused. Organizations responsible for coordinating and offering such afterschool activities are usually specialized and tied to a single enterprise. Many of them hold democratic principles resembling this one stated in a letter to the coaches of a girls' softball association:

> The key message is that the program or activity must be fun if the participants are going to get something out of it and continue to play. We have tried to structure the program on that basis and our #1 goal is that the girls have fun.

Yet sometimes young participants get mixed messages about goals from adults. In a soccer league office at the beginning of the season, as mothers thronged to switch their children onto teams with their friends, one administrator sardonically remarked to another, "What do they think this is, a social club? This is a soccer club." After overhearing this, another mother whispered, "They may not think it's a social club, but I sure do."

Other times it is the parents, more than the administrators, who take the activity seriously. Attending a soccer game of a friend's nine-year-old son in another state, we observed

> The parents were all running up and down the sidelines screaming and shouting at the players and coach. While this looked fairly typical to us, several of the parents felt the need to approach us during the game and offer accounts for their behavior. "Oh, you'll have to excuse us, we get kind of carried away."

Their words suggested that they perceived their intensity, involvement, and competitiveness as deviant relative to the league.

While more talented players may find the increase in competition satisfying, those with weaker skills often find themselves unhappy. In one democratically geared competitive league, a nine-year-old baseball player described his dissatisfaction with his coach's philosophy and practice:

> My coach wants to win too much. It's not fun for me. I'm not getting enough playing time. . . . I never get to start. . . . He only puts me in to play the outfield. I can play other positions, it's not fair. I want to get a chance to pitch. I can do it.

At this level, differences become more apparent between the stated ideals of the program and the actual practices of the participants. Adults usually espouse the commonly held notion of democratic play and skill development, but in practice often emphasize winning. Children have no such ambiguities. Despite the admonitions that "It's only a game," they are quite aware of performance differentials. They easily note that, when faced with crucial situations, the rhetoric of democracy is abandoned and unmitigated competition comes to the fore. The values communicated to youngsters through this level of afterschool participation include a beginning awareness of inequality in talent and the legitimization of how this affects opportunity as well as the introduction of focus on competition and winning.

Meritocratic Organizations sponsoring afterschool activities progressively detach from the vestiges of recreation and democracy as they become more competitive and professionalized. Skill, performance, and talent criteria associated with a meritocracy become increasingly rewarded, while effort, fairness, and democratic decision-making diminish in centrality. As one 12-year-old girl explained,

> I really like being in the performing company. Having tryouts cut out all of the people who can't really dance, who aren't coordinated, and who just don't pay attention. For so many years my dance classes were filled with those kids and they dragged the class down, made us go slowly when they couldn't get the steps or they forgot them. Now we learn much more because we can move faster, the classes are tough, and the rehearsals are

serious. The people with the solos are the ones who are the best and who work the hardest, and you don't have someone's mother complaining . . . that their daughter should have a solo, because they know it's based on who's best.

The shift in character associated with the move from democratic to meritocratic deters some participants from continuing to advance in the activity. One boy dropped out of his quiz bowl team when the level of competition exceeded his abilities. The funneling process also led young people to abandon some activities while concentrating on others. As one 10-year-old boy commented,

I'm not going out for soccer this year. It was boring last year because I didn't get to play enough, and I also had a problem because I was in a play and there was an overlap between the play and soccer. . . . [S]ometimes rehearsals were on Saturday mornings when we had soccer games, and I had to leave games to go to rehearsals, and my coach yelled at me. So I think I'll just stick with acting this year. I don't have any friends in it, but I like it more and I'm better at it. It's more *me.*

The funneling effect of youngsters dropping out of extracurricular activities is thus associated with the rise in commitment these activities require, so that the youngsters have to choose the one to which they will dedicate themselves. These choices then represent a deliberate move on the youngsters' part (as opposed to their parents signing them up) and usually reflect their developing identities.

Beyond whole leagues organized at the more competitive level, coaches or parents occasionally push individual squads into a more competitive posture. Most youngsters enjoyed participating in some of the more competitive afterschool activities. They gave greater concentration and effort, drew a greater sense of self-involvement from them, and received greater attention from the surrounding adult world. Youngsters also noted that the more competitive activities were for older and more talented participants, and appreciated the greater status they derived. Part of growing up for these middle-class youngsters involved being exposed to a variety of diverse extracurricular experiences and acquiring many skills. But they also learned from their recognition of the escalating ladder of extracurricular participation that adults and society value depth involvement. Recreation and breadth, then, were less notable than competition and depth. Other values associated with advancing to this level of afterschool included the stratification of hierarchy and ranking, exclusion, seriousness of purpose, and identification with participation in the activity.

Elite   Almost all competitive afterschool activities can also be performed on an elite level. School plays or local dance troupes can be replaced by participation in adult theater or dance performances, band players can move up to junior symphony, local competitions can give way to regional or national meets, and competitively oriented leagues can be succeeded by elite leagues.

The decision to pursue an elite activity means allocating it a greater amount of time, making it a priority ahead of other activities, and eliminating other projects. Not everyone wants to make this commitment, but for those who do, not only their own but their families' time may have to be shifted and activities rescheduled around what was once an afterschool activity. One 14-year-old ballet enthusiast regularly took a lighter scholastic load so she could spend several hours every day in the studio, while another student left school during the day to attend art classes at the local university. Many families plan their vacations around the demands of their children's activities. Parents have to transport their children to and from these activities daily, adjusting their work, mealtime, and family schedules. One mother expressed her concerns:

> Ever since my son started to play hockey seriously, my days have been ruined. He has to be on the ice at 5 A.M. and then he practices after school. My husband and I haven't had a free weekend in months.

The accoutrements associated with elite afterschool activities are superior to those at the organized and competitive levels, featuring enhanced, often personalized uniforms, better equipment, better arenas, and upgraded transportation. All of this costs more money, frequently a substantial burden for both students and parents. Concomitant with these changes, exclusiveness becomes salient in elite activities. All remaining vestiges of democratic participation are replaced by a philosophy of selectivity. The unconcealed exclusiveness may hurt young people who do not qualify for participation. Others are more sanguine about their failures, putting them aside and turning to other activities.

Once selected for the squad, participants soon discover that the adults treat elite activities seriously and expect adolescents and pre-adolescents will make the commitment. Practices may move from weekly or twice weekly to almost daily and require strenuous concentration. Many participants and parents regard elite activities as a track into some future enterprise. As one 15-year-old girl remarked,

> My goal is to get more speed on my fastball and to stick with softball through the teenage years, which are the tough ones. Then I want to get a scholarship to college. My coach has already told me that I can get one, it's just a matter of how much I improve that will make the difference of if I go to, say, Arizona State or end up at Western Kansas. And then after that, there's no professional softball for women right now, but maybe by the time I get out, there will be.

As the youngsters increasingly specialize, the funnel continues to narrow.

Although elite participants make sacrifices for the team, they never abandon their focus on developing their own careers. Individuals accord the coach authority but never abdicate their own decision-making completely. Decisions are made by the coach for the competitive advancement of the group, overruling parents' authority. Yet participants will only stay with a coach as long as it suits their best interests.

Decisions are no longer made on the basis of fairness or ethical values, but on merit, pragmatics, and outcomes (Berlage, 1982). Coaches and leaders foster a mentality of winning. Young people become introduced to some of the other less idealistic and "fair" aspects of society such as politics, favoritism, and financial considerations.

Berlage (1982) has suggested that the meanings and values embedded in elite afterschool activities are those of the corporate world. These values socialize children to the attitudes and behavior they will encounter in corporate jobs. They are predominantly transmitted through the specialization of roles, the subjugation of individualism and family for the team, the emphasis on sacrifice and self-discipline, the structuring into well-defined units, and the processual model of advancement. Other values associated with the elite level include subordination, interdependence, deferred gratification, professionalism, commercialism, and a strong focus on winning. While these characteristics define the elite level, as time progresses, the characteristics of each tier of afterschool activity tends to diffuse down to the developmentally preceding stage.

Conclusion

In considering the benefits or harm generated by the infusion of adult control and the associated adult-oriented structures and values into the leisure of children and youth, we see elements of both. Our observations suggest that in progressing through the stages of afterschool activities children learn several important norms and values about the nature of adult society. They discover the importance placed by adults on rules, regulations, and order. Creativity is encouraged, but acknowledged within the boundaries of certain well-defined parameters. Obedience, discipline, sacrifice, seriousness, and focused attention are valued; deviance, dabbling, and self-indulgence are not. Coordination with others, the organic model of working toward challenging and complex goals, stands as the ultimate model toward which young people are directed. This anticipatory socialization to the organized, competitive world used to be the more exclusive domain of young boys, giving them what Borman and Frankel (1984) have claimed is a competitive advantage in the adult corporate world. The recent massive growth of institutionalized competitive play has extended into the female realm, however, and this may portend greater gender equalization for this generation's future adults. In fact, as we noted in recent research (Adler, Kless, and Adler, 1993), girls' gender role images appear to be expanding to a much greater extent than boys'. Girls, then, are already beginning to change the world by moving into boys' realms.

At the same time, the earlier imposition of adult norms and values onto childhood may rob children of developmentally valuable play, unchanneled and pursued for merely expressive rather than instrumental purposes. By participating in increasing amounts of adult-organized activities, children are steered away from goal-setting, negotiation, improvisation, and self-reliance toward the acceptance of adult authority and adult pre-set goals. While these afterschool activities may prepare children for the formally rational, hierarchical, and, in Foucault's (1977) words, disciplined adult world, children's spontaneous play may teach different but important social lessons. Adult-organized activities may prepare children for passively accepting the adult world as given; the activities that children organize themselves may prepare them for creatively constructing alternative worlds.

Moreover, entering into these adult-organized activities draws children into junior versions of the existing social order. Not only are the norms and values of adult culture embodied in organized afterschool activities, but the structural inequalities of race, class, and, to a lesser extent, gender are inherent as well. It takes money to support a child going "up the ladder" of afterschool activities, and while families from lower socioeconomic groups participate in these endeavors, they cannot afford them to the same extent as more affluent households. If these experiences prepare youngsters for the corporate work world—partly through their enhanced "cultural capital" (Bourdieu, 1977a) of additional knowledge, skills, and disposition and partly through the "habitus" (Bourdieu, 1977b), the attitude and experience of achievement they acquire—then afterschool activities are yet another route to reproducing social inequities. The channeling of low-income racial group members into inexpensive, segregated activities may also become exacerbated as more extracurricular activities are moved from the realm of public education into the private domain. Finally, while girls' activities have made significant strides toward parity with boys' over the last decade, gender segregation and stereotyping still remain, with patriarchal structures and attitudes continuing to dominate significant portions of the field.

The afterschool phenomenon thus offers a doorway into the youth subculture for adults. In exploiting this opening, adults have seized the opportunity to transform the character of children's play. What they have created, in doing so, is an ironic juxtaposition of work and play: Play has become used as the vehicle for infusing adult work values into children's lives.

REFERENCES

Adler, P. A., S. J. Kless, and P. Adler. 1993. Socialization to gender roles: Popularity among elementary school boys and girls. *Sociology of Education* 65: 169–187.

Berlage, G. 1982. Are children's competitive team sports corporate values? *ARENA Review* 6: 15–21.

Best, R. 1983. *We've all got scars.* Bloomington: Indiana University Press.

Biernaki, P., and D. Waldorf. 1981. Snowball sampling. *Sociological Research and Methods* 10: 141–163.

Borman, K. M., and J. Frankel. 1984. Gender inequalities in childhood social life and adult work life. Pp. 55–63 in *Women in the workplace,* edited by S. Gideonse. Norwood, NJ: Ablex.

Bourdieu, P. 1977a. Cultural reproduction and social reproduction. Pp. 487–511 in *Power and ideology in education,* edited by J. Katabel and A. H. Halsey. New York: Oxford University Press.

Bourdieu, P. 1977b. *Outline of a theory of practice.* Cambridge: Cambridge University Press.

Cahill, S. 1990. Childhood and public life: Reaffirming biographical divisions. *Social Problems* 37: 390–402.

Coakley, J. J. 1990. *Sports in society,* 4th ed. St. Louis: Mosby.

Devereaux, E. 1976. Backyard versus Little League baseball: The impoverishment of children's games. Pp. 37–56 in *Social problems in athletics,* edited by D. M. Landers. Urbana: University of Illinois Press.

Eitzen, D. S., and G. Sage. 1989. *Sociology of North American sports,* 4th ed. Dubuque, IA: William C. Brown.

Fine, G. A., and K. L. Sandstrom. 1988. *Knowing children.* Newbury Park, CA: Sage.

Foucault, M. 1977. *Discipline and punishment.* New York: Pantheon.

Rodman, H. 1990. The social construction of the latchkey children problem. Pp. 163–174 in *Sociological studies of child development.* Vol. 3, edited by N. Mandell. Greenwich, CT: JAI.

Thorne, B. 1986. Girls and boys together. But mostly apart: Gender arrangements in elementary schools. Pp. 167–184 in *Relationships and development,* edited by W. Hartup and Z. Rubin. Hillsdale, NJ: Lawrence Erlbaum.

Webb, H. 1969. Professionalization of attitude toward play among adolescents. Pp. 161–178 in *Aspects of contemporary sport sociology,* edited by G. Kenyon. Chicago: Athletic Institute.

STOP AND THINK   *What did you do for afterschool activities in elementary school? In junior high school and high school? Does the Adlers' developmental model of afterschool activities jibe with your experiences of afterschool activities as you grew up? Does the "funneling" they describe ring true or was your experience of afterschool activities somewhat different?*

Observational Techniques Defined

"Observational techniques" are sometimes referred to as "qualitative methods" and "field research." But both "qualitative methods" and "field research" connote additional methods (e.g., qualitative interviewing and using available data) that we cover elsewhere in this book. Here we focus on "observations" alone. Adler and Adler have clearly asked a lot of questions in the course of their work (hence all the participants' recountings and remarks), but they've also made a lot of observations, including having observed nine-year-old boys in one football league keeping track of their own record for comparative purposes and having seen mothers throng, at the beginning of a soccer season, to a league office to switch their children onto teams with friends. We'd like to stress that a variety of methods *are* and *should be* used in the field, but here we want to focus on those observational methods that are used in the field but have not been covered in other chapters.

Of course, the data from questionnaires, interviews, content analysis, experiments, and so on, *are* all based on observations of one kind or another. For questionnaires and interviews, the observations are about responses to questions. For content analysis, which we'll cover in Chapter 13, the observations are about communication of one sort or another. In experiments, however, observations are frequently of the sort that closely resemble those used by participant and nonparticipant observers. In most cases, experimental observations are of a more controlled sort than are those used in the field. **Controlled (or systematic) observations** involve clear decisions about just what kinds of things are to be observed. In the "Coins in the Kettle" experiment by Bryan and Test (1982), observers focused solely on whether people would or wouldn't contribute coins to a Salvation Army kettle after they'd seen, or not seen, another person do so (the stimulus). Similarly, Harrell and his colleagues made more than 400 observations of children and their parents in 14 supermarkets to see if more attractive children were treated better by their parents than others (cited in Bakalar, 2005). But such controlled observations are not the primary focus of this chapter.

controlled observation, observation that involves clear decisions about what is to be observed.

Adler and Adler's immersion in the natural setting of their subjects distinguishes their work from these other approaches. Rather than merely read about or ask about "afterschool activities," they've gone to and observed those activities in person (though they've obviously also read about and asked about the activities). Moreover, although they have questioned participants and made their presence felt in one way or another (as when they've acted as "parent, friend, counselor, coach, volunteer, and carpooler"), their intervention has been minimal. They have, as they've said elsewhere, tried to create a situation in which "behavior and interaction continue as they would without the presence of the researcher, uninterrupted by intrusion" (1994: 378). We here confine our attention to observational techniques, then, that are used to observe actors who are, at least in the perception of the observer, acting naturally in natural settings.

STOP AND THINK   *Could Adler and Adler have used only questionnaires to study afterschool activities? What kinds of insights might have been "lost" if they had?*

The origins of modern observational techniques are the subject of some current debate. Conventional histories attribute their founding to anthropologists like Bronislaw Malinowski, with his (1922) *Argonauts of the Western Pacific,* and sociologists like Robert Park, who, around the same time, began to inspire a generation of students to study city life as it occurred naturally (see, for example, Wax, 1971, and Hughes, 1960). More recently, however, Shulamit Reinharz (1992) has argued that Harriet Alice Fletcher, studying the Omaha Indians in the 1880s, and Harriet Martineau, publishing *Society in America* in the 1830s, promoted essentially the same techniques that Malinowski and Park received fame for in the 1920s, thus suggesting the earliest "social science" observers might have been American women.

Reasons for Doing Observations

When are observational techniques desirable? As we've suggested, observation can be useful when you don't know much about the subject under investigation. A classic technique in anthropology, the ethnography, or study of culture, is primarily an observational method that presumes that the culture under study is unknown to or poorly known by the observer. Margaret Mead's (1933) classic investigation, *Sex and Temperament,* of gendered behavior among three tribes in New Guinea, for instance, found evidence that sex-role definitions were more malleable than convention held at the time. Similarly, many well-known sociological investigations have been of things less than well known beforehand. More recently, Mears and Finlay (2005) examined how fashion models manage bodily capital and Muir and Seitz (2004) studied rituals within the male collegiate rugby subculture. Little was known by social scientists about any of these topics beforehand. Adler and Adler justify their investigation of adult-controlled afterschool activities, in part, by how new those activities are to the American cultural landscape.

Observational techniques also make sense when one wants to understand experience from the point of view of those who are living it or from the context in which it is lived. Norman Denzin makes a distinction between **thin** and **thick description,** the former merely reporting on an act, the latter providing a sense of "intentions, motives, meanings, context, situations, and circumstances of action" (1989: 39). We see Adler and Adler moving from thin to thick description as they relate first-hand accounts of participants about what it's meant to them to engage in ever-more specialized afterschool activities. In a related study, Cooky and McDonald (2005) use specific observations from athletic events to formulate questions about the meaning of athletic activity for 11- to 14-year old girls.

The goal of understanding experience is clearly not antithetical to the goal of studying a subject that is previously unknown. The two goals frequently go hand in hand. On the one hand, the Adlers' progressive typology of afterschool activities (from recreational to competitive to elite) gives us an interesting new way of understanding those activities. On the other hand, this and other new insights come from focusing on a subject (afterschool activities) that had received very little attention before.

Observational techniques are also useful when you want to study quickly changing social situations. Kai Erikson largely based his account in *Everything in Its Path* (1976) of the trauma visited on an Appalachian community by a disastrous flood in 1972 on descriptions gleaned from legal transcripts and personal interviews. He, as a sociologist, had been asked by a law firm to assess the nature and extent of the damage to the Buffalo Creek community by a flood that had been caused, in part, by the neglect of a coal company. His book presents the legal case he made then, using content analyses and unstructured interviews. When he presented the effects of the favorable judgment on the community he studied, he described a gathering "in a local school auditorium to hear a few announcements and to mark the end of the long ordeal." He interprets the gathering as "a kind of graduation . . . that ended a period of uncertainty, vindicated a decision to enter litigation, and furnished people with sufficient funds to realize whatever plans they were ready to make. But it was also a graduation in the sense that it propelled people into the future at the very moment it was placing a final seal on a portion of the past" (1976: 248–249). Erikson then used direct observation to mark a momentous occasion: the moment (two years after the

---

thin description, barebone description of acts.

thick description, reports about behavior that provide a sense of things like the intentions, motives, and meanings behind the behavior.

> **BOX 11.1**
>
> ## Reasons for Doing Observation
>
> 1. To understand new or little-studied social phenomena
> 2. To develop thick description of social phenomena
> 3. To study quickly changing social situations
> 4. To offer a relatively unfiltered view of human behavior

flood) when Buffalo Creek began, as a community, to heal itself after its calamitous ordeal. If he hadn't been there when he was, this moment of significant change would necessarily have gone unnoticed by himself and his readers. Similarly, if Sudhir Venkatesh (1997) hadn't been around the Saints street gang in the early 1990s, he could never have observed it and told us about the processes it underwent as it turned toward a systematic involvement in local drug economies. (See Box 11.1.)

Another reason for using observational techniques is that they offer a relatively unfiltered view of human behavior. You can, of course, ask people about what they've done or would do under given conditions, but the responses you'd get might mislead you for any number of reasons. People do, for instance, occasionally forget, withhold, or embellish. Thus, you might never be able to figure out whether parents take better care of pretty children than they do ugly ones just by asking them, but Harrell and his colleagues feel they have been able to do so by observing those 400 parent-child interactions in supermarkets (cited in Bakalar, 2005)—and they think they do.

There are other reasons why one might use observational techniques rather than others described in this text. Questionnaire surveys can be very expensive, but participant or nonparticipant observation, although time-consuming, can cost as little as the price of paper and a pencil. Ethically, studying homeless people in their natural settings, as Liebow (1993) did, raises fewer questions than would studying, say, the effects of homelessness by creating an experimental condition of homelessness for a group of subjects. As you can imagine, the reasons for doing observation can have complex interactions with one another, with the salience of some undermining the feasibility of others. Rebekah Nathan (2005), a college professor, became a college student for a year to gain a deep, unfiltered understanding of what it means to be a college student today, partly because she didn't think she'd get the same kind of high-quality data from simply asking her students. (Rebekah Nathan is a pseudonym that the anthropologist adopted for her study.) Ethically, however, Nathan didn't want to accept grants from funding agencies to pay for her tuition, books, housing, etc. because she feared that would give the agency too much influence over what she could report. So she paid for these things out-of-pocket, making the research for *My Freshman Year: What a Professor Learned by Becoming a Student* the most expensive research she'd ever done.

Observer Roles

We suggested earlier that one of the defining characteristics of observational techniques is their relative inobtrusiveness. But this inobtrusiveness varies with the role played by the observers and with the degree to which they are open about their research purposes. Raymond Gold's (1958) typology of researcher roles suggests a continuum of possible roles: the **complete participant,** the **participant-as-observer,** the **observer-as-participant**, and the **complete observer.** It also suggests a continuum in the degree to which the researcher is open about his or her research purpose. As a complete participant, the researcher might become a genuine participant in the setting under study, or might simply pretend to be a genuine participant. When a genuine participant, the complete participant not only risks the fear of being found out but also entertains the danger of over-identifying with other participants, which in some cases can lead to abandoning the research for the joys of participation. One of us, Emily, intended, for instance, to report on a feminist consciousness-raising group in the early 1970s until her participation altered her consciousness, as it were, and she decided that her membership in the group precluded her using it as an object of study. The fate of Emily's project (its abandonment) is just one of the reasons why conventional wisdom has warned against identifying too closely with other participants; another concern has been the bias that tends to be associated with "over-rapport" (see, for example, Hammersley and Atkinson, 1995). Current "critical" (for example, Morrow, 1994) and "feminist" (for example, Reinharz, 1992) perspectives have questioned this concern, sometimes even claiming that understanding other people requires becoming close to them.

In any case, the complete participant, in Gold's typology, is defined by his withholding of the fact that he is, in fact, doing research. Thus, Nathan, for the most part, took on the role of "student" in her (2005) study of students. She collected data, while participating in dormitory life, study groups, classroom tasks, and discussion groups, telling few people, except those who persistently asked, that she was a professor or even that she was doing research.

complete participant role, being, or pretending to be, a genuine participant in a situation one observes.

participant-as-observer role, being primarily a participant, while admitting an observer status.

observer-as-participant role, being primarily a self-professed observer, while occasionally participating in the situation.

complete observer role, being an observer of a situation without becoming part of it.

STOP AND THINK

Some have suggested that it is unethical for a researcher to withhold her research purposes from her subjects (see Chapter 3). Do you think Nathan was unethical in disguising her research purpose from the students she studied? Why or why not?

The opposite of the complete participant role is that of the complete observer: one who observes a situation without becoming a part of it. In principle, complete observers might ask for permission to observe (as Gottman and Levenson [2000] did in their work on how married couples interacted), but they usually don't. Harrell and his colleagues (cited by Bakalar, 2005) anonymously watched the parents and children in a supermarket to code not only the attractiveness of the children but also for whether the parents belted their youngsters into grocery cart seats, the frequency with which parents lost track of their children, and the number of times children were permitted to do dangerous things like standing up in the cart. Unlike

Nathan, who played the role of a student, Harrell and others remained socially invisible to those whom they observed. They didn't tell those they watched that they were observing them. A rather more expensive version of Harrell's basic strategy was employed by observers trained at the National Opinion Research Center (NORC) in preparation for, among others, the work of Robert Sampson and Stephen Raudenbush (1999) on disorder in urban neighborhoods. These observers were driven in a sport utility vehicle at a rate of five miles per hour down every street in 196 Chicago census tracts, videotaping from both sides of the vehicle while speaking into the videotape audio. Martyn Hammersley and Paul Atkinson (1995: 96) note the paradoxical fact that complete observation shares many of the disadvantages and advantages of complete participation. Neither is terribly obtrusive: Harrell, the NORC researchers, and Nathan were not likely to affect the behavior of those they watched. On the other hand, neither complete participation nor complete observation permits the kind of probing questions that one can ask if one admits to doing research. In addition to the question of (not being able to ask scientifically interesting) questions, the lack of openness (about one's research purposes) implicit in pure participation or pure observation has raised, for some, the ethical issue of whether deceit is ever an appropriate means to a scientific end—an issue, you will recall, that also has come up in our discussion of other methods. In describing her work, Nathan acknowledges feeling deceitful and guilty at times, and notes a few times when she "outed" herself as researcher and professor: perhaps most amusingly when one student, whose integrity she admires, confides that he can't find a third professor to write a character reference for an application. After considering the pros and cons, Nathan tells this "friend" who she is and offers to write him the letter he needs (Nathan, 2005, 161).

Both the participant-as-observer and the observer-as-participant make it clear that they are doing research; the difference is the degree to which they emphasize participation or observation. The participant-as-observer is a participant in the setting who admits he or she is also an observer. The observer-as-participant is a self-professed observer who occasionally participates in the situation. Slavin's (2004) observation of illicit drug use in the gay nightclub scene was facilitated by his status as a gay man who had participated in the "scene" before. Slavin thus plays a participant-as-observer role. He claims most of his participant observation was overt and occurred as he hung out with participants and their friends in nightclubs, parties, cases, or homes. Sometimes, though, "key participants chose to make my role covert vis-à-vis their friends or acquaintances, and I left this decision to them" (2004, 269). When that was true, Slavin was primarily an observer at least in relation to those friends or acquaintances. One senses that Adler and Adler were sometimes primarily participants, as when they undertook the diverse roles of "parent, friend, counselor, coach, volunteer, and carpooler." At times, they seem to have been and, perhaps more important, been perceived to be observers, as when they attended a friend's son's soccer game and observed parents shouting at players and coaches, after which observations the parents offered accounts for their behavior. Liebow, in his (1993) study of homeless women, is more explicit about his alternation between being primarily participant and primarily observer in the women's shelter called The Refuge:

> I remained a volunteer at the Refuge [throughout the study], and was careful to make a distinction between the one night [a week] I was an official volunteer and the other nights when I was Doing Research (that is, hanging around). (1993: x)

Adler and Adler divulged their research purposes to their respondents (and report that child respondents enjoyed playing "expert"), especially, of course, when they began to conduct interviews. But,

> While we did not go out of our way to hide the fact that we were researchers (as we did in one of our previous books, *Wheeling and Dealing*), in some cases we didn't go out of our way to divulge it either. Due to the characteristics of the children (i.e., young), we did not think it necessary to announce our intentions at every event. (personal communication)

Their research for *Wheeling and Dealing* (Adler, 1985), incidentally, had been of drug dealers and smugglers and might therefore have been compromised by a frank disclosure of their research interest.

Liebow clearly announced his interest and even records the announcement of his purpose, and his request for permission:

> "Listen," I said at the dinner table one evening, after getting permission to do a study from the shelter director. "I want your permission to take notes. I want to go home at night and write down what I can remember about the things you say and do. Maybe I'll write a book about homeless women." (1993: ix)

One danger of both the participant-as-observer and observer-as-participant roles is that they are more obtrusive than are those of the pure participant or the pure observer and that those being studied might shift their focus to the research project from the activity being studied, much as they might in responding to a questionnaire or an experimental situation. Those being studied might have been aware of being studied in the Adlers' project from time to time, but the verisimilitude, what they elsewhere (1994) call **vraisemblance,** of their writing style tends to dispel one's doubts about the overall validity of their reports.

vraisemblance the verisimilitude, or sense of truth, that a reporter can convey through his or her writing.

STOP AND THINK   *Recall that within their section on "democratic" activities, Adler and Adler claim "sometimes young participants get mixed messages about goals from adults." What about their supporting evidence in this section gives you the feeling this claim is essentially true?*

The danger of obtrusiveness aside, there are times when announcing one's research intention offers the only real chance of access to particular sites. The only other way participant observers Becker, Geer, Hughes, and Strauss (1961), could have gained access to the student culture in medical school for their classic study *Boys in White* would have been to apply and gain admission to medical school themselves. Think of all the science courses they would have had to take!

BOX 11.2

The Continuum of Observer Roles: The Chances of Over-Identification, Being Obtrusive, and Asking Probing Questions Vary by Role

Complete Participant	**Participant-As-Observer & Observer-As-Participant**	**Complete Observer**

$+\longleftarrow$——————— Chances of Over-rapport ———————— $-$

$-$————————————$\rightarrow+$Chances of Being Obtrusive$+\longleftarrow$————— $-$

$-$————————————$\rightarrow+$Chances of asking Probing$+\longleftarrow$————— $-$
Questions

One thing Adler and Adler's study makes clear is that within any given research project, researchers can play several different observational roles. There were times when they were primarily participants, and times when they were primarily observers. In fact, as they've suggested elsewhere (Adler and Adler, 1996), capitalizing on their parent role meant that their engagement in both of these capacities was, depending on the situation, more nuanced still; at times, they were just *parents-in-the-home,* at times, *parents-in-the-school,* at times, *parents-in-the-community,* and sometimes, more explicitly, *parents-as-researchers.* Moreover, as they've suggested (Adler and Adler, 1998: 33–37), there are always potential ethical issues, even when one is open about one's multiple roles and purposes: issues of power (as in a parent's power over a child); issues of guilty knowledge (as when one knows something about another parent's child that one might report to the parent or other authority); and issues of responsibility (as when one moves from the role of normal parent to conduct, say, taped interviews).

Above all, it's clear that different research purposes require different roles. It's hard to imagine, for instance, that Nathan's study of college students or that the Adlers' of drug dealers and smugglers would have worked if they'd announced their research intentions. It is hard to decide which role to adopt in a given situation. You should be guided by methodological and ethical considerations, but there are, unfortunately, no definite guidelines.

Getting Ready for Observation

We've seen how researchers using other techniques (for example, experiments or surveys) can spend much time and energy preparing for data gathering, in formulating research hypotheses, and developing measurement and sampling strategies, for example. Such lengthy preparations are usually not as vital for studies involving observational techniques, where design

FIGURE 11.1

A Potential Codesheet for Recording Observations in the "Care of Pretty/Ugly Children Study" by Harrell and others (cited in Bakalar, 2005)

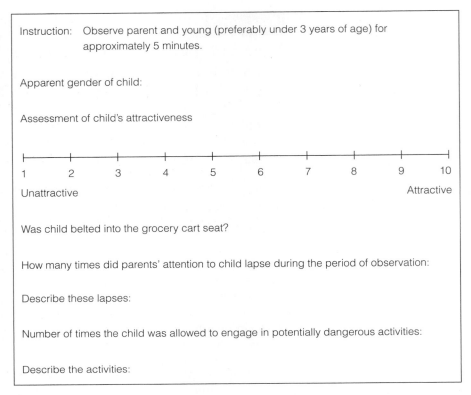

elements can frequently be worked out on an *ad hoc* basis. The exception to this rule, however, is controlled, or systematic, observations, which are defined (not unlike the controlled observations of experiments, mentioned earlier) by their use of explicit plans for selecting, recording, and coding data (McCall, 1984). Harrell and his colleagues (cited in Bakalar, 2005), as you know, studied 400 parents and their young children in 14 supermarkets. Because this team began with a more or less well-defined hypothesis (parents will take better care of pretty children than they will of ugly ones), one can imagine them filling out a codesheet like the one we've constructed for you in Figure 11.1. One doesn't have to imagine such codings of the NORC observations (remember the SUVs with dual camcorders in Chicago?) used by Sampson and Raudenbush (1999). One need only consult the *Systematic Social Observation Coding Manual* (NORC, 1995).

Far more often, however, observers begin their studies with less clearly defined research questions and considerably more flexible research plans. Adler and Adler (1994b) suggest that an early order of business for all observational studies is selecting a setting, a selection that can be guided by opportunistic or theoretical criteria. The selection is made opportunistically if, as the term suggests, the opportunity for observation arises from the researcher's life's circumstances (Riemer, 1977). Adler and Adler chose to study afterschool activities because their own children were engaged in those activities. In fact, they have confided that their general preference is for opportunistic settings:

> We try to research in our own "backyard" (in this case [that is, in the afterschool activities study], literally) so that we can spend as much time in our research setting as our life will allow. . . . We got interested in our children's afterschool activities, and then developed a theoretical interest in them. (personal communication)

Liebow studied shelters for homeless women because, after he'd been diagnosed with cancer and quit his regular job in 1984, he decided to work in soup kitchens and shelters, only to find their clients extraordinarily interesting. Jeffrey Riemer (1977: 474) advocates opportunistic settings for ease of entry and of developing rapport with one's subjects, and for the relatively accurate levels of interpretation they can afford.

On the other hand, Liebow's other book, *Tally's Corner* (1967), was based on a 12-month study of African American men in Washington, D.C., an area that was chosen because of the light it might cast on theories of poverty that were prevalent at the time. Similarly, Hochschild studied airline flight attendants and bill collectors for her book *The Managed Heart* (1983) because they "illustrate[d] two extremes of occupational demand on feeling" (16), the one demanding that one "feel sympathy, trust, and good will," the other, "distrust, and sometimes positive bad will" (137). The settings for *Tally's Corner* and *The Managed Heart,* then, were chosen primarily for their theoretical relevance rather than for their ready access.

STOP AND THINK   *Do you remember Laud Humphreys' (1975) study of brief, impersonal, homosexual acts in public restroooms (mentioned in Chapter 3)? Do you imagine that Humphreys' decision to study public restrooms was informed primarily opportunistically or purposively? What makes you think so? (Humphreys' answer appears in the next Stop and Think paragraph.)*

The suggestion that setting selection can be made on either opportunistic or theoretical grounds might be slightly simplistic, however, insofar as it implies that one chooses only one site, once and for all, and that it is primarily opportunity or theory that informs the choice. Actually, Adler and Adler are not so very unusual in their decision, initially, to immerse themselves in settings that were available, but then to seek out interviews that spoke to ideas they developed on their own:

> [W]e derived our ideas about what to write *about* from our observations of and casual interaction with kids, . . . but to get quotes from kids themselves and to see how they felt or if their experiences jibed or differed from our perspectives, we conducted interviews. . . . These interviews were selected by theoretical [or purposive] sampling to get a range of different types of kids in different activities, age ranges, and social groups. (personal communication)

Especially in seeking their interviews, then, Adler and Adler were guided by the effort to test their own theoretical notions against those of other participants, an approach they recommend elsewhere (Adler and Adler, 1987).

theoretical saturation, the point where new interviewees or settings look a lot like interviewees or settings one has observed before.

By emphasizing their efforts to "get a range of different types of kids," in fact, they suggest they attempted to approximate the ideal of **theoretical saturation** (Adler and Adler, 1994a; Glaser and Strauss, 1967)—that is, the point where new interviewees or settings looked a lot like interviewees or settings they'd observed before. Opportunity, then, might have inspired their early choices; some mix of opportunity and theory influenced their later choices.

STOP AND THINK

Humphreys says he embarked on his study of sex in public restrooms after he'd completed a research paper on homosexuality. He states that his instructor asked about where the average guy went for homosexual action, and that he (Humphreys) articulated a "hunch" that it might be public restrooms. His further comment, "We decided that this area of covert deviance, tangential to the subculture, was one that needed study" (1975: 16), suggests that his decision to study public restrooms was purposively, rather than opportunistically, motivated.

purposive sampling, a nonprobability sampling procedure that involves selecting elements based on the researcher's judgment about which elements will facilitate his or her investigation.

Moreover, Adler and Adler's study suggests what seems to be generally true of sampling in studies using observational techniques: Observations are most often done in a nonrandom fashion. They practiced, as most participant observers do, **purposive sampling,** or sampling based on an intuitive sense of what would offer a reasonably full understanding of the subject matter. Devising a sampling frame of the kinds of things studied by participant and nonparticipant observers (for example, afterschool activities, occupations involving emotion work, homeless women, bathrooms used for homosexual encounters) is often impossible, though again, this might be less true of the (relatively infrequent) studies involving systematic observations. The researchers from NORC did, in fact, select a stratified probability sample of Chicago census tracts "to maximize variation by race/ethnicity and SES" (Sampson and Raudenbush, 1999) for its SUV-driven study of public spaces.

One other decision one should make before entering the field is how much one will tell about yourself and your research. *Disclosure* of your interests (personal and research) can help develop trust in others, but it can also be a distraction from, even a hindrance to the unfolding of, events in the field. Nathan (2005, 6–7), for instance, indicates that she had decided that she would merely reveal that she was a "returning" older student to other students and faculty, unless there came a need for more disclosure. This was of course the truth, but not the whole truth. And she encouraged friends and colleagues to help her think about what she'd say and do. She of course couldn't anticipate the class in which, as a classroom exercise, she ended up revealing lots about her sexual history and listened as other students revealed much of theirs (2005, 162–163). She finally decided that she needed to tell this group about her identity as a researcher and to assure them that she had no intention of revealing their sexual histories in print. In exchange, she got group members to promise not to reveal details of her sexual history to her future students.

We would like to make two more general recommendations about preparing yourself for the field. First, we think potential observers (and, perhaps, especially if they are novices) should review as much literature in advance of their observations as possible. Such reviews can sensitize the researchers to the kinds of things they might want to look for in the field and suggest new settings for study. You should know, however, that our advice on this matter is not universally accepted. In fact, the Adlers tell us that they now try to follow the advice of Jack Douglas (1976), who advises researchers to immerse themselves in their settings first and avoid looking at whatever literature is available until one's own impressions are, well, less impressionable (personal communication). The Adlers suggest that researchers should begin to "compare their typologizations to those in the literature" only once they begin to "grasp the field as members do." Second, whatever decision you make about when to review the literature on your own topic, we strongly recommend spending some time leafing through earlier examples of participant or nonparticipant observation, just to see what others have done. Particularly rich sources of such examples are the *Journal of Contemporary Ethnography* (which was, in earlier avatars, *Urban Life* and *Urban Life and Culture*) and *Qualitative Sociology*.

Gaining Access and Ethical Concerns

Although choosing topics and selecting sites are important intellectual and personal tasks, gaining access to selected sites is, as Lofland, Snow, Anderson, and Lofland (2005) point out, an important *social* task. At this stage, a researcher must frequently use all the social skills or resources and ethical sensibilities she has at her command.

Lofland et al. (2005) suggest that the balance between the need for social skills or resources and ethical sensibilities, however, can depend, as much as anything, on whether or not the observer plans to announce his or her presence as observer. If the observer plans not to reveal the intention to observe, the major issues entailed in gaining (or using) access are often ethical ones (see Chapter 3), especially because of the deceit involved in all covert research. This can be a particularly knotty problem when one seeks "deep cover" in a site that is neither public nor well known, as was the case for Patricia Adler in her study of drug dealers and smugglers (1985), but it exists nonetheless even when one plans to study public places (as Humphreys did while watching impersonal sex in public restrooms) or to study familiar places (as you might do if you decided to study your research methods class today while you continued playing your role as student and as Nathan did when she re-entered the college she teaches at as a slightly older student [2005, especially 158–168]). According to Lofland and Lofland (2005), the decision to engage in covert research and thereafter to establish access, however, is ethically acceptable, if other concerns, such as ensuring lack of harm to those observed and pursuing worthwhile topics in settings that cannot be studied openly, neutralize or overwhelm concerns about deception. Nathan, for instance, found herself, in compensation for failing to reveal her research purpose in the field, suppressing the urge to write about informal conversations she'd had with students about, say, cheating, as she engaged in the act of writing because she realized that reading about such conversations in a book (written by a professor at the students' own university) could generate discomfort and even fear in the students she referred to (2005, 165–166).

Gaining access to, and then getting along in, a site becomes much more a matter of social skill or resources when the observer intends to announce his or her observer role. The degree to which such skills or resources are taxed, however, is generally a function of one's current participation in the setting. Furthermore, for Adler and Adler, who were already participants in many afterschool settings, at least early in their study, their pre-study participation not only made access easy, but also made it relatively easy to decide whom they needed to tell about their research, whom they needed to ask permission from, and whom they needed to consult for clarifications about meanings, intentions, contexts, and so on. Ashley Mears (Mears and Finlay, 2005) found access to models in the Atlanta fashion industry very easy because she, even as an undergraduate and graduate student, already was a model in that industry. Ease of access, then, is certainly one major reason why the Adlers advocate "starting where you are."

STOP AND THINK   *Suppose you wanted to study a musical rock group ("The Easy Aces") on your campus, to which you did not belong. How might you gain access to the group for the explicit purpose of studying the things that keep them together?*

Access particularly taxes social skills or resources when you want to do announced research in a setting outside your usual range of accustomed settings. In such cases, Lofland et al. (2005: 41–47) emphasize the importance of four social skills or resources: connections, knowledge, accounts, and courtesy. If, for instance, you want to study a rock group ("The Easy Aces") comprising students on your campus, you might first discuss your project with several of your friends and acquaintances, in the hopes that one of them knew one of the "Aces" well enough to introduce you. In so doing, you'd be employing a tried and true convention for making contacts—using "who you know" (*connections*) to make contact. In general, it's a good idea to play the role of "learner" when doing observations, but such role-playing can be overdone on first contact, when it's often useful to indicate some *knowledge* of your subject. Consequently, before meeting "The Easy Aces," it would be a good idea to learn what you can about them—either from common acquaintances or from available public-access information. (If, for instance, "Aces" had made a CD, you might want to listen to a copy and refer to their songs, perhaps even admiringly, at your first meeting.) You'd also want to have a ready, plausible, and appealing explanation (an **account**) of your interest in the group for your first encounter. The account needn't be terribly clever, long-winded, or highbrow (something reasonably direct like "I'd like to get an idea of what keeps a successful rock group together" might do), but it might require some thought in advance (blurting out something like "I'm doing a research project and you folks seem like good subjects" might not do). Finally, and perhaps most commonsensically, a bit of common *courtesy* (such as the aforementioned expressions of appreciation for the "Aces'" music) can go a long way toward easing access to a setting.

account, a plausible and appealing explanation of the research that the researcher gives to prospective participants.

gatekeeper, someone who can get a researcher into a setting.

All the aforementioned skills and resources—connections, knowledge, accounts, and courtesy—can be more difficult to muster effectively when there are gender, class, language, or ethnic differences between the researcher and other actors. Convincing **gatekeepers,** or persons who can let you into a setting, can be hard work, as Michelle Wolkomir discovered when she tried to gain access, as a heterosexual Jewish woman, to an ex-gay ministry in a southern, conservative Christian church. She spent five months meeting sporadically with the group leader and the church board to convince them that she would not undermine the security of the men using the ministry to "cure" their homosexuality and that she would not inhibit group talk (2001, 309–310).

Still, the eventual success of Wolkomir's project, and that of other female observers in mixed-gender settings, as well as that of white-male observers, like Liebow, who, for *Tally's Corner* (1967), studied poor, urban, African American males, suggests that even gender, race, class, and sexual-preference divides are not necessarily insurmountable.

Participant and nonparticipant observations, perhaps more than any other techniques we've dealt with in this book, lend themselves poorly to "cookbook" preparations. We can't tell you precisely what to do when you come up against bureaucratic, legal, or political obstacles to your observations, but we can tell you that others have encountered them and that authors like Lofland et al. (2005) provide some general strategies for surmounting them.

Gathering the Data

Perhaps the hardest thing about any kind of observation is recording the data. How does one do it?

STOP AND THINK *Quick. We're about to list three ways of recording observations. Can you guess what they are?*

Three conventional techniques for recording observations are writing them down, recording them mechanically, as on a tape recorder or camera, or digitally (perhaps with an IPod), and recording them in one's memory, to be written down later (only shortly later, one hopes). Given the notorious untrustworthiness of memory, you might be surprised to learn that the third technique, trusting to memory for short periods, is probably the most common. Despite gaining the explicit permission of his subjects, Liebow (1993) waited until he got home at night to write down what he'd observed about homeless women. So did Slavin (2005) in his study of activities in a gay nightclub. Both were cognizant that even their relatively well-trained powers of observation and memory were subject to later distortion (whose aren't?) and so were at pains to record their observations as soon after they'd made them as possible. Their goal, the goal of all observers, was to record observations, including quotations, as precisely as possible, as soon as possible, before the inevitably distorting effects of time and distance were amplified unnecessarily.

Adler and Adler admit to being not "as thorough about this phase of research as others" (personal communication). Occasionally, they would come back home and write "jotted fieldnotes." But one of the advantages of their "team" approach was that when the time came "to make outlines and cull the data for examples," they could brainstorm "together to try to recall examples and talk to each other about what happened and what was said."

Why not record them on the spot? Joy Browne, in her (1976) reflections on her study of used car dealers, gives one view when she advises prospective observers,

> [D]o not take notes; listen instead. A panicky thought, perhaps, but unless you are in a classroom where everyone is taking notes, note taking can be as intrusive as a tape recorder and can have the same result: paranoia or influencing the action. (1976: 77)

The intrusiveness of note-taking, tape-recording, digital-recording, and picture-taking, then, make them all potentially troublesome for recording data.

On the other hand, shorthand notes and tape recordings are certainly appropriate when they can be taken without substantially interrupting the natural flow. Frances Maher and Mary Kay Tetreault (1994), for instance, taped classes of professors to discern varieties of feminist teaching styles. Neither the NORC researchers who used the slow-moving SUV to tape-record the goings-on on Chicago streets (Sampson and Raudenbush, 1999) nor Duneier, when he taped ordinary street conversations by keeping the recorder running on a milk crate under his vending table (Duneier and Molotch, 1999), worried about disturbing the scenes they were recording. In fact, all assumed they weren't. A problem of tape-recording, however, is that one can gather daunting quantities of data: It can take a skilled transcriber three hours to type one hour of tape-recorded conversation (Singleton, Straits, Straits, and McAllister, 1988: 315). And even if you're able to take notes or tape-record, it is still wise to transcribe and elaborate those recordings as soon as possible, while various unrecorded impressions remain fresh enough that you also can record them.

Participant and nonparticipant observers commonly supplement their observations with interviews and available data. Adler and Adler refer to informal interviews with "a range of young people," a Kidsport flyer, and a letter to coaches. Interviewing other participants, known as **informants,** frequently provides the kind of in-depth understanding of what's going on that motivates researchers to use observation techniques in the first place. One concern raised by feminist approaches to observation, in fact, is how to retain as much of the personal perspective of informants as possible (for example, Reinharz, 1992: 52). How, in effect, to give "voice" to those informants? Some observers (for example, Maher and Tetreault, 1994) have even asked permission of their respondents to report their names and have thus abandoned the usual practice of offering confidentiality or anonymity. We recommend, however, that confidentiality or anonymity be offered to

informants, participants in a study situation who are interviewed for an in-depth understanding of the situation.

all interviewees (see Chapters 3, 9, and 10) and also be given to interviewees in the field. Informants who later see their words used to substantiate points they themselves would not have made, for instance, might regret having given permission to divulge their identities in print, no matter how generous the intention of the researcher.

Analyzing the Data

The distinction between data collection and data analysis is generally not as great for participant and nonparticipant observation as it is for other methods (for example, questionnaires, experiments) that we've discussed. (See Chapter 15, where we discuss qualitative data analysis in greater depth than here.) For, although these other techniques tend to focus on theory verification, observational techniques are more often concerned with theory discovery or generation. Thus, Adler and Adler have generated a theory of afterschool activities that speaks to their functions in American society (for example, the preparation of characteristics that are amenable to corporate labor) as well as their classification (for example, as recreational, competitive, and elite). This is not to say that observational studies can't be used for theory verification or falsification. Perhaps the primary theoretical upshot of Harrell and his colleagues' (cited in Bakalar, 2005) study of parent-child interactions in supermarkets is to falsify the notion that the attractiveness of the children has no effect.

Nevertheless, more studies based on observation techniques are like Adler and Adler's than like Harrell and his colleagues'—more concerned, that is, with theory generation or discovery than with theory verification or falsification. Important questions are these: When does the theory building begin and how does it proceed? The answer to the first question—when does theory building begin?—seems to be soon after you record your first observation. Let's say, for instance, that one of Adler and Adler's early observations was of the soccer game in which parents screamed and shouted on the sidelines. The very act of recording that observation, rather than whatever else they might have recorded (for example, meteorological conditions, the presence or absence of hot-dog concessions, the kind of uniform worn by players, and so on), implies that the observer attaches some importance or relevance to the observation. This choice is likely to be based on some preconceived notion of what's (theoretically) salient, a notion that the sensitive researcher will, some time during or after the initial observation, try to articulate to himself. Thus, Jessica Fields (2005) found herself writing "notes on notes" as she wrote up her near-verbatim fieldnotes on her observation of a southern county school board's meetings on the proposed ways sexuality education would be taught in its schools, as did Michelle Wolkomir (2001) in her study of Christian support groups.

Once the researcher does articulate these notions, they become concepts or hypotheses, the building blocks of theory. At this point, the researcher can begin to look for behaviors that differ from or are similar to the ones observed before.

STOP AND THINK *Suppose an early observation led Adler and Adler to the suspicion that parents (and coaches) of children who are in activities described as primarily "for fun" will occasionally help create a competitive atmosphere. What other kinds of observations might they make that would tend to confirm this suspicion? What kinds of observations might they make that would tend to disconfirm it?*

grounded theory, theory derived from data in the course of a particular study.

The pursuit of similarities can lead to the kind of generalizations on which **grounded theory** (Glaser and Strauss, 1967, Glaser, 2005), or theory derived from the data, is based. Hence, the observation, repeated many times, that older children tend to get involved in one activity, shunning the multiple activities of earlier childhood, led Adler and Adler to the concept of a continually narrowing funnel of activities. But this "observation" is really founded as much on differences as on similarities, isn't it? For instance, it depends on the observation of differences between what younger and older children tend to do when they participate in afterschool activities. Moreover, apparently disconfirming observations (like "not everyone wants to make this commitment") are acknowledged to suggest the probabilistic, rather than absolute, nature of the generalizations that emerge.

The process of making comparisons, of finding similarities and differences among one's observations, is the essential process in analyzing data collected during observations. It is also the essential ingredient for making theoretical sense of those observations. Through it, Humphreys (1975) came up with his typology of those who used public restrooms for impersonal sex. Through it, Liebow (1993) discovered the processes that created group solidarity among homeless women in shelters; Erikson (1976) came up with the concept of "collective trauma"; Hochschild (1983) developed the concept of "emotive dissonance" (the strain that develops when one's occupation makes one feign what one is not feeling); and Wolkomir (2001) and Mears and Finlay (2005) used Hochschild's concept of "emotion work" to explain how gay and ex-gay Christians, on the one hand, and fashion models, on the other, develop or maintain dignity.

By now, it might have occurred to you that the analysis of observed data is not so very different from what you do in everyday life to make sense of your world. But the metaphor of everyday thinking, and especially of the simple pursuit of similarities and differences, surely understates the complexities involved in analyzing observed data. The Adlers stress this point when they speak of the varied sources of theoretical insight in their larger opus, as well as in their afterschool piece:

> On the issue of analysis and theory-building, we have taken different tacks in the various papers we have written. For some, the patterns and their implications jumped out at us easily and in a clear-cut manner. For others, we see contradictory patterns and have to mull things over for months and years to try to grasp the underlying themes explaining how and why things work the way they do. We interview all types of participants, alone and in groups, to try to reconcile conflicting

interpretations into one that incorporates all the data. This is not always smooth, as life is contradictory and complex. Usually, we spend 18 months to 2 years on a given article, and begin to mull over ideas at the beginning, working on them throughout the whole time. Anything in our life can pop up and make us think about an issue and analyze or re-analyze our thoughts on it. . . . In the case of the afterschool paper, people kept challenging us to take a side on the issue, determining if it was good or bad. There was a lot of pressure placed on us to condemn afterschool activities (by scholars and reporters), so this made us think about these issues more and rehash them between ourselves. . . . No matter what, we kept seeing both good and bad elements in these activities. Finally, we had to go with this dual approach, despite the pressure to morally condemn these programs. Ultimately we asked, "Why do they all do it if it's so bad?" (personal communication)

STOP AND THINK   *Do you think the Adlers achieved the even-handedness they aspired to in their analysis? Do you think they successfully avoided "condemning afterschool activities?"*

Advantages and Disadvantages of Observational Techniques

The advantages of observational techniques mirror the reasons for using them: getting a handle on the relatively unknown, obtaining an understanding of how others experience life, studying quickly changing situations, studying behavior, saving money (see the subsection labeled "Uses of Observations" earlier in this chapter). We'd like to emphasize, once again, the possibility of gaining an in-depth understanding of a natural situation or social context, including a sense of what might be called an "insider's view" of that situation and context. We reiterate that Adler and Adler might well have used a survey, or even an experiment, to study afterschool activities, but it is very unlikely that they would have been able to tell us such a plausible and detailed story about the meaning of such activities unless they'd actually gone and observed the activities as they naturally occurred.

Adler and Adler (1994a) suggest that one of the chief criticisms of observational research has to do with its validity. They claim that observations of natural settings can be susceptible "to bias from [the researcher's] subjective interpretation of situations" (Adler and Adler, 1994a: 381). They are surely right. But compared with measurements taken with other methods (for example, surveys and interviews), it strikes us that measures used by observers are especially likely to measure what they are supposed to be measuring and, therefore, be particularly valid. We can imagine few combinations of questions on a questionnaire that would elicit as clear a picture of the funneling process that leads some young people to abandon some activities while concentrating on others as the statement by the ten-year-old boy quoted by Adler and Adler:

I'm not going out for soccer this year. It was boring last year because I didn't get to play enough, and I had a problem because I was in a play and there was an overlap between the play and soccer . . . So I think I'll just stick with acting this year. I don't have any friends in it, but I like it more and I'm better at it. It's more *me*. [From the Focal Research earlier in this chapter.]

A final advantage of observation is its relative flexibility, at least compared with other methods of gathering social data. Observation is unusually flexible for at least two reasons: one, because a researcher can, and often does, shift the focus of research as events unfold and, two, because the primary use of observation does not preclude, and is often assisted by, the use of other methods of data gathering (for example, surveys, available data, interviews). Thus, as their study progressed, the Adlers' focus shifted from providing an outline of the characteristics of available afterschool activities to one of assessing a model of developmental progression that young people frequently follow in the course of pursuing extracurricular interests, even as their primary method of data acquisition alternated among observation, interview, and using available data.

Generalizability is a perpetual question for participant and nonparticipant observation alike, however. This question arises for two main reasons. The first has to do with the reliability, or dependability, of observations and measurements. One example will stand for the rest. In Erikson's (1976) study of the effects on a flood on one community, he concludes that the community experienced "collective trauma" because all the individual presentations given by survivors were "so bleakly alike" in their accounts of "anxiety, depression, apathy, insomnia, phobic reactions, and a pervasive feeling of depletion and loneliness" (1976: 198). But it remains a question whether another observer of the same accounts would have detected the bleak sameness that Erikson did, much less come to the conclusion that that sameness was an indicator of "collective trauma," rather than of many individual traumas. One way to check whether one's observations are reliable, of course, is to ask other participants whether they seem right to them. This is what Slavin (2005) did as much as possible in his study of a gay nightclub. But sometimes such a strategy can't work, as, for instance, when one doesn't disclose his or her research purpose to others (as, of course, Nathan didn't in her [2005] study of college students) or when one thinks one might elicit inaccurate interpretations of the situation by asking other participants about it (as, for instance, Liebow (1993: 321) did in his study of homeless women).

The second reason that questions of generalizability arise is a direct result of the nonrandom sampling procedures used by most observers. The statement of the ten-year-old boy quoted by Adler and Adler is, as far as we're concerned, compelling evidence that he was funneling his activities as he grew older. We also have no doubt that the bulk of the information that Adler and Adler collected about children of his age and older was consistent with a similar funneling process. In short, we think funneling occurred within the sample they examined, but whether it occurs among some larger population of children in the United States depends on how generalizable their sample was. And because that sample was not randomly selected (how could it have been?), the question of its generalizability is one that, of necessity, must wait another day (and replication [by one of you?] or, perhaps, the use of another method).

Participant observers are usually their own best critics when it comes to the generalizability of their findings. Adler and Adler, for instance, in the (1998) book, *Peer Power: Preadolescent Culture and Identity,* which summarizes much of their parent-as-researcher work including the work on afterschool activities, are quick to point to the limitations of their observations by acknowledging the special nature of the community they studied:

> The community encompasses a population of about ninety thousand; it is predominantly white and middle- to upper-middle class. Minority groups, of which Hispanics are the largest, constitute less than 10 percent of the population. Like the large, state university located within it, the city is heavily oriented toward research and development, with some small industry. A clean, environmentally concerned town with an active recycling program and a slow-growth and liberal political environment, it is geared toward fitness, promoting outdoor athletics of many kinds, especially individual, nontraditional pursuits. (1998: 5)

STOP AND THINK   *Can you imagine how the generalizability of the "funneling" concept of Adler and Adler might be studied with another method? Would you be inclined to use a survey? A cross-sectional or longitudinal design? A probability or nonprobability sample?*

demand characteristics, characteristics that the observed take on simply as a result of being observed.

A third problem associated with observational methods, especially when they are announced, is the bias caused by **demand characteristics,** or the distortion that can occur when people know (or think) they are being observed. One wonderful example of demand characteristics in action comes from D. L. Rosenhan's classic (1971) study entitled "On Being Sane in Insane Places." Rosenhan was concerned about the accuracy of psychiatric diagnoses and so, with seven accomplices, he gained admission as a pseudopatient to eight psychiatric hospitals. In this first instance, all pseudopatients took the role of complete participants and, although all were as sane as you and us, none was diagnosed as sane once he or she was admitted as a patient. In a second instance, however, Rosenhan told the staff in one hospital that, during a three-month period, one or more pseudopatients would be admitted. Of the 193 patients who were admitted in the next three months, 41 were alleged by at least one member of the staff to be pseudopatients; 23 of them by attending psychiatrists. Because no pseudopatients, to Rosenhan's knowledge, had attempted to be admitted, the results of this second "experiment" suggest that even the threatened presence of observers can distort what goes on in a setting. Thus, despite the effort of most observers to create a situation in which action goes on as if they weren't there, there's little reason to believe that these efforts can be totally successful—especially when their presence is announced.

Particularly when that presence is unannounced, of course, the ethical problem of deceit comes into play. Should Rosenhan and his accomplices, in the first instance mentioned, have disguised their intention to observe when they had themselves admitted as pseudopatients (while, incidentally, claiming initially to possess symptoms none of them actually had)? The range of responses to the ethical dilemma of deceit is almost as wide as there are authors who have written about it, but there can be little doubt that observation, even when it *is* announced, frequently involves the researcher in some measure of deceit.

Three other problems of observational methods are worth emphasizing. First, in comparison with virtually every other technique we present in this book, observation can be extremely time-consuming. The Adlers speak of spending from 18 months to two years on most articles, eight years on the observations for the book, *Peer Power;* Venkatesh spent nearly four years doing the participant observation for his article about a street gang; Liebow saw about nine years pass between the conception and publication of his book on homeless women. Second, and clearly related to the issue of time consumed, is the problem of waiting around for events "of interest" to occur. In almost every other form of research (administering a questionnaire, an interview, an experiment, or collecting available data), the researcher exercises a lot of control over the flow of significant events. In the case of field observations, control is handed over to those being observed—and they might be in no hurry to help events unfold in an intellectually interesting way for the observer. Third, although observation can be fun and enlightening, it can also be terribly demanding and frustrating. Every stage—from choosing appropriate settings, to gaining access to settings, to staying in settings, to leaving settings, to analyzing the data—can be fraught with unexpected (and expected) difficulties. Observation is not for the unenergetic or the easily frustrated.

Summary

Although observation is often used in experiments, the observational techniques we've focused on here—those used by participant and nonparticipant observers—are those used to observe actors who are, in the perception of the observer, acting naturally in settings that are natural for them. Adler and Adler, for instance, studied children, parents, and other adults as they participated in the wide variety of natural settings that have been created for adult-organized afterschool activities. Observational techniques are frequently used for several purposes: to get a handle on relatively unknown social phenomena (a major aim of the Adlers), to obtain a relatively deep understanding of others' experiences (another aim of the Adlers), to study quickly changing situations, to study behavior, to save money, and to avoid ethical problems associated with other techniques.

A sense of the variety of roles a participant or nonparticipant observer can take is suggested by Gold's (1958) classic typology. A researcher can become a complete participant or a complete observer, both of which imply that the researcher withholds his or her research purposes from those being observed. Although both of these roles entail the advantage of relative inobtrusiveness, they also entail the disadvantages of incomplete frankness: for example, of not being able to ask probing questions and of deceiving one's subjects. Both the participant-as-observer and the observer-as-participant do reveal their research purposes, purposes that can prove a distraction for other participants. On the other hand, by divulging those purposes, they earn the advantage, among others, of being able to ask other participants detailed questions about their participation. The Adlers were, alternately, participants-as-observers and observers-as-participants in their study of afterschool activities, and thereby gained the methodological advantage, among others, of being able to ask other participants about the meaning of events for them.

Observers can choose their research settings for their theoretical salience or their natural accessibility, or for some combination of theory and opportunity. The Adlers emphasize the natural accessibility in their own account of why they chose to study afterschool activities. In some instances, it might be possible to sample multiple settings randomly, and thereby maximize their representativeness, but usually the selection of such settings is done in a nonrandom fashion (as it was by the Adlers). Generally, participant and nonparticipant observation begin without well-defined research questions, measurement strategies, or sampling procedures, though systematic observations can involve some combination of these elements. In all cases, however, researchers will want to decide how much of themselves and their research they will want to disclose to others in the research setting. At some point they may also need to employ social skills to gain access to the setting from gatekeepers and to learn what they need to learn from other actors.

Once in a setting, researchers can record observations by writing them down, by using some mechanical device (for example, tape recorder or camera), or by trusting to memory for short periods. Most often observers, like the Adlers, prefer trusting to memory because alternative techniques involve some element of intrusiveness, but in all cases, the effort to record or transcribe "notes" should be made soon after observations are made, improving the chances for precise description and for recording less precise impressions. The analysis of the data collected during observations involves comparing observations with one another and searching for important differences and similarities.

The advantages of participant and nonparticipant observation mirror the purposes for using them and include the likelihood of valid measurement, and of considerable flexibility. Both types of observation are particularly useful for gaining an in-depth understanding of particular settings or situations. On the other hand, disadvantages of observational techniques include almost unavoidable suspicions that they might yield unreliable and nongeneralizable results. Observation can also affect the behavior of those observed, especially when the observation is announced, and contains an element of deceit, especially when the observation is not announced. Observation can also be a particularly time-consuming and sometimes frustrating path to social scientific insight.

Observing Gestures in a Film or TV Program

For this exercise, which we've adapted from ones suggested by Tan and Ko (2004, 117–118), you'll need to find a film or TV program that you can watch with another student to observe mannerisms (gestures) and facial expressions. We want you to discuss your observations with your partner and prepare a two-page essay in which you discuss the expression of emotions through gestures and facial expressions in one scene of the film or text. We'd like you to show off how perceptive your observations are. Be sure your essay mentions the title of the film or TV program and gives a one- or two-sentence synopsis of the characters and action in the scene. Then outline the emotions you observe in the various characters involved. What are the actions/expressions you observe that lead you to your conclusions about these emotions? You need to clearly distinguish the mannerisms and expressions from what they mean. (For instance, you might point out that a character raised his/her voice or motions with an index finger [expressions or mannerisms you observe] and you might say the person looked angry [your conclusion about the emotional meaning of the expressions]). This means you need to distinguish the action you observe from how you interpret what the particular action means. **Please do not write about the storyline beyond your brief synopsis of the action.**

The Institutionalization of Afterschool Activities Revisited

For this exercise, you'll need to locate two similar settings in which children engage in afterschool activities. (Hint: if you don't know where such settings are, you might try to find informants who can help. You could call a local YMCA or recreational center, for instance, and ask about organized activities for children near you.) You should try to play the role of a pure observer in one—that is, you should not announce your research intention. (You might pretend to be an interested observer, which, of course, you are.) In the other, you should try to play the role of participant-observer—that is, you should try primarily to observe, but only after announcing your intention to do so to as many of the participants as you can—perhaps even to the assembled group before it starts the activity. Tell them you are a student, collecting information for a homework assignment. Tell them what school you're from. Tell the truth. Use the same data-recording technique in both settings. If you choose to take notes in the one, take notes in the other. If you choose to tape-record in one, tape-record in the other. If you choose to trust to your memory, short-term, in one, do the same in the other. Your research question in each setting is this: To what extent is competitiveness encouraged in this setting? You should spend at least a half-hour in each setting. Once you leave, you should write up your data: Don't summarize what the participants have done or said; try to record what they've done and said as precisely as you can. Then summarize your overall impressions and respond to the research question. Finally, answer the following questions:

1. Did you notice any differences in the ways your two groups acted toward you or toward one another that you might attribute to your announcement of your research purpose in one and your lack of such an announcement in the other?

2. Which setting did you prefer to observe? Why?

3. Discuss the extent to which you think competitiveness was encouraged in each afterschool setting by children and adults. Give support for your conclusion by summarizing some portions of your field notes.

4. Compare your research findings with those of the Adlers.

EXERCISE 11.3

Participant Observation

Try to replicate what you've done for Exercise 11.2, only this time with two groups you're already a member of (for example, your family [at dinner?], [a meeting of?] a school organization, [a meeting of?] a class at school). (For best results, try to make the settings somewhat comparable. You probably don't want to study your own family at dinner, one time, and a dinner in which you meet your fiancé[e]'s parents at another time.) In one of the groups, do not announce your research purpose; in the other, make a general announcement. Again, for the second group, tell them of your assignment and your plan to take notes on what goes on. Your research question, again, is this: How competitive is this group? Use the same note-taking regime you used in Exercise 11.2 for both of these observations. Write up your notes, again trying carefully to make precise accounts of what went on and what was said, and then answer these questions:

1. Did you notice any differences in the ways these two groups acted toward you or toward one another that you might attribute to the announcement of your research plan in one and the absence of that announcement in the other?

2. Did you feel any different in these two settings?

3. Compare your experiences in the two exercises, 11.2 and 11.3. Which of the four observations felt most comfortable to you? Which was the least comfortable? Why? In which do you feel you gained the greatest understanding of the setting? In which did you gain the least? Why? In which did you learn the most about the setting (a different question)? In which did you learn the least? Why?

4. Which of the groups examined for 11.2 and 11.3 seems to have been most competitive? Give support for your conclusion by summarizing some portions of your field notes.

Using Available Data

LIDIAN EMERSON
In her youth an unusual sense of
the Divine Presence was granted her
and she retained through life
the impress of that high Communion
To her children she seemed in her
native ascendancy and unquestioning
courage a Queen. a Flower in
elegance and delicacy.
The love and care for her husband and
ldren was her first earthly interest
it with overflowing compassion
art went out to the slave. the sick
e dumb creation. She remembered
were in bonds, as bound with them.

© Emily Stier Adler

Introduction

You've probably heard immigration blamed for a lot of what ails us as a nation: the competition for skilled and unskilled jobs, demands on educational and other social services, and violent criminality, to name but three. Suppose you wanted to test one of these assertions. How might you do it? Let's say you wanted to focus on the connection between immigration and criminal violence. You might try participant observation (see Chapter 11), but the chances of seeing anything of interest in any particular immigrant (or non-immigrant) community probably aren't that great. You might use a questionnaire (see Chapter 9) or even qualitative interviews (see Chapter 10) to determine whether immigrants are more likely than non-immigrants are to admit to having committed violent crimes, but the chances of receiving reliable responses (see Chapter 6) are pretty small (for all groups). Who's going to volunteer, after all, that they've committed a violent crime?

A concern with the crime/migrant connection moved Ramiro Martinez and Matthew Lee, the authors of this chapter's focal research. The two concerns, about crime and migration, didn't develop at the same time, though. Martinez asserts that he first became interested in crime, then in migration: "When in graduate school, about ten years ago, I asked somebody a question about Latino (Hispanic) homicides. . . . At the time we knew that homicide was the leading cause of death among young African American males but little attention was paid to Latinos. Only once I began to collect Latino data did the question of immigration status occur to me" (personal correspondence).

available data, data that are easily accessible to the researcher.

Martinez and Lee's paper illustrates the use of **available data** in the social sciences. Available data are, as the term suggests, data that are available to the researcher. Available data are sometimes actually statistics, or summaries of data, that have been computed from data collected by a large organization—e.g., governments (as when they publish census or crime statistics) or businesses (as when they prepare for annual reports). Such statistics, when they have been made available in statistical documents, are called **existing statistics.** If you're sitting at your computer, for instance, you can try the website www.fedstats.gov/, "the gateway to statistics from over 100 U.S. Federal agencies," and see how quickly you can find the population of the country or of your state.

existing statistics, summaries of data collected by large organizations.

Available data are sometimes actual *data*, as opposed to statistics, collected by others, data that the researcher analyzes in a new way. Data that have been collected by someone other than the researcher are called **secondary data.** Secondary data are frequently contrasted with **primary data,** or data that are collected and used by the same researcher. The data collected by the Adlers for the focal piece on afterschool activities are a good example of primary data, as are the data collected by Gray, Palileo, and Johnson for the piece on rape blame attribution. Survey data that have been made available for others to reanalyze are perhaps the best known kind of secondary data. The General Social Survey, conducted now every other year by the National Opinion Research Center (NORC) to collect data on a large number of social science variables, is probably the most popularly used set of survey data by

secondary data, research data that have been collected by someone else.

primary data, data that the same researcher collects and uses.

social researchers, but it's just one of hundreds of sets of survey data that have been subjected to secondary analysis. Yet secondary data don't need to be survey data; they just need to have been collected by someone other than the current researcher. Thus, one of Martinez and Lee's main sources are the "logs" Miami police created to help them solve homicide cases. These became secondary data as soon as Martinez and Lee adapted them for their research purposes. We've chosen to focus on Martinez and Lee's type of secondary data analysis, not because it's especially typical of research based upon available data, but because it's a particularly interesting version of it. The use of available data generally, however, and of secondary survey data in particular, became the dominant mode of social research toward the end of the twentieth century. The increasing popularity is suggested because although Roger (Clark, 1999) found that only about 12 percent of a probability sample of sociology articles published in 1956 involved the analysis of data that had been collected by governments, organizations, or private researchers (i.e., available data), 36 percent did so in 1976, and almost 50 percent did in 1996. Fully 39 percent of articles published in 1996 were based on survey data collected by one group and reanalyzed by the authors of the articles (i.e., secondary data).

A particularly accessible form of available data (for research purposes) is that which is presented in published reports of government agencies, businesses, and organizations. Many studies of crime and violence have relied on national statistics gathered and forwarded to the FBI's *Uniform Crime Reports* (UCR). One such study, reported by Martinez and Lee, was the U.S. Commission on Immigration Reform (1994) investigation that, among other things, compared the city of El Paso's overall crime and homicide rates to those of similar-sized cities in the United States. El Paso is a town along Mexico's border and therefore has a high percentage of migrants within its population. The Commission's finding that El Paso's overall crime and homicide rates were much lower than those in cities of comparable size supports the belief that migrants don't commit more crimes, and particularly more homicides, than nonmigrants. The Commission's evidence, however suggestive, is nonetheless subject to a famous criticism: that it commits the **ecological fallacy.** A fallacy is an unsound bit of reasoning. The ecological fallacy is the fallacy of making inferences about certain types of individuals from information about groups that might not be exclusively composed of those individuals. The Commission wanted to see whether cities with high proportions of migrants had higher crime rates than did cities with lower proportions of migrants, and so using cities as a unit of analysis made a great deal of sense. So far so good. But does its finding that a city with a high proportion of migrants had a lower crime rate than comparable cities with lower proportions of migrants mean that you should conclude that migrants are less likely to commit crimes than nonmigrants? Not legitimately. Such a conclusion risks the ecological fallacy because it might be that the migrants are committing most of the crimes in the cities with low proportions of migrants. The problem is that we have looked at cities as our units of analysis and that, whatever the proportions of migrants, all cities are composed of both migrants and nonmigrants. Another example of the

ecological fallacy, the fallacy of making inferences about certain types of individuals from information about groups that might not be exclusively composed of those individuals.

ecological fallacy would occur if we inferred that individuals who are unemployed are more likely than those who are employed to suffer from mental illness from a finding that a city with a particularly high unemployment rate also has a high rate of mental illness. (It could be, for instance, that it is the employed that are mentally ill.)

STOP AND THINK *You might remember that Émile Durkheim (in his classic sociological study,* Suicide*) found that Protestant countries and regions had higher suicide rates than Catholic countries and regions did. He inferred from this the conclusion that Protestants are more likely to commit suicides than Catholics. Is this a case of the ecological fallacy? Explain.*

Martinez and Lee, you will see, have devised an especially clever way of avoiding the ecological fallacy. To do so, they had to dig a little deeper than many researchers using available data do: in this case, into the "homicide logs" of the Miami Police Department. In using these logs, however, they show us many of the virtues of using available data.

FOCAL RESEARCH

Comparing the Context of Immigrant Homicides in Miami: Haitians, Jamaicans, and Mariels

by Ramiro Martinez, Jr., and Matthew T. Lee[1]

Are new immigrants more crime prone than native-born Americans? This issue has been debated since the early 1900s and continues to fuel controversy on the "criminal alien" problem in contemporary society (Hagan and Palloni, 1998; Short, 1997). Some scholars speculate that immigrants engage in crime more than other population groups by using the growing presence of the foreign born in the criminal justice system as evidence of heightened criminal involvement (Scalia, 1996). Other writers encourage legislation designed to rapidly deport immigrant criminals, such as California's Proposition 187 and the U.S. Congress 1996 Immigration Reform bill, despite any systematic evidence of an "immigrant crime problem" (Beck, 1996; Brimelow, 1995). Still others maintain that large labor pools of young immigrant males willing to work for lower wages than native-born Americans, in particular inner city residents, exacerbate ethnic conflict and contribute to urban disorders (Brimelow, 1995). The consequence of immigration, according to this view, is a host of social problems, especially violent crime (Beck, 1996).

[1] This article is adapted with permission from Ramiro Martinez and Matthew T. Lee's (2000) "Comparing the Context of Immigrant Homicides in Miami: Haitians, Jamaicans, and Mariels," *International Migration Review,* Vol. 34, 794–812.

This study probes the immigration and crime link by focusing on where immigrants live and how immigration might increase or decrease violent crime, at least as measured by homicide in a city that serves as a primary destination point for newcomers. Immigrants are undoubtedly influencing urban crime, but they are also potential buffers in stabilizing extremely impoverished areas (see Hagan and Palloni, 1998). Our primary argument is that while immigration shapes local crime, it might have a lower than expected effect on violent crime rates. The result is areas with high proportions of immigrants but low levels of violence, relative to overall city rates. The key question is whether immigrants have high rates of violence or if immigration lessens the structural impact of economic deprivation and thus violent crime.

It is, of course, preferable to study the impact of immigration and social institutions on violence across time and in diverse settings. While we do not directly test these variables with a multivariate analysis, we provide a starting point for this research agenda by conducting one of the first contemporary studies of immigrant Afro-Caribbean (Haitians, Jamaicans, and Mariel Cubans) homicide in a major U.S. city.

Previous Immigration and Crime Studies

Evidence of high immigrant crime rates is mixed (Shaw and McKay, 1931; see Short, 1997 for review of the literature). Lane (1997) notes that white newcomers (e.g., Irish, Italian) were often disproportionately involved in violent crime in the turn of the century Philadelphia and New York City. Crime rates declined, however, as settlers from abroad assimilated into mainstream society and integrated into the economy (Short, 1997). This finding is particularly instructive, since most immigrant victims and offenders were young males without regular jobs and presumably fewer ties to conventional lifestyles, suggesting that, even a hundred years ago, gender and economic opportunities played a major role in violent crime (Lane, 1997).

More contemporary studies on immigrant violence, however, have been scarce (Short, 1997). Many crime and violence studies rely on national statistics gathered from police agencies and forwarded to the FBI's Uniform Crime Reports (UCR). The UCR typically ignores the presence of ethnic groups such as Latinos or Afro-Caribbeans and directs attention to "blacks" or "whites" in compiling crime data. As a result, comparisons beyond these two groups are difficult to examine in contemporary violence research (Hawkins, 1999).

Furthermore, the handful of recent studies that have explored the connection between immigrants and crime have focused on property crimes or criminal justice decision-making processes (*cf.* Pennell, Curtis, and Tayman, 1989). And, despite two studies (the U.S. Commission on Immigration Reform, 1994, and Hagan and Palloni, 1998) that suggest metropolitan areas with large immigrant populations have lower levels, or at least no higher levels, of violent and property crime than other metropolitan areas, the impression remains that immigrants are a major cause of crime (Hagan and Palloni, 1998: 378).

Research Setting

The city of Miami, Florida, provides a unique opportunity to examine the issue of whether or not immigration influences violent crime. Miami is one of the nation's most ethnically diverse cities and prominently known for its hot weather, natural disasters, waves of refugees, and tourist killings. It is also well known for its record high rates of violence. At various points throughout the 1980s, the violent crime rate was over three times that of similarly sized U.S. cities, and the city was widely characterized as a "dangerous" city in the popular media (Nijman, 1997).

This perception is linked, in part, to the presence of its large immigrant population (Nijman, 1997). After several distinct waves of immigration from Latin America since 1960, Latinos, mostly Cubans, comprised almost two thirds of the entire population in 1990 and had grown to almost one million in surrounding Dade County (Dunn, 1997). In 1980 alone, the number of Cubans increased dramatically as the Mariel boatlift brought 100,000 persons to the United States, and most of these settled in the greater Miami area (Portes and Stepick, 1993). The arrival of the Mariel refugees generated a moral panic in which the newcomers were labeled as deviants allegedly released from Cuban prisons and mental hospitals (Aguirre, Saenz, and James, 1997: 490–494).

Unlike the Mariels, who were welcomed by government officials (at least initially), Haitian immigrants were alternately detained or released by the INS in the early 1980s while fleeing political terror and economic despair in their home country (Portes & Stepick, 1993). Haitians, like the Mariels, however, were also stigmatized in the popular press, not only as uneducated and unskilled boat people, but also as tuberculosis and AIDS carriers (Martinez and Lee, 1998).

Miami was also a leading destination point for another Afro-Caribbean group—Jamaicans. Thousands of Jamaicans emigrated throughout the 1970s and 1980s, after open access to England was closed. The Jamaican population in Miami-Dade County tripled between 1980 and 1990, becoming the second largest Caribbean black group in the area (Dunn, 1997: 336).

As the newcomers converged in Miami, speculation arose that crime, especially drug related types surrounding the informal economy, was growing within immigrant and ethnic minority communities (Nijman, 1997; see also Inciardi, 1992). *The Miami Herald,* for instance, reported that throughout the early 1980s (Buchanan, 1985; Schenker, 1984) Jamaican gangs operated freely in south Florida, controlling a large share of the local crack-cocaine trade and engaging in widespread violence (see also Maingot, 1992).

In sum, it stands to reason that immigration should increase absolute crime rates in Miami if it brings in a large number of people, without a corresponding emigration to surrounding areas (see Hagan and Palloni, 1998). Yet, despite the important empirical and theoretical reasons to anticipate that immigration should influence urban violence, this link has been largely ignored in the social science literature. In this article, we extend previous studies on the social mechanisms of immigration and crime by considering the violence patterns of three distinct immigrant groups in the city of Miami.

Data Collection

Data on homicides from 1980 through 1990 have been gathered directly from files in the Homicide Investigations Unit of the City of Miami Police (MPD). These "homicide logs" contain information on each homicide event, including a narrative of the circumstances surrounding each killing not readily available from other data sources. Most importantly, extensive data on victim and, when known, offender or suspect characteristics such as name, age, gender, race/ethnicity and victim/offender relationship are also noted by the investigators on the logs.

Starting in 1980, homicide detectives began to note whether the victim or offender was white, black, or Latino. As the year progressed, and in response to media inquiries, MPD personnel would also note whether the victim or offender was a Mariel Cuban refugee, Haitian, or Jamaican. These designations were based on a variety of sources. Some Mariels were identified through INS documentation such as a "Parole ID" discovered at the scene of the crime. A temporary tourist visa or INS green card was also frequently found on the victim or at his/her residence. Still other ethnic clues were gathered from witnesses, family member, friends, or neighbors including language (e.g., speaking Creole or English with West Indian accents), providing country of birth on death certificate, or names less than common to African Americans ("Michel Pierre" or "Augustine Seaga"). All of these clues taken together helped provide a more precise definition of victim and offender ethnicity.

Each homicide type is examined, including the level of gun use and circumstances surrounding each killing. The nature of the prior contact and the circumstances of offense type are often distinct and vary by ethnicity (Hawkins, 1999). Types of homicide are therefore placed into five categories. First is "other felony" or a type of homicide committed during the course of a felony other than robbery. The second category consists of "robbery" homicides. Next is "primary nonintimate" or killings among acquaintances, neighbors, friends, or co-workers. Third is "family intimate homicides" or killings between spouses, lovers, estranged partners, and immediate family members. Finally, an "unknown category" is included to account for homicides not cleared with an arrest.

Since our focus is on comparing and contrasting immigrant homicides, we concentrate our account on immigrant Caribbean killings and their characteristics across two time points, 1980 and 1990. Specifically, the average number of homicide suspects (offenders) per 100,000 population from 1980 through 1984 is used for the first, and the average over 1985 to 1990 for the second. Total city homicides are also included as a baseline comparison. We begin by directing attention to suspect rates, but, given the current limitation of missing data, victim and offender relationships will be more fully explored.

We also use the 1980 and 1990 decennial censuses to provide estimates of the population in each city. Reliable population figures for our three immigrant groups are, of course, notoriously difficult to provide. We did, however, employ several strategies. First, Portes and colleagues (1985) estimated about 75 percent of the Mariel population (roughly 125,000) settled in the greater Miami area after the 1980 census was taken across the United States. Of that percentage, most (61.1 percent) moved into the city of Miami, at least as we can best determine (Portes, Clarke, and Manning, 1985: 40). Thus, we estimated roughly 57,281 Mariels were in the city limits between 1980 and 1990.

The number of Haitians and Jamaicans was equally difficult to estimate. Unfortunately, neither the 1980 nor 1990 census provides specific information on Jamaicans and Haitians at the city level, forcing us to use county estimates by Marvin Dunn (1997: 336) on Caribbean-born blacks in 1980 and 1990. The following group-specific rates are therefore based on the number of Caribbean blacks in Miami-Dade County born in Jamaica and Haiti. While these data are useful due to their detail regarding geography and ethnicity, we acknowledge that an undercount problem exists, especially among immigrants. It is therefore likely that the level of immigrant homicides is even lower because the denominator in the calculation of the rates is likely to be larger. With these cautions in mind, we believe that the following results are reliable guides on immigrant homicides.

Results

Between 1980 and 1990, over 1,900 homicides were recorded by the Miami Police Department; 28 percent of these victims were Caribbean immigrants and 15 percent of all suspected killers were identified as Haitian, Jamaican, or Mariel. While the percentage of Mariel victims was higher than Haitians and Jamaicans (78 percent : 22 percent), the number of Mariels settling in the city of Miami (57,281) in the early 1980s was almost twice the number of Caribbean blacks in the entire greater metropolitan Miami area.

To further illustrate the distinct experiences of each group, we examine the homicide offending rate in Table 12.1. These results are especially important because much of the current anti-immigrant sentiment focuses on newcomers as killers rather than as victims. But, with a lone exception, the immigrants were not high-rate killers. In Miami during the early 1980s, the total city homicide offender rate was 69.9. This rate was higher, and in one case substantially higher, than the group specific rates for Mariels (55.5) and Haitians (26.4). One prominent contrast is the Jamaican rate of offending (102.2)—a level one and a half times the total city rate.

The offender data cannot be understood outside of their historical context, as the early 1980s proved to be a time of both social strife and social change in Miami. Social and political frustration among the city's African American population, for instance, erupted into violent riots touched off by questionable police killings in 1980, 1982, and 1989 (Dunn, 1997). In general, however, as the local turmoil subsided throughout the 1980s, the Caribbean violator rate decreased dramatically and quickly, relative to the city as a whole. The Jamaican killer rate displayed the largest decrease of all three groups (84 percent), but the Mariel and Haitian rates also declined sharply by 1990. These decreases are even more noteworthy after

TABLE 12.1

Immigrant Homicide Suspect Rates: 1980 and 1990

	1980	*1990*
Haitian	26.4	12.8
Jamaican	102.2	15.9
Mariel	55.5	10.5
City Total	69.9	49.9

Sources: City population estimates from 1980 and 1990 Bureau of the Census. Haitian and Jamaican population estimates cited in Dunn, 1997: 336. Mariel estimates described in Martinez, 1997: 120 and 121.

considering the substantial Haitian and Jamaican population growth throughout metropolitan Miami. In 1980, less than 13,000 Haitians lived in the country, but ten years later the number quadrupled to over 45,000. Similarly, the Jamaican population tripled over the same time span to more than 27,000 (Dunn, 1997: 336–337).

If the stereotype of Miami in the popular media as a high crime, drug-infested area, overrun by dangerous immigrants, was correct, then immigrant Caribbeans should have rates consistently higher than the city average. Apparently, immigrant murderers, at least in most cases, were engaged in killing at a much lower level than the total city rate, and the lone exception was confined to a single time point.

Table 12.2 describes two variables linked to type of homicide characteristic—gun use and victim-offender relationship. The gun distribution of victims varied in some cases. Gun inflicted homicides for Caribbeans were always higher than for the rest of Miami at both time points, except for the Mariels in 1990, although percentages for Haitians and Jamaicans dropped that same year.

The circumstances surrounding the homicide victim levels also varied. Table 12.2 reports that most Haitian victims were killed during the course of a robbery or by a non-intimate, regardless of the time frame. Jamaicans were usually murdered during the course of a felony, robbery, or by a non-intimate, but some variation occurred. For example, felony and robbery-related homicide increased in 1990, similar to the rest of Miami, while non-intimate killings dropped by half. Mariels, typically, were fatally assaulted by non-intimate acquaintances, friends, or neighbors.

One finding worthy of special attention is that of domestic-related homicides between family members and intimates. A popular myth exists that immigrant Latinos and Afro-Caribbeans are more likely to engage in domestic violence, in part because of "traditional values" or "machismo" that encourage greater control at home, including physical responses to interpersonal problems (see Buchanan, 1985). Contrary to this image, we discover lower family/intimate proportions for immigrants than for the city as a whole.

TABLE 12.2

Characteristics of Immigrant Homicide Offense Type: 1980–1990

Offense	Victim			
	Haitian	*Jamaican*	*Mariel*	*Miami Total*
1980				
Gun	95.2	85.7	86.0	78.3
Felony	8.7	17.9	10.9	11.2
Robbery	39.1	25.0	9.2	17.0
Nonintimate	34.8	50.0	60.9	48.7
Family/Intimate	17.4	3.6	15.5	18.7
Stranger/Unknown	0.0	3.6	3.4	4.4
1990				
Gun	81.6	85.7	74.0	74.1
Felony	22.2	33.3	18.2	22.3
Robbery	31.5	33.3	15.6	18.8
Nonintimate	37.0	26.7	55.8	46.7
Family/Intimate	7.4	6.7	6.5	10.3
Stranger/Unknown	1.9	0.0	3.9	2.0

Note: All numbers in percentages

Discussion and Summary

A popular perception exists that immigrants are responsible for crime and violence in U.S. cities with large foreign-born populations. It is important to bring empirical evidence to bear on the policy debates surrounding this concern. To address this issue, we examined Afro-Caribbean homicides in the city of Miami. The picture that emerged undercuts claims that immigrants significantly influence violent crime.

We found that, almost without exception, Afro-Caribbeans had a lower homicide rate than the city of Miami total population. Comparing the early 1980s to the late 1980s, we also found a strong pattern of declining violence, especially for Jamaicans and Mariel Cubans, while Haitians continuously maintain an overall low rate of violent crime. As immigrant groups became more established, grew in size, and were less dominated by young males, the most violence-prone group, the homicide rate rapidly dissipates.

To illustrate this process, recall the high 1980 Jamaican offending level—the only immigrant rate to exceed the city average. It is possible that a high proportion of unattached and young Jamaican males migrated to south Florida and were employed in the informal economy. As the decade progressed, they left or were removed from the drug industry, and Jamaicans with more conventional ties to society followed the earlier entrants in search of legitimate work. Although our explanation is speculative and data are not available to directly examine this proposition at the city level, we believe this is a plausible account for our findings.

The results of this study have relevance for policies in vogue, such as targeted sweeps by the Immigration and Naturalization Service (INS) or legislation designed to deport criminal aliens. Many of these policies are motivated by the notion that immigrants threaten the order and safety of communities in which they are concentrated (Beck, 1996; Brimelow, 1995). We discovered, however, that, overall, Afro-Caribbeans are involved in violence less than the city of Miami population. Thus, the current INS strategy to remove "criminal aliens" in an effort to combat violent crime is questionable and is a misguided basis for limiting and restricting immigration.

The question of comparability to other areas naturally arises. We stress the caveat that our findings are, perhaps, unique to Miami for a host of demographic and historical reasons. It is possible that crime differences might arise relative to other cities with large immigrant populations (e.g., New York City, Chicago, Los Angeles, San Diego). Thus, local conditions, social class in the home country, and manner of reception might be important determinants of immigrant well-being. We invite further inquiries along these lines.

Our findings suggest that some of the key propositions of social disorganization theory must be revised to account for contemporary relationships among the immigration and homicide patterns found in cities such as Miami. Rapid immigration might not create disorganized communities but instead stabilize neighborhoods through the creation of new social and economic institutions. For example, Portes and Stepick (1993) describe how immigrant-owned small business revitalized Miami by providing job opportunities for co-ethnics. The low rate of involvement of the three immigrant groups we studied demonstrates the value of exploring intra-ethnic distinctions in future research on neighborhood stability, economic conditions, and violent crime. Increased attention to the issues we have raised will likely lead scholars to challenge misleading and deleterious stereotypes and result in more informed theory, research, and policy on urban immigrant violence.

REFERENCES

Aguirre, B. E., R. Saenz, and B. S. James. 1997. Marielitos ten years later: The Scarface legacy, *Social Science Quarterly,* 78: 487–507.

Beck, R. 1996. *The case against immigration.* New York: Norton.

Brimelow, P. J. 1995. *Alien nation.* New York: Random House.

Buchanan, E. 1985. Dade's murder rate was on the rise again in 1984, *Miami Herald,* January 2, B1.

Dunn, M. 1997. *Black Miami in the twentieth century.* Gainesville: University Press of Florida.

Hagan, J., and A. Palloni. 1998. Immigration and crime in the United States. In *The immigration debate.* Ed. J. P. Smith and B. Edmonston. Washington, DC: National Academy Press. Pp. 367–387.

Hawkins, D. F. 1999. African Americans and homicide. In *Issues in the study and prevention of homicide.* Ed. M. D. Smith and M. Zahn. Thousand Oaks, CA: Sage.

Inciardi, J. A. 1992. *The war on drugs II: The continuing epic of heroin, cocaine, crack, crime, AIDS, and public policy.* Mountain View, CA: Mayfield.

Lane, R. 1997. *Murder in America: A history.* Columbus: Ohio State University Press.

Maingot, A. P. 1992. Immigration from the Caribbean Basin. In *Miami Now!* Ed. G. J. Grenier and A. Stepick. Gainesville: University of Florida.

Martinez, R., and M. T. Lee. 1998. Immigration and the ethnic distribution of homicide in Miami, 1985–1995, *Homicide Studies, 2:* 291–304.

Nijman, J. 1997. Globalization to a Latin beat: The Miami growth machine. *Annals of AAPSS,* 551: 164–177. May.

Pennell, S., C. Curtis, and J. Tayman. 1989. *The impact of illegal immigration on the criminal justice system.* San Diego: San Diego Association of Governments.

Portes, A., J. M. Clarke, and R. Manning. 1985. "After Mariel: A survey of the resettlement experiences of 1980 Cuban refugees in Miami," *Cuban Studies,* 15: 37–59.

Portes, A., and A. Stepick. 1993. *City on the edge: The transformation of Miami.* Berkeley: University of California Press.

Scalia, J. 1996. *Noncitizens in the federal justice system, 1984–1994.* Washington, DC: U.S. Department of Justice, Bureau of Justice Studies.

Schenker, J. L. 1984. "Police: Rastafarian gangs setting up shop in Florida," *Miami Herald,* March 20, B8.

Shaw, C. R., and H. D. McKay. 1931. *Social factors in juvenile delinquency. Volume II of Report on the causes of crime.* National Commission on Law Observance and Enforcement, Report No. 13. Washington, DC: U.S. Government Printing Office.

Short, J. F. 1997. *Poverty, ethnicity, and violent crime.* Boulder, CO: Westview Press.

STOP AND THINK   *A good question to ask after reading a piece of research is "What was the unit of analysis, or type of subject, studied by the researchers?" What was Martinez and Lee's unit of analysis?*

STOP AND THINK AGAIN   *Other good questions are "What were their major independent and dependent variables?" "How did they measure those variables?" "How did they expect those variables to be related?" "How were they related?" Try to answer as many of these questions about Martinez and Lee's article as you can.*

Kinds of Available Data

Martinez and Lee take advantage of at least four sources of available data in their study: "homicide logs" from the Miami Police Department for information about offenders and victims and their relationships, decennial census reports for information about the population of Miami, Alejandro Portes et al.'s (1985) estimates of the number of Mariels in the city of Miami, and Marvin Dunn's (1997) estimates of the number of Caribbean-born blacks born in Jamaica and Haiti. The imaginative combination of a variety of data sources is quite common in analyses involving available data. It can make these analyses a great deal of fun, as well as illuminating. Useful in the art of such analyses is understanding the kinds of available data. Three kinds we'll introduce you to are secondary data, existing statistics, and the ever-popular "others."

Secondary Data

Secondary data are data that have been collected by others and have been made available to the current researcher. The "homicide logs," created by the Miami Police while they tried to solve murder cases, but then used by Martinez and Lee to identify the ethnicity of offenders and victims, are an unusual instance of secondary data. Martinez and Lee's access to actual police crime logs is an interesting story, indicating that access to available data isn't always just a matter of finding the right report or downloading the right computer tape. Their accounts of this story are slightly different, but both point to possible stumbling blocks in accessing "available" data. They both agree that, in the beginning, Martinez was a new assistant professor at the University of Delaware with few research contacts. He wanted to study Latino homicides, but had no real access to information about Latinos. "There are few Latinos in Delaware," he reports (personal communication), "but I had a colleague with research projects in Miami. He introduced me to somebody on his research staff who knew the Miami Police Department Chief. Introductions were made and I eventually received permission to enter the police department" and access its homicide logs. His co-author, Lee, suggests that Martinez's account is overly "modest." Lee points out that "police are understandably reluctant to give 'outsiders' unrestricted access to study their files," and that "only by logging countless hours accompanying the detectives on calls, and demonstrating that he was trustworthy, could Ramiro [Martinez] gain the kind of unrestricted access that I [Lee] witnessed when I visited the police department with him: the detectives greeted him enthusiastically and treated him as an insider" (personal communication).

STOP AND THINK *While the information that Martinez and Lee collected from homicide logs certainly counts as secondary data, it is not typical of the kind of secondary data most used by social scientists to produce research articles. What kind of secondary data, do you recall, provides the basis for around 40 percent of all articles published in the social sciences?*

By far the greatest amount of social science research based on the analysis of secondary data, however, employs survey data collected by others. In fact, Clark (1999) estimates that about 40 percent of all research articles published in 1996 were based on such data and that the percentage is increasing. Thus, Wasserman and Richmond-Abbott (2005) used General Social Survey (GSS) data to determine the different ways that men and women use the Internet, while Burdette, Ellison, and Hill (2005) used GSS data to discern why members of conservative Protestant churches tend to be less tolerant of homosexuals than others. Lichter, Qian, and Crowley (2005) analyzed data from the 1990 and 2000 Public Use Microdata Samples (PUMS) of the U.S. Decennial Censuses to determine how child poverty among racial minorities in the U.S. changed during the 1990s and Hogan (2005) analyzed data from the Health and Retirement Survey to show that a Marxist approach to understanding social stratification needs to be rethought. And Hofferth (2005) offers an overview of research in family studies that has relied on secondary analysis of survey data: data from, among others, the National Longitudinal Survey of Youth, National Survey of Families and Households, the Survey of Program Dynamics, the Early Childhood Longitudinal Survey, the Panel Study of Income Dynamics (PSID) Child Development Supplement, the National Survey of Adolescent Health, the National Survey of Child and Adolescent Well-Being, the Health and Retirement Study, the Fragile Families and Child Wellbeing Study, the Early Head Start (EHS) Research and Evaluation Project, and the Welfare, Children, and Families: a Three-City Study. You can find hundreds of such data sets at the Inter-University Consortium for Political and Social Research (ICPSR) at the University of Michigan (http://www.icpsr.umich.edu/) and at the Roper Center for Public Opinion Research (http://www.ropercenter.uconn.edu/), both clearinghouses for survey data, mainly, but not exclusively, from the United States. Again, the point is that a large and increasing percentage of the research articles and books published in the social sciences are based on secondary data, data collected by others, and that secondary data sets are increasingly accessible to those of us interested in using them.

Both the ICPSR and the Roper Center place some restrictions on access to secondary data, either through fees or the requirement of institutional membership. If you are planning a major research project, you might want to consult your professor about what data sets your college or university already has or whether there might be another convenient way for you to gain access to secondary data you've seen referred to in the literature. One of the most delightful sources of easy-access secondary data, however, is the Social Documentation and Analysis (SDA) Archive at the University of California, Berkeley (http://sda.berkeley.edu/archive.htm). Here you can browse a variety of data sets (see Box 12.1) and analyze the data online.

Suppose you have a research project in social psychology in which you're studying the connection between gender and love. You might want to find out whether, for instance, males or females are more likely to say they would rather suffer themselves than let their loved one suffer.

STOP AND THINK   *Quick. Take a guess. Who do you think is more likely to say they'd rather suffer themselves than let loved ones suffer: males or females?*

Well you can use the most recent General Social Survey, which just happens to ask respondents whether they'd rather suffer themselves than let the one they love suffer, as well as what their gender is. One way to analyze this association would be to access the SDA Archive by going to web address http://sda.berkeley.edu/archive.htm.

Some Data Sets Available at SDA Archive (http://sda.berkeley.edu/archive.htm)

GSS Cumulative Datafile (1972–2004)

National Election Study (1948–2004)

Census Microdata: 2000–2003 American Community Surveys (3.8 million cases)

1990 and 2000 U.S. 1 percent PUMS (Public Use Microdata Samples)

National Race and Politics Survey 1991

National Health Interview Survey 1991

Health Studies from Brazil

STOP AND DO

If you have easy access to a computer why don't you try this? Then click on the GSS Cumulative Datafile 1972–2004, or whatever GSS file contains the 2004 survey data. Now "start" the "Frequencies and Crosstabulation (with charts)" action. Then place "sex" in the column box and "agape1" in the row box, before clicking on "run the box." What do you find? Are males or females more likely to say they strongly agree with the statement "I would rather suffer myself than let the one I love suffer?"

It might not surprise you as much as it did us that males are substantially more likely than females (69.1 percent to 57.3 percent) to strongly agree with the statement "I would rather suffer myself than let the one I love suffer," but perhaps you will be interested to know how easily, using secondary survey data, you found out. Think how impressive it would be to add such a finding to your social psychology paper!

The *unit of analysis* for studies based on secondary analysis of survey data is almost always the individual. We've learned something about people from our foray into GSS data on love: that men are more likely than women to say they strongly agree with the statement about suffering for the one they love. Burdette, Ellison, and Hill (2005) used GSS data to discover something else about individual people: that, if they're members of conservative Protestant churches, they tend to be less tolerant of homosexuals than others. People respond to the surveys and people, therefore, tend to be the units of analysis of the research based on those surveys. The units of analysis in Martinez and Lee's study are also individuals. They are focusing on individual people, for instance, when they use the homicide logs to determine how many murders were committed by Mariel Cubans, Haitians, and Jamaicans. And when they use the various estimates of the Mariel, Haitian, and Jamaican populations in Miami, they are simply trying to supply the denominators necessary to calculate homicide rates, much as we needed to use the total number of males and females in the GSS survey to calculate the percentage of each that strongly agreed with the statement about love. In general, a big difference between studies based on secondary analysis of, say, survey data and studies based on existing statistics is the unit of analysis, as we'll see below.

STOP AND THINK   *Emile Durkheim's ([1897] 1951) classic study,* **Suicide,** *is based, in part, on comparisons of suicide rates in more or less Catholic provinces in Europe. He finds the provinces with high percentages of Catholic residents, for instance, have lower suicide rates than the provinces with low percentages of Catholic residents. Can you tell what the unit of analysis of this part of Durkheim's study was?*

Existing Statistics

STOP AND THINK   *Did you ever wonder what is of interest in other countries? One source of available data is the webpage "Zeitgeist Around the World" where the most common queries to Google are posted. In November, 2005, for example, "Harry Potter" was in the top 10 most frequent queries in 13 of 20 countries, while "cricket" was in the top 10 only in India. If you check out the site at www.google.com/intl/en/press/intl-zeitgeist.html, look at the data presented and think about what the unit of analysis is.*

Existing statistics are statistics provided by large organizations. Martinez and Lee employ one of the most frequently employed sources of existing statistics: the U.S. Bureau of the Census to gather information about the overall population of Miami. Census products come in printed reports, on computer tapes, on microfiche and CD-ROM laser disks, and now via websites. But they found that they needed to combine statistics from one source with data or statistics from another (in their case, data from the homicide logs). Most users of existing statistics find the need to make similar combinations. Thus, Lee and Ousey (2005) used data from the offender files of Fox's (2000) Supplementary Homicide Reports, with data from, among other things, the Census of Churches and Church Membership to show that in large urban counties where Blacks have access to significant social institutions, the Black homicide rate is lower than it is in other communities. Similarly, Rey and Barkdull (2005) employed data from Jacobson's (1998) *Membership in International Governmental Organizations* and from the CIA (1993) *World Factbook* to show that nations with two-house legislatures are more likely than ones with one-house legislatures to join international organizations.

In general, the *unit of analysis* in studies based on existing statistics is not the individual. For Durkheim's study of suicide rates by province, the unit of analysis was the province. In Lee and Ousey's study, it was urban counties. In "Zeitgeist Around the World" and in Rey and Bardull's study, it was nations. Statistics, by their nature, summarize information about aggregates of people (countries, regions, states, cities, counties, organizations, etc.) and when statistics are the data or information we have to work with, we're almost surely focused on comparing aggregates of people.

The real trick to doing research with existing statistics (after, of course, you've determined what your research question is) is often to find sources of data about your unit of analysis that permit you to address your research question. Roger, along with two students, April Knights and Ashley Folgo (Clark, Knights, and Folgo, 2005), recently became interested in seeing if we could work out some of the things that have contributed to women's empowerment in society. We'd come across, through our readings for a course, a measure that the United Nations Development Program (UNDP) had devised to measure women's power in countries: the Gender Empowerment Measure (GEM). We decided that this measure, based upon such things as women's access to the parliament of the country and their income compared to men's income, had merit and that we'd use it for our initial study of women's empowerment. Because GEM, which became our dependent variable, was a variable characteristic of countries, we needed to decide what other kinds of variable characteristics of countries might affect the degree of women's empowerment, as measured by GEM. We consulted the literature and found many candidates, two of which were the level of women's labor force participation and when women gained the right to vote. Then we did what is very common in research based on existing statistics: we tried to find sources of statistics that provided information about our units of analysis (in our case, nations) that measured our concepts of interest. We'd found our measure of women's empowerment (GEM) at a UNDP website (http://www.un.org/hdr2002/). We found other indicators by following up leads in literature we read. But some we found by Googling things like "date women's vote." In this way we discovered the Inter-Parliamentary Union's (2004) *Women's Suffrage* webpage (http://www.ipu.org/wmn-e/suffrage.htm) and found the years in which women got the vote in different countries, and the United Nations Statistics Division (2000)*The World's Women 2000: Trends and Statistics* (http://unstats.un.org/unsd/demographic/ww2000/) and found the level of women's labor force participation.

Roger, April, and Ashley actually located other sources that yielded statistics enabling them to measure variables of interest about most countries of the world. And what is true of their work is true of much work involving existing statistics: one searches for information about one's units of analysis (whether they be countries, regions, states, organizations, etc.), often from different sources and combines that information into a single data set about those units (see Box 12.2) Creating the combination often involves listing units of analysis (as Clark, Knights, and Folgo do their countries in the left hand column of Box 12.2) and then listing values of the variables of interest (as they do in subsequent columns).

STOP AND THINK   *Can you use Box 12.2 to determine the year that women in Argentina gained the vote? What was the women's labor force participation rate in Austria in 1999?*

Once you've collected statistics from various sources, you can begin analyzing them using statistical techniques like the ones outlined in Chapter 15's section on quantitative data analysis.

The sources of existing statistics are innumerable. If you have an idea of the kind of topic you'd like to work on, and the unit of analysis, you can find usable data in either of two ways: do a literature review to see what others have used or do a Web search. You could, of course, do both. If you choose to search the Web, findings key words is crucial. Are you interested

---

BOX 12.2

A Portion of Clark, Knights, and Folgo's Dataset on Countries: Finding Data Sources that Complement One Another

Countries	From UNDP (2002) ↓ GEM (2000)	From Inter-Parliamentary Union (2004) ↓ Year Women Got Vote	From United Nations (2000) ↓ Women's Labor Force Participation Rate (1999)
Argentina	65	1947	41
Australia	81	1902	55
Austria	77	1918	50
Bahamas	70	1961	66
⋮	⋮	⋮	⋮

---

in homicide rates in American cities? Try Googling "homicide rates in American cities." You'll get there. If you remember that the FBI warehouses information about crime in the United States in its *Uniform Crime Reports*, you might Google "Uniform Crime Reports." You get right to the *Uniform Crime Reports* website (http://www.fbi.gov/ucr/ucr.htm), and you'll soon be looking at interesting information.

STOP AND DO   *If you're sitting at your computer, why don't you see if you can find the website for the U.S. Census Bureau.*

If you wanted the U.S. Census Bureau, you could Google "U.S. Census Bureau" and find its Web address: http://www.census.gov/.

We won't try to anticipate your needs, but we will provide you with some Web addresses most frequently used for sources of existing statistics (Box 12.3) and the assurance that, even as recently as the last decade, finding existing statistics (which used to mean a trip to the right library and a hunt for the right volume) was never so easy.

Of course, the use of existing statistics constitutes one of the richest research traditions in the social sciences. It provided grist, as we've already pointed out, for Émile Durkheim's classic ([1897] 1964) study, *Suicide*, and for Karl Marx's argument for economic determinism in his ([1867] 1967) *Das Kapital*. It is hard to imagine what these early social scientists would make of the ever increasing and increasingly accessible amounts of statistics available today—and, frankly, there's no way we could do justice to them here—but it's almost a certainty that Marx and Durkheim would have done a lot with them.

BOX 12.3

Some Good Sources of Existing Statistics

Units of Analysis: American states, cities, counties, etc.

- *The Bureau of the Census Home Page* (http://www.census.gov/) allows users to search census data on population. In particular, the bureau's *Statistical Abstract* (http://www.census.gov/statab/www/) provides useful information on crime, sex and gender, social stratification, and numerous other topics.

- *FBI Uniform Crime Reports Page* (www.fbi.gov/ucr/ucr.htm) offers access to recent crime data.

- *Bureau of Labor Statistics Data Home Page* (www.bls.gov/data/) has, among other things, data on men's and women's labor force participation.

- *Department of Education Home Page* (http://www.ed.gov/index.jhtml) leads to much information about educational achievement, etc.

- *The National Center for Health Statistics Home Page* (http://www.cdc.gov/nchs/) is a wonderful source of information about health and illness.

Units of Analysis: Countries

- *The United Nations Yearbook Statistical Division* (http://unstats.un.org/unsd/default.htm) provides basic statistical information, including data on population (in its *Demographic Yearbook),* trade, health, education and many other topics.

- *The International Labor Office* (http://www.ilo.org/) is a great source of information about work worldwide.

- CIA *World Factbook* (http://www.cia.gov/cia/publications/factbook/) provides a wide variety of information on most countries in the world

- *The World Bank* (http://www.worldbank.org) offers basic information and much financial information about countries of the world.

physical traces, physical evidence left by humans in the course of their everyday lives.

erosion measures, indicators of a population's activities created by its selective wear on its physical environment.

accretion measures, indicators of a population's activities created by its deposits of materials.

Other Kinds of Available Data

You don't need to depend on outside agencies to make your "available data" available. In their now classic list of alternative data sources, Eugene Webb and his associates (2000) mention, for instance, **physical traces** and personal records. Physical traces include all kinds of physical evidence left by humans in the course of their everyday lives and can generally be divided into what Webb and his associates call **erosion measures** and **accretion measures.** An erosion measure is one created by the selective wear of a population on its physical environment. Webb and associates cite Frederick Mosteller's (1955) study of the wear and tear on separate sections of the *International Encyclopedia of the Social Sciences* in various libraries as a quintessential example of the use of erosion measures, in this case to study which sections of those tomes received the most frequent examination. An accretion measure is one created by a population's remnants of its past behavior. The "Garbage Project" at the University of Arizona has studied garbage since 1973. By sorting and coding the contents of randomly selected samples of household garbage, William Rathje and his colleagues have been able to learn a great deal, including what percentage of newspapers, cans, bottles, and other items aren't recycled and how much edible food is thrown away (Rathje and Murphy, 1992: 24). Because there are "certain patterns in which people wrongly self-report their dietary habits" (Rathje and Murphy, 1992: 24), the proponents of "garbology" argue that this type of available data can sometimes be more useful than asking people questions about their behavior. One great advantage of using physical traces, like garbage, is that their "collection" is **unobtrusive** (or unlikely to affect the interactions, events, or behavior under consideration).

unobtrusive measures, indicators of interactions, events, or behaviors whose creation does not affect the actual data collected.

STOP AND THINK

Suppose that, like Kinsey and associates (1953), you wanted to study the difference between men's and women's restrooms in the incidence of erotic inscriptions, either writings or drawings. Would you be taking erosion or accretion measures?

personal records, records of private lives, such as biographies, letters, diaries, and essays.

Personal records include autobiographies, letters, diaries, and essays. With the exception of autobiographies, which are often published and available in libraries, these records tend to be more difficult to gain access to than the public records we've already discussed. Frequently, however, personal records can be used to augment other forms of data collection. Roger has used autobiographies, for instance, to augment information from biographies and questionnaire surveys to examine the family backgrounds of famous scientists, writers, athletes, and other exceptional achievers (Clark, 1982; Clark and Ramsbey, 1986; Clark and Rice, 1982). Private diaries, when they are accessible, hold unusually rich possibilities for providing insight into personal experience. R. K. Jones (2000) suggests ways in which these can be used in contemporary medical research, and historians like Laurel Thatcher Ulrich have used them to create classics, such as her (1990) *A Midwife's Tale,* to piece together accounts of times gone by. Letters can be an especially useful source of information about the communication between individuals. Lori Kenschaft (2005), for instance, has studied the ways in which two important nineteenth-century educators, Alice Freeman Palmer and George Herbert Palmer, who happened to be married to each other, affected each other's thinking and American education by carefully studying their written correspondence (available at the Harvard University library). Similarly Karen Lystra's (1989) *Searching the Heart: Women, Men and Romantic Love in Nineteenth-Century America* uses letters almost exclusively to build an argument that nineteenth century conventions of romantic love required people to cultivate an individual, unique, interesting self that they could then present to the beloved other—that is, that romance is an historical and psychological root of individualism.

Kenschaft's and Lystra's works point to one great advantage of available data: the potential for permitting insight into historical moments that are inaccessible through more conventional social sciences techniques (such as interviews, questionnaires, and observation). Other sources of such historical data are newspapers and magazines, and organizational and governmental documents. Beverly Clark (2003) has scrutinized nineteenth- and early twentieth-century newspapers and magazines for insight into the popular reception of (children's) books written by, among others, Louisa May Alcott. Clippings from the *Cleveland Leader* and the *Albany Daily Press and Knickerbocker* indicate that the Cleveland public library required 325 copies of *Little Women* in 1912 to satisfy constant demand; the New York City branch libraries required more than a thousand copies.

Texts, like the diary studied by Thatcher (1990), the letters studied by Kenschaft (2005) and the newspapers and magazines studied by Clark (2003) are amenable to both quantitative and qualitative data analysis. Have a look at the next chapter, on content analysis, for an idea of how others have done these kinds of analyses.

The larger point here is, however, this: You really don't want to be guided in your search for available data by any of the "usual suspects" we've suggested. A much sounder approach is to immerse yourself in a topic by reading the available literature on it. In the course of your immersion, look at the sources other researchers have used to study the topic. If you discover that they've typically used the FBI *Uniform Crime Reports,* you might want to use that too. If they've used the U.N.'s *Demographic Yearbook,* you also might want to look into that. Very often, using earlier research as a resource in this way is all you'll need to find the data you'll want to use.

Advantages and Disadvantages of Using Available Data

Available data offer many distinct advantages over data collected through questionnaires, observation, or experiments. Most striking among these are the savings in time, money, and effort they afford a researcher. The analysis of secondary data or existing statistics might cost little more than the price of your Internet connection (though, depending on your institutional affiliation, there can be a charge of several hundred dollars—still relatively little— for access to and the right to use such data) and the time it takes to figure out how you want to analyze them. Each national census might cost a government billions of dollars, but comparing 150 nations based on the results of those collections might cost you only the price of a library card, or a few keystrokes on your computer.

Martinez and Lee's work with police logs, however, provides a useful caveat to the general rule that available data are cheap and time-saving. In fact, Martinez estimates that he spent roughly five years, off and on, collecting data on 2,500 homicide cases that occurred from 1980 to 1995, and probably spent about $200,000, using grants, salary, and sabbatical supplements to do so. Part of this expense was because he started his work while working in Delaware. He says, "Flying to Miami from Delaware sounded great at first but it was not cheap. You needed airfare, hotel and food monies, as well as money for supplies. You'd manually pull out homicide records from a file room, copy them, and ship them back to your home base. Once you were at home, you needed to store, read and code the data, then have them entered and readied for statistical analysis [usually by a hired student]" (personal communication). Martinez speaks of "maxing out two credit cards" while getting the project going and hoping that a "grant proposal would hit." Eventually, he did "hit" on grants, but, when he didn't, even later on, "it was depressing. At times I felt like quitting. But then I remembered my credit cards."

Matthew Lee suggests that we should think of available data as being on a "continuum of availability" which reinforces the point that, although much available data is, in fact, easily available, some can come at considerable cost to the researcher (personal communication). He points out, however, that sometimes the less available data are the only means to true conceptual breakthroughs. He points out that he came to his interest in the migrant/crime question because he'd been so impressed by Cliff Shaw and Henry McKay's (1931) "idea that social processes like immigration and residential turnover (rather than 'bad' individuals) weaken community levels of social control and thereby facilitate crime." Yet Lee and Martinez found little support for that notion. In fact, immigrants appear to be stabilizing urban communities and suppressing levels of violent crime. Lee believes that one reason why many people see newcomers as a disorganizing influence on "institutions and sensibilities" is that the most easily available data, such as the FBI's *Uniform Crime Reports*, cannot be used to address the validity of conventional wisdom on immigration and crime. And it's hard work to get alternative data. But, Lee points out, "Our experiences suggest that taking a gamble on collecting secondary data that are not easily available can lead to a bigger theoretical payoff" (personal communication).

A related advantage of using available data is the possibility of studying social change over time and social differences over space. Robert Putnam, in his (2000) national bestseller, *Bowling Alone: The Collapse and Revival of American Community*, analyzes available data from many sources (election data, survey data, organizational data) to substantiate his argument that political, civic, religious, and work-related participation have substantially declined in America during the past quarter century, as have informal social connections and volunteer and philanthropic activities. Using data from the National Longitudinal Study of Young Men for 1966–81, Teachman (2005) was able to show that while serving in the armed forces in Vietnam had cost veterans about one year in terms of educational attainment when compared to non-veterans, the difference diminished, though it did not disappear, afterwards. And using U.S. census data collected since 1790, gerontologists have been able to discern that the proportion of the American population that is 65 years or older has grown from 2 percent in 1790 to 4 percent in 1890 to 12.4 percent in 2000. Most individual researchers don't live long enough to observe such trends directly, but, because of available data and the imagination to use them, they don't have to.

STOP AND THINK *Imagine you want to find out how much the size of the "average" American family household has changed over the last two centuries. Would you be more inclined to use a survey, participant observation, an experiment, or U.S. census reports as your data source? Why?*

Available data also mean that retiring and shy people, people who don't like the prospect of persuading others to answer a questionnaire or to participate in a study, can make substantial research contributions. Moreover, the researcher can work pretty much where he or she wants and at whatever speed is convenient. Roger once coded data from a source he'd borrowed from the library while waiting (along with 150 other people who could do no more than knit or play cards or read) in a courthouse to be impaneled on a jury. (At other times, when unable to sleep, he's collected data in the wee hours of the morning—a time when researchers committed to other forms of data collection pretty much have to shut down.) Again, this advantage doesn't always exist. Martinez found it useful, for instance, to cultivate many MPD officers, if only to provide details about cases that were left out of the logs he studied (personal communication).

A final, and most important, advantage of using available data is that the collection itself is unobtrusive and extremely unlikely to affect the nature of the data collected. Unlike researchers who use questionnaires, interviews, observations, or experiments, those of us who use available data rarely have to worry that the process of data collection will affect our findings—only that the original collection by primary researchers may affect those findings.

On the other hand, the unobtrusiveness of using available data is bought at the cost of knowing less about the limitations of one's data than if one collects the data firsthand. One major disadvantage of available data is that one is never quite sure about its reliability. Martinez and Lee, for instance, were evidently concerned that the "homicide logs" they investigated provided adequate definitions of victim and offender ethnicity. Martinez observes, "If a homicide record written in 1980 was not clear on certain things, you were stuck with it." Eventually, though, he had access to an atypical resource for researchers relying on available data. "With time," he reports, "I became acquainted with detectives, especially the 'old-timers' and asked them about certain cases. . . . If I had not had direct access, it would have been very difficult. I was very fortunate" (personal communication).

Most researchers, using available data, are not so fortunate. Cross-national researchers (like Clark, Knights, and Folgo [2005]) using data on female labor force participation, for instance, have been plagued by their awareness that, in many countries, especially in those where women's work outside the home is frowned upon, there is considerable undercounting of that work.

STOP AND THINK   *What kinds of nations, do you think, would be most likely to undercount women's work outside the home? Why?*

Sometimes the best you can do is to be explicit about your suspicion that some of your data are not reliable. You can at least try to clarify the kinds of biases the sponsors of the original data collection might have had, as we did when we used available data on female labor force participation. This kind of explicitness provides readers with a chance to evaluate the research critically. At other times, checks on reliability are possible, even without the capacity to perform the kinds of formal tests of reliability (for example, inter-rater tests—see Chapter 6) that you could use while collecting primary data, or access, such as Martinez's, to the collectors of the original data. Steven

Messner (1999) found that—the conventional wisdom about watching violence on TV and personal acts of violence to the contrary notwithstanding—cities where people watched lots of violent TV programs were likely to have lower violent crime rates than other cities were. Messner was concerned, however, that the "underreporting" of violent crimes in cities where many people were desensitized to crime by the viewing of violent television programming might have been responsible for his unexpected finding that such viewing was negatively associated with the presence of such crimes. But he knew that one kind of violent crime, homicide, was very unlikely to be underreported under any circumstances. So, when he found that the unexpected negative association between viewing violent programs and violent crime rates persisted even for homicide rates, he concluded that it was less plausible that his other unexpected findings were only the result of systematic underreporting.

STOP AND THINK   *Why is homicide less likely than, say, shoplifting to be underreported, whatever the television viewing habits of inhabitants?*

A second disadvantage of using available data is that you might not be able to find what you want. In some cases, there just might not have been anything collected that's remotely appropriate for your purposes. Modern censuses are designed to enumerate the entire population of a nation, so you can reasonably expect to be able to use their data to calculate such things as sex ratios (the number of males per 100 females, used, for instance, by Messner as a control variable in his analysis). But ancient Roman censuses focused on the population of adult male citizens (the only group, after all, owning property and eligible for military service and therefore worth knowing about for a government interested primarily in taxation and military recruitment [Willcox, 1930, cited in Starr, 1987]), and so the data necessary for the computation of, say, sex ratios, not to mention population density or the percentage of the population that's a certain age, might simply not be available. Also, available data might not present data the way you want. Martinez notes, for instance, that crime data from the FBI are presented in racial categories ("black" and "white") and not ethnic ones (e.g., "Mariel," "Haitian," and "Jamaican").

Sometimes, however, you can locate data that very nearly (if still imperfectly) meet your needs, so the issue of validity arises. Numerous cross-national studies (for example, Roger, April, and Ashley in their study of women's power mentioned above) have used the women's labor force participation rate as a rough measure of women's economic status in a nation. A women's labor force participation rate refers to the proportion of women in an area (in this case, a country) who are employed in the formal workforce. The rationale behind its use as a measure of women's economic status is the notion that, if women don't work for pay, they won't have the power and prestige within the family and in the community that access to an independent income can buy. But what if women and men don't always (as we know they don't) receive equal monetary returns for their work? Wouldn't the economic status of women then depend on, among other things, how closely their average income approximated that of men's—something that their labor force participation rate doesn't begin to measure? We're sure you can think of other problems with labor force participation as a valid measure of economic status, but you probably get the idea: Sometimes researchers who employ available data are stuck using imperfectly valid indicators of the concepts they want to measure because more perfect indicators are simply not available.

STOP AND THINK   *Can you think of other problems with using women's labor force participation, or the proportion of women over 16 years old who are in the labor force, as an indicator of women's economic status in an area? (Hint: What other things, besides working at all, contribute to our sense that a person has a high economic status?)*

One peculiar danger of using available data, especially existing statistics, is the possibility of becoming confused about your unit of analysis (or subjects) and committing the ecological fallacy. Much of the research on the migrant-crime connection (e.g., Hagan and Palloni, 1998; U.S. Commission on Immigration Reform, 1994), before Martinez and Lee's study, had looked at the relationship between the proportion of migrants in a city and its crime rates. This research, however plausible, was nonetheless subject to the ecological fallacy because its units of analyses (cities) were not the same as units of analyses (people) researchers wanted to make inferences about. Martinez and Lee avoid the ecological fallacy by obtaining information that enables them to directly compare the homicide rates of various immigrant groups and of immigrants and non-immigrants.

Summary

In contrast with primary data, or data that the researcher collects herself, available data exist before you begin your research. Secondary data, or data that have been collected by someone else, have become increasingly popular in social science research. More and more social scientists are using survey data that have been collected by other researchers and have been made accessible for secondary analysis. Two major repositories for such data are the ICPSR at the University of Michigan and the Roper Center at the University of Connecticut.

A major advantage of using available data is that the data are usually cheap and convenient. Martinez and Lee's experience indicates, however, that there is a "continuum of availability," and that sometimes the least available of available data are well worth exploring. Available data also allow the study of social change over long periods and of social difference over considerable space. Finally, their collection is unobtrusive (or unlikely to affect the interactions, events, or behavior under consideration) even if their original collection might not have been.

Common sources of existing statistics are governmental agencies, like the U.S. Bureau of the Census, or supra-governmental agencies, like the United Nations. These statistics tend to be summaries at some aggregate level of information that has been collected at the individual level. These data have become increasingly available on the Internet.

The main disadvantage of using available data is that, because someone else has collected the data, their reliability might be unknown or unknowable by their user. Moreover, one frequently has to settle for indicators that are known to be imperfectly valid. Finally, especially when using existing statistics, the researcher has to be especially cognizant of the temptation to commit the ecological fallacy—of making inferences about individuals from data about aggregates.

EXERCISE 12.1

Urbanization and Violent Crime: A Cross-Sectional Analysis

This exercise gives you the opportunity to examine one assumption that seems to lie behind Martinez and Lee's analysis: Cities, or more particularly SMSAs, are especially good places to study violent crime. We'd like you to see whether, in fact, population aggregates that are more highly urbanized (that is, have a greater proportion of their population in urban areas) have higher violent crime rates than those that are less highly urbanized.

Let's examine a population aggregate that can be more or less urbanized: states in the United States. Find a recent resource, perhaps in your library, that has data on two variable characteristics of states: their levels of urbanization, and their overall violent crime rate. Collect data on these variables for the first 10 states, listed alphabetically. (Hint: *The Statistical Abstract of the United States* [http://www.census.gov/statab/www/] is a good source of the data you'll want. Most college libraries will have this in their reference section.)

1. Prepare ten code sheets like this (actually, you can arrange all data on one page):

   State: Total Violent Crime Rate: Percent Urban:

2. Go to your library (or the Website) and find the data on crime and urbanization. Using the code sheets you've prepared, collect data on 10 states of your choice.

3. Once you've collected the data, analyze them in the following ways:

   a. Mark the five states with the lowest percent urban as **LU** (low urban) states. Mark the five states with the highest percent urban as **HU** (high urban) states.

   b. Mark the five states with the lowest total violent crime rates as **LC** (low crime) states. Mark the five states with the highest total violent crime rate as **HC** (high crime) states.

   c. Calculate the percentage of **LU** states that you have also labeled **HC**. Calculate the percentage of **HU** states that have been labeled as **HC**.

4. Compare your percentages. Which kinds of states tend to have the highest violent crime rates: highly urbanized or not so highly urbanized states? Interpret your comparison relative to the research question that sparked this exercise.

Life Expectancy and Literacy: A Cross-National Investigation

This exercise asks you to investigate the connection between education and health by examining the research question: Do more educated populations have the longest life expectancies? Investigate this hypothesis using data from 10 countries (your choice) about which you obtain data on the percentage of the population that's literate and life expectancy. [You can find the data in most world almanacs and in a volume of *Human Development Reports,* any year, online. You can search for them by using key words "Human Development Reports" and, then, "Life Expectancy."]

1. Prepare 10 code sheets like this (again, you can arrange all data on one page):

   Country: Literacy: Life Expectancy:

2. Go to your library (or the Website) and find the appropriate data. Using the code sheets you've prepared, collect data on 10 countries of your choice.

3. Once you've collected the data, analyze them using the technique introduced in Exercise 12.1.

   a. Mark the five countries with the lowest literacy rate as **LL** (low literacy) nations. Mark the five countries with the highest literacy rates as **HL** (high literacy) nations.

   b. Mark the five countries with the lowest life expectancy as **LLE** (low life expectancy) nations. Mark the five countries with the highest life expectancy as **HLE** (high life expectancy) nations.

   c. Calculate the percentage of **LL** nations that you have labeled **HLE.** Calculate the percentage of **HL** nations that are labeled **HLE.**

4. Compare your percentages. Which kinds of nations tend to have the highest life expectancy: those with high literacy rates or those with not-so-high literacy rates? Interpret your comparison relative to the research question that sparked this exercise.

5. What is the ecological fallacy? Do you think you succumbed to this fallacy in your interpretation of your analysis in this exercise? Why or why not?

EXERCISE 12.3

For Baseball Fanatics

This exercise gives you the chance to examine an age-old social science dispute: whether hitting or pitching wins ball games. Help shed light on this dispute by, again, turning to a recent almanac, this time to collect winning percentages, team batting averages, and team earned-run averages for all the teams in either the American or the National League for one year. Use data collection techniques similar to those you used for the last two exercises, this time collecting data on three variables, rather than two. Moreover, you should use the data analysis techniques of each exercise, this time examining the relationship between winning percentage and team batting average, on the one hand, and winning percentage and team earned-run average, on the other. Are both batting average and earned-run average associated with winning percentage? Which, would you say, is more *strongly* associated with it? Why?

Write a brief essay in which you state how your analysis bears on this research question.

13

Content Analysis and Comparing Methods

© Bob Daemmrich/Stock, Boston Inc.

Introduction

When Roger was the father of two young children, a boy and a girl, he became interested in the ways in which children's picture books depicted gender. So he enlisted the aid of two students, Rachel Lennon and Leanna Morris, in the early 1990s, and three more students, Monica Almeida, Tara Gurka, and Lisa Middleton, in the early 2000s and found that between the late 1960s and the late 1990s, female characters had become increasingly visible in award-winning picture books and that gender stereotyping had declined. The study that you are about to read, however, asks the question of whether these trends toward increasing female visibility and decreasing gender stereotyping were characteristic of the period between the 1930s and the 1960s as well. The piece illustrates the technique of **content analysis,** a technique that is particularly useful for doing historical investigations. Content analysis is a method of data collection in which some form of communication (speeches, TV programs, newspaper articles, films, advertisements, even children's books) is studied systematically. Content analysis is one form of available data analysis.

content analysis, a method of data collection in which some form of communication is studied systematically.

FOCAL RESEARCH

Two Steps Forward, One Step Back: The Presence of Female Characters and Gender Stereotyping in Award-Winning Picture Books Between the 1930s and the 1960s

by Roger Clark, Jessica Guilmain, Paul Khalil Saucier, and Jocelyn Tavarez[1]

Introduction

Clark, Lennon, and Morris (1993) and Clark, Almeida, Gurka, and Middleton (2003) studied winners and runners-up of the Caldecott Medal (the major national award for children's picture books) from the late 1960s to the late 1990s, and found that depictions of gender had changed among the Caldecotts in the 30 years, with female characters becoming considerably more visible and both male and female characters becoming less stereotyped. The Caldecott Medal was first awarded in 1938, and we wondered

[1] This article is adapted from Clark, Guilmain, Saucier, and Tavarez's article of the same title that appeared in *Sex Roles* 49: 439–449. Reprinted by permission of Plenum Publishing Corporation.

whether the trend towards increasing female visibility and decreasing stereotyping could be safely extrapolated backward in time. We came to our study with two competing sets of expectations. On the one hand, we thought it possible, especially given the trend of the last 30 years, that we could find decreasing visibility of female characters and ever more stereotyping as we examined older and older Caldecotts. This we labeled our "monotonic change" (MC) hypothesis. On the other hand, we entertained a "local variation" (LV) hypothesis. Pescosolida, Grauerholz, and Milkie (1997), after all, had found that the appearance and portrayal of Black characters in children's picture books varied with social and political forces in the twentieth century. During periods of racial conflict (notably during the 1950s and the early 1960s), Black characters virtually disappeared from children's books. We thought it possible that conflict over gender roles could have similar effects on the appearance of female characters in children's books. In particular, we wondered whether, because the Great Depression degraded women's relative status in the public sphere, later 1930s Caldecotts might have tended to depict men and women in traditional roles. We also speculated that, because the Depression actually diminished conflict over women's roles, it might have led to relatively high levels of female-character visibility. We wondered whether women's greater access to the public sphere in the 1940s, particularly during World War II, might have led to a more egalitarian portrayal of gender in the late-1940s children's books. But we thought it possible that the implicit conflict over women's roles, especially during the late 1940s, might have led to relatively few female characters in the books. Because the 1950s were a time when traditional gender roles prevailed in the larger society, and when there was relatively little conflict over gender roles, we hypothesized that Caldecotts might again have depicted men and women in conventional stereotyped ways and contained more female characters. Finally, we wondered whether incipient Second Wave feminist struggles during the 1960s might not have contributed to at least marginally less stereotyping in picture books for this decade, at least in comparison with earlier decades, and to fewer female characters.

Method

One reason for studying Caldecott winners is that the Caldecott is not only the most prestigious award for preschool literature, but it also guarantees its winners phenomenal sales (Clark 1992: 6). For the current project, we closely studied the 20 Caldecott winners and runners-up from 1938 to 1942, the 28 from 1947 to 1951, the 18 from 1957 to 1961, and the 18 from 1967 to 1971.

We specify our units of analysis and some operational definitions as we report our findings. But one set of operational definitions is worth early comment. As Clark et al. (1993) and Clark et al. (2003) did before us, we employ Davis' (1984) refined set of variables for dealing with various aspects of gender-related behavior (see the Appendix to this report for variable definitions). Given these variable traits, and following Clark et al. (1993), we observe that mainstream gender stereotyping entails expectations that female characters will be more dependent, cooperative, submissive, imitative, nurturant, emotional, and passive and that they will be less independent, competitive, directive, persistent, explorative, aggressive, and active than male characters. Both male and female characters can be expected to be creative, but usually in different ways, with male characters often being stereotyped as creative problem solvers (as in getting themselves out of difficult situations) and female characters as being more naturally or domestically creative (as in giving birth to a child or, perhaps, devising a clever cleaning technique).

Following Clark et al. (1993) and Clark et al. (2003), we coded whether each of four possible characters in each book possessed a given trait (say, dependence). We analyzed the book's central character, the most important other character of the same sex, and the two most important characters of the opposite sex, where these characters were deemed sufficiently visible for analysis. If three or four of us independently decided a character possessed a trait, we accorded this trait to the characters. If three or four failed to see the trait, we decided the trait did not exist in the character. If there was a tie (with two of us seeing the trait and two not seeing it), we assigned a missing value to the character for that trait and the character was deleted in analyses of that trait. Overall, the four of us examined 133 characters and obtained an overall intercoder or interrater reliability of 88 percent, with agreements varying between 80 percent (for emotionality) and 99 percent (for activeness).

Results

Our analysis showed almost no support for the monotonic change (MC) hypothesis in the decades before the 1960s. It provided much more support for the local variation (LV) hypothesis. Our results follow, decade by decade.

As shown in Table 13.1, female characters of the late 1930s were more visible than those in any other decade. All of the 1930s books included a central female character, and 45 percent included a central female character. Girls or women were shown in 24 percent of human single-gender illustrations, and female animals were shown in 19 percent of the animal single-sex illustrations. Clark et al. (1993, 2003) have traced a pattern in which the presence of female characters is associated with low gender stereotyping in Caldecotts between the 1960s and the 1990s. But despite the high visibility of female characters in 1930s Caldecotts, we found (Table 13.2) that there is a good deal of gender stereotyping in them. Female characters were more dependent, submissive, imitative, nurturant, emotional, and passively active than male characters. Male characters, on the other hand, were more independent, competitive, directive, persistent, explorative, aggressive, and active than female characters. Male and female characters evinced roughly the same levels of creativity and cooperation. This makes the 1930s the decade in which we found the greatest amount of gender stereotyping in Caldecott winners and runners-up.

As Table 13.1 indicates, we found that female characters in Caldecotts of the late 1940s were more visible than they were in the late-1960s Caldecotts, but less visible than they were in the late 1930s or the late

TABLE 13.1   **Visibility of Female Characters in Caldecott Winners and Runners-Up**

	1930s	1940s	1950s	1960s
Total number of books	20	28	18	18
Percent without female characters[a]	0%	22%	11%	33%
Percent with central female characters[b]	45%	14%	39%	17%
Human single-gender illustrations				
Total number	335	508	163	196
Percent with female characters[c]	24%	16%	40%	10%
Nonhuman single-sex illustrations				
Total number	190	130	101	119
Percent female[d]	19%	20%	10%	1%

[a] Differences between 1930s and 1940s books, significant at .03 level; between 1930s and 1960s books, at .01 level; between 1940s and 1950s books, at .11 level. No other differences statistically significant.

[b] Differences between 1930s and 1940s books, significant at .02 level; between 1930 and 1960s books, at .06 level, between 1940s and 1950s books, at .06 level; between 1950s and 1960s books, at .14 level. No other differences close to being significant.

[c] All differences significant at .05 level.

[d] Differences between 1940s, 1950s, and 1960s books, significant at .05 level. Differences between 1930s and 1940s books, not statistically significant.

1950s. Fully 22 percent of 1940s books had no female character, and only 14 percent had a central female character. Only 16 percent of the human single-gender illustrations were of girls or women, and only 20 percent of animal single-sex illustrations were of females. Consistent with the LV hypothesis, the late-1940s Caldecotts exhibited less gender stereotyping than those of any other decade we examined. Female characters were just as independent, competitive, and aggressive as male characters, and less cooperative, more directive, less submissive, less imitative and more persistent than their male counterparts.

We found that female characters were less visible in the Caldecotts of the late 1950s than they were on the 1930s, but that they were more visible in the 1950s than they were in the Caldecotts of the 1940s or the 1960s. Only 11 percent of the 1950s books had no female character and fully 39 percent had a central character. Forty percent of the human single-gender illustrations were of girls or women. The only indicator on which we found female-character in visibility was for animal single-sex illustrations: only 10 percent of these were of females. Whereas female visibility in 1950s Caldecotts was second only to its counterpart in 1930s Caldecotts, the amount of gender stereotyping was also second only to that found in the 1930s books (see Table 13.2). Female characters were more dependent, submissive, and

TABLE 13.2   Percentage of Characters Exhibiting Behavioral Traits by Gender or Sex

	Late-1930s		Late-1940s	
Traits	Male characters	Female characters	Male characters	Female characters
Dependent	24	62 S*	26	54 S
Independent	90	70 S	93	90 T
Cooperative	81	85 T	83	69 R
Competitive	36	27 S	30	30 T
Directive	48	27 S	36	44 R
Submissive	25	50 S	15	0 R
Persistent	90	50 S*	78	85 R
Explorative	85	46 S*	83	43 S*
Creative	58	57 T	52	33 S
Imitative	8	15 S	18	8 R
Nurturant	43	75 S	44	75 S
Aggressive	35	23 S	17	21 T
Emotional	63	92 S	54	67 S
Active	100	92 S	100	85 S
Passively active	90	100 S	100	85 R

	Late-1950s		Late-1960s	
Traits	Male characters	Female characters	Male characters	Female characters
Dependent	11	50 S	37	29 R
Independent	100	80 S	88	83 T
Cooperative	87	92 T	79	100 S
Competitive	42	8 S	18	17 T
Directive	46	27 S	59	57 T
Submissive	8	27 S	0	0 T
Persistent	85	67 S	73	60 S
Explorative	36	50 R	66	33 S
Creative	64	36 S	40	80 R
Imitative	0	0 T	0	0 T
Nurturant	33	73 S*	56	71 S
Aggressive	13	0 S	27	14 S
Emotional	60	57 T	79	50 R
Active	100	100 T	100	100 T
Passively active	100	100 T	100	100 T

Note: * indicates the difference between male and female characters is significant at the .05 level; S indicates that the difference is in the stereotypical direction; R indicates that the difference is in the reverse stereotypical direction; T indicates that there's essentially no difference (less than a 5% difference) between males and females (i.e., that there is a tie).

nurturant than male characters, and were less independent, competitive, persistent, explorative, creative, and aggressive.

We found considerable evidence of female invisibility in Caldecotts of the late 1960s. Fully 33 percent of the books had no female character, and only 17 percent had a central female character. Moreover, only 10 percent of all single-gender human illustration in these books were, by our count, of girls or women, and only a minuscule 0.8 percent of single-sex animal illustrations were of females. On the other hand, we found the 1960s Caldecotts to be the second-least (to the 1940s books) stereotyped of the ones we examined for this study. Females were less dependent and less emotional than males. And on 7 of the 16 dimensions (e.g., independence), male and female characters had the focal trait in approximately equal measure.

Conclusion

Overall, we found more support for our "local variation" (LV) hypothesis than for the "monotonic change" (MC) hypothesis. Unlike what Clark et al. (1993, 2003) found in award-winners between the late 1960s and the late 1990s, we found that when female characters were most visible between the 1930s and the 1960s, gender stereotyping was most present, and when female characters were least present, gender stereotyping was most absent. This poses one challenge to the MC hypothesis. Another is the nonlinear pattern in both female visibility and gender stereotyping. Female visibility was greatest in the 1930s and the 1950s; gender stereotyping was greatest in these decades too. Female visibility and gender stereotyping were least evident in the 1940s and the 1960s.

One implication of our findings is that gender stereotyping and the presence of female characters in award-winning picture books may reflect different aspects of gender relations in society: that they may not both, as previously supposed (Clark et al., 1993 & 2003), be functions of the relative status of males and females in society. Gender stereotyping may well reflect the relative status of women. It seems to do so in the period from the 1930s to the 1960s, as it did in the period from the 1960s to the 1990s. But the visibility of female characters may reflect something different: perhaps the degree of conflict over women's roles. Perhaps female characters are least likely to appear in periods of conflict over gender roles (like the 1940s and the 1960s), just as Black characters have been least likely to appear during periods of racial conflict (Pescosolido et al., 1997). If so, it may have been the decreasing conflict over gender roles during the period from the 1960s to the 1990s that accounted for the greater degrees of female visibility in picture books (Clark et al., 1993, 2003), rather than the greater relative status of women in society.

Another implication of our work is that the recent "trend" toward female visibility and decreased stereotyping cannot be taken for granted. The trend is probably just as reversible, just as susceptible to antifeminist backlash, today as it was in the period from the late 1930s to the late 1960s.

Appendix: Behavioral Definitions from Davis (1984)

Dependent: seeking or relying on others for help, protection, or reassurance; maintaining close proximity to others.

Independent: self-initiated and self-contained behavior, autonomous functioning, resistance to externally imposed constraints.

Cooperative: working together or in a joint effort toward a common goal, complementary division of labor in a given activity.

Competitive: striving against another in an activity or game for a particular goal, position, or reward; desire to be first, best, winner.

Directive: guiding, leading, impelling others toward an action or goal; controlling behaviors of others.

Submissive: yielding to the direction of others; deference to wishes of others.

Persistent: maintenance of goal-directed activity despite obstacles, setbacks, or adverse conditions.

Explorative: seeking knowledge or information through careful examination or investigation; inquisitive and curious.

Creative: producing novel idea or product; unique solution to a problem; engaging in fantasy or imaginative play.

Imitative: duplicating, mimicking, or modeling behavior (activity or verbalization) of others.

Nurturant: giving physical or emotional aid, support, or comfort to another; demonstrating affection or compassion for another.

Aggressive: physically or emotionally hurting someone; verbal aggression; destroying property.

Emotional: affective display of feelings; manifestation of pleasure, fear, anger, sorrow, and so on via laughing, cowering, crying, frowning, violent outbursts, and so on.

Active: gross motor (large muscle) physical activity, work, play.

Passively active: fine motor (small muscle) activity; alert, attentive, activity but with minimal or no physical movements (for instance, reading, talking, thinking, daydreaming, watching TV).

REFERENCES

Clark, B. L. 1992. American children's literature: Background and bibliography. *American Studies International* 30: 4–40.

Clark, R., M. Almeida, T. Gurka, and L. Middleton. 2003. Engendering tots with Caldecotts: An updated update. In E.S. Adler and R. Clark (Eds.), *How it's done: An invitation to social research* (2nd ed., 379–386). Belmont, CA: Wadsworth.

Clark, R., R. Lennon, and L. Morris. 1993. Of Caldecotts and Kings: Gendered images in recent American children's books by black and non-black illustrators. *Gender & Society* 7: 227–245.

Davis, A. J. 1984. Sex-differentiated behaviors in non-sexist picture books. *Sex Roles* 11: 1–15.

Pescosolida, B., Grauerholz, E., and M. Milkie. 1997. Culture and conflict: The portrayal of Blacks in U.S. children's picture books through the mid- and late-twentieth century. *American Sociological Review* 62: 443-464.

Appropriate Topics for Content Analysis

Although Clark, Guilmain, Saucier, and Tavarez applied content analysis to children's books, content analytic methods can be applied to any form of communication,[2] including movies, television shows, speeches, letters, obituaries, editorials, and song lyrics. In the recent past, Gilbert Shapiro and Philip Dawson (1998) looked at the notebooks of revolutionaries to gain insight into the causes of the French Revolution, Loe (2004) examined newspaper advice columns to make sense of sexuality in the Viagra era, and Holster (2004) studied egg donation related websites to explore the characteristics of the Internet egg donation business. Much earlier, the freed slave Ida Wells (1892, cited in Reinharz [1992]) and G. J. Speed (1893, cited in Krippendorff [2003]) analyzed newspaper articles to show, respectively, the extent to which black men were being lynched in the South and how gossipy newspapers in New York were becoming during the 1880s. More recently, Gormly (2005) studied Pat Robertson's TV program, the *700 Club,* the two weeks following the September 11th attacks to determine its position toward Israel and the Jewish faith and Baker (2005) examined magazine advertisements to distill how they constructed women's sexuality.

Materials Appropriate for Content Analysis

Studies have focused on suicide notes (Bourgoin, 1995, to amend Durkheimian suicide theory), letters (Kenschaft, to examine nineteenth-century efforts to reinvent marriage, 2005), magazines (Friedan, to define the *Feminine Mystique,* 1963, Taylor, to see how they construct the sexuality of young men, 2005), wills (Finch and Wallis, 1993), textbooks (Lewis and Humphrey, 2005), radio programs (Albig, 1938), personal ads (Jaggar, 2005), speeches (Rutherford, 2004), verbal exchanges (Bales, 1950), diaries (Ulrich, 1990), parenting-advice books (Krafchick, Zimmerman, Haddock, and Banning, 2005), films, both conventional (Leslie, 2005) and pornographic (Cowan and Campbell, 1994), e-mail messages (Schleef, 1995), and websites (Holster, 2004). The list is endless. Systematic content analysis seems to date

---

[2] In fact, although content analysis is most often applied to communications, it may, in principle, be used to analyze any *content.* Thus, for instance, Rathje and Murphy's (1992) "garbology" project mentioned in Chapter 12 employed content analysis to study the contents of garbage cans.

back to the late 1600s, when Swedish authorities and dissidents counted the words in religious hymns and sermons to prove and disprove heresy (Dovring, 1973; Krippendorff, 2003; Rosengren, 1981).

Questions Asked by Content Analyzers

Generally the questions asked by content analyzers are the same as those asked in all communications research: "Who says what, to whom, why, how, and with what effect?" (Lasswell, 1965: 12). Clark, Guilmain, Saucier, and Tavarez were interested in what children's books had said about gender to young children from the 1930s to the 1960s and held that content up against liberal feminist standards that advocate that males and females be shown in equal numbers and with similar capabilities and tendencies. They found that, during that time period, the visibility of female characters and the degree of gender stereotyping fluctuated enormously and speculated that, while the visibility of female characters is tied to conflict over women's roles in society, the degree of gender stereotyping is tied to the relative status of women. Evaluating communication content against some standard and describing trends in communication content are two of the general purposes for which content analysis has been used (see Holsti, 1969: 43).

STOP AND THINK *Another frequent use of content analysis is hypothesis testing. For instance, do you think local, regional, or national newspapers are more likely to "headline" international stories on the top of the front page? What is your hypothesis? Why? (You might test this hypothesis in exercise 13.1)*

Units of Analysis

Sampling for content analysis involves many of the same problems that sampling for any type of social research involves. First, you need to determine the units of analysis (or elements), the kinds of subjects you want to focus on. In the focal research, this task was simplified because the authors were replicating earlier studies (Clark et al.'s, 1993 and 2003), whose units of analysis they obviously wanted to use again. And when it came to questions of female visibility, such as how many books had female main characters, or even how many human single-gender illustrations were of females, determining units of analysis was a snap: They were clearly books, in the first instance, and human single-gender illustrations, in the second. These were the units the earlier studies had used.

When it came to questions of gender stereotyping, the authors could also rely on the earlier studies, though it's useful to recall how the authors of the earlier studies, especially how Clark, Lennon, and Morris, 1993, chose their units. They had stumbled (as a result of a systematic literature review) on Albert Davis' (1984) article in which he reported his content analysis of books that had been recommended by liberal feminist organizations. Davis had identified and operationalized the 15 behavioral traits (for example, dependence, independence, and so on) listed in the Appendix to the focal research and used these variables to content-analyze all illustrations and text messages in those books. Clark, Lennon, and Morris decided that Davis' behavioral traits (his variables) were appropriate for their purposes, but that his units of analysis (all illustrations and text messages) were less so. In particular, they thought that the overall presentation of notable characters in books would be more likely to impress young readers than would the details of individual illustrations or text messages.

Therefore, Clark, Lennon, and Morris decided to focus (somewhat arbitrarily) on the two major characters of each gender in each book. (If there weren't two of each, the researchers didn't make them up. They evaluated those characters that were available.) Clark, Lennon, and Morris then decided, after they'd read the book, whether each of the four characters embodied any of the behavioral traits developed by Davis. Clark, Guilmain, Saucier, and Tavarez, in the focal research for this chapter, employed the same strategy.

units of analysis, the units about which information is collected.

units of observation, the units from which information is collected.

Earl Babbie (2003) offers another useful distinction: between **units of analysis**—the units about which information is collected—and **units of observation**—or the units from which information is collected. This, of course, is a distinction without a difference when the unit of analysis and observation are the same. The units of observation in the focal research were always children's books (that is, the authors *always* got information from the books) and in some parts of the analysis (when the authors were determining the gender of the main character of each *book*), the units of analysis were also books. But when the units of analysis and observation are different, remember that the constituents of one's sample are the units of analysis, not the units of observation. Thus, when Clark, Guilmain, Saucier, and Tavarez shifted their focus (their units of analysis) to illustrations or characters, they needed to remember that they were then interested in variable characteristics of illustrations and characters (within books) and not in the books themselves.

STOP AND THINK *Suppose you were interested in whether commercials during children's TV programs were more utopian (that is, focused on improvement rather than on things as they are) than were those for adult programs. What would your unit of analysis be? What would your unit of observation be?*

If you were interested in the difference between commercials during children's and adults' TV programs, your units of observation (or the units from which you collected information) would be children's and adults' TV programs (because commercials from other types, say, adolescent TV programs, would be irrelevant), but the units of analysis (or the units about which you'd get information) would be the commercials attached to those programs. An illuminating recent content analyses (see Box 13.1) used TV programs as both its unit of analysis and unit of observation.

STOP AND THINK *How can you tell from these findings that the units of observation for the Kaiser Foundation Study are TV programs?*

BOX 13.1

Major Findings of the Kaiser Family Foundation Biennial Report (2005) of "Sex on TV"

- 70 percent of all shows have sexual content, up from 56 percent in 1998 and 64 percent in 2002.

- In shows that include sexual content, the number of sexual scenes is also up, to an average of 5.0 scenes an hour, compared to 3.2 scenes in 1998 and 4.4 scenes in 2002.

- Sexual intercourse is depicted or strongly implied in 11 percent of TV shows, up from 7 percent in 1998, but down from 14 percent in 2002.

- Reality shows are the only genre of programming studied in which less than two thirds of shows include sexual content. In 92 percent of movies shown on television, 87 percent of sitcoms, 87 percent of drama series, and 85 percent of soap operas, there is sexual content.

- About half of all scenes with intercourse (53 percent) involve characters who have an established relationship.

- 15 percent of scenes with intercourse present characters having sex when they have just met, compared to 7 percent in 2002.

Sampling

In content analysis, units of analysis can be words, phrases, sentences, paragraphs, themes, photographs, illustrations, chapters, books, characters, authors, audiences, or almost anything you want to study. Once you've chosen the units of analysis, you can sample them with any conventional sampling technique (for example, simple random, systematic, stratified, or cluster—see Chapter 5) to save time or effort. In the analysis of human single-gender illustrations, for instance, Clark, Guilmain, Saucier, and Tavarez might have numbered each one and systematically sampled 25 percent of them by random sampling the first one and picking every fourth illustration after that. As things turned out, however, they decided that, because Clark et al. (1993) had examined all of the single-gender illustrations in the Caldecotts between 1967 and 1971 and between 1987 and 1991, and because Clark et al. (2003) had looked at all of those illustrations in the Caldecotts between 1997 and 2001, they would do the same for all the Caldecotts between 1938 (the year the award was first given) and 1942, between 1947 and 1951, between 1957 and 1961, and between 1967 and 1971.

STOP AND THINK   *How might Clark, Guilmain, Saucier, and Tavarez have collected a stratified sample of picture books? (Hint: Can you think of any characteristic of authors they might have used to divide their sample?) How might they have cluster-sampled illustrations?*

Creating Meaningful Variables

Content analysis is a technique that depends on the researcher's capacity to create and record meaningful variables for classifying units of analysis. Thus, for instance, in classifying books, Clark, Guilmain, Saucier, and Tavarez were interested in knowing whether they had female characters or not and whether they had female central characters or not; in classifying human and animal single-gender illustrations, whether the depicted character was female or not; in classifying main characters, whether they appeared dependent or not, independent or not, and so on. In each of these cases, variables were of *nominal* scale, but variables of ordinal or interval scales also may be used. In analyzing books, for instance, the authors could have asked "how many" major male or female characters appeared, in which case they would have created a *ratio* scale variable.

Actually, their coding sheet, depicted in Figure 13.1, makes it look as though Clark, Guilmain, Saucier, and Tavarez did collect ratio-scale information about single-gender illustrations after all.

STOP AND THINK   *Can you think of a question Clark et al. might have asked about children's books that would have elicited ordinal-scale data?*

FIGURE 13.1

Sample Coding Sheet from "Two Steps Forward, One Step Back" Study.

Book: Goggles   Author: E. J. Keats   Year of Publication: 1969
Publisher: MacMillan: New York

1. Does the book have any female characters?: No
2. Does the book have a central female character?: No
3. Number of human illustrations that depict a single and clearly determinate gender: 16
4. Number of these that depict females: 0
5. Number of animal illustrations that depict a single and clearly determinate gender: 0
6. Number of these that depict females: 0

BEHAVIORS OF CENTRAL CHARACTERS

	Main Character	Supporting Character of Same Gender	Most Important Character of Opposite Gender	Supporting Character of Opposite Gender
ROLE:	PETER	ARCHIE		
Dependent				
Independent	X	X		
Cooperative	X	X		
Competitive	X	X		
Directive	X			
Submissive		X		
Persistent	X	X		
Explorative				
Creative	X	X		
Imitative				
Nurturant				
Aggressive				
Emotional	X	X		
Active	X	X		
Passively Active				

Question 3 in Figure 13.1 asks each coder, for instance, about the number of illustrations that depict characters of clear and singular gender (compared with, for example, those that show characters of both genders or characters of ambiguous gender). But the appearance of interval categories is a sleight-of-hand, made possible because the unit of observation (picture books), in most cases, contains several units of analysis (single-gender illustrations) and because all the researchers were really interested in, after all, was the proportion of all single-gender illustrations in a given sample of books that depicted females.

STOP AND THINK *How do you think Clark et al. used the data from their coding sheets to calculate the proportion of all single-gender illustrations that showed females?*

Clark, Guilmain, Saucier, and Tavarez calculated the proportion of all single-gender illustrations that showed females by dividing the response to item 3 (number of human illustrations that depict a single and clearly determinate gender) into the response to item 4 (number of these that depict females). The unit of analysis here was the human single-gender illustration and the variable was gender (a nominal, not an interval, variable).

As in any measurement operation, the scheme left room for unreliable observations, or the possibility, in this case, that parallel observations of the same phenomenon would yield inconsistent data. Thus, the four coders (in this case, the four authors) disagreed about several books even on the number of unambiguous, single-gender illustrations. The authors handled these disagreements by chatting about them. In all cases, discussion soon made it obvious that one or more of them had simply under- or overcounted cases.

STOP AND THINK *If Clark, Guilmain, Saucier, and Tavarez hadn't used the discussion method described in this paragraph, what could they have done with their four estimates of the number of single-gender illustrations in a book to arrive at a final, single estimate for that book?*

If the authors hadn't used the discussion method, they could have taken the arithmetic average of their four estimates of the number of single-gender illustrations as their final, single estimate for each book.

Clark, Guilmain, Saucier, and Tavarez were tempted to use the discussion method to resolve differences in the classification of behavioral traits as well. But early attempts suggested that many differences here were not only perceptual (Guilmain might honestly disagree with Saucier that a certain character was persistent, despite a common agreement about what Davis [1984] meant by his definition of persistence [see Appendix in the focal research]) but also not easily reconcilable. So when it came to the final classification of characters by whether they possessed a particular behavioral trait, the authors agreed that they'd finally classify a character as possessing such a trait (for example, persistence) only if three coders independently (and without discussion) coded him or her that way. If there was a two-two tie, they agreed not to classify the character by the trait at all (i.e., rather than classifying him or her as persistent or nonpersistent, they simply didn't use the persistence variable to classify the character at all).

Quantitative or Qualitative Content Analysis?

quantitative content analysis, content analysis designed for statistical analysis.

We have discussed the piece by Clark, Almeida, Gurka, and Middleton to give you some appreciation for the potential of a relatively **quantitative** brand of **content analysis:** namely, a brand that was designed with statistical analysis in mind. Quantitative content analysis is the most common kind in the social sciences. But we wouldn't want to leave you with the impression that content analysis must be quantitative to be effective (see also Reinharz, 1992). Actually, we think content analysis is well employed when qualitative social scientists aim for primarily verbal, rather than statistical, analysis of various kinds of communication. Roger did—with another student, Heather Fink—a **qualitative content analysis** of 33 recent picture books that focus on characters on the powerless side of some powerless/powerful social dichotomy (Clark and Fink, 2004). Roger and Heather came to these books with few preconceived notions of what they would "see" (that is, their major purpose was exploration) and so did not try to find ways to "quantify the data," as the previous section suggests they might have. Instead, they found themselves "listening" to themes of oppression and resistance and noting patterns.

qualitative content analysis, content analysis designed for verbal analysis.

Compared with the relatively well-accepted techniques (for example, for quantifying and ensuring the reliability of data) for doing quantitative content analysis, however, guidelines for doing qualitative content analysis are few and far between. (For those interested in such guidelines, however, we recommend Reinharz, 1992, and Strauss, 1987: 28ff.) Heather and Roger tried to discern theme patterns by a process of comparison and contrast that would be congenial to the advocates of grounded theory (see also Glaser, 2005, and Glaser and Strauss, 1967). Thus, we noticed that in picture books where the text celebrated "difference," as the text of hooks and Raschka's (1999) *Happy to be Nappy* does, the art also celebrated differences, as do Raschka's paint-stroke images in *Happy to be Nappy*—images that are of African American girls with nappy hair and cheerful expressions.

Generally speaking, then, quantitative content analyses tend to be of the deductive sort described in Chapter 2—analyses that begin, perhaps, with explicit or implicit hypotheses that the researcher wants to test with data. Explicit in the focal research of this chapter, for instance, is a hypothesis that Caldecott books of later decades are more likely to present characters in less gender-stereotyped ways than their counterparts of earlier decades are. (This hypothesis is, in fact, shown to be false by the data.) Qualitative content analyses, on the other hand, tend to be of the inductive sort, analyses that might begin with research questions, but are then likely to involve observations about individual texts (or portions of texts) and build to empirical generalizations about texts (or portions of texts) in general. Clark and Fink, for instance, began with the general research question "What kinds of presentations are typical of picture books about relatively powerless characters?" and then analyzed the kinds of presentations they discovered as they read, and viewed, such books.

Visual Analysis

A relatively new set of techniques, closely related to those of content analysis, is **visual analysis.** Visual analysis, we would argue, is actually a kind of content analysis. It refers to a set of approaches for analyzing visual images. Visual analysts have studied photographs, video images, paintings, drawings, maps, single images, and collections of images. Some visual analysts (e.g., Bell, 2001) advocate the value of the kind of quantitative content analysis of pictures done by Clark, Guilmain, Saucier, and Tavarez in this chapter's focal research, when they, for instance, calculate the percentage of single-gender pictures in children's books that depict females. Some advocate a more qualitative approach, like the one used by Clark and Fink in the aforementioned study of picture books about relatively powerless characters, one that explicitly involves "listening" to the "overtones and subtleties" of images (e.g., Collier, 2001: 39). Some visual analysts suggest that images can be used as sources of factual information—as records of reality—and some, that they should be read as indicators of the way their creators would have us think about reality, something Asthana (2003) does when he examines Indian music videos for ways in which they construct patriotism. Van Leeuwen and Jewitt's (2001) *Handbook of Visual Analysis* is perhaps the best guide to the variety of approaches taken by visual analysts; *Visual Sociology* is the journal in which one can find the best examples of its uses by sociologists.

STOP AND THINK   *Have a look at the picture on the cover of this book. What do you think we've tried to convey about the book by using it? In answering this question, have you done a quantitative or qualitative visual analysis, would you say?*

Advantages and Disadvantages of Content Analysis

One great advantage of content analysis is that it can be done with relatively little expenditure of time, money, or person power. Clark, Guilmain, Saucier, and Tavarez worked as a team of four on their examination of children's books, all of which they borrowed from local libraries. They did so partly for the fun of working with one another and partly for the enhanced data reliability that comes from cross-checking one another's observations in various ways. In principle, however, their project could have been done by one of them, with no special equipment other than a library card.

Another related plus is that content analysis doesn't require that you get everything right the first time. Frankly, despite appearances to the contrary, the coding sheet shown in Figure 13.1 is really a compilation (and distillation) of several separate efforts made by Clark et al. (1993). Indeed, the authors had made a first pass at all the books in their sample to discern the visibility of female characters (in illustrations, and so on) *before* they discovered Davis' scheme for coding behavior. They then returned to their subjects (the books) to code behavior. This kind of recovery from early oversights is much less feasible in survey or experimental research.

STOP AND THINK   *Compare this to having surveyed 500 people and then deciding you had another question you wanted to ask them.*

Yet another advantage is that content analysis is unobtrusive; our content analysis itself can hardly be accused of having affected the content of the books we read. Nor can the Kaiser Foundation's (2005) study of "Sex on TV" be criticized for having affected the sexual content of the TV programming it examined. This is not to say that content analysis might not affect the content of subsequent communications. In fact, we would guess that the findings of some researchers who first content-analyzed picture books (e.g., Weitzman and colleagues, 1972) did contribute to the kinds of attitude shifts (in award-givers, in publishers, and in writers) that can account for some of the changes later studies (e.g., Clark et al., 1993 and 2003) documented. And, when the Kaiser Foundation introduces its study by referring to the "a third (34 percent) of young women still become pregnant at least once before they reach the age of 20—about 820,000 a year" (2005: 2), it seems to be encouraging a diminution of the number of sexual messages on television.

STOP AND THINK   *Can you imagine whose attitudes would need to be affected so that the "Sex on TV" study would have its desired effect?*

Like using other kinds of available data, content analysis permits the examination of times (and peoples) past—something surveys, experiments, and interviews don't. Clark, Guilmain, Saucier, and Tavarez's analysis of Caldecotts from the 1930s to the 1960s is a good example, as are Shapiro and Dawson's (1998) examination of French revolutionaries' notebooks and Sheldon Olson's (1976) use of the Code of Hammurabi, a list of laws and punishments from ancient Babylonia.

STOP AND THINK   *The Code of Hammurabi consists of 282 paragraphs, each specifying an offense and its punishment. One could code each of the paragraphs by the severity of the punishment associated with an offense, for instance, to gain a sense of what kinds of behavior the Babylonians most abhorred. Based on the punishments associated with stealing from a temple and striking another man, mentioned in the paragraphs below, which of these offenses was held in greater contempt?*
Paragraph 6. If a man steal the property of a god (temple) or palace, that man shall be put to death . . .
Paragraph 204. If a freeman strike a freeman, he shall pay ten shekels of silver.

One of Emily's future projects is an analysis of life expectancy, family structure, and gender roles in previous centuries by doing content analysis of the inscriptions on head stones in New England cemeteries.

Yet another advantage of content analysis is that, because it typically permits one to avoid interaction with human subjects, it also permits the avoidance of the usual ethical dilemmas associated with research involving human subjects and with the time involved in submitting a proposal to an Institutional Review Board, unless one plans to use private documents.

Some disadvantages of content analysis might have occurred to you by now. Because content analysis is usually only applied to recorded communication, it can't very well be used to study communities that don't leave (or haven't left) records. Thus, we can study the legal codes of ancient Babylonia because we've discovered the Code of Hammurabi, but we can't study the codes of societies for which no such discoveries have been made. And Clark, Guilmain, Saucier, and Tavarez can content-analyze children's books with some sense that what children read (or heard) might have predictable consequences for their attitudes. But to tap those attitudes directly, it would be more valid to analyze literature produced *by*, not *for*, children. Peirce and Edwards (1988) actually did content-analyze stories written by 11-year-olds, but this would be pretty hard to do for pre-literate, and certainly pre-verbal, children, the not infrequent audience for picture books. Content analysis can also involve social class biases because communications are more likely to be formulated by educated persons.

There are also questions about the validity of content-analytic measures. For example, is the invisibility of female characters in award-winning children's books of the late 1940s and the late 1960s really a valid indicator of attitudes about women and girls? Perhaps, but maybe not. And even if one were pretty sure that the invisibility was a valid measure, whose attitudes are being measured: those of the authors, the publishers, the award-givers, or the adult purchasers and readers?

STOP AND THINK   *One of us, Emily, and a colleague thought they'd found evidence of significant social change when they content-analyzed 20 years of the "Confidential Chat" column (which published reader contributions) in the Boston Globe and found the column increasingly dealing with women's work outside the home between 1950 and 1970. They were discouraged when a reviewer of their research report for a scholarly journal suggested that their results might reflect changes in the editorial climate at the newspaper, rather than changes in the attitudes of the readership. When they interviewed the editor of the column, they found, indeed, he and the former editor published only about 10 percent of the letters he received and that changes in content were likely to reflect his sense of what his readership (mostly women) wanted to read. Emily and her co-author never resubmitted their paper because they now felt it failed to say anything significant about social change. What do you think of this decision?*

Finally, content analysis doesn't always encourage a sensitivity to context that, say, literary criticism often does. A male character, for instance, might be portrayed as aggressive, but the point of the story might be to show how foolhardy that aggressiveness can be. Is it the aggressiveness or the author's implicit condemnation of the aggressiveness that will affect a young listener? Questions of validity plague those who do content analysis just as much as they plague those who employ other methods of data collection.

STOP AND THINK *Suppose your main concern was whether authors endorsed, rather than simply portrayed, male aggressiveness? What questions might you ask of your books to code such an endorsement? Must content analysis be as insensitive to context as the previous paragraph suggests?*

Validity concerns are less likely to be legitimate in the case of qualitative content, and visual, analyses, where the researchers focus on things like context and author (artist) intention and the verbal analysis is more likely to take such things into account. Thus, a project that focused on the research question, "What attitudes do authors take toward aggressiveness?," is more likely to identify such attitudes validly than one that begins with the hypothesis "Books about male characters are more likely to be about aggressive characters than books about female characters."

Comparing Methods

We've now covered the major methods of data collection in the social sciences: questionnaires, interviews, observations, and available data (including content analysis). We believe that no one method is inherently better than others. Instead, we believe that some methods are more or less useful for studying specific topics and in specific situations.

Researchers should consider the following issues when deciding whether to observe something, to ask questions about it, or to use some available data as the data collection method:

- Which method(s) will allow the collection of valid information about the social world?
- Which method(s) will allow the collection of reliable information about the social world?
- To what extent will the results be generalizable?
- To what extent are these data collection methods ethical?
- To what extent are these data collection methods practical?

Each method of data collection has advantages and disadvantages. A brief overview of these is found in Table 13.3.

STOP AND THINK *Assume our research question is "What factors contribute to alcohol consumption among college students?" If we were interested in studying this on your campus, assuming we had the funding and the time to spend a semester on this question, what method of data collection should we select?*

We could, like many researchers, administer a voluntary, anonymous questionnaire in a group setting because it is likely to be ethical, fairly practical, and, if there is a good response rate, produce results that are generalizable. However, we might worry about validity—will students tell the truth? Do they remember accurately? Will they over- or underestimate usage?

TABLE 13.3   **Advantages and Disadvantages of Methods of Data Collection**

Method	Advantages	Disadvantages
Asking Questions		
Group-administered questionnaire	Inexpensive Time efficient Can be anonymous, which is good for "sensitive" topics Good response rate	Not all populations possible Must have literate sample Must be fairly brief Need permission from setting or organization
Mailed questionnaire	Fairly inexpensive Completed at respondent's leisure Time efficient Can be anonymous, which is good for "sensitive" topics	Response rate may be low Need names and addresses Must have literate sample Can't explain questions
Individually administered questionnaire	Good response rate Completed at respondent's leisure	Most expensive questionnaire Need names and addresses Must have literate sample
Internet-based questionnaire	Very inexpensive and time efficient Can be completed at respondent's leisure	Not all populations possible Random samples might not be possible Low response rate
Structured phone interview	Good response rate Little missing data Can explain questions and clarify answers Can use "probes" Don't need names or addresses Good for populations with difficulty reading/writing	Expensive Time consuming Respondents might not feel anonymous Some interviewer effect
Structured in-person interview	Good response rate Little missing data Good for populations with difficulty reading/writing Can assess validity of data Can explain questions and answers and use "probes"	Expensive Time consuming Need names and addresses Interviewer effect Confidential, not anonymous
Semi-structured interview	Allows interviewer to develop rapport with study participants Good response rate Questions can be explained and modified for each participant Useful for discussing complex topics because it is flexible and allows follow-up questions Can be used for long interviews Useful when the topics to be discussed are known	Time consuming Expensive Requires highly skilled interviewer Interviewer effect Data analysis is time consuming

TABLE 13.3 *(continued)*

Method	Advantages	Disadvantages
Unstructured interview	Allows interviewer to develop rapport with study participants Good response rate Questions can be explained and modified for each participant Participant can take lead in telling his or her own story Useful for exploratory research on new topics Useful for discussing complex topics and seeing participant's perspective Can be used for long interviews	Time consuming Expensive unless small sample Requires highly skilled interviewer Interviewer effect Data analysis is time consuming May not have comparable data for all participants
Focus group interview	Allows participants to have minimal interaction with interviewer and keep the emphasis on participants' points of view Typically generates a great deal of response to questions Often precedes or supplements other methods of data collection	Attitudes can become more extreme in discussion Possibility of "group think" outcome Requires highly skilled interviewer Data less useful for analysis of individuals
Observation		
Participant and nonparticipant observation	Permits researcher to get handle on the relatively unknown Permits researcher to obtain understanding of how others experience life Permits the study of quickly changing situations Relatively inexpensive Permits valid measurement especially of behavior Relatively flexible, at least insofar as it permits researcher to change focus of research	Results are often not generalizable Being observed sometimes elicits characteristics (demand characteristics) in observed simply because of observation Especially in unannounced observations, requires deceit Time consuming
Available Data and Content Analysis		
Available data	Savings in time, money, and effort Possibility of studying change and differences over space Good for shy and retiring people, and for those who like to work at their own convenience Data collection is unobtrusive	Inability to know limitations of data Might not be able to find information about topic Sometimes one has to sacrifice measurement validity Possibility of confusion about units of analysis Danger of ecological fallacy

Continued

TABLE 13.3   *(continued)*

Method	Advantages	Disadvantages
Content analysis	Savings of time, money, and effort Don't need to get everything right the first time Good for shy and retiring people, and for those who like to work at their own convenience Data collection is unobtrusive	Can't study communities that have left no records Validity issues, especially resulting from insensitivity to context of message examined

We could decide to hire students to work as complete participant observers. After all, we'd expect that if they don't drink much themselves and record the information periodically, they could make accurate observations of other students' behavior. On the other hand, we don't know how representative their observations are, and we'd face the ethical issue that the students being observed have not been informed and have not given consent.

We could decide to use available data and analyze bottles in the trash or kegs waiting to be returned to distributors. Such indicators are practical and ethical (because trash is public), but we'd need to worry about reliability—thinking about when and how frequently we examined the data. Also we might worry about validity as the bottles for some alcoholic beverages consumed by students might have been returned for the deposit, others might be in dorm rooms or in recycling bins, and some might have been left by nonstudents. At best, the data tell us about the campus as a whole rather than about individual student use.

Multiple Methods

Each individual method has advantages and disadvantages, so it is always useful to employ more than one method of data collection whenever it is practical. In studying alcohol use on campus, if the resources were available, we could select more than one technique. A questionnaire, for example, could be coupled with the analysis of available data or a focus group or both. Together, they would give a better overview of the topic than any one of the methods.

Using multiple methods, or what has been called triangulation (Denzin, 1997), allows the strengths of one method to compensate for the weaknesses of another and increases the validity of the study's conclusions.

BOX 13.2

Examples of Multiple Methods

Research questions: Do nurses treat terminally ill patients differently from those who are expected to recover?

Possible Methods: Being a participant observer in a hospital for at least several weeks coupled with semi-structured, in-person interviews with samples of nurses, patients, and patients' families and an analysis of a sample of patient records.

Research questions: How do two communities in the same urban area differ? Do they receive different levels of services? Do residents in one community feel more socially isolated than the other? Does one community experience higher levels of crime than the other?

Possible Methods: Available data (such as the annual reports from local agencies and organizations, editorials, and news stories in local newspapers, National Crime Victim Survey data, and the FBI *Uniform Crime Reports*), along with observations in public places as a complete observer, and structured phone interviews conducted with a random sample of residents.

Research questions: What do high school students think about careers in law enforcement? Do attitudes differ by race?

Possible Methods: Interview high school guidance counselors concerning student views, do an Internet-based survey of a sample of high school students, be a complete observer at career fairs that include law enforcement opportunities, and use available data from law enforcement recruitment efforts.

Of course, using additional methods means increases in time and cost, the need to obtain access more than once, and the obligation to reconcile discrepant findings. These considerations must be weighed against the increases in validity or generalizability that the use of multiple methods brings. Box 13.2 has examples of alternate methods. An increase in validity and cost accrues if more than one method is used.

Summary

Content analysis is an unobtrusive, inexpensive, relatively easy, and flexible method of data collection that involves the systematic and—typically, but not necessarily—quantitative study of one kind of available data—some form of communication. Content analysis can be used to test hypotheses about communication, to compare the content of such communication with some standard, or to describe trends in communication, among other things. Clark, Guilmain, Saucier, and Tavarez simultaneously looked at trends in the depiction of gender in children's picture books and measured these trends against an egalitarian standard.

An early step in content analysis is to identify a unit of analysis. This is the unit about which information is collected and can be words, phrases, or paragraphs; stories or chapters; themes; photographs; illustrations; authors; audiences; or almost anything associated with communication. In Clark, Guilmain, Saucier, and Tavarez's study, the unit of analysis was sometimes whole books, sometimes illustrations, and sometimes characters.

Once a unit of analysis has been identified, any conventional sampling technique can be employed to ensure representativeness. Sometimes, as in Clark, Guilmain, Saucier, and Tavarez's study, investigating a whole population is also a feasible route to this goal.

The quality of a content analysis depends on the creation and meaningful usage of variables for classifying units of analysis. In studying main characters, for instance, Clark, Guilmain, Saucier, and Tavarez used a set of variables, created by Davis (1984), to examine various aspects of gender-related behavior. Explicit and reliable coding schemes for recoding variation are most important. Checks on measurement reliability can be pretty simple, but are always appropriate. Clark, Guilmain, Saucier, and Tavarez used discussion to check the reliability of some of their classification efforts and statistics to check others.

On the down side, content analysis cannot tell us much about communities that have left no recorded communications. Moreover, content analysis is frequently criticized on validity grounds, mainly because those criticizing aren't sure whether the content analyzers are measuring what they think they are. Validity concerns are less well founded, however, in the case of qualitative content and qualitative visual analyses.

We ended this chapter by summarizing the advantages and disadvantages of the major methods of data collection in the social sciences and by suggesting the possibility that, under certain circumstances, it is useful to consider using multiple methods of data collection.

EXERCISE 13.1

How Newsworthy Is the News?

This exercise offers a chance to observe, and speculate about, differences in the treatment of a given news story in daily newspapers. Your research questions in this exercise is this: Does the treatment of a given news story vary from one newspaper to another and, if so, what might account for this variation?

1. Select a current national or international news story. What story did you choose?

2. Select four national, regional, or local newspapers (either from a newsstand or your library). Get copies of these papers on the same day. List the names of the newspapers and the date.

   Date:

   Newspaper 1:

   Newspaper 2:

   Newspaper 3:

   Newspaper 4:

3. Decide how you will measure the importance given a story in a newspaper. (Hint: You might consider one of the following options, or combinations thereof: What page does the story appear on? Where does the story appear on the page? How big is the typeface used in the headline? How many words or paragraphs are devoted to the story? Is it illustrated? What percentage of the total words or paragraphs in the paper is devoted to it? Does the newspaper devote an editorial to the story?) Describe the measure you will use to determine the relative importance given your story in each of the papers.

4. Report your findings. Which paper gave the story greatest play, by your measures? Which paper, the least? Interpret your findings by trying to account for differences in terms of characteristics of the papers you've examined. Please submit copies of the stories with the assignment.

EXERCISE 13.2

Analyzing Website Commentaries on School Shootings

This exercise adapts an idea from Professor Molly Dragiewicz of George Mason University to give you the chance to analyze website commentaries on school shootings in the United States.

1. Do a Web search using a search engine or a subject guide such as Google or Yahoo, using the terms "school shootings" and "commentary." Find four commentaries about school shootings that have taken place in the past decade and print them out.

2. Your research questions are the following:

   a. Do website commentaries tend to focus on the perpetrators or the victims of school shootings?

b. Do these commentaries tend to focus most on gender, race, or class in their interpretations of shootings in American suburban schools?

3. Prepare four code sheets like this:

Commentary: Author: Publication date:

a. Would you say that the commentary focuses more on the perpetrators or the victims of school shootings?

b. If the commentary focuses on the perpetrators, in your view, does it focus on the race, gender, or class of these perpetrators? (Yes or no)

c. If so, which of these characteristics seem most important to the author?

d. If commentary focuses on the victims, in your view, does it focus on the race, gender, or class of these victims? (Yes or no)

e. If so, which of these characteristics seem most important to the author?

4. Write a paragraph summarizing your findings. (You might summarize, for instance, whether the commentaries seem more concerned with the perpetrators or the victims of the shootings. Then you might summarize whether race, gender, or class seems to be the "lens" through which the commentaries tend to be made.)

5. Include the copies of the commentaries as an appendix to the exercise.

EXERCISE 13.3

The Utopianism of TV Commercials

It has been said that commercials for children's TV programs are utopian—that they tap into utopian sentiments for something different, something better; that they express utopian values of energy, abundance, and community (see, for example, Ellen Seiter: 1993: 133)—rather than extolling the virtues of things as they are. But are commercials for children's programs really all that different in this regard from those for adult programs? This is your research question for this exercise.

1. Select a children's TV show and an adults' TV show on commercial stations. (Hint: Saturday mornings are particularly fruitful times to look for children's commercial TV programs.) Pick shows of at least ½ hour in duration (preferably one hour).

2. Indicate the shows you've chosen and the times you watched them.

> Children's Show:
> Time watched:
> Adults' Show:
> Time watched:

3. Describe the measures you will use to determine whether a commercial appealed to utopian sentiments or values. (For example, you might ask whether the commercial makes an explicit appeal to the viewer's desire for a better life [for instance, by showing an initially depressed-looking character fulfilled by the discovery of a soft drink or an initially bored breakfaster enlivened by the acquisition of a new cereal] or if it makes implicit appeals to, say, the viewer's wish for energy [is there dancing or singing associated with a product?], abundance [is anyone shown to be constrained by considerations of price?] or community [do folks seem to get along better after acquiring brand X?].) Show an example of the code sheet you will use, including a space for the product advertised in the commercial and a place for you to check if you observe the presence of what it is you're using to measure an appeal to utopian sentiments or values.

4. Use your coding scheme to analyze the first five commercials you see in both the children's and the adults' show.

5. Report your findings. In which kind of programming (for children or adults) did you find a higher percentage of commercials appealing to utopian sentiments or values? Interpret your findings.

EXERCISE 13.4

Gender Among the Newberys: An Internet Exercise

The American Library Association (ALA) awards the Newbery medal for adolescent fiction. For this exercise, you'll want to use the ALA website (www.ala.org) to answer questions about the gender of characters in Newbery winners and honor books over the last decade. (Hint: one way of finding such a list, after getting on the ALA site, is to select the "search website" option, type "Newbery" at the request line, request information about one of the recent winners, then request the list of other winners and honor books at the bottom of that page.)

1. Find a list of Newbery award winners and honor books for the last 10 years.

2. Begin your content analysis of the titles of these books by listing the titles that suggest to you that there is a female character in the book.

3. Now list those books whose titles suggest to you that there is a male character in the book. (The two lists might overlap. For instance, the title *Lizzie Bright and the Buckminster Boy,* an earlier title, suggests to us that there are at least two characters in the book, one male and one female. It would, therefore, make both your lists, if it fell into the period of your study.)

4. Does your brief content analysis of recent Newbery award and honor books suggest that more attention has been paid to males or females?

14

Applied Social Research

© Mary Kate Denny/PhotoEdit, Inc.

Introduction

By now you've read quite a few social research studies. Our hope is that you found at least some of the studies to be interesting. You also might have found some of them focused on topics that are of interest to you personally or professionally. Perhaps you've even thought that one or two of the studies could be of use to you in a current or future work setting. If so, then you've probably identified the kind of research we'll be considering in this chapter—**applied research.**

applied research, research intended to be useful in the immediate future and to suggest action or increase effectiveness in some area.

We've explored a variety of topics in this text: creating research questions and hypotheses, considering ethical and practical issues, selecting populations, sampling strategies and research designs, and choosing one or more methods of data collection. Before covering data analysis, we want to expand on our discussion of applied research—and in so doing, provide the details of an exemplary research project.

Basic and Applied Research

All social research is designed to increase our understanding of human behavior and can be useful to individuals, groups, or the whole society, but some work is more *immediately* useful than other research. As we noted in Chapter 1, **basic research** seeks to create new knowledge for the sake of that knowledge, whereas applied research is designed to help solve immediate social problems and answer broad research questions about settings in complex social, and political environments. Although practical applications can usually be derived from basic research, those projects are designed primarily to provide greater understanding of our social world and to develop or test theories. In applied work, theory is used instrumentally to identify concepts and variables that will produce practical results (Bickman and Rog, 1998: xii) and the research is designed to provide organizations such as schools, legislatures, communities, social service agencies, health care institutions, and the like with practical information. As a great deal of research can be seen as falling somewhere along the continuum between these two poles, identifying the major goal or emphasis is helpful in determining whether a research project is primarily basic or applied work.

basic research, research designed to add to our fundamental understanding and knowledge of the social world regardless of practical or immediate implications.

Social Problems and Social Solutions

In the twentieth century in the United States, awareness of social problems and programs directed toward solving them became increasingly common. The efforts before World War I focused on literacy, occupational training, and controlling infectious diseases, with projects designed to determine if funding initiatives in these areas were worthwhile (Rossi and Freeman, 1993: 9). Through World War II and into the 1960s, federal human service projects in health, education, housing, income maintenance, and criminal justice became widespread, as did the efforts to demonstrate their effectiveness.

By the end of the 1970s, many interested parties started questioning the continued expansion of government programs, partly because of changes in ideology, but also because of disenchantment with the outcomes and implementation of many of the programs advocated by public officials, planners, and politicians (Rossi and Freeman, 1993: 24). Despite disappointments with earlier efforts, both old and newly identified social problems remained, and programs continued to be created and implemented. With many millions of program dollars spent on local, state, and federal levels, policy makers wanted answers: Are the programs effective? Should they be continued? Modified or eliminated entirely? Applied research attempts to answer these questions.

STOP AND THINK *What do you think are the most serious social problems that we face as a society? Violence in our schools and communities? Unemployment? Poverty? Racism? AIDS? Obesity? Teenage pregnancy? Homelessness? Drug and alcohol abuse? Are you aware of programs nationally or in your state or local community designed to work toward solving any of these problems? If you are familiar with a program, do you think it is effective? Are those who are served by such programs involved in deciding if they are useful?*

You may believe that all of the problems mentioned warrant attention. In fact, there have been programs addressing each of them. Sometimes we read or hear about efforts to improve things, but many such efforts are ignored by the media. We rarely learn about the *effectiveness* of programs and policies designed to address social problems; even more rarely do we hear how those affected by the programs and policies feel about the situation or what their concerns are.

In the past three decades, one social concern that's received widespread public attention is youth drug use. Over the years, public policies defined this as a critical area for action, with a variety of programs proposed. Drug Abuse Resistance Education (DARE) was one of the American anti-drug education programs developed in the mid-1980s. Implemented in thousands of communities, DARE reached millions of students in a matter of years. In 2006 alone, the DARE program reached 36 million school children around the world, with 26 million in the U.S. alone (DARE, 2006). Central to its continued use is the *assumption* that the program is effective in helping students resist pressures to use tobacco, alcohol, cocaine, heroin, and similar drugs. Considering the widespread implementation of DARE and the amount of money spent on this program, clearly it is important to test the assumption about its effectiveness.

STOP AND THINK *Did you take part in an anti-drug education program through your elementary or secondary school or through a religious or community youth group? If so, was it a DARE program? If not, what did your program involve?*

Sociologists on the faculty of Indiana University, Kokomo, were given the chance to ask and answer an important research question: Is Kokomo's DARE program effective? As in other studies, the social and political climate, chance factors, and the professional training and interests of the social scientists provided the impetus for an interesting and important project.

In the mid-1980s, the first police officer recruited and trained to teach the DARE curriculum in the Kokomo schools had been a student in one of Professor Richard Aniskiewicz's sociology courses. DARE had program evaluation as one of its funding requirements, so the officer recommended Professor Aniskiewicz to the local police and schools as a possible evaluator. Aniskiewicz recruited Earl Wysong, and in 1987, after receiving approval and some funding from the local school system, they began an evaluation of the local DARE program that lasted more than seven years.

One article that resulted from their work is the basis for the focal research selection that follows. Because it is applied rather than basic research, Wysong, Aniskiewicz, and Wright tried to answer a research question with practical and political implications. The authors worked in the "real world" context of applied research—a context that can make research simultaneously interesting and frustrating. The kind of applied work that Professors Wysong, Aniskiewicz, and Wright did is evaluation research. The following focal research tells us about several things: the evaluation research process, the social and political context of such research, and the success of DARE.

FOCAL RESEARCH

Truth and DARE: Tracking Drug Education to Graduation and as Symbolic Politics

by Earl Wysong, Richard Aniskiewicz, and David Wright[1]

Introduction

The period following former President Reagan's 1986 "War on Drugs" campaign witnessed a rapid expansion of anti-drug programs at federal, state, and local levels with active support from corporations, the mass media, and community agencies (Jensen, Gerber, and Babcock, 1991; Males, 1992). Federal drug-related expenditures for law enforcement, treatment, and education rose from $2.6 billion in 1985 to nearly $13 billion in 1993 (U.S. Office of Management and Budget, 1990; 1992). To coordinate public and corporate policies aimed at reducing both the supply of and demand for drugs, a "National Drug Control Strategy" was initiated (National Drug Control Strategy, 1990). Schools quickly became the locus of such efforts, and by 1990 more than 100 school-based drug education programs were being promoted (Miley, 1992; Pereira, 1992) with wide variations reported

---

[1] This report is an abbreviated version of the article by Wysong, Aniskiewicz, and Wright, which appeared in *Social Problems* 41 (August, 1994): 448–472. It is reprinted with permission of the Society for the Study of Social Problems.

in their approaches, duration, costs, and reputed effectiveness (Bangert-Drowns, 1988; Pellow and Jengeleski, 1991).

Within the diverse array of drug prevention curricula in the United States, DARE (Drug Abuse Resistance Education) has become the largest and best known school-based drug education program (Rosenbaum, Flewelling, Bailey, Ringwalt, and Wilkinson, 1994). In 1990, DARE programs were in place in more than 3,000 communities in all 50 states and were reaching an estimated 20 million students (U.S. Congress: House, 1990). The DARE curriculum involves the presentation of anti-drug lessons by uniformed police officers who have taken 80 hours of specialized training (U.S. Department of Justice, 1991). With training costs of more than $2,000 per officer and program costs reaching as much as $90,600 per year for each full-time officer-instructor, DARE is expensive to implement and maintain (Miley, 1992; Pope, 1992). In 1993, total nationwide expenditures for DARE programs from all sources were estimated to be $700 million (Cauchon, 1993). The costs of adopting DARE are typically covered by federal, state, and foundation grants, by corporate support, and by local education and law enforcement agencies (U.S. Department of Justice, 1991). However, at this time no institutionalized funding base exists to sustain the program on a permanent basis (Congressional Record, 1992).

Despite its high cost and the absence of laws or formal policies mandating its implementation, DARE has gained widespread acceptance and support. Popularity notwithstanding, DARE's size, claims, and expense have generated some questions about its effectiveness. Short-term evaluations conducted within weeks or months following completion of the DARE program indicate it does produce "anti-drug" effects among students, especially in terms of creating and/or reinforcing anti-drug attitudes and drug-coping skills (Aniskiewicz and Wysong, 1990; Clayton, Cattarello, Day, and Walden, 1991; Harmon, 1993). However, claims for the program extend beyond the short-term. DARE is now being represented as a "long-term solution" to the problem of drug use (U.S. Department of Justice, 1991: i), but studies of the program's long-term effectiveness are virtually nonexistent. Longitudinal studies are underway at DARE Regional Training Centers (U.S. Department of Justice, 1991), and DARE AMERICA, a nonprofit agency in Los Angeles, is now funding an eight-year study in six U.S. cities (DARE AMERICA, 1991). In addition to the issue of efficacy are the less frequently addressed, but equally important, questions concerning DARE's emergence as a potent political force and the implications of this for program evaluation.

Assessment Framework

We view the issues of DARE's efficacy and political potency as interrelated. The present study analyzes data collected from 1987 to 1992 as part of a multiyear DARE assessment in Kokomo, Indiana (a medium-sized midwestern city). We explore DARE's effects on high school seniors five years after their initial exposure to the program using quantitative measures to determine if students' exposure to DARE produces lasting effects related to the program's objectives.

As an extension of our previous work (Aniskiewicz and Wysong, 1990), we also utilize a macro-level process perspective to explore DARE's political potency, organizational support structure, and the implications of these factors for program evaluation and implementation. This approach views DARE as a socially constructed form of symbolic politics growing out of and embedded within the larger Drug War of the late 1980s and early 1990s. It calls attention to the role played by stakeholders in developing, legitimating, and expanding the program as well as the nature and extent of DARE's organizational support structure.

Project DARE

DARE originated in 1983 as a joint venture between the Los Angeles Police Department (LAPD) and the Los Angeles United School District (LAUSD). It is the best known of the "new generation" of drug prevention programs and uses a combination of factual/informational, cognitive/affective decision-making, and psychosocial life skills approaches. Uniformed officers teach a standardized anti-drug curriculum that attempts to shape students' attitudes and social skills to help them resist peer and media pressures to try drugs including tobacco and alcohol.[2] The DARE "core" curriculum is designed for fifth and sixth graders. It typically consists of 17 in-class weekly lessons, each approximately 45 to 60 minutes long, with active student involvement, including questions and answers, role playing, and group discussion (U.S. Department of Justice, 1991: 7).

Kokomo was the first city in Indiana to implement the DARE program (Miley, 1992). During 1987–1988, all Kokomo fifth graders received the full DARE core curriculum, while all seventh graders received a shortened 11-week version of the curriculum. At that time, we conducted short-term program assessments for both grades using only the "DARE SCALE," an instrument developed in Los Angeles (Nyre, 1984; 1985), which supposedly provides a general measure of DARE's effectiveness in the areas of drug-related attitudes, knowledge, and anti-drug coping skills. In a pre- and post-test design, we found significant increases in anti-drug attitudes among the fifth but not the seventh graders. While both groups are part of a larger study, the former seventh graders (as 1992 seniors) are used as the basis for evaluating DARE's long-term effectiveness in this inquiry.

---

[2] DARE's objectives regarding student drug use embrace the same "zero tolerance/no-drug/drug free" orientations that guided national drug policies since the 1980s. DARE makes no distinctions between legal and illegal drugs and advocates total abstinence as the only acceptable approach to all types and categories of drugs. The program appears to assume that drug experimentation (with any drug at any level) equals drug abuse or will inevitably lead to it. DARE's objectives are not only unrealistic, but possibly counterproductive because they are obviously unattainable.

Impact Evaluation: Objectives, Sample, Design

The impact dimension of the study uses data from a multipart questionnaire completed by Kokomo High School (KHS) seniors in 1991 (non-DARE group, no DARE exposure) and in 1992 (DARE group, exposed to DARE as 7th graders). Several comparisons are made to assess whether and the extent to which DARE exposure is associated five years later with DARE's primary objectives of preventing/reducing/delaying drug use, including measures of self-reported drug use rates and of drug attitudes, drug knowledge, and drug-resisting coping skills. If DARE exposure in the seventh grade produces lasting anti-drug effects, then measures of the DARE group's characteristics in these areas should be significantly different from the non-DARE group.

The questionnaire was completed by samples of 1991 and 1992 KHS seniors nearly evenly divided by gender. Subjects were chosen each year by distributing questionnaires to all students in a random selection of classes (for example, English, science, and so on) and asking them to participate on a voluntary basis. Asking students not to put their names on the survey and using ballot-box style procedures to ensure confidentiality and anonymity, responses were obtained for 331 1991 seniors (class size = 511) and 334 1992 seniors (class size = 474). Of the later group, 288 were identified as having completed DARE as seventh graders. This sub-population is referred to as the 1992 "DARE group" in the comparisons developed throughout the study.

The questionnaire included both the Drug Use Scale and the DARE Scale.[3] The Drug Use Scale is used to compare the 1991 and 1992 KHS senior groups on five measures of drug use:[4]

1. *Lifetime Prevalence.* Percentages of students reporting use (any amount) of each type/category of drug during their entire lives.

2. *Recency of Use (Past Year).* Percentages of students reporting use (any amount) of each type/category during the past 12 months.

3. *Recency of Use (Past Month).* Percentages of students reporting use (any amount) of each type/category during the past 30 days.

4. *Grade Level at First Drug Use.* The mean grade at first use for each type/category of drug.

5. *Frequency of Use.* Percentages of students reporting drug use by number of occasions for three time periods—lifetime, past year, and past 30 days.

---

[3] See the appendix to this article for a more complete description of the instrument and examples of questions from both scales.

[4] The Drug Use Scale categories of hallucinogens, cocaine, sedatives/tranquilizers, and narcotics are used to report use rates among KHS seniors for combinations of specific drugs reported separately in the national senior survey. For example, our questions on hallucinogens asked students to respond within that category if they had used any related drugs (such as LSD, PCP, and so on). By contrast, the national senior survey instrument from which this scale was adapted asks for use rates on each specific drug (Johnston, O'Malley, and Bachman, 1992).

FIGURE 14.1

Drug Use Scale: Lifetime Prevalence

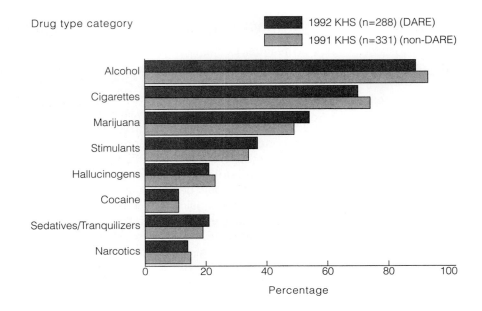

The DARE Scale consists of 19 questions with five response categories for each item (strongly agree to strongly disagree). Scores were computed by calculating the percentage of students responding with "appropriate" answers (disagree or strongly disagree) to each item along with the overall mean percentage of "appropriate" responses for all 19 items. According to this coding and scoring approach, the higher the individual item percentages and overall mean score, the more the results are supposedly indicative of anti-drug attitudes, greater drug knowledge, and enhanced drug-resistant coping skills.

Results: Impact Evaluation

Drug Use Scale

Comparisons of data from the Drug Use Scale for 1991 non-DARE and 1992 DARE seniors reveal that self-reported drug use rates among both groups are very similar for (1) Lifetime Prevalence (see Figure 14.1), (2) Recency of Use (Figure 14.2), (3) Grade Level at First Drug Use (Figure 14.3) and (4) Frequency of Use (data not shown). For some drugs and periods, drug use rates are higher for the 1992 DARE group than for the 1991 non-DARE group. However, for other drugs and periods, the rates are reversed. Figure 14.3 illustrates the similarities between the two groups regarding Mean Grade Level at First Drug Use.[5] A separate set of eight *t*-tests revealed no significant differences between the two groups on this measure for each drug type/category.

---

[5] The "numbers reporting first use" sub-figure illustrates that aside from alcohol, the mean grade level figures reflect drug use among relatively small groups of students compared to the total samples.

FIGURE 14.2

Drug Use Scale: Recency
of Use (Last 30 Days)

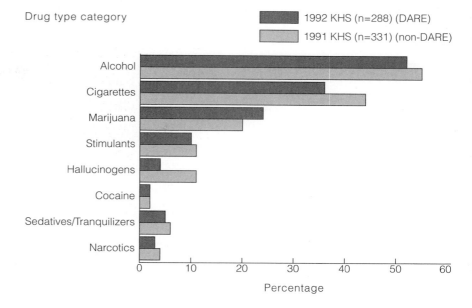

FIGURE 14.3

Drug Use Scale: Mean
Grade Level at First Use

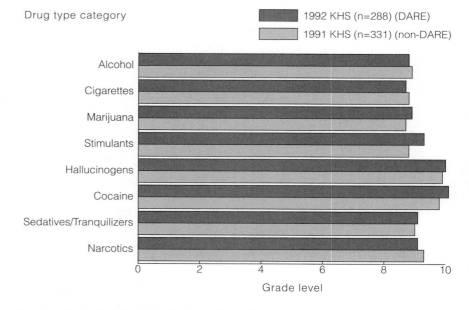

Number Reporting First Use by Drug Type/Category

	Alcohol	Cigarettes	Marijuana	Stimulants	Hallucinogens	Cocaine	Sedatives/ Tranquilizers	Narcotics
1991 KHS	279	143	175	88	71	36	63	41
1992 KHS	259	137	142	75	64	61	54	29

Notes: The Alcohol category includes those who tried more than a few sips; cigarettes includes those who smoked on a daily basis; and the other categories include those who first tried (any level) the substances. See footnote 5 in the text.

TABLE 14.1   **DARE Scale Results: ANOVA of "Appropriate" Response Means (1991 Seniors, 1992 Seniors, and 1992 Seniors as 7th Graders)**

DARE SCALE	1991 KHS (non-DARE Group)	1992 KHS (DARE Group)	1992 KHS (as 7th Graders)	F-value	Sig. F
"Appropriate" Response Means	82	82	79		
St. Dev:	12.86	12.21	11.10	.313	.732*
Cases:	n = 331	n = 288	n = 596		

Note: * = nonsignificant

Prevalence and recency data were further analyzed using chi-square tests comparing drug use among DARE and non-DARE seniors for all drug types/categories, time periods, and use levels. While too lengthy to report in detail, the results reveal no significant differences in drug use rates or patterns between the two groups in each area (prevalence, recency, frequency) for each drug type/category—with two exceptions. The results were significant for comparisons of the 1991 and 1992 groups' hallucinogen use over the last 30 days in terms of use/no use categories (sig. = .0004) and levels of use (sig. = .02). In both instances *higher* use rates were recorded for the 1992 DARE group.

DARE Scale

Table 14.1 summarizes the DARE Scale results and reinforces the basic findings of the previous section: DARE exposure appears to produce no significant long-term effects in areas related to the program's primary objectives. As Table 14.1 indicates, the mean percentages of "appropriate" responses for all 19 scale items were similar for all three groups with no statistically significant differences apparent in the groups' mean scores.[6]

Despite the similarities, there are some interesting differences on individual items. For example, there was a sharp decline in positive attitudes toward the police and a growing unwillingness to condemn peers' consumption of alcohol. However, on the other items the "Appropriate" response results were higher for the 1992 DARE seniors than the results recorded as seventh graders. Despite these potentially confusing trends, the DARE and non-DARE senior groups had virtually identical response percentages for all 19 items, indicating that the differences between student responses as seventh graders and later as seniors were not due to DARE exposure.

[6] While too lengthy to report here, in a separate analysis of "appropriate" mean response percentages for each DARE scale question by each of the three groups identified in Table 14.1, we found similar results for most of the 19 items.

Finally, the comparisons of results of the DARE Scale with those from the Drug Use Scale call into question the utility of the DARE Scale as a meaningful predictor of drug use. For example, despite the fact that high percentages of both senior groups gave "Appropriate" responses for most items on the scale, substantial majorities in these groups reported using various drugs (especially alcohol and cigarettes). These results are consistent with a number of findings in the drug education literature showing that while anti-drug information is easily imparted, producing changes in drug-related attitudes and behavior is much more difficult and problematic (Bangert-Drowns, 1988; Pellow and Jengeleski, 1991). Moreover, the results suggest that rather than measuring any meaningful attitudinal or behavioral changes, the DARE Scale simply assesses the extent to which DARE encourages students to uncritically recall and repeat information. In this sense, the scale appears to be a very self-serving measure which by the design of its questions and coding procedures elicits the *appearance* of positive results thereby overstating the efficacy of the DARE program.

Micro-Level Process

A micro-level process evaluation of a focus group interview with six KHS seniors, evenly divided by gender and recruited from informal community contacts, supported the quantitative findings. The first part of the interview concentrated on recollections of their DARE experiences as seventh graders; the second part addressed student perceptions of DARE's long-term impact. DARE was judged by all group members as having had no lasting influence on their peers' drug-related attitudes or behaviors, but they did not see it as a waste of time *in the lower grades.* Students thought that DARE had a short-term anti-drug effect and could be seen as giving students the message that adults cared about them.

Symbolic Politics and Organizational Support for DARE

Programs are never implemented in a vacuum or under ideal conditions (Weisheit, 1983). Our approach views social problems and related public policies as linked to political, social, and cultural processes (Gamson, Croteau, Hoynes, and Sasson, 1992). In the case of DARE, this orientation directs us to explore (1) the links between the media and the politically constructed Drug War and DARE's emergence as a form of symbolic politics, (2) the role of various stakeholders in promoting, legitimating, and expanding the program, and (3) the program evaluation implications of DARE's evolution into a major political force.

Edelman's (1964) concepts of "symbolic politics/action" are useful in understanding the links between the Drug War and DARE's emergence as a popular programmatic policy response. They focus attention on the role and interests of political elites and the media in constructing the Drug War and in promoting policies to deal with it. They alert us to the importance of the public reassurance features of ameliorative programs in generating political and public support (which might be more important than their actual substantive effects). The reassurance value of such programs is linked to the extent to which the programs are grounded in widely respected and legitimate institutions and cultural traditions. Programs imbued with potent symbolic qualities (such as DARE's links to schools and police) are virtually assured of widespread public acceptance (regardless of actual effectiveness), which in turn advances the interest of political leaders who benefit from being associated with highly visible, popular symbolic programs.

In the mid-1980s, illicit drug use was defined as a "drug crisis" and a major social problem by the mass media and national political leaders (Jensen, Gerber, and Babcock, 1991). President Reagan officially declared a national War on Drugs (Boyd, 1986), and as the 1980s ended, the Bush administration extended and systematized the Drug War in the "National Drug Control Strategy."

Public attitudes became more supportive of anti-drug programs and political interest in drug education increased. With a positive track record in Los Angeles and its powerful symbolic affirmation of traditional values and institutions, DARE was well positioned to take advantage of this emerging context by offering a convenient individual-level programmatic "solution" to the drug threat. Since expanding the DARE program offered public reassurance in the face of the constructed drug threat, DARE can be seen as a clear example of symbolic politics and action.

As political support for DARE increased, the program shifted from being a local success story to being a national symbol in the war on drugs. This resulted in mutual support and reinforcement among direct stakeholders (such as staff and directly supportive organizations) and *indirect stakeholders* (such as political supporters). Both groups benefited from the reflected approval, support, and legitimacy associated with a program linking a popular cause to traditional authority structures.

The stakeholders collaborated to embed the program within a complex organizational support structure to ensure its continued existence and growth. One important feature of this structure is DARE AMERICA, a nonprofit corporation with a large annual budget ($1.3 million in 1990) (Pope, 1992). Another dimension of the support structure is an interrelated network of organizational ties linking DARE programs to federal, state, and local agencies (such as the U.S. Department of Justice [U.S. Department of Justice, 1991], the National Institute on Drug Abuse [Clayton et al., 1991], the U.S. Department of Education [Rogers, 1990]), and to private corporations, including McDonald's, Security Pacific National Bank, and Kentucky Fried Chicken (Pope, 1992).

Implications for Program Evaluation

DARE's existence as a potent form of symbolic politics has important implications for evaluation. It increases the prospects for tension as negative evaluations threaten the interests of powerful stakeholders. Furthermore, it underscores the importance of framing results in a process perspective offering alternative interpretations for observed outcomes.

While our negative results alone could create the potential for conflict with DARE stakeholders, this prospect is magnified by recent criticisms (Pereira, 1992) and concerns about the expense of the program (Pope, 1992). Our results could make the case for cutting the program's budget, constituting a potential threat to stakeholders and eliciting their wrath. This does not mean evaluators should anticipate stakeholder pressures by discounting the validity or importance of their results. However, when the issue under consideration is important and the stakes are high, program evaluators must be the first to call attention to the complex and inexact nature of the evaluation process. Our obligation is to provide the DARE program and its supporters not only with our findings, but also with an interpretation of factors that may have influenced our results.

Alternative interpretations of our "no effects" findings essentially cluster around three main issues. The first is the influence of contextual factors. While the National Drug Control Strategy was built around the complementary goals of reducing both the demand for and the supply of drugs, the exposure of the local seventh graders to DARE occurred before the implementation of several important facets of the strategy, such as the Partnership for Drug-Free America and the passage of major federal anti-drug laws. Therefore, our DARE group did not get the reinforcing effects of coordinated mass media campaigns, community programs, and federally mandated anti-drug policies until after their initial exposure to DARE.

The second issue is the disjunction between key DARE program assumptions/ideals and the manner in which the program was implemented. DARE views drug use as an individual decision and choice that is heavily dependent upon knowledge, attitudes, and social skills. Consequently, DARE assumes it is possible to intervene in these areas via systematically organized anti-drug lessons that will produce long-lasting attitudinal and behavioral effects. DARE implementation includes the ideas that anti-drug effects are maximized by early intervention and the uniform presentation of a standardized curriculum.

The DARE implementation we evaluated differs from the assumptions in several ways. The group was exposed to the program two years later than recommended, they did not receive the full 17-week program, and uniform presentation and "curriculum fidelity" did not occur in at least some sessions for some students. While these disjunctions could be a threat to DARE's long-term effectiveness, they may also be trivial and perhaps inconsequential factors.

The third set of issues to consider in conjunction with our findings involves the program design limitations related to areas not explicitly considered by the DARE curriculum. DARE's mandatory nature in participating schools means that resistant students will have DARE imposed upon them. Other studies confirm that "when socially deviant youths are required to participate in the school setting in peer-led denunciation of activities they value, they are more likely to become alienated than converted" (Baumrind, 1987: 32). "Classroom climate" (including pedagogical styles and classroom teachers' level of enthusiasm for DARE) can also be influential in conditioning student receptiveness to DARE (Aniskiewicz and Wysong, 1990). DARE's implicit embrace of "free will/user accountability" principles means limiting the scope and emphasis to the context within which students' choices are made and ignoring important structural factors such as students' socioeconomic background, family influences, and employment prospects.

Discussion

After tracking DARE for five years, our data point in the direction of no long-term program effects for preventing or reducing adolescent drug use. Despite the massive scope of the Drug War, the existence of powerful DARE stakeholders, and the potential controversies that our findings might generate, we believe our evaluation of DARE represents a fair and accurate assessment of the program's long-term effects *for the group we studied.* We believe our results must be viewed as suggestive rather than definitive. We explicitly recognize the limitations of our findings given the local nature of our sample and the complexities of conducting longitudinal evaluation on a popular program addressing adolescent drug use. However, because our efforts have gone far beyond a one-dimension impact evaluation approach, we believe the study deserves widespread attention.

Our work should help stimulate a reconsideration of several issues related to the etiology of drug use among young adults. For example, should drug prevention efforts be expanded beyond those programs aimed at changing attitudes or improving social skills through sophisticated persuasive communication programs? We need a more complete consideration of the appropriate balance between approaches focusing primarily on the psychosocial dimension of drug use and those addressing cultural and structural factors related to it (for example, those that work towards reducing inequalities, alienation, and social isolation among adolescents). To the extent that our evaluation helps to stimulate a more multifaceted discourse on the nature and merits of various psychological and structural drug prevention policies and programs, then DARE, regardless of its long-term effects, will have made an important contribution.

REFERENCES

Aniskiewicz, R., and E. Wysong. 1990. Evaluating DARE: Drug education and the multiple meanings of success. *Policy Studies Review* 9: 727–747.

Bangert-Drowns, R. L. 1988. The effects of school based substance abuse education: A meta-analysis. *Journal of Drug Education* 18: 243–264.

Baumrind, D. 1987. Familial antecedents of adolescent drug use: A developmental perspective. In *Etiology of drug abuse: Implications for prevention,* eds. Coryl L. Jones and Robert J. Battjes, 13–44. Rockville, MD: National Institute for Drug Abuse.

Boyd, G. M. 1986. Reagan's advocate 'crusade' on drugs. *New York Times,* September 15: A1, B10.

Cauchon, D. 1993. Studies find drug program not effective. *USA Today,* October 11: 1–2.

Clayton, R. R., A. Cattarello, L. E. Day, and K. P. Walden. 1991. Persuasive communication and drug prevention: An evaluation of the DARE Program. In *Persuasive communication and drug abuse prevention,* eds. L. Donohew, H. I. Sypher, and W. J. Bukowski, 295–313. Hillsdale, NJ: Lawrence Erlbaum.

Congressional Record. 1992. Senate Joint Resolution 295. Joint Resolution designating September 10, 1992, as National D.A.R.E. Day. April 30: S5915–S5916.

D.A.R.E. AMERICA. 1991. *Request for proposal.* D.A.R.E. America, Inc., 606 South Olive, Suite 1206, Los Angeles, CA.

DeJong, W. 1987. A short term evaluation of project DARE (Drug Abuse Resistance Education): Preliminary indications of effectiveness. *Journal of Drug Education* 17: 279–294.

Edelman, M. 1964. *The symbolic uses of politics.* Chicago: University of Illinois Press.

Gamson, W., D. Croteau, W. Hoynes, and T. Sasson. 1992. Media images and the social construction of reality. In *Annual Review of Sociology.* Eds. J. Blake and J. Hagan, 373–393. Palo Alto, CA: Annual Reviews.

Harmon, M. A. 1993. Reducing the risk of drug involvement among early adolescents. *Evaluation Review* 17: 221–239.

Jensen, E. J., J. Gerber, and G. M. Babcock. 1991. The new war on drugs: Grass roots movement or political construction? *Journal of Drug Issues* 21: 651–667.

Johnston, L. D., P. M. O'Malley, and J. G. Bachman. 1992. Smoking, drinking, and illicit drug use among American secondary school students, college students and young adults, 1975–1991. Volume 1. Rockville, MD: National Institute on Drug Abuse.

Males, M. 1992. Drug deaths rise as the war continues. *In These Times,* May 20–26: 17.

Miley, S. L. 1992. DARE beats drugs; truth is, funding's in a 'crisis.' *Indianapolis Star,* May 18: D-1.

National Drug Control Strategy. 1990. *The White House, January 1990.* Washington, DC: U.S. Government Printing Office.

National Drug Control Strategy. 1992. *The White House, January 1992.* Washington, DC: U.S. Government Printing Office.

Nyre, G. F. 1984. An evaluation of project DARE (Drug Abuse Resistance Education). Los Angeles. Evaluation and Training Institute.

Nyre, G. F. 1985. Final evaluation report, 1984–1985: Project DARE (Drug Abuse Resistance Education). Los Angeles. Evaluation and Training Institute.

Pellow, R. A., and J. L. Jengeleski. 1991. A survey of current research studies on drug education programs in America. *Journal of Drug Education* 21: 203–210.

Pereira, J. 1992. The informants: In a drug program, some kids turn in their own parents. *Wall Street Journal,* April 20: 1, A4.

Pope, L. 1992. DARE-ing program at risk. *Los Angeles Daily News,* September 20: G-6, G-7.

Rogers, E. M. 1990. Cops, kids, and drugs: Organizational factors in the spontaneous diffusion of Project D.A.R.E. Paper presented at the Conference on Organizational Factors in Drug Abuse Prevention Campaigns. Bethesda, MD.

Rosenbaum, D. P., R. L. Flewelling, S. L. Bailey, C. L. Ringwalt, and D. L. Wilkinson. 1994. Cops in the classroom: A longitudinal evaluation of Drug Abuse Resistance Education (DARE). *Journal of Research in Crime and Delinquency* 31: 3–31.

U.S. Congress: House of Representatives. 1990. Oversight hearing on drug abuse education programs. Committee on Education and Labor. Subcommittee on Elementary, Secondary and Vocational Education. 101st Congress. 1st session. Serial No. 101–129. Washington, DC: Government Printing Office.

U.S. Department of Justice Bureau of Justice Assistance. 1991. *PROGRAM BRIEF: An introduction to DARE* (2nd ed.). Washington, DC: Bureau of Justice Assistance.

U.S. Office of Management and Budget. 1990. *Budget of the U.S. government, Fiscal Year 1991.* Washington, DC: U.S. Government Printing Office.

U.S. Office of Management and Budget. 1992. *Budget of the U.S. government, Fiscal Year 1993.* Washington, DC: U.S. Government Printing Office.

Weisheit, R. A. 1983. The social context of alcohol and drug education: Implications for program evaluations. *Journal of Alcohol and Drug Education* 29: 72–81.

Appendix: The Questionnaire

Students were given a 15-page, voluntary, and anonymous questionnaire. The instructions stated that they would be asked about their opinions on a number of topics, and that they should view their answers as "votes" on a wide range of important issues. The specified interest was in group results, not individual scores and students were assured that there were no right or wrong answers.

The DARE Scale consisted of 19 questions with five answer categories. The following are examples of the questions:

1. There is nothing wrong with smoking cigarettes as long as you do not smoke too many.

5	4	3	2	1
strongly disagree	disagree	unsure	agree	strongly agree

2. Students who drink alcohol are more grown up.

5	4	3	2	1
strongly disagree	disagree	unsure	agree	strongly agree

3. It is safe to take medicine that a doctor orders for another person.

5	4	3	2	1
strongly disagree	disagree	unsure	agree	strongly agree

4. Police officers would rather catch you doing something wrong than try to help you.

5	4	3	2	1
strongly disagree	disagree	unsure	agree	strongly agree

5. If your best friend offers you a drug, you have to take it.

5	4	3	2	1
strongly disagree	disagree	unsure	agree	strongly agree

6. It is okay to drink alcohol or use drugs at a party if everyone else is.

5	4	3	2	1
strongly disagree	disagree	unsure	agree	strongly agree

The Drug Use Scale consisted of 12 multipart questions. Questions focused on tobacco, alcohol, marijuana, hashish, psychedelics, cocaine, crack, tranquilizers, barbiturates, amphetamines, and narcotics (heroin, morphine, and the like). The following are some examples of questions:

1. Have you ever smoked cigarettes?
____ 1. Never—**Go to Question 2.**
____ 2. Once or twice
____ 3. Occasionally, but not regularly
____ 4. Regularly in the past
____ 5. Regularly now

1a. How frequently have you smoked cigarettes during the past 30 days?
____ 1. Not at all
____ 2. Less than 1 cigarette per day
____ 3. 1 to 5 cigarettes per day
____ 4. About ½ pack per day
____ 5. About 1 pack per day
____ 6. About 1½ packs per day
____ 7. 2 or more packs per day

2. Have you ever used smokeless tobacco?
____ 1. Never—**Go to Question 3.**
____ 2. Once or twice
____ 3. Occasionally, but not regularly
____ 4. Regularly in the past
____ 5. Regularly now

3. Have you ever had any beer, wine, wine coolers, or liquor to drink?
____ 1. No—**Go to Question 4.**
____ 2. Yes

4. On how many occasions have you had alcoholic beverages to drink?

4a. in your lifetime?	**4b. during the last 12 months**	**4c. during the last 30 days?**
____ Zero occasions	____ Zero occasions	____ Zero occasions
____ 1–2 occasions	____ 1–2 occasions	____ 1–2 occasions
____ 3–5 occasions	____ 3–5 occasions	____ 3–5 occasions
____ 6–9 occasions	____ 6–9 occasions	____ 6–9 occasions
____ 10–19 occasions	____ 10–19 occasions	____ 10–19 occasions
____ 20–39 occasions	____ 20–39 occasions	____ 20–39 occasions
____ 40 or more	____ 40 or more	____ 40 or more

5. On how many occasions (if any) have you used marijuana (grass, pot) or hashish (hash, hash oil)?

5a. in your lifetime?	5b. during the last 12 months	5c. during the last 30 days?
____ Zero occasions	____ Zero occasions	____ Zero occasions
____ 1–2 occasions	____ 1–2 occasions	____ 1–2 occasions
____ 3–5 occasions	____ 3–5 occasions	____ 3–5 occasions
____ 6–9 occasions	____ 6–9 occasions	____ 6–9 occasions
____ 10–19 occasions	____ 10–19 occasions	____ 10–19 occasions
____ 20–39 occasions	____ 20–39 occasions	____ 20–39 occasions
____ 40 or more	____ 40 or more	____ 40 or more

6. On how many occasions (if any) have you used LSD ("acid") or other psychedelics (like mescaline, peyote, psilocybin, or PCP)?

6a. in your lifetime?	6b. during the last 12 months	6c. during the last 30 days?
____ Zero occasions	____ Zero occasions	____ Zero occasions
____ 1–2 occasions	____ 1–2 occasions	____ 1–2 occasions
____ 3–5 occasions	____ 3–5 occasions	____ 3–5 occasions
____ 6–9 occasions	____ 6–9 occasions	____ 6–9 occasions
____ 10–19 occasions	____ 10–19 occasions	____ 10–19 occasions
____ 20–39 occasions	____ 20–39 occasions	____ 20–39 occasions
____ 40 or more	____ 40 or more	____ 40 or more

7. Amphetamines are sometimes prescribed by doctors to help people lose weight or to give people more energy. They are sometimes call uppers, speed, bennies, dexies, pep pills, and diet pills. On how many occasions (if any) have you taken amphetamines on your own—that is, without a doctor telling you to take them?

7a. in your lifetime?	7b. during the last 12 months	7c. during the last 30 days?
____ Zero occasions	____ Zero occasions	____ Zero occasions
____ 1–2 occasions	____ 1–2 occasions	____ 1–2 occasions
____ 3–5 occasions	____ 3–5 occasions	____ 3–5 occasions
____ 6–9 occasions	____ 6–9 occasions	____ 6–9 occasions
____ 10–19 occasions	____ 10–19 occasions	____ 10–19 occasions
____ 20–39 occasions	____ 20–39 occasions	____ 20–39 occasions
____ 40 or more	____ 40 or more	____ 40 or more

8. How difficult do you think it would be for you to get each of the following types of drugs, if you wanted some? (please circle your answer for each question)

	1	2	3	4	5
8a. cigarettes?	probably impossible	very difficult	fairly difficult	fairly easy	very easy
8b. alcohol (beer, wine, liquor)?	probably impossible	very difficult	fairly difficult	fairly easy	very easy

8c. cocaine or "crack"?	1 probably impossible	2 very difficult	3 fairly difficult	4 fairly easy	5 very easy

Evaluation Research

evaluation research, research specifically designed to assess the impact of a specific program, policy, or legal change.

There are several kinds of applied research. One of the best known is **evaluation research**—research specifically designed to assess the impact of programs, policies, or legal changes. Often the focus of an evaluation is whether the program, policy, or law has succeeded in effecting intentional or planned change. An example of this kind of work is the Adler and Foster experiment (in Chapter 8) on the effect of the school reading project on caring.

Evaluation research is not a unique research *method*. Rather, it is research with a specific *purpose*. Rossi and Freeman (1993: 15) call evaluation research "a political and managerial activity, an input into the complex mosaic from which emerge policy decisions and allocations for the planning, design, implementation, and continuance of programs to better the human condition."

Outcome Evaluation

outcome evaluation, research that is designed to "sum up" the effects of a program, policy, or law in accomplishing the goal or intent of the program, policy, or law.

Wysong, Aniskiewicz, and Wright's work is a good example of an outcome evaluation, which is the most common kind of evaluation research. **Outcome evaluation,** also called impact or summative analysis, seeks to estimate the effects of a treatment, program, law, or policy and thereby determine its utility.

As in the research by Wysong and his colleagues, outcome evaluations typically begin with the question "Does the program accomplish its goals?" or the hypothesis "The program (independent variable) has a positive effect on the program's objectives (dependent variable)." The researcher first determines *what* the goal of the program is and *how* the program was implemented and designs a project useful to practitioners and policy makers. At the end of an outcome evaluation, the researcher should be able to answer questions about a program's success and to speculate about how to improve it.

Wysong and colleagues focused on a local version of a large national program, but a wide variety of programs, policies, or laws can be evaluated. Programs and policies can be evaluated regardless of whether they serve many or few people; whether they last for days, weeks or years; if their geographic target area is large or limited; and whether they have broad or narrow goals. Examples of the many hundreds of research questions that have guided outcome evaluations in the past decade include:

- Do client-centered programs lead to a decrease in adolescent pregnancies? (McBride and Gienapp, 2000)

- Do comprehensive, whole-school intervention programs designed to decrease bullying in schools actually work? (Smith, Schneider, Smith & Ananiadou, 2004)
- Do marriage education programs originally designed for white, middle class, well-educated couples strengthen couple relationships in disadvantaged populations? (Dion, 2005).

Cost-Benefit Analysis

cost-benefit analysis, research that compares a program's costs to its benefits.

cost-effectiveness analysis, comparisons of program costs in delivering desired benefits based on the assumption that the outcome is desirable

In the focal research, Wysong and his colleagues focused on whether DARE accomplished its goals and, in so doing, implied that it might be better to expend the resources consumed by the program in other ways. An evaluation can be more explicitly a **cost-benefit analysis,** which is a study designed to weigh all expenses of a program (its costs) against the monetary estimates of the program's benefits (even putting dollar values on intangible benefits), or a **cost-effectiveness analysis,** which estimates the approach that will deliver a desired benefit most effectively (at the lowest cost) without considering the outcome in economic terms. Comparing program costs and benefits can be helpful to policy makers, funding agencies, and program administrators in allocating resources. With large public expenditures on health care, correctional facilities, social welfare programs, and education in the United States, saving even small percentages of costs add up.

Typical questions asked in cost-benefit and effectiveness analyses are these: How effective is the program? How expensive is it? Is it worth doing? How does this program compare with alternative programs? These analyses can compare outcomes for program participants with those not in programs or with those in alternative programs.

Collection and interpretation of data is often a complex task when evaluating social programs. For example, with more than 100 marriage education curriculums in the U.S. alone (Dion, 2005), there are many options to consider. In doing a cost-benefit analysis of any marriage education program, direct expenditures (such as materials, salaries, and the costs for the sites) and indirect costs (such as child care expenses) need to be considered. Both the direct benefits of an effective program (such as fewer relationship breakups and greater levels of individual and couple happiness) and the indirect effects (for example, decreased poverty due to decreased divorce and the impact of staying together on the couple's children) will need to be measured.

Evaluations for Other Purposes

stakeholders, people or groups that participate in or are affected by a program or its evaluation, such as funding agencies, policy makers, sponsors, program staff, and program participants.

The tasks an evaluation sets out to accomplish are determined by the stage of the program (whether it is a new, innovative program or an ongoing one) and the needs and interests of the **stakeholders** (those involved in or affected by the program in some way). In some instances, before a program

needs assessment, an analysis of whether a problem exists, its severity, and an estimate of essential services.

or project is designed, a needs assessment is conducted to determine the needs for various forms of service. A **needs assessment** evaluates the existence of a problem, its severity, and the number of people affected. It can also include an analysis of the services or programs that could be implemented to ameliorate the problem and estimate the required duration. For example, if school administrators, elected officials, or parents were concerned about math or reading skills of middle school students, a needs assessment would first collect data about the current skill levels of students. If the analyses and subsequent decision-making process were supportive of a programmatic solution, then a new curriculum might be designed.

formative analysis, evaluation research focused on the design or early implementation stages of a program or policy.

If a new program is funded and there is time in the early stages of design or implementation to make improvements, program staff and developers can benefit from a **formative analysis.** This kind of analysis occurs when an evaluator involved in the beginning stages of program development and implementation provides program staff with an evaluation of the current program and suggestions for improving the program design and service delivery. Suppose, for example, a high school principal sought assistance with a new anti-bullying program. A formative analysis would carefully review the program's goals and its current instructional materials (including comparisons of the program with programs and materials used in other school systems) and collect data on the ongoing program. Administrators, teachers, and students might be interviewed, teacher training and program sessions might be observed, and the first students involved in the program might be asked to complete a questionnaire or to participate in focus groups. After analysis of the data, evaluators should be able to offer advice, including suggestions for staff development, for modifying instructional materials or techniques, and for making adjustments in program management.

process evaluation, research that monitors a program or policy to determine if it is implemented as designed.

A related kind of evaluation is **process evaluation,** also called an implementation study, which is research to determine if a program was implemented as designed. Variations in program delivery occur because of differences in program deliverers, recipients, sites, and changes over time (Sheirer, 1994: 42). It's especially important for the program staff responsible for delivering subsequent versions of a program to know exactly how the initial version was implemented. For instance, if a state implemented a new victim assistance program, a process evaluation could ascertain the actual staffing patterns, administrative arrangements, and primary program activities. The evaluators could also determine the cost of the program, the services provided, staff responsibilities, number of participants, and the extent to which the program differed from site to site. Monitoring a program's quality helps to see whether a program includes any unplanned aspects or substandard procedures. In process evaluation, the goal is not to see whether there is an effect or how big the effect is but, rather, to try to pinpoint how something works and what aspects of the program contribute to the effect.

Sociologists can do important work for their local communities as applied sociologists. Box 14.1 describes Kristy Maher's view of practicing sociology in Greenville, NC.

> **BOX 14.1**
>
> ## Practicing Sociology as a Vocation
>
> Kristy Maher believes in practicing sociology in a way that uses her knowledge and skills to effect a positive change in her local community of Greenville, NC. Her first project was a needs assessment of the health care needs of the local Hispanic population and her second was an outcome evaluation of a prenatal outreach program for high-risk pregnancies. The first resulted in the hospital system's hiring full-time translators for the Hispanic community and the second was helpful in securing continuing funding for the successful program. Since then she has worked on a variety of projects, including serving as the medical sociologist in a community health-care assessment and as an evaluator for a program aimed at providing quality of health-care access for children in low-income schools. Maher (2005) says that she believes in practicing sociology as a *vocation*, which means using her expertise as a medical sociologist, a researcher, and a statistician to better her community.

Designing Evaluation Research

As in all kinds of research, the evaluation researcher needs to select research strategies. Making specific choices for study design, measurement, data analysis, and the like depends on the specific evaluation, including intended audience, resources available, ethical concerns, and the project's time frame.

Typically, many parties will have an interest in an evaluation, including program administrators, legislators, client groups, interest groups, and the general public. Different constituencies often want different information. In our focal research on DARE, Wysong and associates were not paid as project consultants, but did see the Kokomo Center School Corporation as a client because the organization had requested the evaluation and contributed resources. The school department provided the researchers with access to the schools for data collection and contributed $1,500 toward the expenses associated with the early phases of the research. However, Wysong and associates saw the study as having additional clients: the students, families, and community of Kokomo, professional colleagues, and the social sciences.

The objective of most evaluation research is to determine if a program, policy, or law accomplished what it set out to do, so one of the study's concepts or variables must be the goal or desired outcome of the program, policy, or law. A typical evaluation research question asks whether the program accomplished its goal(s). In our focal research example, Wysong and his colleagues asked the question "Is DARE effective?" That is, did the DARE curriculum produce "anti-drug" effects among students?

If a researcher constructs a hypothesis, typically the program, policy, or law will be the independent variable and the goal(s) or outcome(s) will be

---

BOX 14.2

Evaluation Issues

When evaluating a program, policy, or law, the following must be decided:

- What is the desired outcome?
- Is the outcome to be short- or long-term?
- Are attitudinal changes sufficient, or is it also essential to study behavioral change?
- How should change be determined?
- Should all aspects of the program be studied or only certain parts?
- Should all the targets of an intervention be studied or only some of them?

---

the dependent variable(s). Although Wysong, Aniskiewicz, and Wright did not state an explicit hypothesis, the implied hypothesis was that students who participated in DARE developed stronger anti-drug attitudes, better drug-coping skills, and lower drug use than did those who did not participate in DARE.

If a program has broad or amorphous goals, it can be difficult to select one or more specific dependent variables. Therefore, a critical challenge for evaluators is to articulate appropriate dependent variables and select valid measurement techniques. Even when program personnel help define a specific goal, it still might be hard to determine the desired outcome precisely and how to measure it. Some of the issues the evaluation researchers must confront are specified in Box 14.2.

In an earlier project Aniskiewicz and Wysong (1990) had focused on the initial implementation of DARE and short-term attitudinal changes, finding significant increases in anti-drug attitudes and drug information and knowledge among fifth grade students at the end of the one-year program. However, because the researchers felt the ultimate goal of DARE was to prevent or delay student use of drugs (both legal and illegal), they wanted to conduct a long-term study focusing on *behavioral outcomes*. Five years after the first students in Kokomo participated in DARE, Wysong and his colleagues did the research presented in this chapter using "preventing/reducing/delaying drug use" as a dependent or outcome variable.

STOP AND THINK   *In the focal research, Wysong and his associates used two different self-report methods to collect data about preventing/reducing/delaying drug use. The primary data collection method was an anonymous, group-administered questionnaire, and a secondary method was a focus group interview. What do you think are the advantages and disadvantages of the two data collection methods for this variable?*

As we discussed in previous chapters, questionnaires, focus groups, and observation all have advantages and disadvantages. Wysong and colleagues wanted to identify differences in attitudes and behavior among students. Attitudes are usually best studied by asking people questions, whereas behavior is typically better studied by observing. However, because drug use is not a behavior that lends itself easily to observation, the researchers selected self-report methods and used a group-administered questionnaire as the primary data collection method. The group-administered questionnaire is a good choice for large, literate samples that, like the students in Kokomo, can be found in accessible, group settings. This kind of questionnaire typically has a good response rate and allows anonymous collection of data, an advantage when researchers ask about sensitive topics like drug use. The questionnaire provided quantitative data that were augmented by the qualitative data of the focus group interview of six Kokomo High School seniors. The focus group had the advantage of being flexible and providing answers from the interaction among participants.

STOP AND THINK   *Look at the appendix of the focal research for examples of the questions included in the questionnaire. What do you think of the validity of the DARE scale and the DRUG USE scale?*

STOP AND THINK AGAIN   *An evaluation of a DARE program in Ohio did a survey of teachers and principals (RMBSI, 1995). The researchers found that personnel in schools with DARE gave the program "high marks." For example, 90 percent of the 400 staff members who completed the survey reported that the program had made a positive difference in students' attitudes about drugs; 75 percent said it delayed students' use of illegal substances (RMBSI, 1995: 2). What do you think of these data as indicators of DARE's impact?*

Wysong and associates did not ask teachers or administrators to comment on student behavior. Asking school personnel to report on student attitudes and drug use is too indirect and makes an enormous compromise in validity. In the focal research, students reported on their own attitudes and behaviors, a technique that we believe was more appropriate. It is important to recognize, however, that using self-report data about attitudes and behavior depends on accurate memory and honest reporting. Asking students about behavior that occurred five years earlier can be problematic because events and activities can be remembered incorrectly. In addition, even with anonymity, respondents might not always be truthful about their attitudes or drug use—perhaps underestimating for fear of discovery or overstating to "look cool."

STOP AND THINK   *Evaluation researchers almost always have an explanatory purpose and typically test a causal hypothesis about the effect of an independent variable such as a program, law, or policy on a desired outcome. Can you remember some of the study designs discussed in Chapters 7 and 8? If you have a study with an explanatory purpose, what is a particularly useful study design?*

causal hypothesis, a testable expectation about an independent variable's effect on a dependent variable.

We talked about experiments in Chapter 8 as useful for testing **causal hypotheses.** Built into the experimental design are ways to decrease systematic differences between the experimental and control groups before introducing the program or policy. For this reason, we'd have more confidence that observed differences between the groups after they had received program services were not simply the result of those differences or other factors, but of the introduction of the independent variable.

The Adler and Foster research on the school reading project discussed in Chapter 8 is an example of evaluation research using an experimental design. The hypothesis tested was that participating in a 10-week program of reading and discussing books with themes of caring increased students' support for the value of caring for others. A large, team-taught seventh-grade class was selected, students were assigned to experimental and control groups, the independent variable was "introduced," the dependent variable was measured both before and after the program, and the pre- and post-program results were compared. The experimental design handled many of the possible challenges to **internal validity.** That is, there was support for the assumption that the control group was exposed to all the other factors that the experimental group was exposed to *except* for the program, and that the small differences in support for the value of caring were likely to have been the result of the independent variable (participation or nonparticipation in the innovative program).

internal validity,
agreement between a study's conclusions about causal connections and what is actually true.

STOP AND THINK
The controlled experiment assumes that assigning people to experimental and control groups is possible. Can you think of evaluation research situations where a researcher either couldn't or wouldn't want to put people in a control group?

Not all situations lend themselves to using the true experimental model. Outside a laboratory, the evaluator might not be able to control all aspects of the design. A well-known field study of the impact of police intervention on wife abuse in Minneapolis (Berk, Smyth, and Sherman, 1988) was supposed to be a true experiment. Police were to randomly assign domestic violence calls to one of three treatments (mediating the dispute, ordering the offender to leave the premises for several hours, or arresting the abuser). Despite the intended design, however, in about one-third of the calls, officers used individual discretion, not random assignment, to determine treatment. Without a pretest measurement or random assignment, there was no way to estimate prior differences between the experimental and control groups.

Sometimes practical concerns of time or money limit design choices. Evaluation research can take five or more years from design to the final report—often too much time for administrators and policy makers (Rossi and Wright, 1984). In other situations, researchers know from the beginning that assigning participants to experimental or control groups is not possible or desirable. For example, when studying the impact of high school restructuring on student achievement, Valerie Lee and Julia Smith (1995) could not determine which school a student attended, nor could they control when a school system would make significant changes. In studying the impact of "no fault" divorce laws on husbands' and wives' post-divorce economic statuses, Lenore Weitzman (1985) could not regulate where a couple would live when filing for divorce nor when couples would want to end their marriages. And clearly, in situations where there is an ethical responsibility to offer services to the most needy—such as AIDS treatment studies and child abuse programs—using random selection as the criterion for assignment to treatment or no-treatment groups is not acceptable. Instead, it is ethical to give treatment and services to the whole sample and have no control group.

STOP AND THINK   *The DARE research was not an experiment. What aspects of the experiment design were missing from this research?*

The study of DARE in this chapter was limited by the realities of the situation. The Kokomo school administration made the decision to implement DARE for *all* fifth and seventh graders. Therefore, random assignment of students into control and experimental groups was not possible. In addition, because, as we've already noted, the outcomes being studied—attitudes toward drugs and self-reports of drug use—are best handled with anonymous questionnaires, no names or other identifications were used. Without a way to identify individual responses at different times, individual changes in attitudes and behavior over time could not be tracked. In retrospect, Wysong (personal communication) felt that it would have been advantageous to have had the ability to create experimental and control groups with random assignment. But, without the possibility of doing a true experiment, the researchers selected alternate strategies.

In earlier work, the researchers had compared the aggregated before and after scores for all of the fifth and seventh graders on the DARE Scale. In the focal research study, the researchers used a modified version of a quasi-experimental design. The two groups were the 1991 and 1992 seniors at Kokomo High School. The "experimental" group was composed of the former seventh graders who took the DARE program, and the comparison group was the former sixth graders who did not participate in DARE.[7] With only one data collection, the answers of these two groups of seniors on the DARE Scale and Drug Use Scale were compared.[8] Although it is likely that the two grades of students had been alike as sixth and seventh graders and, aside from DARE participation, had similar school experiences, the researchers cannot really know if this assumption is true. Although their design raises concerns about internal validity, it was a "do-able" and ethical option.

All Kokomo students in one grade were enrolled in the DARE program, and the students in another grade were not, regardless of personal preferences.

[7] Only students who had not moved from Kokomo in the years between 1987 and 1990 or 1991 were in the sample. The former sixth and seventh graders were asked as seniors to identify the particular school they attended during the 1987–1988 school year and to identify the particular months of that year that they attended if they were not in the school for the entire school year. The indicator of DARE participation was having been in seventh grade in the specified schools for the months of the DARE program.

[8] Some of you might be wondering what happened to the fifth graders who participated in DARE in 1987–1988. After all, they, not the seventh graders, had experienced short-term changes in the anti-drug attitudes immediately after DARE intervention. Two years after the focal research study, Wysong and Wright (1995) studied these former fifth graders when they were seniors. Once again, they found that the attitudes toward drugs and self-reported drug use among DARE-exposed seniors were very similar to those of non-DARE seniors.

However, when members of a sample make choices about participating in a program or an experience, the issue of self-selection leads to other validity concerns. A well-known study, the Coleman Report (National Center for Education Statistics, 1982), which compared the academic achievement of students in public and private high schools, has been vulnerable to the criticism that differences in school admission criteria and parent and student characteristics confound any simple comparisons between groups of students in different kinds of schools (Rossi and Wright, 1984).

Another common design for evaluation is the panel study—a design that uses before and after comparisons of a group that receives program services or that is affected by policy changes. If, for example, a school intended to change a discipline code by increasing the use of in-school suspension and decreasing out-of-school suspension, student behavior for the months preceding the change could be compared with behavior afterward. If student behavior improved after the policy was introduced, school personnel might decide to keep the new policy, even if they weren't sure about the extent to which other factors (time of year, a new principal, a new course scheduling pattern, and so on) contributed to the differences in behavior. A panel study can be useful if a comparison group is not available, even though this design is better at documenting changes in the group under study than in identifying the causes of changes.

Cross-sectional studies are sometimes used in evaluating programs or policies, especially if other designs are problematic. In a cross-sectional design, data about the independent and dependent variables are collected at the same time. Say, for example, that a city has been using a series of public service announcements about AIDS prevention in all buses for the past six months. Officials might commission a survey of city residents to determine the effectiveness of the ads. Interviewers could survey a random sample of respondents asking if they had seen the ads, their opinions of the ads, and then include a series of questions about AIDS and AIDS prevention. If the survey's data show that a higher percentage of those who remember seeing and liking the ads are more knowledgeable about AIDS and its prevention than those who either don't remember or didn't like the ads, the city officials could conclude that the ad campaign was successful.

STOP AND THINK *Why would we need to be cautious in concluding that the ad campaign was successful?*

In our hypothetical city with our imagined ad campaign, with this relatively inexpensive and easy-to-do study, we could determine that a *correlation* exists between two variables. Although the connection between noticing ads and knowledge is suggestive of causation, we don't have evidence of the time order. For example, it's possible that those already most interested in and best informed about AIDS prevention would be the people to pay the most attention to the ads. And those who are least informed about AIDS to begin with could be the ones who didn't read the public service announcements on the topic. In other words, the proposed independent variable could actually be the dependent variable!

Evaluation research can be judged by many standards: usefulness for intended users, degree of practicality, meeting ethical responsibilities, balancing client needs against responsibilities for public welfare, and issues of validity and reliability. For a statement of program evaluation standards, see the American Evaluation Association's Guiding Principles for Evaluators posted on their website at http://www.eval.org/Publications/GuidingPrinciples.asp.

In most evaluation projects, the researchers—like Wysong, Aniskiewicz, and Wright in their evaluation of DARE—are typically "outsiders" to the organization under study. As an outsider, the evaluator is seen as being able to be detached and objective. In most of these projects, the researcher uses what William Foote Whyte (1997: 110) calls the "professional expert model," where the researcher is "called upon by a client organization (or gets himself or herself invited) to evaluate existing policies and to suggest policy changes to meet new objectives." Typically when evaluators are hired, they are expected to give priority to the perspectives and needs of their clients or the funding agency. One of the limitations of traditional evaluation research is that research participants and other stakeholders are given very limited opportunities to get involved and, as a result, they might not be invested in the outcome.

Some researchers want more equity for research participants. For example, psychologist Anne Brodsky (2001) and a team of students were invited to evaluate job training and education programs for low-income women at a community center. During the two-year project, the research team learned a great deal about the center and the research process and provided feedback about the programs to the center. However, Brodsky (2001: 332) was troubled by the difficulty in balancing the needs of the center and those of the researchers, noting that the team was asking research questions that were more theoretical in nature and probably more important in the long run to her career and her students' education than to the center's effectiveness. In the more activist and participatory kinds of applied research to which we now turn, the process, objectives, and priorities are different.

Participatory Action Research

In the past few decades, a growing body of activist research from different disciplines, historical traditions, and geopolitical environments has emerged with an explicit value system and view of stakeholders. Some of the names for the recent kinds of activist research are empowerment evaluation, empowerment research, participatory research, collaborative research, action research, critical action research, and participatory action research.

Explicit Goals

In our discussion, we'll concentrate on participatory action research, although much of the commentary applies to the other forms as well.

participatory action research (PAR), research done by community members and researchers working as co-participants, most typically within a social justice framework to empower people and improve their lives.

Participatory action research or **PAR** is different from more conventional sociological research in at least two ways: the purpose of the work and the view of study participants, including the researcher. PAR is defined as research in which "some of the people in the organization or community under study participate actively with the professional researcher throughout the research process from the initial design to the final presentation of results and discussion of action implications. . . . The social purpose underlying PAR is to empower low status people in the organization or community to make decisions and take actions that were previously foreclosed to them" (Whyte, 1997: 111–112). One essential aspect of this research is that participants take roles that were formerly occupied by social researchers, challenging some fundamental notions about who can be a researcher (Kemmis and McTaggart, 2003). The other essential part of this work is the *action* that comes from the research—action that can help the community that is participating, rather than the researcher and the research community.

Some point out that PAR can have conservative goals, as when corporations, foundations, and government bureaucracies use participation in research to get people at the grassroots to comply with the organization's goals, such as profit-making (Feagin and Vera, 2001: 165). However, PAR usually aims to have an emancipatory cast. Feagin and Vera (2001) argue, in fact, that the roots of participatory action research are in the very beginnings of U.S. sociology—in the efforts of female sociologists, such as Jane Addams, working out of and in the Hull House settlement in Chicago in the 1890s. Addams sought the advice of women at all class levels, brought them onto the stage at gatherings, and provided them with research data to help local residents "understand community patterns in order to make better decisions" (Feagin and Vera, 2001: 66).[9]

The practical significance of research and its potential impact are critical to this growing activist tradition. Rejecting the view of research as a value-free endeavor, participant action researchers see themselves as combining popular education, community organizing, and issue-based research in the service of the community (Brydon-Miller, 2002: 2). Researchers in this tradition argue that human beings can't be value neutral because *not* stating values is another way of supporting the status quo (Serrano-Garcia, 1990: 172). They argue that all action requires research, and that research is impossible without action because once we start uncovering the social construction of reality we affect it (Serrano-Garcia, 1990: 172). A common theme in participatory work is a focus on social justice. Peter Park (1993: 2) describes PAR as seeking to "bring about a more just society in which no groups or classes of people suffer from the deprivation of life's essentials, such as food, clothing, shelter, and health, and in which all enjoy basic human freedoms and dignity. . . . Its aim is to help the downtrodden be self-reliant, self-assertive, and self-determinative, as well as self-sufficient."

---

[9] Feagin and Vera indicate that Addams very clearly saw herself as a sociologist and that her "demotion" to mere reformer and social worker, by subsequent generations of sociologists, is more a reflection of the loss of the original emancipatory vision of sociology than of Addams' lack of concern with empirical research.

Participation and the Researcher's Role

Doing participatory research means working in partnership with those in the community being studied to obtain and use the knowledge that is generated to empower the community. These research projects are "done *with* the community and not *to* the community" (Nyden, Figert, Shibley, and Burrows, 1997: 7). PAR does not aim to control the setting but to educate practitioners in ways that will help them to more fully understand the nature and consequences of their actions (Kemma and McTaggart, 2003: 363).

In the more traditional research perspectives, the researchers are sometimes seen by community organizations and social service agencies as coming in to judge how well they have been performing their jobs. Often in nonparticipatory evaluation research, the researcher treats the members of organizations and communities being studied as passive subjects, and the relationship between researcher and researched is hierarchical, with the researcher being above others. But in participatory action work, the relationship is more equal because the researcher acts collaboratively with community and organization members throughout the research process—from selecting the problem to discussing the action implications of the study's findings (Whyte, Greenwood, and Lazes, 1991: 20).

In PAR, the researcher must be involved intimately in the life of a community and its problems and be in dialogue with other researcher collaborators while eschewing hierarchy (Esposito and Murphy, 2000: 181). The research groups are usually composed of professionals and ordinary people, with all viewed as authoritative sources of knowledge and experience. The success of the project is not seen as depending on expertise related to technical skill but, rather, on the ability to communicate and be responsive (Esposito and Murphy, 2000: 181).

An example of activist research was conducted by Mary Brydon-Miller and the Community Accessibility Committee in western Massachusetts. The initial objectives of this project on accessibility included encouraging participants to see themselves and each other as legitimate experts on disability, developing their potential for advocacy efforts to achieve social change, and creating a sense of community and ownership of the research process (Brydon-Miller, 1993). Brydon-Miller conducted lengthy, individual interviews with members of the local disabled community and organized a workshop for the participants to identify important accessibility-related problems and issues. One outcome of the workshop was the decision to focus on a local shopping mall as an accessibility target. Over a five-year period, the group evaluated the mall's accessibility, communicated problems to the mall management, filed a complaint with the Architectural Barriers Board, and followed through after the initial decision against the mall was appealed and re-appealed. The outcome was successful: The mall was renovated after the case became the first accessibility-related case to be heard by the Massachusetts Supreme Court (Brydon-Miller, 1993). By focusing on community concerns and making action possible, the result was a fundamental structural transformation that improved the lives of the people involved (Brydon-Miller, 1993: 136). Brydon-Miller (2002) sees this work as research, a form of social activism, and a political act.

Researchers in the activist traditions can select from all the methods of social inquiry. The critical issue is that members of the community are involved in deciding which methods to use, developing or adapting the research instruments, and carrying out the research and analysis. The more naturalistic methods of inquiry, such as participant observation and informal interviewing (perhaps in a more reciprocal way) are often selected because the research focus is usually on the participants and their interpretations of the social context. Each aspect of the research is seen as an opportunity for interaction and discussion. In this perspective, information is not transmitted from others to researchers but is co-created (Esposito and Murphy, 2000: 182). Kemmis and McTaggart (2003: 375) argue that in participatory action and related kinds of research, the researchers may make sacrifices in methodological and technical rigor in exchange for better immediate face validity so that the evidence makes sense within its context. They argue that this trade-off is acceptable because a goal of the research is transformative social action, where participants live with the consequences of the transformations.

A study by Sullivan et al. (2005) illustrates the goals of PAR and its methodology. The project began in order to understand the cultural context of domestic violence in a community with multiple minority groups and to examine access to and satisfaction with services for battered women. Community-based organizations, government agencies, advocates, survivors of domestic violence, and activists working to end domestic violence were involved in all phases of the research process.

A research team composed of university researchers, Health Department personnel, and advocates from the partner agencies was formed; an advisory group with members from the larger domestic violence communities was selected; and a community activist who was also a survivor of domestic violence was hired to help coordinate the work (Sullivan et al., 2005: 980). Decision making was shared by researchers and community partners at each step of the process. For example, while the original plan was to conduct focus groups and individual interviews, some community advocates argued that women in their communities would be reluctant to talk about domestic violence in groups, but others disagreed. The resolution was suggested by one of the advocates: let the women themselves chose the kind of interview. Recruitment of a sample of women aged 18 and older who had experienced domestic violence by an intimate partner or family member living in their households was also a joint process, with advocates in some communities spending a great deal of time finding participants. Similarly, the data collection methodology, including the specific questions asked of participants, was first designed by brainstorming in team meetings.

The researchers report that those with the most limited participation in the research process were the immigrant and ethnic minority women participants: "Although these informants reported having enjoyed taking part in the focus groups and interviews, the structure of the project did not facilitate them being further involved. In particular, constraints arose because of the demands on informants' lives as related to possible ongoing abuse, parenting responsibilities, and employment" (Sullivan et al., 2005: 990). However, the community-based agencies did participate in the research process,

> ### BOX 14.3
>
> ## Planning Participatory Action Research
>
> - Find a community and one or more agencies or organizations to be your partner(s).
> - With your community partners, make joint decisions on a research topic or question and the time frame you have in mind.
> - Find out about the requirements for approval of your project from the Institutional Review Board at your college or university and apply for approval in a timely manner.
> - Think about your stake in the research, your expertise, and what you want to contribute.
> - With your community partners, estimate the time and the costs of doing the project, locate possible resources, and put together an action plan.

and then took a leadership role in implementing the kinds of projects that best fit the research findings. The power of the women's stories galvanized them. "Even advocates who had worked in the field for years learned new information from listening to the women's stories. Hearing, reading, and rereading the stories also developed a strong sense of responsibility to the participants" (Sullivan et al., 2005: 991).

Steps to begin participatory action research are described in Box 14.3. If you want to get more involved, check out www.parnet.org for an archive of articles and a discussion forum for students, faculty, community activists, and others interested in PAR.

Final Considerations

A Middle Ground?

> Although "flattening" the research hierarchy may be an ideal in community research, it is neither honest nor helpful to portray the roles of researcher and participant as equal. Some power inequality is inherent, and . . . we need to acknowledge the power and resource differences that typically exist and, thereby, accept additional responsibilities as researchers that are not necessarily balanced by parallel or complementary responsibilities on the part of participants. (Bond, 1990: 183)

Although at first glance it appears that researchers must choose between the more conventional and the more activist kinds of applied work, there might be a middle ground. Looking at grass-roots groups, some have noted that with relatively few paid workers and volunteers, many organizations do not have the human resources or time to commit to empowerment research (Lackey, Moberg, and Balistrieri, 1997). As a result, groups might be willing to work in collaboration with researchers to obtain information about effectiveness of programs for their own use. In addition, they sometimes prefer more conventional evaluation work to attract funding or help generate recognition for their work (Lackey, Moberg and Balistrieri, 1997).

Leonard Krimerman (2001) argues that PAR will not replace mainstream or expert-directed work, but rather that it is possible and sensible for the practitioners from both traditions to collaborate. He believes that social scientists can play a major role in the invention and extension of key concepts: initiating identification of and response to oppressive conditions; contributing macro-level policy analyses; providing the cross-national analysis. On the other hand, he notes that any research focused on the poor, the disenfranchised, or the marginalized that does not include their voices and contributions will be flawed or incomplete (Krimerman, 2001).

Using a combined approach, described by some as engaged evaluation or community-based research, the researcher and the organization or group work together and share responsibilities to assess a program or project. Evaluation is typically different when done for internal rather than external purposes. In internal evaluations, questions are asked about "our program," whereas external evaluation asks about "your program" (Fetterman and Eiler, 2000). The researcher working *with* the community can bridge the gap between these approaches.

The researcher and community group can split the work. The researcher can take responsibility for considering all possible stakeholder groups and their agendas and take a broker role between the groups, insisting that all viewpoints be heard and adding credibility and visibility to views that otherwise might be ignored (Hills, 1998). This new kind of evaluation could provide a space for discussion and communication among the various stakeholders, encouraging self-evaluation where the participants themselves engage in discussion and systematic analysis of the issues, collect data, and reach conclusions (Hills, 1998).

A model of work that balances conventional evaluation research perspectives with more participatory viewpoints can be found at the Center for AIDS Prevention Studies (CAPS) in its attempts to stimulate collaboration among academic researchers, public health professionals, and community-based organizations. One CAPS project was a study of 11 partnerships between community-based organizations and university-based researchers. The study found that the community-based organizations took the lead in developing the research question, delivering the program, and collecting the data; the academic researchers took the lead in developing the instrument, consent procedures, and data analysis; together the groups trained the staff, interpreted the data, and disseminated the findings (CAPS, 2000). In the past decade, CAPS has supported more than two dozen collaborative research and evaluation projects using their model of community collaborative research (CAPS, 2006).

Politics and Applied Research

Regardless of approach, all research is political in nature and is affected, to some degree, by the social, political, and economic climates that surround the research community. Knowledge is socially constructed, so the choice of research topics and questions and all methodological decisions are related to the social and political context.

In evaluation research, the specific choice of research project is affected not only by societal values and the priorities of funding agencies, but also by the perspectives of various constituencies and program stakeholders. Evaluations are typically conducted in field settings, within ongoing organizations such as school systems, police departments, health care facilities, and social service agencies. Such organizations are very political: There is "turf" to protect, loyalties and long-standing personal relationships to look after, and careers and economic survival to think about. These are not the ordinary, newsworthy politics of the legislative variety—intellectual issues, public morals, votes, and the next election. Rather, these politics are the kind that are internal to organizations, politics spelled with a lower case "p," the politics of authority, sexual relationships, and small groups (Chambers, Wedel, and Rodwell, 1992: 11). Organizational participants can either help or hinder evaluation research efforts depending on their assessment of the politics of the research.

Participatory action researchers are aware of the politics of setting and, even though they have professional expertise, usually work as nonprivileged members of a team. Communities make an initial judgment about researchers and choose whether or not to invite them into the setting. Because of shared responsibility, the working group will determine the form and interpretation of the research project.

Beyond Our Control

Organizations are not "neutral territory" and in most cases, the researcher is an outsider, working in someone else's sphere. The results of any assessment will have the potential to affect the organization and the individuals under study. As a result, a special challenge in evaluation research can be obtaining the cooperation of program staff for access to data. Sometimes this is especially challenging. When one of us, Emily, was hired by a special education professor and researcher to help with the evaluation of special needs educational policies in several communities, feasibility became a concern. The research plan was for Emily to observe school meetings at which individual education plans for special-needs students were designed by school personnel in consultation with parents. There was no difficulty in getting cooperation at three of the schools selected, but, at the fourth school, the principal kept "forgetting" to communicate when meetings were canceled or had been rescheduled because of "unavoidable problems and crises." Several times, on arriving for a meeting at this particular school, Emily was told that the meeting had already been held—the day or even the hour before. The message was clear: The principal did not want an observer. Because the schools selected were a purposive, rather than random sample, the meetings at another school were substituted and observed.

The sentiment that evaluators aren't welcome isn't unusual. As the following excerpt from an editorial in a magazine for youth program administrators makes clear, evaluators and their work are often distrusted by program personnel:

> As part of the price of doing business, youth service managers put up with a stream of over-schooled but often under-educated "evaluators" typically drawn from academia or a Beltway Bandit consulting firm. Much of the supposed evaluation time is spent by the staff teaching the evaluators the realities of the rough and tumble world of youth work. When evaluation results are made available years later they speak to a staff, funding mix, and business climate that no longer exist. In evaluating the actual helpfulness of evaluators in steering the nation toward better, more cost effective programs for youth, it is reasonable to wonder if this whole mini-industry isn't just another example of white collar welfare masquerading as help for the disadvantaged. (Vanneman, 1995: 2)

Evaluation researchers must realize that those who work in the programs might feel that the time and money spent on evaluation would be better spent on programs. They must be sensitive to the fact that the program staff has a job to do that might be made more difficult by ongoing evaluation.

Participant action researchers can be invited in, but also might be asked to leave. They have less control over (and responsibility for) the research than evaluation researchers. They must be willing to contribute their expertise and work as team members rather than in a traditional hierarchy.

Having an Impact

As education and action are explicit parts of their work, participatory action researchers can usually see the impact of the work. For example, at the end of the project described by Brydon-Miller (1993), the mall had been renovated and was much more accessible to disabled patrons. As a result of the work on domestic violence described by Sullivan (2005), culturally appropriate programs were developed to provide education and skill-building through support groups. As one member of the team, the participatory action researcher can know that the efforts will continue because a group of knowledgeable and committed community participants remain.

More conventional evaluation faces the challenge of implementing change after the research is completed. This is not a new story. In 1845, for example, the Board of Education of the city of Boston initiated the use of printed tests to assess student achievement. The resulting evaluation of test scores showed them to be low, but rather than analyze the causes of the poor performance (Traver, cited in Chambers et al., 1992: 2), the board decided to discontinue the test! Unfortunately, this example, more than a century and a half old, still has relevance.

Even if evaluators are trusted, allowed to collect data, and do a credible job of evaluating a program or policy, their conclusions might have little impact. Ideological and political interests can sometimes have more influence on decisions about the future of social interventions than evaluative feedback. Even if a program is shown to be ineffective, it might be kept if it fits with prevailing values, satisfies voters, or pays off political debts. Often changes in social programs are very gradual because frequently no single authority can institute radical change (Shadish, Cook, and Leviton, 1991: 39).

The DARE case provides an interesting example of how programs that receive negative evaluations can continue to flourish. Evaluations of DARE in Kokomo and elsewhere (Ennett, Tobler, Ringwalt, and Flewelling, 1994; Perry, et al., 2000) indicated that local DARE programs are less than successful. In fact, an analysis of 11 peer-reviewed evaluations of DARE judged the program to be ineffective (West and O'Neal, 2004). However, some stakeholders—in this case, organizations with direct and indirect involvement—have accumulated sufficient resources and legitimacy to protect the program.

The initial reaction of local policy makers in Kokomo to the "no effects" evaluation ranged from silence and indifference to hostility (Wysong and Wright, 1995); other DARE evaluations have met similar fates (Vanneman, 1995: 2). One unexpected outcome in Kokomo was a decision by the local school board to approve random drug testing for students in grades 8 through 12. The irony was that the superintendent of schools cited the estimates of drug use in the Wysong data as contributing to the new policy (Hubbard, 1996).

When evaluations are bought and paid for by agencies that are also stakeholders, an even more difficult political situation can arise. For example, at least one completed analysis of DARE (commissioned by the National Institute of Justice and conducted by the North Carolina Research Triangle Institute) was initially denied release by the funding agency, perhaps to restrict or discredit its negative findings (Wysong and Wright, 1995). The DARE evaluation in the focal research was conducted by researchers working as interested scholars rather than as paid consultants to the DARE program or the school department; the bulk of the funding for the study's expenses ($11,000) came from university faculty grants. These factors made possible the widespread dissemination of the study's findings, regardless of the reaction of local or national stakeholders. Not all evaluation researchers are so fortunate.

Outside evaluators usually have an advisory role, "closest to that of an expert witness, furnishing the best information possible under the circumstances; it is not the role of judge and jury" (Rossi and Freeman, 1993: 454). Evaluators can provide reports with significance for policy making (Newman and Tejeda, 1996), discuss the larger context of their work, argue forcefully for their positions, and work toward disseminating their findings widely, but they rarely have the power to institute changes in programs or policies. The recent changes in the DARE program and the ongoing evaluation discussed in Box 14.4 illustrate the effect that stakeholders can have on public policy and evaluation research.

BOX 14.4

DARE: Evaluating a Moving Target

Having an Impact on DARE

In study after study, researchers like Wysong and his colleagues have concluded that DARE is ineffective. But DARE either ignored evaluation efforts or tried to discredit the studies as flawed or the work of groups favoring the decriminalization of drugs (Zernike, 2001: A1).

Finally, by the late 1990s, the chorus of voices yelling "no effect" started to have an effect. Because implementation decisions about DARE are made on the local level, that's where some of the battles were waged. The Seattle Police Department dropped the program in 1998 (Deveny, 1998). In Massachusetts, by 1999, five towns had dropped DARE, although 93 percent of the state's municipalities remained loyal to it (Barry, 1999). In 2000, making a politically unpopular decision, Salt Lake City's mayor, Rocky Anderson, arguing that DARE was better at public relations than in preventing long-term drug abuse, dropped the program and saved the city the annual $300,000 DARE budget (KSL, 2000).

At the federal level, officials were frustrated by DARE's shortcomings but did not want to drop it. As former Assistant Attorney General Laurie Robinson said, "A decision was made in [the Department of Justice], sitting around Janet Reno's conference table, that we should mend it, not end it" (Boyle, 2001: 1). On the legislative front, the Safe and Drug-Free Schools and Communities Act was passed, requiring programs receiving Department of Education funding to show effectiveness; DARE did not make the list (Weiss et al., 2005). By 2005, DARE was dropped by many municipalities, including Cincinnati and Chattanooga (Associated Press, 2005).

However, DARE had a big budget, well-paid executives, and many political allies. Earl Wysong (personal communication) found that the 2003 tax return DARE America filed (obtained from GuideStar, a database on nonprofit organizations) indicated that it received more than $4 million in revenue from government sources and contracts annually and had well-paid directors and employees. With the stakes quite high, DARE wasn't going to go away. As part of the deal brokered by Janet Reno's office, the Robert Wood Johnson Foundation agreed to provide $13.7 million to fund revisions of the DARE curriculum and an evaluation of the new program (Boyle, 2001: 19).

In 2001, the revised program, Take Charge of Your Life, was introduced to approximately 1900 students in 83 high schools and 122 middle schools from six U.S. cities. The 10-lesson curriculum was taught by DARE officers in the seventh grade with a seven-lesson "booster" curriculum given in the ninth grade. The University of Akron's Institute for Health and Social Policy is doing the evaluation by collecting data through 2006, comparing program participants to a control group not enrolled in the

(continued)

BOX 14.2 (*continued*)

curriculum (ASAP, 2003). In 2002, preliminary evaluation results indicated 6 percent higher scores for DARE curriculum participants in decision-making skills and a 4 percent lower rate of using inhalants (ASAP, 2003). However, the next reports did not include any information on behavioral differences. Instead, the progress report for year 4 notes that "results to date show that the Take Charge of Your Life program has had the strongest impact on normative beliefs" (ASAP, 2006).

The more things change, the more they stay the same?

Summary

Applied research is research with a practical purpose. It can be undertaken by a researcher with the participation of one or more stakeholders. In evaluation research, the researcher or a specific stakeholder develops a research question or hypothesis, most typically a causal hypothesis with a program or its absence as the independent variable and the goal of the program as the dependent variable. Evaluation research can provide reasonably reliable and valid information about the merits and results of particular programs that operate in particular circumstances. Necessary compromises in programs and research methodologies can mean that the users of the information will be less than fully certain of the validity of the findings. Because it is better to be approximately accurate than have no information at all, researchers might have to settle for practical rather than ideal evaluations.

Participatory action researchers and others in the activist tradition work collaboratively with participants. With the goals of education and social justice, outside professionals and those who are participants in the setting decide collaboratively on the research questions, the study's methodologies, and ways to apply the findings.

Applied work is done in settings with political and social contexts. In each project, there will be multiple stakeholders. At the end of each study, the analyses and their practical implications can be valuable to program staff, clients, other stakeholders, other researchers, and the general public. Applied research can help to create, modify, and implement programs and activities that make a difference in people's lives.

EXERCISE 14.1

Designing Evaluation Research

Find a description of a social program in a daily newspaper or use the following article of a hypothetical program. (If you select a newspaper article, attach it to the exercise.)

"Pets Are Welcome Guests"

Residents of the Pondview Nursing and Rehabilitation Center have a series of unusual guests once a week. The VIPs (Volunteers Interested in Pondview) have organized a "Meet the Pets Day" at the local facility. Each week, one or more volunteers bring a friendly pet for short one-to-one visits with residents. On a typical day, a dozen or so owners will bring dogs, cats, and bunnies to Pondview, but sometimes companions include hamsters and gerbils.

Last week, Buffy, a spirited golden retriever with a wildly wagging tail, made her debut at Pondview. In a 15-minute visit with Mrs. Rita Williams, an 85-year-old widow recovering from a broken hip, Buffy managed to bestow at least several "kisses" on the woman's face and hands. Mrs. Williams said she has seen more sedate pets, but wasn't at all displeased with today's visit.

Margaret Collins, facility administrator, said that the program had been adapted from one she had read about in a nearby city. She was glad that the VIPs had organized the new program. "It gives the residents something to look forward to," she said. "I think it makes them more alert and attentive. If it really does aid the residents' recoveries and results in their improved mental health, we'll expand it to several days a week next year."

Design a research project to evaluate either "Meet the Pets Day" or the program described in your local newspaper by answering the questions that follow.

1. What is an appropriate research question that your evaluation should seek to answer or hypothesis that you would test?

2. Describe the social program that is being offered.

3. Who are the program's participants?

4. In addition to the program's participants, who are the *other* stakeholders?

5. What is the goal or intended outcome of the program?

6. Describe how you would decide if the program was successful in meeting its goal by designing an *outcome evaluation study.* Be sure to describe the study design you would use, who your sample would include, and how you would measure the dependent variable.

7. Comment on the internal validity of the study you designed.

8. What are the ethical considerations you would need to consider if you were interested in doing this study?

9. What are the practical issues (time, money, and access) that you would need to consider if you were interested in doing this study?

Participatory Action Research

Imagine that you would like to do a piece of applied research that focuses on a group in your community, such as a shelter for homeless people, a literacy program, a neighborhood preservation association, a local boys or girls club, a volunteer services for animals organization, or any group of interest.

Visit one such group or organization in your community. See if you can meet either formally or informally with members of the staff and with clients. Ask them what they feel their most pressing needs are and what problems the organization or group has had in meeting these needs.

Based on what you find out, write a short paper describing a possible participatory action research project, which includes the following:

1. A description of the organization, agency, or group.

2. A listing and description of the major stakeholders.

3. A possible research question that could be answered if you were to work collaboratively with the major stakeholders.

4. The benefits of doing the work you're proposing.

5. Any practical or ethical concerns you think you'd run into in doing the project.

Evaluating Your Course

The course you are taking this semester is an educational program designed to help students understand and use social research methods. As such, it can be evaluated like any other program. For this exercise, select one aspect of your course (such as the textbook, the instructional style of the teacher, the frequency of class meetings per week, the length of each class session, the lectures, the exercises, etc.) as an independent variable. Design an evaluation research project that could be conducted to test the effectiveness of the aspect of the course you've selected on the dependent variable of student learning.

Answer the following questions.

1. What aspect of the course are you focusing on as your independent variable?

2. Who are the significant stakeholders in this setting?

3. How would you define and measure the dependent variable "student learning of social research methods"?

4. What would you do to determine the effectiveness of the aspect of the course that you are interested in?

5. Using your imagination, what are some results that you think you'd get from your evaluation?

6. Comment on the practical and ethical considerations you would need to consider if you wanted to do this study.

Drug Use and Attitudes

Work with another student to complete a survey of attitudes toward drugs with a quota sample of 20 college students.

What to Do

1. Review the appendix in the focal research in this chapter and select at least five of the questions from the DARE Scale or the Drug Use Scale to include on an anonymous questionnaire. Construct an additional question that asks the respondent if he or she participated in a DARE or another drug-education program while in elementary or secondary school. Construct several questions asking about background characteristics (age, gender, and so on) that you think are important. Write an introduction to the questionnaire, instructions for completion, and order the questions appropriately. Make at least 20 copies of the questionnaire to distribute.

2. Check with your instructor about the need to seek approval from the Human Subjects Committee, the Institutional Review Board, or the committee at your college or university that evaluates the ethics of research projects. If necessary, obtain approval before completing the rest of the exercise.

3. Approach available students in person or by phone. Tell them that you are conducting a short, anonymous survey of attitudes toward drugs for your research methods class. If a student is willing to be in the study, first ask if he/she remembers having participated in an anti-drug education program (like DARE) while in elementary or secondary school. You will be aiming for a total sample that has approximately 10 people who have been in an anti-drug program in the past and approximately 10 who have not.

4. Select a method of having students return the questionnaire so that the members of the sample can be anonymous. (For example, students can return it to you at an on-campus address or mail it to you using a self-addressed, stamped envelope that you provide.)

5. You and your partner should find 20 people to complete the questionnaire. Keep track of how many "turn downs" you get. Aim for 10 students who have participated in an anti-drug program and 10 students who have not.

6. Once you receive the 20 completed questionnaires, analyze the data using frequency tables as follows:

   a. Separate the questionnaires of respondents who participated in an anti-drug education program from those who did not.

   b. Construct separate frequency tables for the two groups of respondents for at least four of the questions on the drug use and behavior questions.

   c. Compare the answers for the two groups.

   d. Compare your results to those reported in the focal research article.

What to Write

Write a report of your research that includes the following:

1. Describe the approach you used for finding and encouraging students to be part of your survey.

2. How many did you approach? How many agreed? How many actually responded? Calculate your response rate.

3. Describe how the questionnaires were returned so that you could guarantee anonymity to sample members.

4. Present the results of your survey and how the results compare with those of Wysong and colleagues.

5. What are some possible concerns about the validity of your results?

6. If you were to repeat this survey, based on your experience, is there anything you'd do differently?

7. Attach all 20 completed questionnaires to your report as an appendix.

15

Quantitative and Qualitative Data Analysis

© Royalty-Free/CORBIS

431

Introduction

Perhaps you've asked yourself, as we've discussed all the ways of collecting data, "Once I've collected all the data, how do I organize all these facts so that I can share what I've found?"

We'd now like to introduce you to the most rewarding part of the research process: analyzing the data. After all the work you've done collecting your information, there's nothing quite like putting it together in preparation for sharing with others—a little like finally cooking the bouillabaisse after shopping for all the individual ingredients. But like the excitement of cooking bouillabaisse, the excitement of data analysis needs to be balanced by having a few guidelines (or recipes) and principles to follow. In this chapter, we'll introduce you to quantitative and qualitative data analyses. Rather like an elementary cookbook, this chapter offers rudimentary data-analytic recipes, and some insights into the basic delights and principles of data analysis. Bon appetit!

Quantitative or Qualitative?

What is the difference between quantitative and qualitative data? One answer is that quantitative data results from quantitative research and qualitative data results from qualitative research. This sounds like question begging, and of course it is, but it carries an important implication: The distinction drawn between quantitative and qualitative data isn't as important (for data-analytic purposes) as the distinction between the strategies driving their collections. Sure, you can observe, as James Dabbs (1982: 32) and Bruce Berg (1989: 2) have, that the notion of qualitative refers to the essential nature of things and the notion of quantitative refers to their amounts. But, as we've hinted before (for example, in chapters on qualitative interviewing and observation techniques), there's really nothing about qualitative research that precludes quantitative representations (see Morrow, 1994: 207). Patricia and Peter Adler (see Chapter 11) did sometimes count the number of times that children attended different kinds of after-school activities. Similarly, research that might seem primarily quantitative never totally ignores essences. Clark, Guilmain, Saucier, and Tavarez (in Chapter 13) used qualitative judgments about which picture book characters exhibited, for instance, nurturant behaviors before they used quantitative techniques to decide whether male or female characters were more likely to exhibit nurturant qualities. Indeed, although the mix of quantitative and qualitative approaches does vary in social science research, almost all published work contains elements of each. Consequently, the distinction we draw (and have used to organize this text) is not one of mutually exclusive kinds of analysis but, rather, of kinds that, in the real world, stand side by side, and, in an ideal world, would always be used to complement one another.

An Overview of Quantitative Data Analysis

quantitative data analysis, analysis that tends to be based on the statistical summary of data.

However artificial the distinction between quantitative and qualitative research and data might be, there's no denying that the motives driving quantitative and qualitative researchers are distinguishable. Although there are exceptions, **quantitative** researchers normally focus on the relationships between or among variables, with a natural science-like view of social science in the backs of their minds. Thus, in the fashion of the physicist who asserts a relationship between the gravitational attraction and distance between two bodies, Clark, Guilmain, Saucier, and Tavarez (in Chapter 13) focused on whether male or female characters were more likely to exhibit, for instance, nurturant qualities in children's books and Ramiro Martinez and Matthew Lee (in Chapter 12) studied whether migrant communities were more likely to have high crime rates than nonmigrant communities were. Projects like these engaged in other natural science-like activities as well: for example, looking at aggregates of units (whether these were characters in picture books or communities), more or less representative samples in some studies (see Chapter 5 on sampling) of these units, and the measurement of key variables (see Chapter 6 on measurement). For data-analysis purposes, the distinguishing characteristic of this type of study is attention to whether there are associations among the variables of concern: Whether, in general, as one of the variables (say, exposure to themes of caring) changes, the other one (students display caring attitudes) also changes. To demonstrate such an association, social science researchers generally employ *statistical* analyses. Thus, we will aim toward a basic understanding of statistical analyses in the part of the chapter on quantitative data analysis.

STOP AND THINK

Recall Martinez and Lee's study (in Chapter 12) of migration and crime in Miami. Did the migrants they studied generally have higher or lower crime rates than nonmigrants do? Do you see that the focus of this article, like that of others using quantitative data, is on variables (in this case whether a community is migrant or not and crime rate) and not really on people, groups, or social organizations?

An Overview of Qualitative Analyses

qualitative data analysis, analysis that tends to result in the interpretation of action or representations of meanings in the researcher's own words.

The strategic concerns of qualitative researchers differ from those of quantitative researchers and so do their characteristic forms of data analysis. Generally, **qualitative** researchers tend to be concerned with the interpretation of action and the representation of meanings (Morrow, 1994: 206). Rather than pursuing natural science-like hypotheses (as Durkheim advocated in his *Rules of Sociological Method* [1938]), qualitative researchers are moved by the pursuit of empathic understanding, or Max Weber's *Verstehen* (as described in his *Theory of Social and Economic Organization* [1947]), or an in-depth description (for example, Geertz' [1983] *thick description*). Sandra Enos (in the focal research in Chapter 10) aspired to an empathic understanding of the meaning of motherhood for women inmates and of the differences in those meanings for different inmates. Adler and Adler (in the focal research in Chapter 11) provide a thick description of what it means to participate in afterschool activities. Qualitative researchers pursue single cases or a limited number of cases, with much less concern than quantitative researchers have about how well those cases "represent" some larger universe of cases. Qualitative researchers look for interpretations that can be captured in words rather than in variables and statistical language. Later in the chapter, we'll seek a basic understanding of how qualitative researchers reach their linguistic interpretations of data.

Quantitative Data Analysis

Quantitative data analysis presumes one has collected data about a reasonably large, and sometimes representative, group of subjects, whether these subjects (or units of analysis) are individuals, groups, organizations, social artifacts, and so on. Ironically, the data themselves don't always come in numerical form (as one might expect from the term *quantitative* data).

A brief word to clarify this irony: You'll recall that the focus of quantitative data analysts isn't really the individual subjects being studied (for example, school children or communities) but, rather, the variable characteristics of those subjects (for example, whether they were exposed to caring literature or whether they have high crime rates). Remember that variables come in essentially three levels of measurement: nominal, ordinal, and interval-ratio. The most common level of measurement is the nominal level (all variables are at least nominal), because all variables (even marital status) have categories with names (like "married" and "single"). Thus, an individual adult might be married or single, and the information you'd obtain about that person's marital status might, for instance, be "single." But, "single" is obviously not a number. Nonetheless, it is information about marital status that when, taken together with data about marital status from other subjects, can be subjected to numerical or quantitative manipulation, as when we count the number of single people in our sample.

This is just a long way of saying that the descriptor "quantitative" in "quantitative data analysis" should really be thought of as an adjective describing the kind of analysis one plans to do, rather than as an adverb modifying the word "data." When you do a quantitative analysis, the data might or might not be in the form of numbers.

Sources of Data for Quantitative Analysis

Data appropriate for quantitative analysis can come from many sources, including a survey conducted by the researcher herself. When the data are collected by the researcher, an important first step often is **coding** the data. Coding is the process by which raw data are given a standardized form. In quantitative analyses, this frequently means making data computer usable. In most cases, quantitative coding involves assigning a number to each observation or datum. Thus, when coding gender, you might decide to

coding, the process by which raw data are given a standardized form. In quantitative analyses, this frequently means making data computer usable. In qualitative analyses, this means associating words or labels with passages and then collecting similarly labeled passages into files.

assign a value of "1" for each "female," and a "2" for each male. The assignment of a number is often pretty arbitrary (there is no reason, for instance, that you couldn't assign a "1" for females and a "0" for males), but it should be consistent within your data set.

As we suggested in Chapter 12, more and more published research in the social sciences is now produced by researchers who have analyzed secondary survey data—data collected by large research organizations and made available to other researchers, sometimes for a fee. According to one study (Clark, 1999), about 40 percent of all articles published in sociology toward the end of the 1990s were based on analyses of secondary survey data. You might recall that the General Social Survey is the most popular of these sets of survey data. One of the great advantages of secondary survey data is that they usually come pre-coded and ready for analysis via one of the software packages social scientists use for statistical analyses. The General Social Survey comes, in an abridged version, for instance, along with abridged versions of a number of other data sets, in *The SPSS Guide to Data Analysis* (Norusis, 2005). This guide introduces students to statistical analysis using the software package, SPSS, or the Statistical Package of the Social Sciences. General Social Survey data are so widely available that you can even visit a website, The SDA Archive (http://sda.berkeley.edu/archive.htm) and do your own analysis of GSS data from 1972 to 2004, and other data sets. In some of the embedded Stop and Do sections that follow, we'll be asking you to do so, so you can see what can be done with General Social Survey data online.

In any case, we will be using General Social Survey data to illustrate some of the kinds of quantitative analysis most frequently used by social scientists in our discussion of elementary quantitative analyses.

Elementary Quantitative Analyses

Perhaps the defining characteristic of quantitative analyses is their effort to summarize data by using statistics. Statistics, being a branch of mathematics, is an area of research methods that some social science students would just as soon avoid. But the advantages of a little bit of statistical knowledge are enormous for those who would practice or read social science research, and the costs of acquiring that knowledge, we think you'll agree, aren't terribly formidable, after all. Our primary goal in the section on quantitative analysis is to provide you with the kind of overview that will give students who have never had a course in statistics, and those with more experience, a feeling for how statistics advance the practice of social research.

To organize and limit our discussion of social statistics, however, we'd like to begin with two basic sets of distinctions: one distinction, between **descriptive** and **inferential** statistics, relates to the generalizability of one's results; the other distinction among **univariate, bivariate,** and **multivariate** statistics relates to how many variables you focus on at a time. Descriptive statistics are used to describe and interpret data from a sample. We find that about 55 percent of the people sampled in the GSS from 1972 to 2004 were married, so that modal (most common) marital

descriptive statistics, statistics used to describe and interpret sample data.

inferential statistics, statistics used to make inferences about the population from which the sample was drawn.

univariate analyses, analyses that tell us something about one variable.

bivariate analyses, data analyses that focus on the association between two variables.

multivariate analyses, analyses that permit researchers to examine the relationship between variables while investigating the role of other variables.

status of the GSS sample is "married" (data and analysis done using GSS data from the SDA Archive, 2005). The mode is a descriptive statistic because it describes only the data one has in hand (in this case, data on 46,510 respondents). Inferential statistics, on the other hand, are used to make estimates about characteristics of a larger body of data (population data—see Chapter 5 on sampling). We find, using the GSS data again, that men are significantly more likely than women to have been employed full time (SDA Archive, 2005). The word "significantly" in this context means that there's a good chance that, in the larger population from which the GSS sample has been drawn (basically adult Americans), that men are more likely to be employed full time than women. We used an inferential statistic to draw this conclusion. In the rest of this chapter, we'll focus on descriptive statistics, but we'll briefly introduce inferential statistics, too.

Much powerful quantitative research can be done with an understanding of relatively simple univariate (one variable) or bivariate (two variable) statistics. Univariate statistics, like the mode (the most frequently occurring category), tell us about one characteristic. The fact that the modal marital status is "married" in the GSS sample tells us about the typical marital status of that sample. Bivariate statistics, on the other hand, tell us something about the association between two variables. When we wanted to find out whether men or women have been more likely to employed full time—that is, that there is a relationship between "gender" and "work status"—we used bivariate statistics. Multivariate statistics permit us to examine associations while we investigate the role of additional variables. We hypothesized that the association between people's gender and their likelihood of being employed full time might have been accounted for in terms of their level of education. We used GSS data (from the SDA Archive, 2005) and found that the association between gender and employment status could not be accounted for in terms of education by using multivariate statistical methods. In what follows, we'll focus on univariate and bivariate statistical analyses, but we encourage those of you who are interested in pursuing the logic and practice of multivariate analysis to see the website for this text on the Thomson website at http://www.thomsonedu.com/sociology/.

mode, the measure of central tendency designed for nominal level variables. The value that occurs most frequently.

median, the measure of central tendency designed for ordinal level variables. The middle value when all values are arranged in order.

mean, the measure of central tendency designed for interval level variables. The sum of all values divided by the number of values.

Univariate Analyses

Measures of Central Tendency

Perhaps the most common statistics that we use and encounter in everyday life are ones that tell us the *average* occurrence of something we're interested in. In 2005 David Ortiz, the Red Sox designated hitter, sometimes got a hit and sometimes didn't, but his batting average for the year was .300, meaning that he got a hit about 3 out of every 10 times he came to bat. In most distributions, values hover around an average, or central tendency. The most common of these are the **mode,** the **median,** and the **mean.**

In Chapter 6, we observed that the mode, median, and mean are designed for nominal, ordinal, and interval level variables, respectively. The mode, designed for nominal level variables, is that value or category that occurs most frequently. Have a look at the following data about five students' gender, age, and height. Because more students are female than male, the modal gender for the sample is "female."

Student	A	B	C	D	E
Gender	Male	Female	Female	Female	Female
Age	18	18	17	19	20
Height	Tall	Tall	Short	Short	Medium

STOP AND THINK *Look at those data again to see if you can decide what the mode is for "age" in this sample.*

Although the mode has been designed for nominal level variables, like gender, it can be calculated for ordinal and interval level variables (like age) because ordinal and interval level variables, whatever else they are, are also nominal. As a result, the mode is an extremely versatile measure of central tendency: It can be computed for any variable.

STOP AND THINK *We say a variable is unimodal when, like "gender" and "age" in the sample of students above, it has only one category that occurs most frequently. We say it is bimodal when it has two categories that occur most frequently. Check out the data for the variable "height." Is it unimodal or bimodal?*

Now, for those of you with easy Internet access:

STOP AND DO *The mode is easy enough to calculate when, as in our example of students, there are five cases. But let's see if you can apply the principles to a much larger sample? Let's look at data from the General Social Survey on the SDA Archive. Access the Archive by going to http://sda.berkeley.edu/archive.htm. Now hit on the most recent GSS Cumulative Data File, and see if you can figure out what the modal highest educational degree has been for respondents to this survey. (Hint: You'll want to browse the codebook for "education" and then look at the information for "Rs highest degree," a variable called DEGREE. Hit on DEGREE and see what you can see.)*

The median, the measure of central tendency designed for ordinal level variables, is the "middle" case in a rank-ordered set of cases. The variable "height," when measured as it is for our students above in categories like "short," "medium," and "tall," is an ordinal level variable. The height for the Students A through E is reported to be "Tall," "Tall," "Short," "Short," and "Medium," respectively. If you arrange these data from shortest to tallest, they become "Short," "Short," "Medium," "Tall," "Tall." The third, or middle case, in this series is "Medium," so the median is "Medium." Though it is designed for ordinal level variables, the median also can be used with interval variables, because interval variables, whatever else they are, are also ordinal variables.

STOP AND THINK *Can you calculate the median age for the sample above?*

The five students mentioned above have ages of 18, 18, 17, 19, 20. Arranging these in order, from lowest to highest, they become 17, 18, 18, 19, 20. The third, or middle, case in this series is 18, so the median is 18 years of age.

STOP AND THINK *Why wouldn't you want to try to calculate the median "gender" for our sample?*

The median is pretty easy to calculate (especially for a computer) when you have an odd number of cases (provided, of course, you have at least ordinal level information). What you do with an even number of cases is locate the median halfway between the two middle cases. If, for instance, the "number of years of education" for eight cases were 9, 9, 2, 9, 9, 7, 9, 12, you'd want to arrange them in order so that they would become: 2, 7, 9, 9, 9, 9, 9, 12. Then, you'd take the two middle cases (9 and 9) and find that the median for these cases is (9 + 9)/2 = 9.

STOP AND THINK *What would the median age for a sample be if the sample ages were 19, 18, 17, 19, 18, 19?*

The mean, the measure of central tendency designed for interval level variables, is the sum of all values divided by the number of values. The mean years of education for the first seven cases mentioned in the previous paragraph is

$$\frac{9 + 9 + 2 + 9 + 9 + 7 + 9}{7} = 7.7 \text{ years.}$$

STOP AND THINK *What is the mean of the sample in the previous Stop and Think?*

How does a researcher know which measure of central tendency (mode, median, or mean) to use to describe a given variable? Beyond advising you not to use a measure that is inappropriate for a given level of measurement (such as a mean or a median for a nominal level variable like gender), we can't give you hard and fast rules. In general, though, when you are dealing with interval level variables (such as age, years in school, or number of confidantes), variables that *could* be described by all three averages, the relative familiarity of reading audiences with the mean makes it a pretty sound choice. Thus, although you could report that the modal number of years of education among the first seven cases listed above is 9 or that their median is 9, we'd be inclined to report that the mean is 7.7 years, other things being equal. See Box 15.1 for an example of a time when things are not equal.

Variation

Measures of central tendency can be very helpful at summarizing information but, as Stephen Jay Gould eloquently argued (1997), they can also be misleading. Gould gave his own bout with a rare form of cancer, abdominal mesothelioma, as an example. He'd been diagnosed with this disease as a 40-year-old and learned, almost immediately, that mesothelioma was incurable and that people who had it had a median life expectancy of eight months. Gould reported that his understanding of statistics provided him almost immediate consolation (talk about knowing about statistics being personally useful!) because he knew that if all the journals were reporting life expectancy in terms of the median, then it was very likely that life expectancy for individuals was strongly skewed. Otherwise, the journals would

Another Factor in Choosing An Average: Skewness

skewed variable, an interval-level variable that has one or a few cases that fall into extreme categories.

Although our suggestion that you use the mean as your measure of central tendency for describing interval-level variables works some of the time, it can yield pretty misleading "averages" for those many real-world interval-level variables that are skewed. A **skewed variable** is an interval-level variable that has one or a few cases that fall into extreme categories. Although we made up the following distribution of incomes for 11 people, it is not completely unlike the real distribution of incomes (and wealth) in the U.S. and the world, inasmuch as it depicts few very high-income people and many lower-income people.

$20 million

$1,926,363.60 The Mean Income

$1 million

$100,000

$50,000 — Median (the one in the middle; five above, five below)

$10,000 — Mode (occurs most frequently)

This income variable is skewed because it contains at least one extreme case (the one falling into the $20 million category. The one falling in the $1 million category is "out there" too.). Because of this case, the mean of $1,926,363.60 provides a sense of average that somehow seems to misrepresent the central tendency of the distribution. In this case, as in the case of most skewed variables, we would advise reporting an average, like the median, that is less sensitive to extreme cases.

Do you agree that the median provides a better sense of this distribution's central tendency than the mean?

be reporting a mean life expectancy. Moreover, he reckoned, the extreme values of life expectancy after diagnosis were almost surely on the high end, perhaps even many, many years after diagnosis. (One or two life expectancies of zero, after all, couldn't skew a distribution with a median of eight months that much.) Gould also realized that his own personal life expectancy was almost surely going to be on the high end of the scale. He was, after all, young, eager to fight the disease, living in a city (Boston) that offered excellent medical treatment, gifted with a supportive family and "lucky that my disease had been discovered relatively early in its course" (Gould, 1997: 50).

Gould's larger point is a crucial one: human beings—indeed, anything we're likely to study as social scientists—are not measures of central tendency, not means nor medians nor modes. All things worth studying entail variation. In his case, he needed to be able to place himself in a region of variation based on particulars of his own case. Similarly, if we're interested in the educational level, income, or wealth of a group of people, averages will only get us so far: We need to know something about the variation of these variables.

We can discover a lot of important information about variation by describing the distribution of a sample over various categories of a variable. One of the most commonly used techniques for such descriptions is the **frequency distribution**, which shows the frequency (or number) of cases in each category of a variable. To display such a distribution, the categories of the variable are listed, and then the number of times each category occurs is counted and recorded. Box 15.1, with its display of hypothetical data about the income of 11 people, provides an example of a frequency distribution.

frequency distribution, a way of showing the number of times each category of a variable occurs in a sample.

Let's imagine how we might create a frequency distribution for data about the gender of a hypothetical sample of respondents. Suppose we had 20 people in our sample and we found that there were 17 females and 3 males. The frequency distribution for this variable, then, is displayed in Table 15.1.

One thing this frequency distribution demonstrates is that the overwhelming majority of our respondents were female. The table has three rows. The first two rows display categories of the variable (gender). The third row displays the total of number of cases appearing in the table.

The middle column shows the number of cases in each category (17 and 3). The number is called the *frequency* and is often referred to by the letter *f*. The total of all frequencies, often referred to by the letter *N,* is equal to the total number of cases in the sample examined (in this case, 20). The third column shows the percentage of the total number of cases that appears in each category of the variable. Because 17 of the 20 cases are female, the percentage of cases in this category is 85 percent.

The variable in this example, gender, is a nominal-level variable, but frequency distributions can be created for variables of ordinal and interval level as well. Let's confirm this by working with an interval-level variable: the number of close friends our respondents have. Suppose our 20 respondents told us that they had 1, 0, 0, 0, 2, 1, 1, 1, 3, 0, 2, 2, 2, 1, 1, 1, 0, 0, 1 and 1 close friends, respectively.

Create the outline of an appropriate table on a separate piece of paper, using Table 15.2 as a guide.

Now complete the distribution. First, count the number of respondents who report having "0" close friends and put that number (frequency) in the appropriate spot in the table. Do the same for categories "1," "2," and "3."

TABLE 15.1

Frequency Distribution of Gender of Respondent

Gender	Frequency (f)	Percentages
Female	17	85
Male	3	15
Total	N = 20	100

TABLE 15.2

Frequency Distribution of Number of Close Friends

Number of Close Friends	Frequency (f)	Percentages
0		
1		
2		
3		
Total	N = 20	100

Do the frequencies in each of these categories add up to 20, as they should? If not, try again. Once they do, you can calculate the percentage of cases that fall into each category. (Calculate percentages by dividing each *f* by the N and multiplying by 100. We count 6 respondents with 0 close friends, so the percentage of respondents with 0 close friends is 6/20 × 100 = 30.) Into which category does the greatest number of respondents fall? (We think it's category "1," with 9 respondents.) In which category does the smallest number of respondents fall? What percentage of the respondents said they had 3 close friends?

Frequency distributions, simple as they are, can provide tremendous amounts of useful information about variation. One of our students, Angela Lang, did her (2001) senior thesis on Americans' attitudes toward income inequality. Some of her sources had suggested that the American public is basically apathetic toward income inequality in the country and that a majority, in fact, felt that income inequality provided an "opportunity . . . for those who are willing to work hard" (Ladd and Bowman, 1998: 36). Lang was therefore surprised to find, upon examining General Social Survey data from 1996, that, when asked whether they agreed or disagreed with the statement "Differences in income in America are too large," Americans gave responses that are summarized in the frequency distribution depicted in Table 15.3.

Most striking to Lang was the percentage of the 1468 respondents who answered the question and either strongly agreed or agreed with the statement "Differences in income in America are too large:" 32.3 percent and 32.5 percent, respectively. You can spot these percentages in the column

TABLE 15.3

A Frequency Distribution of Responses to the Statement "Differences in Income in America Are Too Large"

Response	Frequency	Percent	Cumulative Percent
Strongly Agree	464	32.3	32.3
Agree	467	32.5	64.8
Neither	178	12.4	77.2
Disagree	169	11.8	89.0
Strongly Disagree	117	8.1	97.1
Can't Choose	41	2.9	100.0
Total	1468	100.0	

Source: General Social Survey, 1996 (see Davis and Smith, 1998).

labeled "percent." Using the cumulative percentage column, which presents the percentage of the total sample that falls into each category and every category above it in the table, Lang was also able to discern that 64.8 percent of this sample either strongly agreed or agreed with the statement. This was a much higher level of agreement than she expected based on her reading. Notice that the median and the mode for this variable were both "agree." The point, though, is that reporting that the median response was "agree" doesn't convey nearly as much information about the level of agreement as saying "About 65 percent of respondents either agreed or strongly agreed with the statement, whereas only about 20 percent disagreed or strongly disagreed."

STOP AND THINK   *Can you see where the figure of 20 percent comes from?*

STOP AND DO   *Over the years the General Social Survey has asked about 24,000 respondents whether, if their party nominated a woman for President of the United States and she were qualified for the job, they would vote for her. What percentage, would you guess, has said, "Yes"? What percentage has said, "No"? Check your answers against the actual percentage by going to http://sda.berkeley.edu/archive.htm, and finding the frequency distribution for the variable FEPRES.*

Measures of Dispersion or Variation for Interval Scale Variables

Examining frequency distributions, and their associated percentage distributions, is a pretty good way of getting a feel for dispersion or variation in nominal or ordinal level variables. If, for instance, you're looking at gender and discern that 100 percent of your sample is female and 0 percent is male, you know that there is no variation in gender in your sample. If, on the other hand, you find that 50 percent of your sample is female and 50 percent is male, you know that there is as much variation as there could be over two categories. Statisticians have actually given us more elegant ways of describing variation for interval level variables. Again, though, we start with the observation that the mean, perhaps the most frequently used measure of central tendency for interval level variables, can hide a great deal about the variable's *spread* or *dispersion*. Thus, a mean of 3 could describe both of the following samples:

Sample A: 1, 1, 5, 5
Sample B: 3, 3, 3, 3

measures of dispersion, measures that provide a sense of how spread out cases are over categories of a variable.

Inspection shows, however, that Sample A's values (varying between 1 and 5) are more spread out or dispersed than are those of Sample B (*all* of which are 3). To alleviate this problem, researchers sometimes report **measures of dispersion** for individual variables. The simplest of these is the **range:** the difference between the highest value and the lowest. The range of Sample A would be 5 − 1 = *4*, while that of Sample B is 3 − 3 = *0*. The two ranges tell us that the spread of Sample A is larger than the spread of Sample B.

range, a measure of dispersion or spread designed for interval-level variables. The difference between the highest and lowest values.

STOP AND THINK

Calculate the range of Samples C and D:

Sample C: 1, 1, 5, 5; Sample D: 1, 3, 3, 5

Having calculated the range, can you think of any disadvantage of the range as a measure of spread or dispersion?

standard deviation, a measure of dispersion designed for interval-level variables and that accounts for every value's distance from the sample mean.

There are several measures of spread that, like the range, require interval-scale variables. The most commonly used is the **standard deviation.** The major disadvantage of the range is that it is sensitive only to extreme values (the highest and lowest). Samples C and D in the last Stop and Think exercise have ranges of 4, but the spreads of these two samples are obviously not the same. In Sample C, each value is two "units" away from the sample mean of 3. In Sample D, two values (the 1 and the 5) are two "units" away, but two values (the 3's) are zero units away. In other words, the average "distance" or "variation" from the mean is greater in Sample C than in Sample D. The average variation is 2 in Sample C, but less than 2 in Sample D. The standard deviation is meant to capture this difference (one that isn't caught by the range) and to assign higher measures of spread to samples like Sample C than to those like Sample D. And, in fact, it does.

It does so by employing a computational formula that, in essence, adds up the "distances" of all individual values from the mean and divides by the number of values—a little like the computation of the mean in the first place. That's the essence. In fact, the computational formula is

$$s = \sqrt{\frac{\Sigma(X - \bar{X})^2}{N}}$$

where s stands for standard deviation
$\bar{X}$ stands for the sample mean
X stands for each value
N the number of sample cases

Although this formula might look a bit formidable, it's not very difficult to use. We'll show you how it works for Sample D.

First notice that the computational formula requires that you compute the sample mean ($\bar{X}$). For Sample D, the mean is 3. Then subtract this mean from each of the individual values, in turn ($X - \bar{X}$). For Sample D (whose values are 1, 3, 3, 5), these differences are -2, 0, 0, and 2. Then square each of those differences [$(X - \bar{X})^2$]. For Sample D, this results in four terms: 4, 0, 0, 4 (-2 squared $= 4$). Then sum these terms [$\Sigma(X - \bar{X})^2$]. For Sample D, this sum is 8 ($4 + 0 + 0 + 4 = 8$). Then the formula asks you to divide the sum by the number of cases [$\Sigma(X - \bar{X})^2 / N$]. For Sample D, this quotient is 2 ($8 / 4 = 2$). Then the formula asks you to take the square root of this quotient. For Sample D, this is the square root of 2, or about 1.4. So the standard deviation of Sample D is about 1.4.

STOP AND THINK   *Now try to calculate the standard deviation of Sample C. Is the standard deviation of Sample C greater than (as we expected it would be) or less than the standard deviation of Sample D?*

We hope you've found that the standard deviation of Sample C (we calculate it to be 2) is greater than the standard deviation of Sample D. In any case, we're pretty sure you'll see by now that the standard deviation does require variables of interval or ratio scale. (Otherwise, for instance, how could you calculate a mean?) You might also see why a computer is helpful when you are computing many statistics.

The standard deviation has properties that make it a very useful measure of variation, especially when a variable is normally distributed. The **normal distribution**, a distribution that is symmetrical and bell-shaped. graph of a **normal distribution** looks like a bell, with its "hump" in the middle and cases diminishing on both sides of the hump (see Figure 15.1). A normal distribution is symmetric. If you folded it in half at its center, one half would lie perfectly on the other half. One of the nice mathematical properties of a variable that has a normal distribution is that about 68 percent of the cases would fall between one standard deviation above the mean (the center of the distribution) and one standard deviation below the mean.

The standard deviation becomes less intuitively useful when a variable's distribution does not conform to the normal distribution, however. Many socially interesting variables do not so conform. Most of these nonconformers, like income in Box 15.1, are skewed. Figure 15.2 shows another skewed distribution: the distribution of the number of children respondents said they had in the 1996 General Social Survey. As you might expect, most of the 2889 who answered this question said they had zero, one, or two children. But just enough reported that they had 5, 6, 7, and 8 (or more), that the distribution has a "tail" . . . this time to the right. Statisticians have provided another very useful statistic that we won't try to show you how to compute: skewness. We discussed the concept of skewness earlier, but skewness is so important that it also merits its own statistic. If the computed skewness statistic is zero, the variable is very likely to be nearly normally distributed. The skewness statistic for number of children, as depicted in

FIGURE 15.1

Normal Distribution

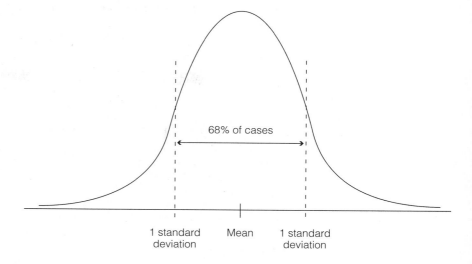

68% of cases

1 standard   Mean   1 standard
deviation   deviation

FIGURE 15.2

Number of Children
Reported by General
Social Survey Respondents.
Mean = 1.8.
Skewness = 1.0.

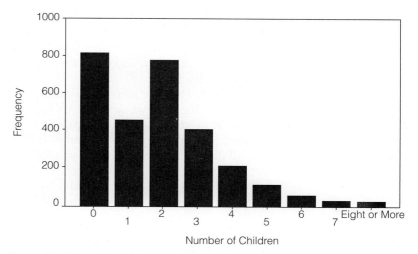

Source: The General Social Survey, 1996 (see Davis and Smith, 1998). Graph was
produced using SPSS.

Figure 15.2, is actually about 1, indicating that the "tail" of the distribution
is to the right. Skewness values can be both positive and negative (when the
"tail" is to the left) and can vary from positive to negative infinity. But when
skewness gets near 1 (or –1), as it does in the case of number of children, or
in the case of life expectancy at diagnosis, as it did for mesothelioma when
Gould was diagnosed with the disease, one can no longer pretend that one
is dealing with a normal distribution. As Gould's case indicates, however,

distributions don't have to be normal to be interesting, enlightening, and even life-affirming . . . especially if one has some appreciation of statistical variation.

Bivariate Analyses

Analyzing single variables can be very revealing and provocative. (Think of Gould's musing about the "life expectancy at diagnosis" variable.) But sometimes the real fun begins only when you examine the *relationship* between variables. A relationship exists between two variables when categories of one variable tend to "go together" with categories of another. You might have found, if you tried to do the last Stop and Do exercise, that about 85.9 percent of adult Americans queried by the General Social Survey since 1972 have said they would vote for a woman if she were nominated by their party and if she were qualified. But do you think this percentage has gone up, down, or remained the same over time since 1972? Let's suppose you think it might have generally gone up since 1972, perhaps because the women's movement since the late 1960s has made it more acceptable to Americans to envision women in leadership positions. If you think of this hypothesis in the variable language introduced in Chapter 2 and the passage of time and people's willingness to vote for a woman for president as variables, then you're saying that you expect the passage of time and people's willingness to vote for a woman to "go together." One way of depicting this expectation is shown in Figure 15.3.

crosstabulation, the process of making a bivariate table to examine a relationship between two variables.

One way of showing such a relationship is to **crosstabulate,** or create a bivariate table for, the two variables. Crosstabulation is a technique designed for examining the association between two nominal level variables and therefore, in principle, applies to variables of any level of measurement (because all ordinal and interval-ratio variables can also be treated as nominal level variables).

Bivariate tables (sometimes called contingency tables) provide answers to questions such as "Has the percentage of adult Americans who are willing to vote for a woman president increased since 1972?" Or, more generally to questions such as, "Is there a difference between sample members that fall into one category of an independent variable and their counterparts in other categories in their experience of another characteristic?"

STOP AND DO

Let's see what's actually happened to the percentage of adult Americans who say they'd be willing to vote for a woman president over time by going back to http://sda.berkeley.edu/archive.htm. This time, after you've gotten into the most recent GSS Cumulative Data File, and "started" the "Frequencies or crosstabulation (with charts)," put FEPRES in the "row" box and YEAR in the "column" box. You'll have crosstabulated the variable "Would you Vote for a Female President?" with the variable "Year of Survey." Examine the resulting table and see whether the percentage of adult Americans who say they would vote for a woman president has gone up, down, or remained the same since 1972.

Isn't it interesting how the percentage of American adults who say they would vote for a qualified woman for president inched up after 1972? At

FIGURE 15.3

Our Hypothesis About the Relationship Between Time and the Willingness to Vote for a Woman President

TABLE 15.4

Whether Respondent Would Vote for a Woman for President by Year of Survey, 1972 or 1998

Would Respondent Vote for Woman?	*Yes*	*No*	*Total*
1972	1,129 (73.6%)	404 (26.4%)	1533 (100.0%)
1998	1,687 (93.6%)	116 (6.4%)	1803 (100.0%)
	Cramer's V = .27	p < .05	

73.6 percent in 1972, it rose, with minor exceptions, pretty steadily until it reached 93.6 percent in 1998. In fact, if we just focused on the years 1972 and 1998, as categories of the independent variable, "Year of Survey," and on "Yes" and "No," as categories of the dependent variable, "Would Respondent Vote for a Qualified Woman for President?," the crosstabulation of these two variables could be reduced to the bivariate (or contingency) table, Table 15.4. (We've calculated a couple of other statistics, "Cramer's V" and "p," and put them in Table 15.4 as well. We'll be getting to them soon.)

This table demonstrates that, as we hypothesized, the year of the survey is related to or associated with whether respondents answered they would vote for a qualified woman for president; it makes a difference what year it was.

Let's create a contingency table using data that Roger collected with students and colleagues (Clark, Keller, Knights, Nabar, Ramsbey, & Ramsbey, 2006) for a recent elaboration of the study of children's picture books presented in Chapter 13. In this elaboration, the researchers were interested to see, among other things, whether award-winning picture books created by Black illustrators are more likely to make female characters central than award-winning picture books created by non-Black illustrators. They collected data on, among other things, Caldecott award and honor books from 2000 to 2004 (created, as it turns out, by non-Black illustrators and authors) and Coretta Scott King award and honor books from 2000 to 2004 (created by Black illustrators and authors). The researchers hypothesized that the King books would make females more central to their books than the Caldecotts, because the King committee, unlike the Caldecott committee, is enjoined to pick "educational" books and because women have been central to the survival of the African American community in the U.S.

Here are data on two variables for the 36 relevant award-winning books from 2000 to 2004: the type of award (Caldecott or King) and whether there is a female central character:

Award	Central Female?	Award	Central Female?
1. King	Yes	19. Caldecott	No
2. Caldecott	Yes	20. Caldecott	No
3. King	Yes	21. King	No
4. Caldecott	No	22. Caldecott	No
5. Caldecott	Yes	23. Caldecott	No
6. King	Yes	24. King	No
7. King	Yes	25. King	Yes
8. Caldecott	No	26. Caldecott	No
9. Caldecott	No	27. King	No
10. King	Yes	28. Caldecott	No
11. Caldecott	No	29. King	No
12. Caldecott	Yes	30. Caldecott	Yes
13. Caldecott	No	31. King	No
14. King	No	32. Caldecott	No
15. King	No	33. Caldecott	Yes
16. Caldecott	No	34. King	No
17. Caldecott	No	35. Caldecott	No
18. King	Yes	36. Caldecott	No

Note that 15 of these 36 books were Coretta Scott King award or honor books and that 21 were Caldecott award or honor books. Of the 15 King books, seven (or 46.7 percent) have female central characters and eight (or 53.3 percent) do not. Of the 21 Caldecott books, five (or 23.8 percent) have female central characters and (76.2 percent) do not. We'll put this information into a bivariate (or contingency) table, Table 15.5.

STOP AND THINK *What percentages would you compare from the table to show that King books were more likely to have female central characters than Caldecott books?*

Note that the way we've formulated our hypothesis forces us to look at the relationship between type of book award and presence of central female character in a certain way. Using the variable language introduced in Chapter 2, we expected "type of book award" to affect "presence of central female character," rather than "presence of central female character" to affect "type of book award." In other words, "type of book award" is our independent variable and "presence of central female character" is our dependent variable. Given the formulation of our hypothesis and given the analysis of Table 15.5, we're in a good position to say that King books are more likely than Caldecott books to have central female characters.

Measures of Association

Before we leave the topic of bivariate analyses, let's note a few more important points. Table 15.4 contains some curious new symbols that are worth

TABLE 15.5

Do Caldecotts or Kings from 2000 to 2004 Have More Female Central Characters?

Award	Does Books Have Female Central Character?		
	No	Yes	Total
Caldecott	16 (76.2%)	5 (23.8%)	21 (100.0%)
King	8 (53.3%)	7 (46.7%)	15 (100.0%)
	Cramer's V = .24	p > .05	

measures of association, measures that give a sense of the strength of a relationship between two variables.

mentioning, mainly because they illustrate a whole class of others. One such class is called **measures of association,** of which "Cramer's V" is a specific example. Measures of association give a sense of the strength of a relationship between two variables, of how strongly two variables "go together" in the sample. Cramer's V can vary between 0 and 1, with 0 indicating that there is absolutely no relationship between the two variables, and 1 indicating that there is a perfect relationship. A perfect relationship exists between two variables when change in one variable is always associated with a predictable change in the other variable. The closer Cramer's V is to 0, the weaker the relationship is; the farther from 0 (closer to 1), the stronger the relationship.

STOP AND THINK

You'll note, from Table 15.5, that Cramer's V for the relationship type of award and the presence of female main characters is .24 and, from Table 15.4, that Cramer's V for year of survey and the willingness of respondents to vote for a qualified woman for president is .27. Which relationship is stronger?

Measures of Correlation

Cramer's V is a measure of association that can be used when both variables are nominal level variables (and have two categories each). Statisticians have cooked up literally scores of measures of association, many of which can be distinguished from others by the levels of measurement for which they're designed: some for nominal, some for ordinal, and some for interval level variables. A particular favorite of social science researchers is one called Pearson's r, designed for examining relationships between interval level variables. Pearson's r falls in a special class of measures of associations: It's a **measure of correlation.** As such, it not only provides a sense of the strength of the relationship between two variables, it also provides a sense of the *direction* of the association. When variables are intervally (or ordinally) scaled, it is meaningful to say that they can go "up" and "down," as well as to say that they vary from one category to another.

measures of correlation, measures that provide a sense not only of the strength of the relationship between two variables, but also of its direction.

Suppose, for instance, that you had data about the education and annual income of three people. Suppose further that the first of your persons had 12 years of education and made $15,000 in income, the second had 13 years

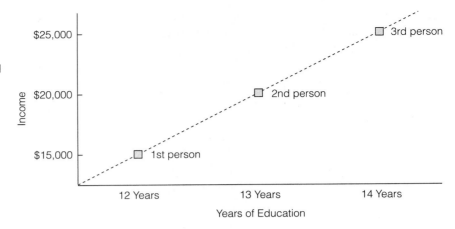

of education and made $20,000, and the third had 14 years of education
and made $25,000. In this case, not only could you say that the two vari-
ables (education and income) were related to each other (that every time education changed, income changed), but you could also say that they were
positively or *directly* related to each other—that as education rises so does income. You might even be tempted to graph such a relationship, as we have
done in Figure 15.4.

One feature of our graphical representation of the relationship between
education and income is particularly striking: All the points fall on a straight
line. One striking feature of this line is that it rises from the bottom left of
the graph to the top right. When such a situation exists (that is, when the
data points from two variables fall on a straight line that rises from bottom
left to top right), the correlation between two variables is said to be perfect
(because all points fall on a line) and positive (because it goes from the bottom left up to the top right), and the Pearson's r associated with their relationship is 1.

STOP AND THINK   *Suppose your sample included three individuals, one of whom had 14 years of education and made $15,000, one of whom had 13 years of education and made
$20,000, and one of whom had 12 years of education and made $25,000. What
would a graphical representation of this data look like? Can you imagine what the
Pearson's r associated with this relationship would be?*

If, on the other hand, the data points for two variables fell on a line that
went from the "top left" of a graph to the "bottom right" (as they would for
the data in the Stop and Think exercise), the relationship would be perfect
and negative and the Pearson's r associated with their relationship would be
−1. Thus, Pearson's r, unlike Cramer's V but like all other measures of correlation, can take on positive *and* negative values, between 1 and −1.

In general, negative values of Pearson's r (less than 0 to −1) indicate that
the two variables are *negatively* or *indirectly* related, that is, as one variable
goes up in values, the other goes down. Positive values of Pearson's r (greater

FIGURE 15.5

Idealized Scattergrams of Relationships of r = .70 and –.70

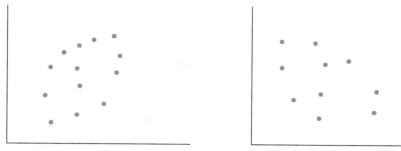

r about equal to .70   r about equal to – .70

than 0 to 1) indicate that the two are directly related—as one goes up, the other goes up. In both cases, r's "distance" from 0 indicates the strength of the relationship: The farther away from 0, the stronger the relationship. Thus, a Pearson's r of .70 and a Pearson's r of –.70 indicate relationships of equal strength (quite strong!) but opposite directions. The first (r = .70) indicates that as one variable's values go up, the other's also go up (pretty consistently); the second (r = –.70) indicates that as one's values go up, the other's go down (again pretty consistently). Figure 15.5 provides a graphical representation (known as a *scattergram*) of what relationships (with many data points) with r's of .70 and –.70 might look like. Notice that, although in neither case do the points perfectly conform to a straight line, in both cases they all hover around a line. r's of .30 and –.30 would, if you can imagine them, conform even less well to a straight line.

STOP AND THINK *Would you expect the association between education and income for adults in the United States to be positive or negative?*

STOP AND DO *Try doing a correlation analysis for the association between years of education and income for adults in the United States in 1998. Go back to* http://sda. berkeley.edu/archive.htm, *enter the most recent GSS Cumulative Data File, start the "correlation matrix" action, enter the variable names "educ" and "rincom98" into the first two input boxes before hitting "run correlations." Check the correlation in the middle of the page. Look at the intersection of a row headed by "educ" and a column headed by "rincom98" (or a row headed by "rincom98" and a column headed by "educ") and read the correlation coefficient (the Pearson's r).*

We did the analysis in the Stop and Do and found what we hope you did as well, that the Pearson r for the association between education and income in the 1998 GSS sample was .29. As it turns out, the association is a positive one (the positive "sign" of .29 indicates a direct relationship), but the relationship is nowhere near a perfect one (.29 is closer to 0 than to 1)—so one would not expect the data points to conform very well to a straight line.

STOP AND THINK *Suppose the Pearson's r for the relationship between years of education and incomes had proven to be –.67, instead of .29. How would you describe the relationship?*

We could tell you more about correlation analysis, but any standard statistics book could do so in more detail. We think you'll find correlation analysis a particularly handy form of analysis if you ever want to study the associations among interval level variables.

Inferential Statistics

We draw your attention to a final detail about Table 15.4 before we conclude our discussion of bivariate relationships. On the same row as Cramer's V in the table is a funny-looking set of symbol, "p < .05." This symbol, standing for "probability is less than .05," enables the reader to make an inference about whether a relationship, like the one shown in the table for the sample (of 3336 respondents), exists in the larger population from which the sample is drawn. This "probability" is estimated using what we've called inferential statistics, in this case chi-square, perhaps the most popular inferential statistic used in crosstabulation analysis. There are almost as many inferential statistics as there are measures of association and correlation, and a discussion of any one of them stands outside the scope of this chapter. Inferential statistics are properly computed when one has a probability sample (see Chapter 5). They are designed to permit inferences about the larger populations from which probability samples are drawn.

You might recall from Chapter 5 however, that, even with probability sampling, it is possible to draw samples whose statistics misrepresent (sometimes even badly misrepresent) population parameters. Thus, for instance, it is possible to draw a sample in which a strong relationship exists between two variables even when no such relationship exists in the larger population. Barring any other information, you'd probably be inclined to infer that a relationship also exists in the larger population. This kind of error—the one where you infer that a relationship exists in a population when it really doesn't—is called a Type I error. Such an error could cause serious problems, practically and scientifically.

Social scientists are a conservative lot and want to keep the chances, or probability, of making a Type I error pretty small—generally lower than 5 times in 100, or "p < .05." When the chances are greater than 5 in 100, or "p > .05," social scientists generally decline to take the risk of inferring that a relationship exists in the larger population.

STOP AND THINK *When "p < .05," social scientists generally take the plunge and infer that a relationship, like the one in the sample, exists in the larger population. Would social scientists take such a plunge with the relationship between year of survey and respondents' willingness to vote for women for president as displayed in Table 15.4?*

The probability of making a Type 1 error, the "p," is related to the size of the sample. The larger the sample, the easier it is to achieve a p of less than .05, and therefore it is easier to feel comfortable with an inference that a relationship exists in the larger population. Consequently, it is not surprising that the association between year of survey and respondents' willingness to vote for women for president as displayed in Table 15.4 is associated with a p less than .05, even though the strength of the association (as measured by a Cramer's V of .29) is not terribly strong. The number of cases in the sample is 3,336, after all.

Measures of association, measures of correlation, and inferential statistics are not an exotic branch of statistical cookery. They are the "meat and potatoes" of quantitative data analysis. We hope we've presented enough about them to tempt you into learning more about them. We'd particularly recommend Marija Norusis' *The SPSS Guide to Data Analysis* (2005), which, in addition to offering a pretty good guide to SPSS (the statistical package we used to generate several tables and figures), offers a good guide to the kinds of statistics we've referred to here.

Qualitative Data Analysis

> Qualitative data are sexy. They are a source of well-grounded, rich descriptions and explanations of processes in identifiable local contexts. . . . Then, too, good qualitative data are more likely to lead to serendipitous findings and to new integrations. . . . Finally, the findings from qualitative studies have a quality of "undeniability." Words, especially organized into incidents or stories, have a concrete, vivid, meaningful flavor that often proves far more convincing to a reader . . . than pages of summarized numbers. (Miles and Huberman, 1994: 1)

As we indicated at the beginning of this chapter and as the quotation from Matthew Miles and Michael Huberman (1994) suggests, a major distinction between qualitative and quantitative data analyses lies in their products: words, "especially organized into incidents or stories," on the one hand, and "pages of summarized numbers," on the other. Even though, as Miles and Huberman argue, the former are frequently more tasty to readers than the latter, the canons, or recipes, that guide qualitative data analysis are less well defined than those for its quantitative counterpart. As Miles and Huberman elsewhere observe, "qualitative researchers . . . are in a more fluid—and a more pioneering—position" (1994: 12). There is, however, a growing body of widely accepted principles used in qualitative analysis. We turn to these in this chapter.

If the outputs of qualitative data analyses are usually words, the inputs (the data themselves) are also usually words—typically in the form of extended texts. These data (the words) are almost always derived from what the researcher has observed (for example, through the observation techniques discussed in Chapter 11), heard in interviews (discussed in Chapter 10), or found in documents (discussed in Chapter 13). Thus, the data analyzed by Adler and Adler were their observations and interviews surrounding children's afterschool activities, and the data analyzed by Enos were the words of incarcerated women. The outputs of both data analyses were the essays themselves (see Chapters 10 and 11).

Social Anthropological versus Interpretivist Approaches

Although there might be general agreement about the nature of the inputs and outputs of qualitative analyses (words), there are nonetheless some fundamental disagreements among qualitative data analysts. One basic division involves theoretical perspectives about the nature of social life. To illustrate, and following Miles and Huberman (1994: 8ff.), look at two general approaches: a social anthropological and an interpretative approach. Social anthropologists (and others, like grounded theorists and life historians) believe that there exist behavioral regularities (for example, rules, rituals, relationships, and so on) that affect everyday life and that it should be the goal of researchers to uncover and explain those regularities. Interpretivists (including phenomenologists and symbolic interactionists) believe that actors, including researchers themselves, are forever interpreting situations, and that these, often quite unpredictable, interpretations largely affect what goes on. As a result, interpretivists see the goal of research to be their own (self-conscious) "accounts" of how others (and they themselves) have come to these interpretations (or "accounts"). The resulting concerns with lawlike relationships, by social anthropologists, and with the creation of meaning, by interpretivists, can lead to distinct data-analytic approaches. Given this difference, our discussion in this chapter will generally be more in line with the beliefs and approaches of the social anthropologists (and those with similar perspectives) than it is with those of the interpretivists. Nevertheless, we believe that much of what we say applies to either approach, and that their differences can be overstated. Thus, although the Adlers clearly sought to discover lawlike patterns in the experience of afterschool activities (remember their concept of "funneling"?), they were nonetheless sensitive to variation in the interpretations or accounts given by children, and adults, of those activities.

STOP AND THINK   *When the Adlers pursued lawlike patterns in the experience of afterschool activities, they were doing what the social anthropologists advocate. What about when they looked to the accounts given by children and adults of those activities?*

Does Qualitative Data Analysis Emerge from or Generate the Data Collected?

In qualitative field studies, analysis is conceived as an *emergent* product of a process of gradual induction. Guided by the data being gathered and the topics, questions, and evaluative criteria that provide focus, analysis is the fieldworker's *derivative ordering* of the data. (Lofland and Lofland, 1995: 181)

In ethnography the analysis of data is not a distinct stage of the research. It begins in the pre-fieldwork phase, in the formulation and clarification of research problems, and continues into the process of writing up. (Hammersley and Atkinson, 1983: 174)

These quotations from John Lofland and Lyn Lofland (1995) and Martyn Hammersley and Paul Atkinson (1983) seem to point to another debate concerning qualitative analysis: whether qualitative analysis *emerges from* or *generates* the data collected. At issue seems to be, to paraphrase the more familiar chicken-or-egg dilemma, the question of which comes first: data or ideas about data (for example, theory). Lofland and Lofland stress the creative, after-the-fact nature of the endeavor by describing it as a "process of gradual induction"—of building theory from data. Hammersley and Atkinson emphasize how data can be affected by conceptions of the research topic, even before one enters the field. These positions, however different in appearance, are, in our opinion, both true. Rather than being in conflict, they are two sides of the same coin. The Adlers, in discussing their own practice in studying afterschool activities (see Chapter 11), have told us about their efforts to immerse themselves in their setting first, *before* examining whatever literature is available on their topic, to better "grasp the field as members do" without bias. But *something* compelled their interest in afterschool activities, and whatever that was surely entailed certain preconceptions. Sandra Enos (the author of the focal research report in Chapter 10) is much more explicit about her preconceptions and, in conversation, has confided that one of her expectations before observing mother-child visits was that she'd see mothers attempting to create "quality time" with their children during the visits. Still, she remained open to the "serendipitous findings" and "new integrations" mentioned by Miles and Huberman. When she found mothers casually greeting their kids with "high-fives," rather than warmly embracing them, as she'd expected, she dutifully took notes. Because Enos has been explicit about various stages of her data analysis, and because she's been willing to share her field notes with us, we use her data to illustrate our discussion of qualitative data analysis in this chapter.

The Strengths and Weaknesses of Qualitative Data Analysis Revisited

Before we actually introduce you to some of the techniques that are increasingly associated with qualitative data analysis, we want, once again, to suggest one of its great strengths. As Gary Fine and Kimberly Elsbach (2000) suggest, qualitative data can produce theories that "more accurately describe real-world issues and processes than do quantitative data." Because qualitative data usually come from some combination of the observation or interview of real-world participants, they are more likely to be grounded in the immediate experiences of those participants than in the speculations of researchers. Thus, rather than simply handing out a set of closed-ended questions to children, and parents of children, who participated in organized afterschool activities, Patricia Adler and Peter Adler (see Chapter 11) watched them as they participated in those activities. Consequently, they didn't need to worry, quite as much, about whether they were asking the right questions (and supplying the right answers, as they would if they'd used closed-ended questions). They could simply observe participants as they became more and more deeply involved, in most cases, in relatively specialized afterschool lives, and ask them what those lives meant to them. The resulting capacity of qualitative data to yield insights into the longer-term dynamics of behavior is also a great advantage in producing theories that "describe real-world . . . processes." We've mentioned earlier (see, for instance, Chapter 11) that one of the disadvantages of qualitative data is the questionable generalizability of the theories they lead to. Because Adler and Adler studied a predominantly white, upper-middle-class subpopulation, for instance, the extent to which their findings may be typical is unclear. In general, though, we feel that the richness and accuracy of theories generated from qualitative data analysis are well worth the trouble, especially if their limitations are kept in mind and, perhaps, investigated further through future research.

Are There Predictable Steps in Qualitative Data Analysis?

Quantitative data analysis often follows a fairly predictable pattern: Researchers, after they have coded their own data or acquired computer-ready available data, almost always begin by doing some sort of univariate analyses, perhaps followed by bivariate and multivariate analyses. The "steps" involved in executing good qualitative analyses, however, are much less predictable, and a lot "more fluid," as Miles and Huberman suggest. Miles and Huberman prefer to see qualitative analysis as consisting of "concurrent flows of activity," rather than as "steps." They dub these flows "data reduction," "data display," and "conclusion drawing/verification," and depict the interactions among these flows as we do in Figure 15.6.

This depiction of qualitative data analysis involves several notable assumptions. First, even data collection is not immune to the "flows" of data analysis, being itself subject to the (perhaps tentative) conclusions a researcher entertains at various stages of a project (note the arrow from "conclusions" to "data collection"). Enos had expected inmate mothers to greet their children with warm embraces (even before she entered the field), and so focused attention on the gestures of greeting and, as you'll see, collected data about them. Second, Figure 15.6 suggests that data reduction and data display (both discussed later) are not only products of data collection but of one another and of conclusions, even if such conclusions are not yet fully developed. Thus, Enos' coding of her data (a data reduction process described later) not only contributed to her final conclusions about mothering in prison, but might have been itself a product of earlier, perhaps more tentative, conclusions she'd reached. Finally, the figure emphasizes that the conclusions a researcher draws are not just products of earlier flows (data reduction and data display) but continually inform both of the other "flows," as well as the data collected in the first place.

We'd like to emphasize that one danger of the model depicted in Figure 15.6 is that it could give the impression that qualitative data analysis

FIGURE 15.6

Components of Data
Analysis: Interactive Model
from Miles and Huberman:
12, 1994. Used with
permission.

invariably follows a predictable pattern of data collection and transcription, followed by data reduction (particularly coding and memoing), followed by data displaying, followed by conclusion drawing and verification. It's not that simple. We think, however, that the advantages of such a model outweigh its dangers. One important advantage is that it gives new (and even experienced) practitioners a sense of order in the face of what otherwise might feel like terribly uncharted and, therefore, frightening terrain. Angela Lang (in Clark and Lang, 2001: 1) speaks of the anxiety she experienced when confronted with the demands of generating theory from qualitative data for an undergraduate project: "I soon realized I had to reinvent my creative side . . . I was nervous that I would discover that I wasn't creative at all." But, once she'd finished the project, she reflected on the model implicit in Figure 15.6 as a source of inspiration and comfort. She said that, by doing qualitative data analysis, she "had gained a tremendous amount of confidence in [her]self." And that, "If I were to do a qualitative project again, I'd probably try to organize my work in much . . . [the same way]. I'd probably plan to create fieldnotes, code data, memo, do data displays, review literature, write drafts of the final paper . . . and hope that the ideas and research questions began to float into my head, as they did this time" (Clark and Lang, 2001: 5–6).

Data Collection and Transcription

We've discussed at some length the process of collecting qualitative data in earlier chapters: through qualitative interviewing (Chapter 10), observation techniques (Chapter 11), and studying the content of texts (Chapter 13). So far, however, we haven't shown you any actual data. We'd like to present a small portion of Enos' fieldnotes, as she composed them on the evening after an interaction with one mother in the prison parenting program. These notes were based on observation and conversation in the off-site parenting program, where women got together with their children. The excerpt in Figure 15.7 is from nearly 150 pages of field notes Enos took before she engaged in the qualitative interviewing she reports on in the essay within Chapter 10.

FIGURE 15.7

A Segment from Enos' Field Notes

I spoke to a woman who had remained silent in the group. She was SITTING ALONE AT A TABLE. I ASKED if I could talk with her. She has six children, two of her children will be visiting today. Her mother has always had her kids. Last year her mother passed away and now her sister is taking care of the kids. When she's released, she'll have responsibility for the whole bunch. She says it was hard to see your kids while you were in jail but that if you were DCYF[1] involved sometimes it was easier. (This has to do with termination of rights.) She remarked that it was important not to be a "stranger" to your kids while you were in jail. It made everything worse. I asked about telling your kids where you are. She said she just straight out told them and they accepted. "Mommy's in jail. That's it." She thought having such direct experience with jail that maybe her kids would "avoid" it when they got older. I asked her when she'd be released.

[1] Department of Children, Youth & Families.

STOP AND THINK   *One of Enos' expectations had been that women would greet their children with hugs. How did this woman greet her son? How did he greet her?*

The physical appearance of these fieldnotes brings up the issue of computer usage in qualitative data analysis. Enos did all her written work, note collection, and analysis with an elementary word-processing program—a practice that is extremely common these days. Enos did not use one of the increasingly available software packages (for example, Kwalitan, MAX, QUALPRO) designed to facilitate the processing of qualitative data, until she entered the qualitative interview phase of her work when she used a package called HyperResearch. To those students who think they might be interested in software for analyzing qualitative data, we recommend Weitzman and Miles (1995), Richards and Richards (1994), Barry (1998), and Dohan and Sanchez-Jankowski (1998) for discussions of what's available. We believe that much good qualitative analysis can be done without the use of anything more sophisticated than a word processor—and are inclined to point out that much has been achieved with nothing more sophisticated than pen and paper. (See Box 15.2 for some sense of the advantages and disadvantages of computer-assisted qualitative data analysis software.)

FIGURE 15.7

(continued)

In a few months, but she'd be going to a drug treatment program. She's been in prison many times before and this time she's going to do things right. She's made a decision to change, to do something new.

She is very happy with this program. She likes the fact that there are no officers around, that kids can run and scream, that there is an extended time out of the joint and that she can spend more time with her kids.

Her children arrived, a boy five years old or so and his older sister. They walked over to us. The mother greeted her son, "Hey, Dude! How you doing?" He didn't say anything. The woman asked for the children's coats and hung them on the chair backs. She introduced me to the kids, who were quiet and beautiful, very well cared for. I complimented her on these kids and excused myself so she could be alone with them.

Before we return to Enos' transcribed fieldnotes, however, we'd like to point to one trend in computer-assisted qualitative data analysis software (CAQDAS) that shows great promise: the direct analysis of audio-recorded data using computer software. The transcription of audiotapes, as mentioned in Chapter 10, is enormously time consuming, sometimes taking, in the case of interviews, up to four hours for an hour's worth of interview. Gibson et al. (2005) and Hutchinson (2005) have used Atlas.ti and GoldWave software, respectively, to analyze sound files, and found them useful. But Gibson et al. (2005), in particular, claim that there remain substantial obstacles to the direct integration of audio materials into qualitative analysis: e.g., absence of good enough recording technologies and the insufficient development of interfaces between recording devices and computers. But this is one technology that, in the near future, could make the analysis of audio data faster and cheaper than before. For now, though, back to Enos' fieldnotes . . .

The physical appearance of these fieldnotes might be the first thing that strikes you. Enos uses only half a page, the left column, for her fieldnotes in anticipation of a need to make notes about the notes later on. (See the data reduction section later.) Leaving room for such notes about notes is excellent practice. Enos didn't have an interview guide at this stage of her investigation, but she evidently had certain questions in mind as she approached

BOX 15.2

The Debate About CAQDAS (Computer-Assisted Qualitative Data Analysis Software)

Advocates of CAQDAS claim that it will

- Speed up and enliven coding (see the section on "coding" later in this chapter)

- Offer a clear-cut structure for writing and storing memos (see the section on "memoing" later in this chapter)

- Provide a "more complex way of looking at the relationships in the data" (Barry, 1998: 2.1)

Opponents see the possibility that CAQDAS will

- Distance analysts from their data

- Lead to qualitative data being used quantitatively

- Generate an unhealthy sameness in the methods used by analysts

- Continue to favor word-processed text over other forms of data such as sketches, maps, photos, video images, or recorded sound (Dohan and Sanchez-Jankowski, 1998: 483)

Barry (1998) gives a relatively balanced appraisal of these positions, and offers specific insights into two of the most popular varieties of CAQDAS, NUDIST, and Atlas.ti. You can read it on the Web at www.socresonline .org.uk/socresonline/3/3/4.html

this woman. Her notes indicate that she asked about how the woman explained her incarceration to her children, and about her expectations for release and her current child-care arrangements.

Enos' interest in the aforementioned questions indicates that she's probably engaged in a fair amount of what Miles and Huberman (1994: 16) call **anticipatory data reduction**—that is, decision making about what kinds of data she'd like to collect. Admittedly, her research strategy (more or less "hanging out" at an off-site parenting program for female prisoners) is pretty loose at this stage.

There is quite a bit of debate about the relative merits of "loose" and "tight" (those that entail explicit conceptual frameworks and standard instruments) research strategies in field research. Those advocating relatively loose strategies or "designs" (for example, Wolcott, 1982) point to the number of questions, and answers, that can be generated through them. Susan Chase, for instance, in the study mentioned in Chapter 10, was primarily interested in getting female school superintendents to tell their life stories and was willing to use any question that stimulated such stories. She was willing to "hang out" during these interviews, though she does confess that she

anticipatory data reduction, decision making about what kinds of data are desirable that is done in advance of collecting data.

eventually found asking female superintendents to describe, generally, their work histories, was a pretty good way to induce "lively, lengthy, and engrossing stor[ies]" (Chase, 1995: 8).

But the advantages of relatively tight strategies, especially for saving time spent in the field, are worth noting. Enos developed an appreciation of tight strategies during the course of her fieldwork. In personal communication, she's indicated how tempting it is to pursue every idea that occurs in a setting:

> So many questions occurred to me after I left the site. I would consider what women had said to me and then try to figure out if that was related to race/ethnicity, to class, to previous commitments to prison, to number of children, to street smarts, to sophistication about the system, and so on and so on. (Enos, personal communication)

In the end, she realized that she couldn't pursue all those questions, that she needed, in her words, "to keep on track." She claims that a most important task was to continually sharpen her focus as the fieldwork went on and that this meant "finally coming to the conclusion that you can't look at everything" (personal communication).

By the time Enos entered the interview stage of her project, she'd developed a very tight design indeed, having, among other things, composed the interview guide shown at the end of her report in Chapter 10. This guide led Enos to ask very specific questions of her respondents, about their age of first incarceration, and so on.

Data Reduction

data reduction, the various ways in which a researcher orders collected and transcribed data.

Qualitative **data reduction** refers to the various ways a researcher orders collected data. Just as questioning 1,000 respondents in a closed-ended questionnaire can generate more data on most variables than the researcher can summarize without statistics, fieldnotes or transcripts also can generate volumes of information that need reduction. Some researchers can do a reasonably good job of culling out essential materials, or at least can tell compelling stories, without resorting to systematic data reduction procedures in advance of report writing. But most can't, or don't. Especially in long-term projects with ongoing data collection, most researchers engage in frequent data reduction. In the following section, we'll focus on two common data reduction procedures: coding and memoing. In doing so, we'll be emphasizing one thing implicit in Figure 15.6: Data reduction is affected by the kinds of data you collect but is also affected by, and affects, the conclusions you draw.

Coding

Coding, in both quantitative and qualitative data analysis, refers to assigning observations, or data, to categories. In quantitative analysis, the

FIGURE 15.8

Enos' Coding of Fieldnotes

Fieldnotes	*Codes*
I spoke to a woman who had remained silent in the group. She was sitting alone at a table. I asked if I could talk with her. She has six children, two of her children will be visiting today. Her mother has always had her kids. Last year her mother passed away and now her sister is taking care of the kids. When she's released, she'll have responsibility for the whole bunch. She says it was hard to see your kids while you were in jail but that if you were DCYF involved sometimes it was easier. (This has to do with termination of rights.) She remarked that it was important not to be a "stranger" to your kids while you were in jail. It made everything worse. I asked about telling your kids where you are. She said she just straight out told them and they accepted. "Mommy's in jail. That's it." She thought having such direct experience with jail that maybe her kids would "avoid" it when they got older.	women as birthgivers, not caretakers motherhood management permanent substitute temporary substitute

researcher usually assigns the data to categories of variables she has already created, categories that she frequently gives numbers to in preparation for computer processing. In qualitative analysis, coding is more open-ended because both the relevant variables and their significant categories are apt to remain in question longer. In any case, coding usually refers to two distinct, but related, processes: The first is associating words or labels with passages in one's fieldnotes or transcripts; the second is collecting similarly labeled passages into files. The goal of coding is to create categories (words or labels) that can be used to organize information about different cases (whether situations, individuals, or so on). When you don't have much of a framework for analysis, as Enos didn't initially, the codes (words or labels), at least initially, can feel a little haphazard and arbitrary. They can nonetheless be useful guides to future analysis. Thus, Enos' own coding of the initial paragraph of her notes (from Figure 15.7) looked something like what appears in Figure 15.8.

Here we see Enos, in her first informal interview, developing codes that will appear again and again in the margins of her fieldnotes: codes that permit her to characterize how social mothering is managed while a biological mother is in prison. The codes "permanent substitute" and "temporary substitute," referring to this particular woman's mother and sister, respectively, become categories for classifying the kinds of mothering that all other children receive.

STOP AND THINK   *Suppose you wanted to compare all the data that Enos coded with the words "temporary substitute." What might you do to facilitate such a comparison?*

Assigning a code to a piece of data (that is, assigning "temporary substitute" to this woman's sister or "permanent substitute" to her mother) is just the first step in coding. The second step is physically putting the coded data together with other data you've coded in the same way. There are basically two ways of doing this: the old-fashioned way, which is to put similar data into a file folder, and the new-fashioned way, which is to do essentially the same thing using a computer program that stores the "files" in the computer itself. Old-fashioned manual filing is simple enough: you create file folders with code names on their tabs, then place a copy of every note you have coded in a certain way into the folder with the appropriate label. In these days of relatively cheap photocopying and the capacity of computers to make multiple copies, you can make as many copies of each page as there are codes on it, and slip a copy of the page into each file with the corresponding code. (In the bad old days, this process required relatively cumbersome manipulations of carbon paper and scissors.) Thus, Enos might have placed all pages on which she placed the codes "temporary substitute" into one folder and all pages on which she wrote the code "permanent substitute" into another. Later in her analysis, when she wanted to see all the data she'd coded in the same way, she'd just have to pull out the file with the appropriate label.

The logic of the new-fashioned, computer-based filing is the same as that of the old-fashioned one. The newer way requires less paper and no actual file folders (and can therefore result in a great savings of money and trees), but it does generally require special computer software (so the initial outlay can be substantial). "Code-and retrieve programs," for instance, like Kwalitan, MAX, and NUDIST, help you divide notes into smaller parts, assign codes to the parts, and find and display all similarly coded parts (Miles and Huberman, 1994: 312). Computer-based filing also makes revising codes simpler. For those who are interested, we recommend Weitzman and Miles (1995), Richard and Richards (1994), Barry (1998), and Dohan and Sanchez-Jankowski (1998) for discussions of what's available.

Types of Coding

One purpose of coding is to keep facts straight. When you code for this purpose alone, as you might be doing if you were to assign a person's name to all paragraphs about that person, you'd be engaged in purely *descriptive coding*. Descriptive codes can emerge from the data, and usually do from when "loose" research designs are employed. Initially Lang (Clark and Lang, 2001: 11), who looked at respondents' open-ended reflections to an image of Sojourner Truth, found herself coding the reflections as "complex" or "simple," with no particular idea where she might be going with those codes. Sometimes, and especially when research designs are of moderate "tightness," as in Levendosky, Lynch, and Graham-Bermann's (2000) interview study of women's perceptions of abuse, the codes can be suggested by participants themselves. Responses, in this study, to the question, "How do you think that violence you have experienced from your partner has affected your parenting of your child?," led naturally to seven categories, such as "no impact on parenting" and "reducing the amount of emotional energy and time available," into which all accounts could be placed. Sometimes, especially when a very "tight" research design is employed, descriptive codes actually exist before the data are collected. Thus, Dixon, Roscigno, & Hodson (2004: 3–33) report that even before their coders read the 133 book-length ethnographies of workplaces in the U.S. and England, they'd been prepared to look for and code evidence of strikes, union organization, internal solidarity, and a legacy of workplace activism. The coders found 21 workplaces (out of the 133) that had had strikes during the period of observation, 67 that were unionized, 78 that exhibited average to strong levels of solidarity, and 16 that had a legacy of frequent strike activity.

Because the basic goal of coding is to advance your analysis, you'd eventually want to start doing a more *analytic* type of coding. Enos' simultaneous development of the two codes "permanent substitute" and "temporary substitute" is an example of analytic coding. Kathy Charmaz calls this preliminary phase of analytic coding *initial coding* and defines it as the kind through which "researchers look for what they can define and discover in the data" (Charmaz, 1983: 113). This is also a point when one's early, provisional conclusions (see Figure 15.6) can inform one's choice about codes.

Eventually, according to Charmaz, initial coding gives way to *focused coding,* during which the initial codes themselves become the subject of one's coding exercise. At this stage, the researcher engages in a selective process, discarding, perhaps, those codes that haven't been terribly productive (for example, Enos never used the "women as birthgivers" code after its first appearance) and concentrating or elaborating on the more enlightening ones. After Enos saw all the ways in which women found temporary substitutes, for instance, she began to use variations in those ways—for example, relatives, foster care, friends—as codes, or categories, of their own.

Memoing

memos, more or less extended notes that the researcher writes to help herself or himself understand the meaning of codes.

data displays, visual images that summarize information.

All data analyses move beyond coding once the researcher tries to make sense of the coded data. In quantitative analyses, the researcher typically uses statistical analyses to make such sense. In qualitative analyses, other techniques, such as **memoing** and **data displaying,** are used. Memos are more or less extended notes that the researcher writes to help understand the meaning of coding categories. Memos can vary greatly in length (from a sentence to a few pages) and their purpose, again, is to advance the analysis of one's data (notes). An example (labeled distinctly as a theoretical note) of a memo appears about 100 pages after Enos' write-up of the encounter with

FIGURE 15.9

Illustration of Memo from Enos

Theoretical note
Consider the work on the breakdown
of the family and compare that to
the efforts of these families to
keep their children in their midsts.
These children are being raised by
persons other than their birth
mothers but they are well dressed,
taken care of, appear healthy, and
so forth. If they are on welfare,
there appears to be another source
of support.

the woman whose children had been (permanently) cared for by her mother until the mother died, but were now being (temporarily) cared for by her sister. It appears to be a reflection on all the material that's been coded "temporary" and "permanent substitute" throughout the fieldnotes. This memo is shown in Figure 15.9.

STOP AND THINK   *Enos tells herself to think about the literature on family breakdown, something that the mother's imprisonment can also lead to. What do you think are some of the more familiar causes of family breakdown?*

Notice that Enos makes meaning by drawing an analogy between family separations brought about by imprisonment and those brought about by divorce, separation, death, and so on. She's focusing her attention on what happens to the care of children when they lose daily access to a parent (in this case, the mother). This has clearly become a central concern for future consideration. With this memo as a springboard, it's not surprising that child care and relationships with family members are a key issue in the research she reports on in Chapter 10.

Because fieldwork and coding can be very absorbing, it's easy to lose track of the forest for the trees. The practice of memoing, of regularly drawing attention to the task of analysis through writing, is one way that many qualitative researchers force themselves to "look for the big picture." Memos can be thought of as the building blocks of the larger story one eventually tells. Some memos, such as Enos', still need some refinement (for example, through the reading of the literature on family breakdown), but others qualify for inclusion in the final report without much alteration. Some, especially those that are memos about linkages among memos, can provide the organizing ideas for the final report. Memos are a form of writing and have the potential for creative insight and discovery that writing always brings. One can easily imagine that Hannah Frith and Celia Kitzinger's (1998) key theoretical question came from a memo about whether qualitative data are transparent reflections of reality or context-dependent constructions of participants. In their own analysis of young women in focus groups talking about how they refuse men sex, Frith and Kitzinger decided that the talk needed to be seen as generative and perhaps reflective of group constructions. The women spoke of their efforts to protect the feelings of the man they'd refused. Frith and Kitzinger saw the women's talk as "constructing" men as needy, and perhaps weak, and themselves as knowledgeable agents, in opposition to traditional stereotypes of men as sexual predators and women as vulnerable prey. The talk was not a "transparent" window on what went on during the refusals, Fritz and Kitzinger conclude. Roger (Clark, 2001b) was also made aware of such group constructions when he asked faculty at a particularly successful urban middle school to talk, in focus groups, about how they'd managed to make differences in test performance between high-income and low-income students "disappear." His memos led to this awareness as he tried to interpret the focus-group interactions in which one faculty member would come up with an explanation and others would agree vigorously. Thus, for instance, one faculty member mentioned at one point that she maintained high standards for all of her students, both high- and low-income. When others voiced immediate agreement (and still others independently wrote of high standards in response to an open-ended question), Roger didn't necessarily feel he'd "seen" what went on in individual classrooms, but he did feel he had evidence of a common definition of faculty values at this school.

Memos can be potent theoretical building blocks, but they also can be vital methodological building blocks. Enos tells us, for instance, that memos were a particularly useful way for her to figure out what she needed to "learn next" during her study, and how to actually learn it (personal communication). One memo (presented in Figure 15.10) shows Enos deciding to adopt a more formal interview strategy than she'd used before. This memo constituted a transition point between the fieldwork and the interview stages of Enos' study.

Data Displays

If memos, in their relative simplicity, are analogous to the relatively simple univariate examinations of quantitative analyses, data displays can be seen as a rough analogue to the more complex bivariate and multivariate analyses used there. Miles and Huberman (1994) and Williamson and Long (2005) are particularly adamant about the value of data displays for enhancing both the processes of data reduction, admittedly a means to an end, and conclusion making and verifying, the ultimate goals of data analysis.

A data display is a visual image that summarizes a good deal of information. Its aim is to present the researcher herself with a visual image of what's going on and help her to discern patterns. Miles and Huberman's figure (our Figure 15.6) itself can be seen as an example of such a display, summarizing the interaction among various processes in qualitative analysis.

The crucial feature of data displays is that they are tentative summaries of what the researcher thinks he knows about his subject. Therefore data

FIGURE 15.10

Methodological Memo
from Enos

Methodological note
While it has been easy to use my role as researcher and "playmate" to gain some information about mothers and children in prison, it has also been frustrating from a data collection perspective. I am involved in some long and wide-ranging conversations with the women and trying to remember the tone, the pacing, and the content of discussion while engaging in these as an active member. I can disappear for a few minutes to an adjoining room to scribble notes but this is difficult.
I focus on the points that are most salient to my interests but can get sidetracked because so much is of interest to me. A more regimented conversation in an atmosphere that is more controlled than this one would probably yield "better" notes.

displays permit him to examine what he thinks by engaging in more data examination (and reduction) and, eventually, to posit conclusions. Enos, for instance, tends to use the space on the right half of her fieldnote pages not only for coding but also for data displays, often in the form of exploratory typologies about the kinds of things she's seeing. Thus, next to notes she's taken about one mother, "Pam," and her interaction with her children, "Tom" (a baby) and "Joy" (a 3-year-old), we find a display (in Figure 15.11) about how mothers seem to manage their weekly interactions with their children in the parenting program.

Enos' coding of the first paragraph (referring to Pam's attempt to normalize her interaction with Tom and Joy) with a label that she's used before ("attempt at normalization") clearly has pushed her, by the time she rereads the second paragraph, to play with a list of ways she's seen mothers interact with their children at the parenting program. (Nothing in the second paragraph, in fact, is likely to have sparked this reflection.) This typology is the kind of data display that Miles and Huberman advocate. It pictorially summarizes observations Enos has made about the varieties of mothering behavior she's seen, and its presence facilitates her evaluation of its adequacy for subsequent presentation. Something like this display guided Adler and Adler (see their report in Chapter 11) when they laid out the various stages (recreational, competitive, and elite) of commitment to afterschool activities shown by the children they studied.

The forms in which data displays occur are limited only by your imagination. Figure 15.6 is a flow chart; Enos' display is a more conventional-looking

FIGURE 15.11

Enos' Emergent Typology of Mothering Behavior in the Parenting Program

There were a number of infants visiting today. One of them, Tom, 2, is the son of Pam who also has a three year old girl. Pam plays actively with her children, is with them on the carpeted area, holding the baby and roughhousing with her daughter. Joy is an active and very outgoing kid who asks to have her hair fixed by her mom who is struggling to feed the baby and pay attention to her daughter.

attempt at normalization

The baby is tiny for his age and pale. These children are being taken care of by Pam's boyfriend's mother. Pam has a problem with cocaine and got high the day before Tom was born. She says that before he was born all she was interested in was getting high and the "chase." During her pregnancy with Joy, she quit smoking and was drug free.

Setting management
Mother Behavior
1) no playing with child
2) active playing
3) sporadic conversation
4) steady conversation
5) steady interaction
6) kid in other's care

2 All names are pseudonyms.

typology. Miles and Huberman (1994) also advocate using various kinds of matrices, something that Roger and a student co-author, Heidi Kulkin, used in preparing their findings about 16 recent young-adult novels about non-White, non-American, or non-heterosexual main characters. Heidi and Roger found themselves most interested in themes of oppression and resistance in such novels and needed a quick way of summarizing the variation they'd discovered. The matrix shown in Table 15.6 worked. Preparing this matrix was helpful in two ways. First, it revealed some things they definitely wanted to say in their final analysis. It gave them the idea, for instance, of organizing their written presentation into four parts: stories about contemporary American characters, historical American characters, non-American characters, and historical non-American characters. Reviewing the table also led them to valid observations of between-group differences for their sample. Thus, for instance, they observed that authors writing about contemporary American characters were less likely to emphasize the importance of formal education and extraordinary self-sacrifice than were those writing about other subjects.

Preparing the matrix, then, led Heidi and Roger to some observations they directly translated into their final report (Clark and Kulkin, 1996). It also helped them in another, perhaps more important, way: It warned them

TABLE 15.6   **Content Analytic Summary Matrix Used in Preparation of Clark and Kulkin's (1996) "Toward Multicultural Feminist Perspective on Fiction for Young Adults"**

Place	Contemporary	Historical
United States	No. of books: 4	No. of books: 3
	Varieties of oppression: racism, classism, cultural imperialism, heterosexism	Varieties of oppression: sexism, racism
	Strategies of resistance: escape, resort to traditional custom, communion with similarly oppressed others	Strategies of resistance: escape, cooperation, extraordinary self-sacrifice, learning, writing
Outside United States	No. of books: 4	No. of books: 5
	Varieties of oppression: sexism, classism, racism, imperialism	Varieties of oppression: all the usual suspects
	Strategies of resistance: education, communion with similarly oppressed persons, creative expression, self-reliance, self-sacrifice	Strategies of resistance: escape, communion with similarly oppressed others, writing, self-definition, education, self-sacrifice

against saying certain things in that report. Thus, for instance, although each mode of oppression and resistance listed in each "cell" of the matrix was prominent in some novel that fell within that cell, each was not equally evident in all the novels of the cell. Heidi and Roger knew, then, that their final written presentation would have to be much more finely attuned to variation within each of their "cells" than the matrix permitted.

Sometimes, after coding, a researcher will have essentially comparable and simple information about cases. Such was Lang's case after she'd coded respondents' reflections on images of Sojourner Truth. Initial coding had led her to observe that some respondents wrote relatively complex reflections. A memo, which consciously located all reflections in the early-morning class in which they'd been received, led Lang to ask respondents whether they considered themselves "morning," "afternoon," or "night" people. Lang then organized the two sets of information (about complexity and whether people considered themselves "morning" people, etc.) using a kind of display that Miles and Huberman call a meta-matrix or "monster-dog." Table 15.7 is one of the monster-dogs that Lang created, albeit a relatively small and tame one. Lang's monster-dog, like all good examples of its breed, is a "stacking up" of "single-case displays" (Miles and Huberman, 1994: 178),

TABLE 15.7   Example of Lang's Monster-Dog Data Displays: Elaborateness of Response vs. When Respondent Feels Most Awake

Respondent	Elaborate	Morning	Afternoon	Night
A				
B		X		
C	X		X	
D			X	
E	X	X		
F		X		
G	X			
H	X	X		
I			X	
J	X			X
K				X
L		X		
M	X	X		
N	X	X		
O				X
P	X	X		

Note: X indicates presence of a trait; no X indicates absence.
Source: Adapted from Clark and Lang, 2001.

one on top of the other, so that the researcher can make comparisons. As Lang (in Clark and Lang, 2001: 14) reports, "I looked at several Monster Dogs, each displaying different sets of variables. I was soon able to see helpful patterns. For instance, self-described morning people looked as though they gave more elaborate responses to the photograph of Sojourner Truth than [others] did. When I crosstabulated these two variables, there was in fact clearly a relationship."

Data displays of various kinds, then, can be useful tools when you analyze qualitative data. Miles and Huberman would take this as too weak an assertion. They claim

> You know what you display. Valid analysis requires, and is driven by, displays that are focused enough to permit a viewing of a full data set in the same location, and are arranged systematically to answer the research questions at hand. (Miles and Huberman, 1994: 91–92)

Although we think this position is overstated, if only because we know of good qualitative analyses that make no explicit reference to data displays, we think you could do worse than try to create data displays of your own while you're engaged in qualitative data analysis. When you do, you might well look to Miles and Huberman (1994) for advice.

Conclusion Drawing and Verification

The most important phase of qualitative data analysis is simultaneously its most exciting and potentially frightening: the phase during which you draw and verify conclusions. Data displays are one excellent means to this end because they force you to engage in two essential activities: pattern recognition and synthesis. Coding, of course, is a small step in the ongoing effort to do these things, and memoing is a larger one. But, you can hardly avoid these activities when you're engaged in creating a data display.

Enos emphasizes that one very important resource for verifying conclusions, especially for researchers who are doing announced field observations or qualitative interviews, is the people being observed or interviewed. These people, she reminds us, can quickly tell you "how off-base you are": "This was very helpful to me, as was asking questions they could theorize about, like 'It seems that white women use foster care a lot more than black women do. Is that right?' [And, if they agree, 'Why is that the case?']" (personal communication)

Elliot Liebow (1993) advocates, however, an "Ask me no questions, I'll tell you no lies" principle, especially relative to participant observation. In his study of homeless women, he consciously decided not to bring the kind of theoretical questions to the women that Enos recommends. In doing so, he claimed he was less concerned with the "lies" he might receive in return for his questions than with "contaminating the situation with questions dragged in from the outside" (Liebow, 1993: 321).

Whatever you decide to do about the issue of "bouncing ideas" off those you're studying, we think you'll find that a vital phase of conclusion drawing and verification comes when you compose your final report. The very act of writing, we (and others [for example, Lofland and Lofland, 1995]) find, is itself inseparable from the analysis process. As a result, we divide the remainder of the section on drawing and verifying conclusions into a discussion of two phases: the pre-report writing and the report-writing phase.

Pre-Report Writing

Lofland and Lofland (1995) point out that the growing use of computers for qualitative data analysis has helped reinforce two points. The first is that qualitative analysis can be facilitated by the kinds of things computers do, like storing and retrieving information quickly. The new programs that facilitate data coding testify to these strengths. The second is that good qualitative analysis requires what computers aren't really much good at. One of these tasks, and something that really distinguishes qualitative research from other brands of social research, is deciding what problems are important. Enos began her study of women in prison with an open mind about what the real problems were. The same was true of Adler and Adler (see Chapter 11) as they approached children's afterschool activities, and Clark and Kulkin (1996) as they began to read nontraditional novels for young adults. These authors all came to their settings with open eyes, open ears, and, particularly, open minds about what they'd find to be the most important problems.

STOP AND THINK *Contrast this openness with the hypothesis-driven approaches of Martinez and Lee (see Chapter 12), and Clark, Guilmain, Saucier, and Tavarez (see Chapter 13).*

Finding out just what problem you should address requires something more than what computers are good at: the imagination, creativity, and moral insight of the human brain.

Imagination and creativity are necessary even after you've chosen your topic and are well into the data analysis. One way to stimulate such imagination and creativity is to use the kinds of data displays advocated by Miles and Huberman. But, as Lofland and Lofland (1995: 202–203) point out, other practices can create fertile conditions for qualitative analysis. One practice is to *alter data displays*. If you've created a data display that, somehow, isn't useful, try a different form of display. We can imagine, for instance, Miles and Huberman building the flow chart that we've reproduced in Figure 15.6, indicating the various interrelationships that might exist among data collection, data reduction, data display, and conclusions, out of a simple typology of (with no indication of relationships among) activities involved in qualitative analysis: data collection, data reduction, data display, and conclusions. Note that altering data displays involves conclusion verification and dismissal, both important goals of data analysis.

Another pre-report writing activity, advocated by Lofland and Lofland, and before them by Glaser and Strauss (1967), is that of *constant comparison*. The key question here is this: how is this item (for example, person, place, or thing) similar to or different from itself at a different time or comparable items in other settings? One can, again, easily imagine how such comparisons led Enos to her eventual interest (see Chapter 10) in how substitute mothering differs for women from different racial backgrounds, or the Adlers (Chapter 11) to their focus on how the meaning of afterschool activities changed with the age of children, or to Clark and Kulkin's interest in how themes of oppression and resistance vary by geographical or historical context, or to Lang's (see Clark and Lang, 2001) belief that "morning" and other people might differ in their capacity to give complex responses early in the morning. Constant comparison is also implicit in Janet Mancini Billson's (1991) progressive verification method for analyzing individual and joint interviews (see Chapter 10).

Lofland and Lofland (1995: 202–203) actually recommend several other pre-report writing techniques for developing creative approaches to your data, but we'd like to highlight just one more of these: *speaking and listening to fellow analysts*. Conversations, particularly with friendly fellow analysts (and these needn't, of course, be limited to people who are actually doing qualitative data analysis but can include other intelligent friends) can not only spark new ideas (remember the adage "two heads are better than one?") but also provide the kind of moral support we all need when ideas aren't coming easily. Friendly conversation is famous not only for showing people what's "right in front of their noses," but also for providing totally new insights. Lang is explicit about the value of sharing her ideas with other students in a data-analysis class devoted to this purpose: "The discussion was similar to having someone proofread a paper for you. Sometimes outsiders help you find better ways to phrase sentences, etc. In this case, they helped me articulate to myself what I was most interested in. The class felt like one big peer review session" (Clark and Lang, 2001: 11–12).

Qualitative Report Writing

A qualitative report writer on report writing had this to say:

> When I began to write, I still had no order, outline, or structure in mind . . . I plunged ahead for many months without knowing where I was going, hoping and believing that some kind of structure and story line would eventually jump out. My original intention was to write a flat descriptive study and let the women speak for themselves—that is, to let the descriptive data and anecdotes drive the writing. In retrospect, that was a mistake. Ideas drive a study, not observations or unadorned facts . . . I shuffled my remaining cards again and discovered that many cards wanted to be grouped under Problems in Day-to-Day Living, so I undertook to write that chunk. The first paragraph asked the question, How do the women survive the inhuman conditions that confront them? After I wrote that paragraph it occurred to me that most situations, experiences, and processes could be seen as working for or against the women's survival and their humanity—that the question of survival might serve as the opening and organizing theme of the book as a whole and give the enterprise some kind of structure. (Liebow, 1993: 325–26)

Lang's reflections on her writing of an undergraduate paper, based on qualitative data, made related, but slightly different points:

> I finally sat down and began writing the paper. It was now that the project actually began to make sense and fit together. One of my biggest concerns going into this step was how I would incorporate the literature reviewed with the rest of my findings. I'd found literature, I knew, that was somewhat relevant, but nothing that spoke directly to my findings that morning people would write more elaborately in the morning. As I wrote, I discovered that my worries were misplaced. Things did flow. The fact that the literature I reviewed didn't speak directly to what I'd found became a virtue; I could simply say that I was addressing a hole in what the literature already said about sleep patterns. Once I began to put it all into words, the paper "practically wrote itself." That was neat. (Clark and Lang, 2001: 17)

Most authorities on qualitative data analysis attest to the difficulty of distinguishing between report writing and data analysis (see Hammersley and Atkinson, 1983: 208; and Lofland and Lofland, 1995: 203). Data collection, data reduction, data display, and other pre-report writing processes all entail an enormous amount of writing, but, whatever else it is, *report writing is a form of data analysis.* Liebow's confession that he came to the report-writing phase of *Tell Them Who I Am,* his touching portrayal of the lives of homeless women, with "no order, outline, or structure in mind" could probably have honestly been made by many authors of qualitative social science—and, frankly, by many authors of quantitative social science. Lang might have had a better sense of "order" going into her short essay than Liebow did going into his book, but writing clearly helped her to refine her ideas, too. The discovery, through writing, of one vital organizing question or another is probably the rule rather than the exception for qualitative analyses.

The writing involved in pre-report phases and the report-writing phase of qualitative analysis can be distinguished, notably by their audiences. Early phases involve writing in which the researcher engages in conversations with herself. We're privy to Enos' codes, her memos, and her typologies, for instance, only because she has generously permitted us to peer over her shoulder, as it were. (Lang's self-conscious renderings [in Clark and Lang, 2001] are exceedingly unusual.) Enos' intended audience was Enos and Enos alone. Formal report writing, on the other hand, is much more clearly a form of communication between author and some other audience. This is not to say that the author herself won't gain new insights from the process. We've already said that she will. But it does imply that at least a nodding acquaintance with certain conventions of social science writing is useful, even if you only intend to "play" with those conventions for effect. As a result, we'd like to introduce you to a few such conventions, and do so in our Appendix A on Report Writing.

Liebow's confession suggests a few guidelines we don't mention in Appendix A. One is that, even if you're not sure just where a particular segment of a report will fall within the whole, as long as you've got something to say, write. Don't insist on seeing the forest, before you write, especially if you've got a few good trees. Liebow wrote and wrote, long before he'd gotten a sense of his main theme. He wrote with the kind of hope and belief that an experienced writer brings: that a story line will emerge. But in writing what he took to be a minor theme, the "Problem of Day-to-Day Living," he first glimpsed what would be the organizing principle of his whole book, "the question of survival." Once he discovered this principle, he found that much of what he'd written before was indeed useful.

Another principle, implicit in Liebow's confession, is that you needn't look for perfection as you write. Especially if you have access to a word-processor and time, you can write relatively freely, in the certain knowledge that whatever you write can be revised. Don't assume you need to say what you want to say in the best (the most grammatical, the most concise, the most stylish) way possible the first time around. Assure yourself that you can always go back, revise, and rethink as necessary. Such assurances are terribly liberating. Just remember to keep your implicit promise to yourself and revise whenever you can.

Summary

The most common goal of quantitative data analysis is to demonstrate the presence or absence of association between variables. To demonstrate such associations, and other characteristics of variables, quantitative data analysts typically compute statistics, often with the aid of computers.

Statistics can be distinguished by whether they focus on describing a single sample, in which case they are descriptive statistics, or on speaking to the generalizability of sample characteristics to a larger population, in which case they are inferential statistics. Statistics can also be distinguished by the number of variables they deal with at a time: one (as in univariate statistics), two (bivariate statistics), or more than two (multivariate statistics).

The most commonly used univariate statistical procedures in social data analysis are measures of central tendency, frequency distributions, and measures of dispersion. Measures of central tendency, or average, are the mode, the median, and the mean. Measures of dispersion include the range and standard deviation. All the univariate statistical procedures described in this chapter focus on describing a sample and are therefore descriptive, rather than inferential, in nature.

We demonstrated bivariate by focusing on crosstabulation, one of many types of procedures that allow you to examine relationships between two variables. In passing, we discussed descriptive statistics, like Cramer's V and Pearson's r, used to describe sample relationships, and inferential statistics, like chi-square, used to make inferences about population relationships.

The canons for analyzing qualitative data are much less widely accepted than are those for analyzing quantitative data. Nonetheless, the various processes involved in such analysis are broadly accepted in general terms: data collection itself, data reduction, data display, and conclusion drawing and verification. These processes are interactive in the sense that they do not necessarily constitute well-defined and sequential steps and insofar as all the individual processes can affect or be affected by, directly or indirectly, any of the others.

Word processors are very useful tools in data collection. Soon after each field or interview session, it is useful to word-process fieldnotes or transcripts. It's often a good idea to leave plenty of room in your notes for coding, comments, memos, and displays.

Data reduction refers to various ways a researcher orders collected and transcribed data. We've talked about two subprocesses of this larger process: coding and memoing. Coding is the process of associating words or labels with passages in your fieldnotes and then collecting all passages that are similarly coded into appropriate "files." These files can be traditional file folders or computer files. Codes can be created using predefined criteria, based on a conceptual framework the researcher brings to his data, or using postdefined criteria, that "emerge," as it were, as the researcher reviews his fieldnotes. Codes can be merely descriptive of what the researcher's seen or analytic, in which case the codes actually advance the analysis in some way. Memos, unlike codes, are more or less extended notes that the researcher writes to help create meaning out of coding categories. Memos are the building blocks of the larger story the researcher eventually tells about her data.

Data displays are visual images that present information systematically and in a way that allows the researcher to see what's going on. The forms in which data can be displayed are limitless, but we've discussed four types in this chapter: charts, typologies, matrices, and "monster-dogs."

Data displays are particularly useful for drawing conclusions that are then subject to verification or presentation. We discussed a few other pre-report techniques for developing such conclusions: altering data displays, making comparisons, and talking with and listening to fellow analysts. We also described report writing itself as an important stage in the analysis process.

EXERCISE 15.1

Analyzing Data on the Web

Visit the GSS Cumulative Data file by going to http://sda.berkeley.edu/archive.htm and hitting on the most recent GSS Cumulative Data File. Do your own univariate and bivariate analyses of marital status (called "marital" in the codebook) and labor force status ("wrkstat").

1. Do you remember what we said the modal marital status (married, divorced, separated, or never married) is among adults in the U.S? Why don't you check for yourself? Start the "Browse codebook" action, hit the "standard codebook," hit the "sequential variable list" and click on "Personal and Family Information." Find the variable named "marital" and hit on it. Which, in fact, has been the modal marital status among adults in the United States?

2. Now, consider the variable "labor force status" ("wrkstat" in the codebook). Which labor force status would you expect to be most common among adults in the United States: working full time, working part time, temporarily not working, retired, in school, keeping house, or something else? State your guess. Now find the variable named "wrkstat." Which, in fact, is the modal labor force status among adults in the United States? Were you right?

3. Now consider both marital status and labor force status: Which marital status would you expect to have the greatest representation in the full-time labor force? State your guess in the form of a hypothesis. Now return to the original GSS menu (where you found the "browse codebook" option before). This time, click "frequencies and crosstabulation"

and click "start." Where the menu asks for "row" variable, type "wrkstat." Where it asks for "column" variable, type "marital." Then "run the table." Which marital status does, in fact, have the greatest percentage of its members employed "full-time"? Were you right?

EXERCISE 15.2

Interpreting a Bivariate Analysis

After Lang (2001) had analyzed Americans' General Social Survey responses to the question of how they felt about income differences in America, she decided to look at how this feeling was related to other variable characteristics of the respondents. Look at the table, below, from her paper and write a paragraph about what it shows about the association between respondents' satisfaction with their own financial situation and their sense of whether income differences in America are too large.

TABLE

Crosstabulation of Satisfaction with Present Financial Situation and Response to Whether Differences in Income in America Are Too Large

Satisfaction with Financial Situation	Differences in Income in America Are Too Large					
	Strongly Agree	Agree	Neither	Disagree	Strongly Disagree	Total
Pretty well satisfied	27.6%	29.7%	16.3%	13.7%	12.6%	100%
More or less satisfied	30.0%	37.5%	13.2%	13.0%	6.3%	100%
Not satisfied at all	44.7%	30.7%	8.5%	8.5%	7.7%	100%

Source: Lang (2001: 25)

EXERCISE 15.3

Interpreting Another Table in Context

This exercise gives you another chance to use the skills you've developed in this chapter to interpret quantitative analyses used by other researchers. Return to Adler and Foster's "A Literature-Based Approach to Teaching Values to Adolescents" in Chapter 8. Focus on the analyses presented in the first table of the article. Review the article well enough so that you can answer the following questions about the table.

1. What research question is addressed in the table?

2. What is the "independent" variable in the table?

3. What are the "dependent" variables in the table?

4. Discuss, in your own words, how the authors demonstrate that a relationship exists or doesn't exist between the independent variable and one of the dependent variables.

5. What answer does this demonstration suggest about the research question that was addressed in the table?

Reconstructing a Qualitative Data Analysis

Return to either Adler and Adler's essay (in Chapter 11) or Enos' essay (in Chapter 10). See if you can analyze one of these articles in terms of concepts developed in this chapter.

1. See if you can find two portions of transcribed field notes or interviews that made it into the report. (Hint: Your best bet here is to locate quotations that might have been taken from those notes.) Write these out here.

2. See if you can imagine two codes or categories that the author or authors may have used to organize her/their field notes. (There are no wrong answers here. We aren't actually told, in either article, what the codes were. You're looking here for categories that come up in the analysis and that seem to have linked two or more segments of the field notes.) Briefly discuss why you think these codes might have been used.

3. See if you can imagine a memo that the author or authors might have written to organize her/their memos. (Again, there are no wrong answers here.) Briefly discuss why you think this memo might have been used.

4. See if you can imagine and draw a data display that the author or authors might have used to organize their essay. (Again, there are no wrong answers, though a very good answer might be the presentation of a typology.)

Data Collection and Coding

For this exercise, you'll need to collect data either through observation or qualitative interview. If you prefer to do observations, you'll need to locate a social setting to which you can gain access twice for at least 20 minutes each time. You can return to one of the settings you observed for Exercises 11.2 or 11.3 (a children's afterschool setting or a group you're already a member of), or you can pick some other easily observable activity (for example, a condiment table at a cafeteria, an elevator on campus, a table at a library where several people sit together). You can choose to tell others of your intention to do research on them or not. You can work out some research questions before you go to the setting the first time or not. You can take notes at the setting, tape-record what's going on, or you can choose to do all your note-taking later.

If you prefer to do two interviews, we recommend that you do something like the work suggested in Exercise 10.1, pick an occupation you know something, but not a great deal, about (such as police officer, waiter, veterinarian, letter carrier, high school teacher, and so on) and find two people who are currently employed in that occupation and willing to be interviewed about their work. Conduct 20-minute semi-structured interviews with each of these people, finding out how they trained or prepared for the work, the kinds of activities they do on the job, about how much time is spent on each, which activities are enjoyed and which are not, what their satisfactions and dissatisfactions with the job are, and whether or not they would like to continue doing their work for the next 10 years.

Whatever the setting or whoever the interviewee, write up your notes afterward, preferably using a word processor and preferably leaving room for coding and memoing (that is, use wide right margins).

After writing up your notes, spend some time coding your data by hand. Remember: This means not only labeling your data but also filing away similarly labeled data (in this case, in file folders you make or purchase).

EXERCISE 15.6

Memoing and Data Displaying

Some time shortly after you've recorded your notes (for Exercise 16.4) and finished coding them (at least for the first time), reread both the notes and your coding files.

Spend about 20 minutes or so writing memos (either word processing them or hand-writing them) at the end of your notes. You can, of course, spend more than 20 minutes at this activity. The important thing is to develop some facility with the process of memoing.

Once you've written your memos, spend some time trying to create a data display that helps you understand patterns in the observations or interviews you've done. This display can be in the form of a chart, a typology, a graph, a matrix, or mini-monster-dog. The purpose is to help you "see" patterns in the data that you might not have already seen.

EXERCISE 15.7

Report Writing

Having completed Exercises 15.5 and 15.6, write a brief report (no more than two or so typed pages) about your project. You should consult Appendix A on Report Writing for guidance here. You can omit a literature review (though, of course, if you found a couple of relevant articles, they might help), but you might otherwise be guided by the list [of items to include in a report] provided in Appendix A. (In particular, consider giving your report a title; providing a brief abstract, an overview, and a sense of your methods; and highlighting your major findings and your conclusions.) Good luck!

Writing the Research Report

The research report can be communicated in a book, a written paper, or an oral presentation. The intended audience(s) can be the teacher of a class, supervisor at work, an agency that funded the research, an organization that gave access to the data, or the general public. Although the report might differ for each audience, each version should describe the research process and the research conclusions in an orderly way. Students wanting a detailed discussion of the "nuts and bolts" of technical writing, including coverage of footnoting, referencing, and writing style, should consult one of the many excellent guides, such a Turabian's (1996) *Manual for Writers of Term Papers, Theses, and Dissertations,* Pyrczak and Bruce's (2005) *Writing Empirical Research Reports: A Basic Guide for Students of the Social and Behavioral Sciences,* or Lester and Lester, Jr.'s (2006) *Writing Research Papers in the Social Sciences.*

We'd like to tell you about some organizational conventions of report writing, even though we feel a bit presumptuous in doing so. After all, what will work for book-length manuscripts probably won't work equally well for article-length reports. Moreover, generally speaking, research reports involving qualitative data are less formulaic than are those involving quantitative data. Nonetheless, because report writing does involve an effort to reach out to an audience, and because social-science report writing frequently involves an effort to reach out to a social science audience, it's not a bad idea to have some idea of the forms expected by such audiences.

We'd like to suggest, then, a checklist of features that often appear in article-length research reports. (Books tend to be much less predictable in their formats.) Audiences can include policy makers or practitioners, who typically will expect practical information about either policy or practice and the general public who is interested in facts, new ideas, or information on their particular interests (Silverman, 2004: 369). Many of you, however, will write for an academic audience, so we'll focus on what that audience will expect most of the time. These features are the following:

1. *A title* that tells what your research is about. We think the title should give the reader some idea of what you have to say about your subject. Four good examples are from reports presented in Chapters 9, 10, 11, and 13:

   - "Explaining Rape Victim Blame: A Test of Attribution Theory" (Gray et al., Chapter 9)

   - "Managing Motherhood in Prison" (Enos, Chapter 10)

   - "The Institutionalization of Afterschool Activities" (Adler and Adler, Chapter 11)

- "Two Steps Forward, One Step Back: The Presence of Female Characters and Gender Stereotyping in Award-Winning Picture Books Between the 1930s and the 1960s"

2. *An abstract,* or a short (usually more than 100 but less than 150 words) account of what the paper does. Such accounts should state the paper's major argument and describe its methods. We haven't included abstracts for any of our focal research articles because we wanted to keep the length of the text manageable. But we've been very self-conscious about not following this convention and think we should give you at least one good example of an abstract, this one from Adler and Adler's original version of "The Institutionalization of Afterschool Activities":

> This article identifies and describes a phenomenon that has arisen over the course of the last generation: an institutionalized "afterschool" period marked by children's involvement in adult-organized and -supervised activities. We trace the historical development of this period and examine its socializing influences on children. Children experience passage through an "extracurricular career" that begins with a recreational ambiance but progresses into competitive and finally elite activities as they grow older and become more skilled. Along this route, their leisure activities become less spontaneous and more rationalized, focused, and professionalized. Adults' incursions into children's play thus represent a means for them to reproduce the existing social structure and to socialize young people to the corporate work values of American culture. (Adler and Adler, 1994b: 309)

Although this abstract is a little longer than many and not focused on method, we think it does a wonderful job of encapsulating key points of the article. Abstracts frequently come before the main body of a report—just after the title—but are generally (and ironically) written *after* the rest of the report's been put together, so the author knows what points to summarize. Here again, however, there are no rules: Drafting an abstract can also be a useful heuristic earlier in the process—to help the writer focus (in a process sometimes called "nutshelling"). Notice, however, that in this abstract, Adler and Adler have actually engaged in a final act of conclusion drawing—the ultimate goal, after all, of data analysis itself.

3. *Overview of the problem* and *literature review.* These two features are sometimes treated distinctly, as they are by Gray et al.'s piece in Chapter 9, but are sometimes combined, as in Adler and Adler's paper in Chapter 11. The basic function of these features is to justify your research, although they're also good places to give credit to those who have contributed ideas, like theories, conceptual definitions or measurement strategies, that you're using in some way. The literature review, in particular, should justify the current research through what's already been written or not written about the topic. Enos does this with two kinds of "literatures": literature that has discussed the "imprisonment boom" in

the United States and has pointed to the special problems of women in prison, and a literature that implies that family forms should have certain "normative" aspects, such as mothers who engage in full-time mothering. If you've never written a research report before, we think you could do worse than study the ways in which the authors of articles in this book have used literature reviews to justify their projects. Literature reviews are an art form, and social science literature reviews are a specialized genre of the larger one. As an art form, there are few hard-and-fast rules, but there are some. One is this: Never simply list the articles and books you've read in preparation for your report, even if you include some of what they say while you're listing them. Always use your literature review to build the case that your research has been worth doing.

Literature reviews themselves offer an excellent chance to engage in analysis. Whether your review occurs before, after, or throughout your field research, it will compel you to make decisions about what your study's really about. (See Chapter 4 for further tips on preparing your literature review.)

4. *Data and Methods.* This section involves the description and explanation of research strategies. Authors generally lay out the study design, population, sample, sampling technique, method of data collection, and any measurement techniques that they used here. They also sometimes discuss practical and ethical considerations that influenced their choice of method. Authors also sometimes justify methodological choices by discussing things such as the validity and reliability of their measures in this section.

5. *Findings.* Being able to present findings is usually the point of all your previous analysis. You might want to subdivide your findings section. In findings sections, researchers always incorporate data or summaries of data. If the project involved quantitative data analyses, this is where researchers display their statistical results, tables, or graphs (see Gray et al., Chapter 9; Martinez and Lee, Chapter 12; or Wysong et al., Chapter 14 for examples). If the research involved qualitative analyses, this is where the authors present summary statements or generalizations, usually supported by anecdotes and illustrations from their interview or field notes. Qualitative researchers usually spice up and document their accounts with well-selected quotations in this section (see Thomson et al., Chapter 7, Enos, Chapter 10, and Adler and Adler, Chapter 11, for examples). In both kinds of studies, writers generally offer interpretations of their data in this section.

6. *Summary and Conclusions.* This section should permit you to distill, yet again, your major findings and to say something about how those findings are related to the literature or literatures you've reviewed earlier. It's very easy to forget to draw such relationships (that is, between your work and what's been written before), but once you get started, you might find you have a lot more to say that could interest other social scientists than you anticipated.

7. *References*. Don't forget to list the references you've referred to in the paper! Here you'll want to be guided by certain stylistic conventions of, perhaps, a formal guide to stylistic conventions (for example, American Sociological Association, 1997). If you're preparing a paper for a course or other academic setting, you might even ask the relevant faculty member what style she prefers.

8. *Appendices*. It's sometimes useful to include a copy of a questionnaire, an interview schedule, or a coding scheme if one was used. Sometimes appendices are used to display elements of a sample (e.g., particular states, countries, or organizations) that were used in the research and sometimes, even the raw data (especially when the data were difficult to come by and other researchers might be interested in using them for other purposes). In general, researchers tend to use appendices to display information that is germane, but not central, to the development of the paper's main argument or that would require too much space in the article proper.

There are few absolutes here, however. Perhaps the most important rule is to give yourself oodles of time to write. If you think it might take about 25 hours to write a report, give yourself 50. If you think it might take 50, give yourself 100. Writing's hard work, so give yourself plenty of time to do it.

Report writing is an art form. Although guidance, such as what we've presented in this section, can be useful, this is one of those human activities where experience is your best guide. Good luck as you gather such experience!

Random Number Table

3250	2881	2326	9108	3702	6844	9827	3210	4895
7961	7952	3660	1503	4644	7656	3856	1672	0695
7739	5850	5998	9821	9476	2165	7662	5742	5048
0417	0262	7442	0873	6101	8592	2862	5346	7284
5816	4578	4061	7262	6328	1355	9386	5446	0666
7843	1922	3038	8442	5367	5952	6782	4206	4630
4912	7287	8526	9602	4545	3755	4159	3288	2900
2105	4791	5097	7386	6040	3575	4143	4820	5547
3263	3102	3555	4204	9411	0088	4873	1407	0858
4619	0105	9946	3234	1265	9509	0227	6671	3992
4460	7800	4476	3613	1252	8579	0611	6919	4208
6041	6103	3043	4634	1191	7886	4579	6301	4488
6718	8181	1425	6396	5312	4700	8077	1604	0870
8253	0766	2292	0146	6302	4413	6170	9764	3377
2061	5261	7862	8368	6533	9481	4098	1313	8527
4712	2096	6000	4173	8920	6913	3092	2028	2678
3344	6985	0656	2416	8367	5673	3272	8878	8714
5965	7229	0064	8356	7767	0960	5365	4980	9568
1203	8216	0354	1005	8063	3853	6732	0012	5391
6619	7927	0067	5559	2929	8706	6366	1111	9997
5545	8811	0833	1259	1344	0579	2735	9419	8533
8446	2198	8304	0803	5947	6322	9623	4419	8007
6017	2981	0880	6879	0193	6721	9130	2341	4528
7582	0139	5100	9315	5571	4508	1341	3450	1307
4118	7411	9556	9907	0612	0601	9545	8084	7117
8443	2486	1121	6714	8133	9113	4613	2633	6828
0231	2253	0931	9757	3471	3080	6369	6270	6296
8393	2801	8322	2871	6504	4972	1608	4997	2211
6092	5922	5587	8472	4256	2888	1344	4681	0332
8728	4779	8851	7466	9308	3590	3115	0478	6956
8585	4266	8533	9121	1988	4965	1951	4395	3267
9593	6571	7914	1242	8474	1856	0098	2973	2863
9356	6955	4768	1561	6065	3683	6459	4200	2177
0542	9565	7330	0903	8971	5815	2763	3714	0247
6468	1257	2423	0973	3172	8380	2826	0880	4270
1201	1959	9569	0073	8362	6587	7942	8557	4586
3616	5075	8836	1704	9143	6974	5971	0154	5509
4241	9365	6648	1050	2605	0031	3136	5878	9501
1679	5768	2801	5266	9165	5101	7455	9522	4199
5731	9611	0523	7579	0136	5655	6933	7668	8344
1554	7718	9450	9736	4703	9408	4062	7203	8950
1471	3213	7164	3788	4167	4063	7881	5348	4486
9563	8925	5479	0340	6959	3054	9435	6400	7989
3722	8984	2755	2560	1753	1209	2805	2763	8189
4237	1818	5266	2276	1877	4423	7318	1483	3153
2859	9216	5303	2154	3243	7359	1393	9909	2482
8303	7527	4566	8713	0538	7340	6224	3135	8367

4150	4056	1579	9286	4353	3413	6618	9524	9284
1507	1745	7221	9009	4769	9333	6803	0198	7879
1365	3544	3753	1530	7384	9357	2231	6166	3217
7414	5143	8912	6611	3396	7969	6559	9463	9114
9658	7935	7850	4297	4908	9653	9540	8753	3920
3169	0990	6016	0631	8451	2488	8724	3625	1805
7294	6007	0887	9667	5211	0082	2522	7055	4431
0514	8862	2713	9944	7880	5101	7535	1200	2583
7016	0451	2039	5827	2820	4979	1997	0920	0988
0389	0734	3322	0822	7189	7229	1031	2444	5827
4903	9574	0313	6611	6589	4437	8070	4480	6481
3746	6799	3082	6987	3766	8230	4376	7150	6963
7695	8918	3141	8417	7200	6066	4097	8883	5254
8423	2024	8911	9430	1711	6618	6768	0251	8926
5009	0421	9412	0227	4881	2652	8863	2900	1878
3265	7124	8628	0386	7584	3412	9002	7013	2898
2263	6500	7451	5922	8289	3552	5546	7389	1959
4629	5201	3214	6479	0313	4100	0596	9274	0415
6833	0952	8529	2108	2219	8249	2184	6631	7768
1991	4762	7349	3282	6394	3786	5792	0337	7781
9628	1897	4009	1946	2765	3552	9865	0398	4612
5860	8540	7938	2799	6545	0922	9124	4352	2057
3394	0297	8151	6445	6994	4566	0323	6185	3117
6976	4713	7478	9092	7028	8140	2490	9508	9481
5359	7743	8669	8469	5126	1434	7695	8129	3184
7127	6156	8500	7593	0670	9534	3945	9718	0834
8690	6983	6359	3205	6167	4362	5340	4104	3004
5705	2941	2505	3360	5976	2070	8450	6761	7404
7210	4415	4744	2061	5102	7796	2714	1876	3398
0777	2055	8932	0542	1427	8487	3761	3793	0954
0270	9605	7099	9063	2584	5289	0040	2135	6934
7194	7521	3770	6017	4393	3340	8210	6468	6144
6029	0732	9672	6507	1422	3483	2387	2472	6951
6631	3518	5642	2250	3187	8553	6747	6029	1871
6958	4528	6053	9004	6039	5841	8685	9100	4097
1589	6697	9722	9343	3227	3063	6808	2900	6970
2004	8823	1038	1638	9430	7533	8538	6167	1593
2615	5006	1524	4747	4902	7975	9782	0937	5157
6487	6682	2964	0882	2028	2948	6354	3894	2360
7238	7432	4922	9697	7764	7714	4511	3252	6864
0382	2511	7244	7271	4804	9564	8572	6952	6155
8918	3452	3044	4414	1695	2297	6360	2520	5814
9331	0179	6815	5701	0711	5408	0322	8085	0080
1065	9702	9944	9575	6714	9382	8324	3008	9373
6663	5272	3139	4020	5158	7146	4100	3578	0683
7045	5075	6195	5021	1443	2881	9995	8579	7923
6483	7316	5893	9422	0550	3273	5488	2061	1532
4396	9028	4536	8606	4508	3752	9183	5669	5927
5766	7632	9132	7778	6887	8203	6678	8792	3159
2743	8858	5094	2285	2824	9996	7384	4069	5442
3330	8618	1974	2441	0457	6561	7390	4102	2311
4791	2725	4085	3190	9098	2300	6241	7903	9150
6833	9145	2459	2393	7831	2279	9658	5115	6906
9474	9070	4964	2035	7954	7001	8688	7040	6924
8576	0380	5722	6149	4086	4437	9871	0850	3248

Check List for Preparing a Research Proposal

A good research proposal should impress someone (e.g., a teacher, a faculty committee, an Institutional Review Board [IRB], or a funding agency) with the worthiness and feasibility of a project, and with the thought that has gone into its design. The following items usually, but not always, appear in a research proposal.

I. A Title   A title that captures the theme or thesis of the proposed project in a nutshell.

II. A Statement of the Project's Problem or Objective   In this section you should answer questions like: What exactly will you study? Why is it worth studying? Does the proposed study have practical significance?

III. Literature Review   In general, a good review literature justifies the proposed research. (See Appendix A for a discussion of literature reviews in research reports. Literature reviews in research proposals should do the same things that literature reviews in research reports do.) The literature review normally accomplishes this goal (of justification) by addressing some of the following questions: What have others said about this topic and related topics? What research, if any, has been done previously on the topic? Have other researchers used techniques that can be adapted for the purposes of the proposed study? In a literature review, one normally cites references that appear in the proposal's reference section (see below) using a style that is appropriate to one's discipline (e.g., American Sociological Association style for sociology, American Psychological Association style for psychology and education, etc.). It is often appropriate to end the literature review with a statement of a research question (or research questions) or of a hypothesis (or hypotheses) that will guide the research.

IV. Methods   In a methods section, you should answer questions like: Whom or what will you study to collect data? How will you select your sample? What, if any, ethical considerations are relevant? What method(s) of data collection will you use—a questionnaire, an interview, an observation and/or available data? You might also, depending on the nature of the study (e.g., whether it is quantitative or qualitative), want to answer questions such as: What are the key variables in your study? How will you define and measure them? Will you be borrowing someone else's measures or using a modified form of measures that have been used before? What kind of data analysis, or comparisons, do you intend to do, or make, with the data you collect?

V. Plan for sharing your findings   In this section, you will want to answer questions like these: Will you write up your results in the form of a paper (or book) to be shared with others? What kinds of reporting outlets might you use: a journal or magazine, a conference or other audience? (If you might prepare a book manuscript, what publisher might be interested in the material?) In general, you want to address the question: with whom will you share your findings?

VI. Budget   What are the major foreseeable costs of your project and how will you cover them?

VII. References   In this section, you want to provide a list of the references (books, articles, websites) already cited in the proposal, using some standard style format appropriate for one's discipline (e.g., American Sociological Association style for sociology, etc.).

Glossary

access   the ability to obtain the information needed to answer a research question.

account   a plausible and appealing explanation of the research that the researcher gives to prospective participants.

accretion measures   indicators of a population's activities created by its deposits of materials.

anonymity   when no one, including the researcher, knows the identities of research participants.

antecedent variable   a variable that comes before both an independent variable and a dependent variable.

anticipatory data reduction   decision making about what kinds of data are desirable that is done in advance of collecting data.

applied research   research intended to be useful in the immediate future and to suggest action or increase effectiveness in some area.

authorities   socially defined sources of knowledge.

available data   data that are easily accessible to the researcher.

basic research   research designed to add to our fundamental understanding and knowledge of the social world regardless of practical or immediate implications.

biased sample   a sample that is not representative of the population from which it is drawn.

bivariate analyses   data analyses that focus on the association between two variables.

case study   a research strategy that focuses on one case (an individual, a group, an organization, and so on) within its social context at one point in time, even if that one time spans months or years.

causal hypothesis   a testable expectation about an independent variable's effect on a dependent variable.

causal relationship   a nonspurious relationship between an independent and dependent variable with the independent variable occurring before the dependent variable.

closed-ended question   a question that includes a list of predetermined answers

cluster sampling   a probability sampling procedure that involves randomly selecting clusters of elements from a population and subsequently selecting every element in each selected cluster for inclusion in the sample.

coding   the process by which raw data are given a standardized form. In quantitative analyses, this frequently means making data computer usable. In qualitative analyses, this means associating words or labels with passages and then collecting similarly labeled passages into files.

cohort   a group of people born within a given time frame or experiencing a life event, such as marriage or graduation from high school, in the same time period.

cohort study   a study that follows a cohort over time.

complete observer role   being an observer of a situation without becoming part of it.

complete participant role   being, or pretending to be, a genuine participant in a situation.

composite measure   a measure with more than one indicator.

concepts   words or signs that refer to phenomena that share common characteristics.

conceptual definition   a definition of a concept through other concepts. Also a theoretical definition.

conceptualization   the process of clarifying what we mean by a concept.

confidence interval   a range of values within which the population parameter is expected to lie.

confidence level   the estimated probability that a population parameter will fall within a given confidence interval.

confidentiality   when no third party knows the identities of the research participants.

construct validity   how well a measure of a concept is associated with a measure of another concept that some theory says the first concept should be associated with.

content analysis   a method of data collection in which some form of communication is studied systematically.

content validity   how well a measure covers the range of meanings associated with a concept.

contingency question   a question that depends on the answers to previous questions.

controlled observation   observation that involves clear decisions about what is to be observed.

convenience sampling   a nonprobability sampling procedure that involves selecting elements that are readily accessible to the researcher.

cost-benefit analysis   research that compares a program's costs to its benefits.

cost-effectiveness analysis   comparisons of program costs in delivering desired benefits based on the assumption that the outcome is desirable.

cross-sectional study   a study design in which data are collected for all the variables of interest using one sample at one time.

crosstabulation   the process of making a bivariate table to examine a relationship between two variables.

cover letter   the letter accompanying a questionnaire that explains the research and invites participation.

coverage errors   errors that result from differences between the sampling frame and the target population.

data displays   visual images that summarize information.

data reduction   the various ways in which a researcher orders collected and transcribed data.

deductive reasoning   reasoning that moves from more general to less general statements.

demand characteristics   characteristics that the observed take on simply as a result of being observed.

dependent variable   a variable that is seen as being affected or influenced by another variable.

descriptive statistics   statistics used to describe and interpret sample data.

descriptive study   research designed to describe groups, activities, situations, or events.

dimensions   aspects or parts of a larger concept.

double-blind experiment   an experiment in which neither the subjects nor research staff who interact with them knows the memberships of the experimental or control groups.

ecological fallacy   the fallacy of making inferences about certain types of individuals from information about groups that might not be composed exclusively of those individuals.

element   a kind of thing a researcher wants to sample. Also called a sampling unit.

empirical generalization   a statement that summarizes a set of individual observations.

erosion measures   indicators of a population's activities created by its selective wear on its physical environment.

ethical principles in research   the set of values, standards, and principles used to determine appropriate and acceptable conduct at all stages of the research process.

evaluation research   research specifically designed to assess the impact of a specific program, policy, or legal change.

exhaustiveness   the capacity of a variable's categories to permit the classification of every unit of analysis.

existing statistics   summaries of data collected by large organizations.

experimental design   a study design in which the independent variable is controlled, manipulated, or introduced in some way by the researcher.

experimenter expectations when expected behaviors or outcomes are communicated to subjects by the researcher.

explanatory research research that seeks to explain the cause of a phenomenon, and typically asks "what causes what?" or "why is it this way?"

exploratory research ground-breaking research on a relatively unstudied topic or in a new area.

face validity the degree to which a measure seems to be measuring what it's supposed to be measuring.

feasibility whether it is practical to complete a study in terms of access, time, and money.

field experiment an experiment done in the "real world" of classrooms, offices, factories, homes, playgrounds, and the like.

focus group interview a type of group interview where participants converse with each other and have minimal interaction with a moderator.

formative analysis evaluation research focused on the design or early implementation stages of a program or policy.

frequency distribution a way of showing the number of times each category of a variable occurs in a sample.

gatekeeper someone who can get a researcher into a setting or facilitate access to participants.

generalizability the ability to apply the results of a study to groups or situations beyond those actually studied.

grounded theory theory derived from data in the course of a study.

group-administered questionnaire questionnaire administered to respondents in a group setting.

group interview a data collection method with one interviewer and two or more interviewees.

history the effects of general historical events on study participants

honest reporting the ethical responsibility to produce and report accurate data.

hypothesis a testable statement about how two or more variables are expected to be related to one another.

in-person interview an interview conducted face-to-face.

independent variable a variable that is seen as affecting or influencing another variable.

index a composite measure that is constructed by adding scores from several indicators.

indicators observations that we think reflect the presence or absence of the phenomenon to which a concept refers.

individually administered questionnaire questionnaire that is hand-delivered to a respondent and picked up after completion.

inductive reasoning reasoning that moves from less general to more general statements.

inferential statistics statistics used to make inferences about the population from which the sample was drawn.

informants participants in a study situation who are interviewed for an in-depth understanding of the situation.

informed consent the principle that potential participants are given adequate and accurate information about a study before they are asked to agree to participate.

informed consent form a statement that describes the study and the researcher and formally requests participation.

institutional review board (IRB) the committee at a college, university, or research center responsible for evaluating the ethics of proposed research.

internal validity agreement between a study's conclusions about causal connections and what is actually true.

Internet questionnaire questionnaire that is sent by email or posted on a website.

interobserver (or interrater) reliability method a way of checking the reliability of a measurement strategy by comparing results obtained by one observer with results obtained by another using exactly the same method.

intersubjectivity agreements about reality that result from comparing the observations of more than one observer.

interval scale measure a level of measurement that describes a variable whose categories have names, whose categories can be rank-ordered in some sensible way, and whose adjacent categories are a standard distance from one another.

interview A data collection method in which respondents answer questions asked by an interviewer.

interview guide the list of topics to cover and the order in which to cover them that can be used to guide less structured interviews.

interview schedule the list of questions read to a respondent, most typically in a structured interview.

interviewer effect the change in a respondent's behavior or answers that is the result of being interviewed by a specific interviewer.

keywords the terms used to search for sources in a literature review.

laboratory research research done in settings that allows the researcher control over the conditions, such as in a university or medical setting.

literature review the process of searching for, reading, summarizing, and synthesizing existing work on a topic or the resulting written summary of a search.

longitudinal research a research design in which data are collected at least two different times, such as a panel, trend, or cohort study.

mailed questionnaire questionnaire mailed to the respondent's residence or workplace.

matching assigning sample members to groups by matching members of the sample on one or more characteristics and separating the pairs into two groups with one group randomly selected to become the experimental group.

maturation the biological and psychological processes that cause people to change over time.

mean the measure of central tendency designed for interval level variables. The sum of all values divided by the number of values.

measurement the process of devising strategies for classifying subjects by categories to represent variable concepts.

measurement error the kind of error that occurs when the measurement we obtain is not an accurate portrayal of what we tried to measure.

measures of association measures that give a sense of the strength of a relationship between two variables.

measures of correlation measures that provide a sense not only of the strength of the relationship between two variables, but also of its direction.

measures of dispersion measures that provide a sense of how spread out cases are over categories of a variable.

median the measure of central tendency designed for ordinal level variables. The middle value when all values are arranged in order.

memos more or less extended notes that the researcher writes to help herself or himself understand the meaning of codes.

mode the measure of central tendency designed for nominal level variables. The value that occurs most frequently.

multidimensionality the degree to which a concept has more than one discernible aspect.

multistage sampling a probability sampling procedure that involves several stages, such as randomly selecting clusters from a population, then randomly selecting elements from each of the clusters.

multivariate analyses analyses that permit researchers to examine the relationship between variables while investigating the role of other variables.

mutual exclusiveness the capacity of a variable's categories to permit the classification of each unit of analysis into one and only one category.

natural experiment a study using real-world phenomena that approximates an experimental design even though the independent variable is not controlled, manipulated, or introduced by the researcher.

needs assessment an analysis of whether a problem exists, its severity, and an estimate of essential services.

nominal scale measure a level of measurement that describes a variable whose categories have names.

nonparticipant observation observation made by an observer who remains as aloof as possible from those observed.

nonresponse errors errors that result from differences between nonresponders and responders to a survey.

nonprobability sample a sample that has been drawn in a way that doesn't give every member of the population a known chance of being selected.

normal distribution a distribution that is symmetrical and bell-shaped.

objectivity the ability to see the world as it really is.

observational techniques methods of collecting data by observing people, most typically in their natural settings.

observer-as-participant role being primarily a self-professed observer, while occasionally participating in the situation.

open-ended question a question that allows respondents to answer in their own words.

operational definition declarations of the specific ways in which the absence, presence, or the degree of presence of a phenomenon will be determined in a specific instance.

operationalization the process of defining specific ways to infer the absence, presence, or degree of presence of a phenomenon.

ordinal scale measure a level of measurement that describes a variable whose categories have names and whose categories can be rank-ordered in some sensible way.

outcome evaluation research that is designed to "sum up" the effects of a program, policy, or law in accomplishing the goal or intent of the program, policy, or law.

panel attrition the loss of subjects from a study because of disinterest, death, illness, or inability to locate them.

panel conditioning the effect of repeatedly measuring variables on members of a panel study.

panel study a study design in which data are collected about one sample at least two times where the independent variable is not controlled by the researcher.

parameter a summary of a variable characteristic in a population.

participant-as-observer role being primarily a participant, while admitting an observer status.

participatory action research (PAR) research done by community members and researchers working as co-participants, most typically within a social justice framework to empower people and improve their lives.

participant observation observation performed by observers who take part in the activities they observe.

passive consent when no response is considered an affirmative consent to participate in research; also called "opt out informed consent," this is sometimes used for parental consent for children's participation in school-based research.

personal inquiry inquiry that employs the senses' evidence.

personal records records of private lives, such as biographies, letters, diaries, and essays.

physical traces physical evidence left by humans in the course of their everyday lives.

pilot test a preliminary draft of a set of questions that is tested before the actual data collection.

placebo a simulated treatment of the control group that is designed to appear authentic.

population the group of elements from which a researcher samples and to which she or he might like to generalize.

posttest the measurement of the dependent variable that occurs after the introduction of the stimulus or the independent variable.

posttest-only control group experiment an experimental design with no pretest.

process evaluation research that monitors a program or policy to determine if it is implemented as designed.

predictive validity how well a measure is associated with future behaviors you'd expect it to be associated with.

pretest the measurement of the dependent variable that occurs before the introduction of the stimulus or independent variable.

pretest-posttest control group experiment an experimental design with two or more randomly selected groups (an experimental and control group) in which the researcher controls or "introduces" the independent variable and measures the dependent variable at least two times (pretest and posttest measurement).

primary data data that the same researcher collects and uses.

probability sample a sample that gives every member of the population a known (nonzero) chance of inclusion.

protecting study participants from harm the principle that participants in studies are not harmed, physically, psychologically, emotionally, legally, socially, or financially as a result of their participation in a study.

purposive sampling a nonprobablity sampling procedure that involves selecting elements based on the researcher's judgment about which elements will facilitate his or her investigation.

qualitative content analysis content analysis designed for verbal analysis.

qualitative data analysis analysis that results in the interpretation of action or representation of meanings in the researcher's own words.

qualitative interview a data collection method in which an interviewer adapts and modifies the interview for each interviewee.

quantitative content analysis content analysis designed for statistical analysis.

quantitative data analysis analysis based on the statistical summary of data.

quasi-experiment an experimental design that is missing one or more aspects of a true experiment, most frequently random assignment into experimental and control groups.

questionnaire a data collection instrument with questions and statements that are designed to solicit information from respondents.

quota sampling a nonprobability sampling procedure that involves describing the target population in terms of what are thought to be relevant criteria and then selecting sample elements to represent the "relevant" subgroups in proportion to their presence in the target population.

random assignment a technique for assigning members of the sample to experimental and control groups by chance to maximize the likelihood that the groups are similar at the beginning of the experiment.

range a measure of dispersion or spread designed for interval level variables. The difference between the highest and lowest values.

rapport a sense of interpersonal harmony, connection, or compatibility between an interviewer and an interviewee.

ratio scale measure a level of measurement that describes a variable whose categories have names, whose categories may be rank-ordered in some sensible way, whose adjacent categories are a standard distance from one another, and one of whose categories is an absolute zero point—a point at which there is a complete absence of the phenomenon in question.

reliability the degree to which a measure yields consistent results.

requirements for supporting causality the requirements needed to support a causal relationship include a pattern or relationship between the independent and dependent variables, determination that the independent variable occurs first, and support for the conclusion that the apparent relationship is not caused by the effect of one or more third variables.

research costs all monetary expenditures needed for planning executing, and reporting research.

research question a question about one or more topics or concepts that can be answered through research.

research topic a concept, subject, or issue that can be studied through research.

researchable question a question that can be answered with research that is feasible.

respondent the participant in a survey who completes a questionnaire or interview.

response rate the percentage of the sample contacted that actually participates in a study.

sample a number of individual cases drawn from a larger population.

sampling distribution the distribution of a sample statistic (such as the average) computed from many samples.

sampling error any difference between the characteristics of a sample and the characteristics of the population from which the sample is drawn.

sampling frame or study population the group of sampling units or elements from which a sample is actually selected.

sampling variability the variability in sample statistics that occurs when different samples are drawn from the same population.

scales indexes in which some items are given more weight than others in determining the final measure of the concept.

scientific methods a way of conducting research following rules that specify objectivity, logic, and communication among a community of knowledge seekers, and the connection between research and theory.

screening question a question that asks for information before asking the question of interest.

secondary data research data that have been collected by someone else.

selection bias a bias in the way the experimental and control or comparison groups are selected that is responsible for preexisting differences between the groups.

self-administered questionnaire a questionnaire that the respondent completes by him or herself.

self-report method another name for questionnaires and interviews because respondents are most often asked to report their own characteristics, behaviors, and attitudes.

semi-structured interview interview with an interview guide containing primarily open-ended questions that can be modified for each interview.

simple random sample a probability sample in which every member of a study population has been given an equal chance of selection.

skewed variable an interval level variable that has one or a few cases that fall into extreme categories.

snowball sampling a nonprobability sampling procedure that involves using members of the group of interest to identify other members of the group.

social theory an explanation about how and why people behave and interact in the ways that they do.

Solomon four-group design a controlled experiment with an additional experimental and control group with each receiving a posttest only.

split-half method a method of checking the reliability of several measures by dividing the measures into two sets of measures and determining whether the two sets are associated with each other.

spurious relationship a noncausal relationship between two variables.

stakeholders people or groups that participate in or are affected by a program or its evaluation, such as funding agencies, policy makers, sponsors, program staff, and program participants.

standard deviation a measure of dispersion designed for interval level variables that accounts for every value's distance from the sample mean.

statistic a summary of a variable in a sample.

stimulus the experimental condition of the independent variable that is controlled or "introduced" by the researcher in an experiment.

stratified random sampling a probability sampling procedure that involves dividing the population into groups or strata defined by the presence of certain characteristics and then random sampling from each stratum.

study design a research strategy specifying the number of cases to be studied, the number of times data will be collected, the number of samples that will be used, and whether or not the researcher will try to control or manipulate the independent variable in some way.

structured interview a data collection method in which an interviewer reads a standardized list of questions to the respondent and records the respondent's answers.

survey a study in which the same data, usually in the form of answers to questions, are collected from all members of the sample.

systematic sampling a probability sampling procedure that involves selecting every kth element from a list of population elements, after the first element has been randomly selected.

telephone interview an interview conducted over the telephone.

test-retest method a method of checking the reliability of a test that involves comparing its results at one time with results, using the same subjects, at a later time.

testing effect the sensitizing effect on subjects of the pretest.

theory an explanation about how and why something is as it is.

theoretical saturation the point where new interviewees or settings look a lot like interviewees or settings one has observed before.

thick description reports about behavior that provide a sense of things like the intentions, motives, and meanings behind the behavior.

thin description bare-bones description of acts.

time expenditures the time it takes to complete all activities of a research project from the planning stage to the final report.

trend study a study design in which data are collected at least two times with a new sample selected from a population each time.

units of analysis the units about which information is collected, and the kind of thing a researchers wants to analyze. The term is used at the analysis stage, whereas the terms element or sampling unit can be used at the sampling stage.

units of observation the units from which information is collected.

univariate analyses analyses that tell us something about one variable.

unobtrusive measures indicators of interactions, events, or behaviors whose creation does not affect the actual data collected.

unstructured interview a data collection method in which the interviewer starts with only a general sense of the topics to be discussed and creates questions as the interaction proceeds.

validity the degree to which a measure taps what we think it's measuring.

variable a characteristic that may vary from one subject to another or for one subject over time. A concept that varies.

vignettes scenarios about people or situations that the researcher creates to use as part of the data collection method.

visual analysis a set of techniques used to analyze images.

voluntary participation the principle that study participants choose to participate of their own free will.

vraisemblance the verisimilitude, or sense of truth, that a reporter can convey through his or her writing.

Acker, J., K. Barry, & J. Esseveld. 1991. Objectivity and the truth: problems in doing feminist research. In *Beyond methodology: Feminist scholarship as lived research,* 133–153, edited by M. M. Fonow & J. A. Cook. Bloomington: Indiana University Press.

Adler, E. S. 1981. The underside of married life: Power, influence and violence. In *Women and crime in America,* 300–319, edited by L. H. Bowker. New York: Macmillan.

Adler, E. S., & P. J. Foster. 1997. A literature-based approach to teaching values to adolescents: Does it work? *Adolescence* 32 (Summer): 275–286.

Adler, E. S., & J. S. Lemons. 1990. *The elect: Rhode Island's women legislators, 1922–1990.* Providence: League of Rhode Island Historical Societies.

Adler, P. A. 1985. *Wheeling and dealing.* New York: Columbia University Press.

Adler, P. A., & P. Adler. 1987. *Membership roles in field research.* Newbury Park, CA: Sage.

Adler, P. A., & P. Adler. 1994a. Observational techniques. In *Handbook of qualitative research,* 377–392, edited by N. K. Denzin & Y. S. Lincoln. Thousand Oaks, CA: Sage.

Adler, P. A., & P. Adler. 1994b. Social reproduction and the corporate other: The institutionalization of afterschool activities. *Sociological Quarterly* 35: 309–328.

Adler, P. A., & P. Adler, 1996. Parent-as-researcher: The politics of researching in the personal life. *Qualitative Sociology* 19(1): 35–58.

Adler, P. A., & P. Adler. 1998. *Peer power. Preadolescent culture and identity.* New Brunswick, NJ: Rutgers University Press.

Albig, W. 1938. The content of radio programs—1925–1935. *Social Forces* 16: 338–349.

Allen, S. 2006. Harvard to study katrina's long-term psychological toll. *Boston Globe* 269(6): A4.

Altman, L. E. 2005. Nobel prize for discovering bacterium causes ulcers. *New York Times,* October 4, D3.

American Sociological Association. 1997. *American Sociological Association style guide,* 2nd edition. Washington, DC: American Sociological Association.

American Sociological Association. 1999. Code of Ethics and Policies and Procedures of the ASA Committee on Professional Ethics. Washington, DC: ASA. http://www.asanet.org/galleries/default-file/Code%20of%20Ethics.pdf).

Andersen, E. 1999. *Code of the street.* New York: Norton.

Angrist, J. D., & J. H. Johnson IV. 2002. Effects of work-related absences on families: Evidence from the Gulf War. *Industrial & Labor Relations Review* 54(1): 41–58.

Aniskiewicz, R., & E. Wysong. 1990. Evaluating DARE: Drug education and the multiple meanings of success. *Policy Studies Review* 9: 727–747.

Aquilino, W. S. 1993. Effects of spouse presence during the interview on survey responses concerning marriage. *Public Opinion Quarterly* 57: 358–376.

Aquilino, W. S. 1994. Interview mode effects in surveys of drug and alcohol use. *Public Opinion Quarterly* 58: 210–242.

ASAP. 2003. The University of Akron's adolescent substance abuse prevention study: A longitudinal evaluation of the new curricula for the DARE middle (7th) and high school (9th grade) programs: Take charge of your life, progress report for year 2. December, 2003. Carnevale Associates. http://www.asapstudy.org/media/Dec1ProgressReport.pdf

ASAP. 2006. The University of Akron's adolescent substance abuse prevention study. A longitudinal evaluation of the new curricula for the DARE middle (7th) and high school (9th grade) programs: Take charge of your life, year four progress report, January, 2006. Carnevale Associates. http://www.asapstudy.org/media/DAREProgressReport-2006.pdf

Asher, R. 1992. *Polling and the public.* Washington, DC: Congressional Quarterly.

Associated Press. 2005. Chattanooga police drop DARE program, The Associated Press State & Local Wire, July 27, 2005. Information from Chattanooga Times Free Press.

Asthana, S. 2003. Patriotism and its avatars: Tracking the national-global dialectic in Indian music videos. *Journal of Communication Inquiry* 27: 337–353.

Atchley, R. C. 2003. *Social forces and aging,* 10th edition. Belmont, CA: Wadsworth.

Atkinson, R. The Life Story Interview. 2001. In *Handbook of interview research,* 121–140 edited by J. F. Gubrium & J. A. Holstein. Thousand Oaks, CA: Sage Publications.

Auerbach, J. 2000. Feminism and federally funded social science: Notes from inside. *Annals of the American Academy of Political and Social Science* 571: 30–41.

Babbie, E. 2003. *The practice of social research,* 10th edition. Belmont, CA: Wadsworth.

Bakalar, N. 2005. Ugly children may get parental short shrift. *New York Times,* May 3, http://www.nytimes.com/2005/05/03/health/03ugly.html.

Baker, C. N. 2005. Images of women's sexuality in advertisements: A content analysis of black-and white-oriented women's and men's magazines. *Sex Roles* 52: 13–27.

Bales, K. 1999. Popular reactions to sociological research: The case of Charles Booth. *Sociology* 33: 153–168.

Bales, R. F. 1950. *Interaction process analysis.* Reading, MA: Addison-Wesley.

Barbour, R. S., & J. Kitzinger, editors. 1998. *Developing focus group research.* Thousand Oaks, CA: Sage.

Barry, C. 1998. Choosing qualitative data analysis software: Atlas/ti and Nudist compared. *Sociological Research Online* 3. Retrieved on June 20, 2001, from www.socresonline.org.uk/socresonline/3/3/4.html.

Barry, E. 2002. After day care controversy, psychologists play nice. *Boston Globe.* Page E1 September 3, 2002.

Barry, E. 1999. Study adds doubts to DARE program. *Boston Globe,* August 2, 1999, pp. A1, B12.

Bart, P. B., & P. H. O'Brien. 1985. *Stopping rape: Successful survival strategies.* New York: Pergamon.

Bartels, L. 2000. Question order and declining faith in elections; Report to the Board of Overseers, National Election Studies. Retrieved June 19, 2001, from www.umich.edu/~nes/resources/techrpts/tech-abs/tech-ab60.htm.

Baruch, Y. 1999. Response rate in academic studies—A comparative analysis. *Human Relations* 52: 421–438.

Bates, B., & M. Harmon. 1993. Do "instant polls" hit the spot? Phone-in versus random sampling of public opinion. *Journalism Quarterly* 70: 369–380.

Baumrind, D. 1964. Some thoughts on ethics of research: After reading "Milgram's behavioral study of obedience." *American Psychologist* 19: 421–423.

Becker, H. S., B. Geer, E. C. Hughes, & A. L. Strauss. 1961. *Boys in white: Student culture in medical school.* Chicago: University of Chicago Press.

Belinfante, A. 2005. Telephone subscribership in the United States (Data Through March 2005). 2005. http://www.fcc.gov/Bureaus/Common_Carrier/Reports/FCC-State_Link/IAD/subs0305.pdf.

Bell, P. 2001. Content analysis of visual images. In *Handbook of visual analysis,* 10–35, edited by T. van Leeuwan & C. Jewitt. Thousand Oaks, CA: Sage.

Bennett, L. A., & K. McAvity. 1994. Family research: A case for interviewing couples. In *Psychosocial interior of the family,* edited by G. Handel & G. G. Whitchurch. New York: Aldine de Gruyter.

Benney, M., & E. C. Hughes. 1956. Of sociology and the interview. *American Journal of Sociology* 62: 137–142.

Berg, B. L. 1989. *Qualitative research methods of the social sciences.* Boston: Allyn & Bacon.

Berk, R. A., G. K. Smyth, & L. W. Sherman. 1988. When random assignment fails: Some lessons from the Minneapolis spouse abuse experiment. *Journal of Quantitative Criminology* 4: 209–223.

Best, J. 2001. Promoting bad statistics. *Society* March/April: 10–15.

Bickman, L., & D. J. Rog. 1998. Introduction: Why a handbook of applied social research methods? In *Handbook of applied social research methods,* ix–xix, edited by L. Bickman & D. J. Rog. Thousand Oaks, CA: Sage.

Billson, J. M. 1991. The progressive verification method toward a feminist methodology for studying women cross-culturally. *Women's Studies International Forum* 14: 201–215.

Bishop, G. F. 1987. Experiments with the middle response alternative in survey questions. *Public Opinion Quarterly* 51: 220–232.

Bock, J. 2000. Doing the right thing? Single mothers by choice and the struggle for legitimacy. *Gender & Society* 14: 62–86.

Bonanno, G. A., C. Rennicke, & S. Dekel. 2005. Self-enhancement among high-exposure survivors of the September 11th terrorist attack: Resilience or social maladjustment? *Journal of Personality and Social Psychology* 88: 984–998.

Bond, M. A. 1990. Defining the research relationship: Maximizing participation in an unequal world. In *Researching community psychology,* 183–185, edited by P. Tolan, C. Keys, F. Chertock, & L. Jason. Washington, DC: American Psychological Association.

Booth, R. E., S. K. Koester, C. S. Reichardt, & J. T. Brewster. 1993. Quantitative and qualitative methods to assess behavioral change among injection drug users. *Drugs and Society* 7: 161–183.

Bosk, C. L., & R. G. De Vries. 2004. Bureaucracies of mass deception: Institutional review boards and the ethics of ethnographic research. *The Annals of the American Academy of Political and Social Science* 595 (Sept): 249–263.

Bourgoin, N. 1995. Suicide in prison. Some elements of a strategic analysis. *Cahiers Internationaux de Sociologie* 98: 59–105.

Boyle, P. 2001. A DAREing rescue. *Youth Today* 19 (April): 1 and 16–19.

Bradburn, N. M., & S. Sudman. 1988. *Polls and surveys.* San Francisco: Jossey-Bass.

Brief, A. P., T. Buttram, J. D. Elliot, R. M. Reizenstein, & L. R. McCline. 1995. Releasing the beast: A study of compliance to use race as a selection criteria. *Journal of Social Issues* 51: 177–193.

Brodsky, A., 2001. More than epistemology: Relationships in applied research with underserved communities. *Journal of Social Issues* 57(2): 323:335.

Brophy, B. 1999. Doing it for science. *U.S. News & World Report* (March 22, 1999). Retrieved on July 21, 2001 from www.usnews.com/usnews/issue/990322/nycu/22nurs.htm.

Browne, J. 1976. The used car game. In *The research experience,* 60–84, edited by M. P. Golden. Itasca, IL: F. E. Peacock.

Bryan, J. A., & M. A. Test. 1982. Models and helping: Naturalistic studies in aiding behavior. In *Experimenting in society,* 139–146, edited by J. W. Reich. Glenview, IL: Scott, Foresman.

Brydon-Miller, M. 1993. Breaking down the barriers: Accessibility self-advocacy in the disabled community. In *Voices of change, participatory research in the United States and Canada,* edited by P. Park, M. Brydon-Miller, B. Hall, & T. Jackson. Westport, CT: Bergin & Garvey.

Brydon-Miller, M. 2002. What is participatory action research and what's a nice person like me doing in a field like this? Paper presented at the 2002 SPSSI Convention June 28–30 Toronto, Canada Downloaded from http://www.scu.edu.au/schools/gcm/ar/w/Brydon-Miller.pdf.

Budiansky, S. 1994. Blinded by the cold-war light. *U.S. News & World Report,* 116: 6–7.

Bullock, K. 2005. Grandfathers and the impact of raising grandchildren. *Journal of Sociology & Social Welfare* 32(1): 43–59.

Bulmer, M. 1982. *Social research ethics: An examination of the merits of covert participant observation.* New York: Holmes & Meier.

Burdette, A. M., C. G. Ellison, & T. D. Hill. 2005. Conservative Protestantism and tolerance towards homosexuals: An examination of potential mechanisms. *Sociological Inquiry* 75: 177–196.

Bureau of Labor Statistics (BLS). 2001. *National Longitudinal Surveys: Mature Women User's Guide 2001.* http://www.bls.gov/nls/mwguide/2001/nlsmwg0.pdf.

Bureau of Justice Statistics (BJS). 2005. Criminal Victimization 2004. http://www.ojp.usdoj.gov/bjs/cvictgen.htm.

Bureau of Justice Statistics (BJS). 2004. *National Crime Victims Survey.* Retrieved on February 15, 2006 from http://www.ojp.usdoj.gov/bjs/abstract/cv04.htm.

Burt, M. R. 1980. Culture myths and supports for rape. *Journal of Personality and Social Psychology* 38: 27–230.

CAPS. 2000. The legacy project: Lessons learned about conducting community-based research. Retrieved on March 11, 2006 from the website of Center for AIDS Prevention Studies at http://www.caps.ucsf.edu/capsweb/publications/LegacyS2C.pdf.

CAPS. 2006. Community collaborative research at CAPS. Retrieved on March 11, 2006 from http://www.caps.ucsf.edu/capsweb/commres.html.

Campbell, D. T., & J. C. Stanley. 1963. *Experimental and quasi-experimental designs for research.* Chicago: Rand McNally.

Carmines, E. G., & R. A. Zeller. 1979. *Reliability and validity assessment.* Beverly Hills, CA: Sage.

CBS News. 2004. Presidential poll-o-rama. Nov. 1. Retrieved on December 13, 2005 from http://cbsnews.cbs.com/stories/2004/10/28/politics/main652020.shtml.

CDE. 2001. *Character education.* Information found on the California Department of Education website on July 19, 2001 from http://www.cde.ca.gov/character/aboutpg.html.

Chambers, D. E., K. R. Wedel, & M. K. Rodwell. 1992. *Evaluating social programs.* Boston: Allyn & Bacon.

Charmaz, K. 1983. The grounded theory method: An explication and interpretation. In *Contemporary field research: A collection of readings,* 109–126, edited by R. M. Emerson. Boston: Little, Brown.

Chase, S. E. 1995. *Ambiguous empowerment: The work narratives of women school superintendents.* Amherst: University of Massachusetts Press.

Chito Childs, E. 2005. Looking behind the stereotypes of the "angry black woman": An exploration of black women's responses to interracial relationships. *Gender & Society* 19(4): 544–561.

Clark, B. L. 2003. *Kiddie lit.* Baltimore: Johns Hopkins University Press.

Clark, R. 1982. Birth order and eminence: A study of elites in science, literature, sports, acting and business. *International Review of Modern Sociology* 12: 273–289.

Clark, R. 1999. Diversity in sociology: Problem or solution? *American Sociologist* 30: 22–43.

Clark, R., M. Almeida, T. Gurka, and L. Middleton. 2003. Engendering tots with Caldecotts: An updated update. In E. S. Adler and R. Clark (Eds.), *How it's done: An invitation to social research* (2nd ed., 379–386). Belmont, CA: Wadsworth.

Clark, R., & H. Fink. 2004. Picture this: A multicultural feminist analysis of picture books for children. *Youth & Society* 36: 102–125.

Clark, R., A. R. Folgo, & J. Pichette. 2005. Have there now been any great women artists? An investigation of the visibility of women artists in recent art history textbooks. *Art Education* 58: 6–13.

Clark, R., P. J. Keller, A. Knights, J. A. Nabar, T. B. Ramsbey, & T. Ramsbey. 2006. Let me draw you a picture: Alternate and changing views of gender in award-winning picture books for children. Unpublished manuscript.

Clark, R., A. Knights, & A. Folgo. 2005. Measuring and predicting women's public empowerment: A cross-national analysis. Unpublished paper.

Clark, R., & H. Kulkin. 1996. Toward a multicultural feminist perspective on fiction for young adults. *Youth & Society* 27(3): 291–312.

Clark R., & A. Lang. 2001. *Balancing yin and yang: Teaching and learning qualitative data analysis within an undergraduate data analysis course.* Presented at the Eastern Sociological Society Meetings. Philadelphia. March.

Clark, R., R. Lennon, & L. Morris. 1993. Of Caldecotts and Kings: Gendered images in recent American children's books by black and non-black illustrators. *Gender & Society* 7: 227–245.

Clark, R., & T. Ramsbey. 1986. Social interaction and literary creativity. *International Review of Modern Sociology* 16: 105–117.

Clark, R., & G. Rice. 1982. Family constellations and eminence: The birth orders of Nobel Prize winners. *Journal of Psychology* 110: 281–287.

Collier, M. 2001. Approaches to analysis in visual anthropology. In *Handbook of visual analysis,* 35–60, edited by T. van Leeuwan & C. Jewitt. Thousand Oaks, CA: Sage.

Converse, J. M., & H. Schuman. 1974. *Conversations at random: Survey research as interviewers see it.* New York: Wiley.

Cooky, C., & M. G. McDonald. 2005. "If you let me play": Young girls' insider-other narratives of sport. *Sociology of Sport Journal* 22: 158–177.

Corrigan, J., K. Smith-Knapp, & C. Granger. 1998. Outcomes in the first 5 years after traumatic brain injury. *Archives of Physical Medical Rehabilitation* 79: 298–305.

COSSA. 2005. *COSSA Washington Update* 24(4): 1–49. http://www.cossa.org/volume24/24.4.pdf.

Couper, M. P., & S. E. Hansen. 2001. Computer-assisted interviewing. In *Handbook of Interview Research*, 557–575, edited by J. F. Gubrium & J. A. Holstein. Thousand Oaks, CA: Sage Publications.

Cowan, G., & R. R. Campbell. 1994. Racism and sexism in interracial pornography: A content analysis. *Psychology of Women Quarterly* 1918: 323–338.

Cramer, K. M., M. D. Gallant, & M. W. Langlois. 2005. Self-silencing and depression in women and men: Comparative structural equations models. *Personality and Individual Differences* 39: 581–592.

Crawford, S., M. Couper, & M. Lamias. 2001. Web surveys: Perceptions of burden. *Social Science Computer Review* 19: 146–162.

Cuddy, A. J. C., M. L. Norton, & S. T. Fiske. 2005. This old stereotype: The pervasiveness and persistence of the elderly stereotype. *Journal of Social Issues* 61(2): 267–285.

Culver, L. 2004. The impact of new immigration patterns on the provision of policy services in midwestern communities. *Journal of Criminal Justice* 32(4): 329–344.

Curtin, R., S. Presser, & E. Singer. 2000. The effects of response rate changes on the index of consumer sentiment. *The Public Opinion Quarterly* 64(4): 413–428.

Curtin, R., S. Presser, & E. Singer. 2005. Changes in telephone survey nonresponse over the past quarter century. *Public Opinion Quarterly* 69: 87–98.

DARE. 2006. About DARE. http://www.dare.com/home/about_dare.asp.

Davidson, J. O., & D. Layder. 1994. *Methods, sex and madness*. New York: Routledge.

Davis, A. J. 1984. Sex-differentiated behaviors in non-sexist picture books. *Sex Roles* 11: 1–15.

Davis, D. W. 1997. Nonrandom measurement error and race of interviewer effects among African Americans. *Public Opinion Quarterly* 61: 183–206.

Davison, J. 2004. Dilemmas in research: Issues of vulnerability and disempowerment for the social worker/researcher. *Journal of Social Work Practice* 18: 379–393.

De Meyrick, J. 2005. Approval procedures and passive consent considerations in research among young people. *Health Education* 105(4): 249–258.

Delva. J., M. P. Smith, R. L. Howell, D. Harrison, D. Wilke, & L. Jackson. 2004. A study of the relationship between protective behaviors and drinking consequences among undergraduate college students. *Journal of American College Health* 53(1): 19–26.

Denzin, N. 1970. *Sociological methods*. Chicago: Aldine.

Denzin, N. 1989. *The research act: A theoretical introduction to sociological methods*. Englewood Cliffs, NJ: Prentice-Hall.

Denzin, N. K. 1997. Triangulation in educational research. In *Educational research, methodology, and measurement: An international handbook*, 318–322, edited by J. P. Keeves. New York: Pergamon.

Desrochers, S. J. M. Hilton, & L. Larwood. 2005. Preliminary validation of the work-family integration-blurring scale. *Journal of Family Issues* 26: 442–466.

Deutscher, I. 1978. Asking questions cross-culturally: Some problems of linguistic comparability. In *Sociological methods: A sourcebook*, edited by N. K. Denzin. New York: McGraw-Hill.

DeVault, M. 1999. *Liberating method: Feminism and social research*. Philadelphia: Temple University Press.

Deveny, J. P. 1998. DARE—Cut it or keep it? Retrieved on August 2, 2001, from the Portland Police Department website at www.portlandpd.com/darecutit.html.

Diallo, A., G. De Serres, A. Beavogui, C. Lapointe, & P. Viens. 2001. Home care of malaria-infected children of less than 5 years of age in a rural area of the Republic of Guinea. *Bulletin of the World Health Organization* 79: 28–38.

Dickert. N., & J. Sugarman. 2005. Ethical goals of community consultation in research. *American Journal of Public Health*. 95(7): 1123–1127.

Diener, E. 2001. Satisfaction with life scale. Retrieved on August 12, 2001, from http://www.psych.uiuc.edu/~ediener/hottopic/hottopic.html.

Diener, E., R. A. Emmons, R. J. larsen, & S. Griffin. 1985. The satisfaction with life scale. *Journal of Personality Assessment* 49: 71–75.

Dillman, D. A., M. D. Sinclair, & J. R. Clark. 1993. Effects of questionnaire length, respondent-friendly design and a difficult question on response rates for occupant-addressed census mail surveys. *Public Opinion Quarterly* 57: 289–304.

Dion, M. R. 2005. Healthy marriage programs: learning what works. *The Future of Children* 15(2): 139–156.

Divett, M., N. Crittenden, & R. Henderson. 2203. Actively influencing consumer loyalty. *Journal of Consumer Marketing* 20(2): 109–126.

Diviak, K. R., S. J. Curry, S. L. Emery, & R. Mermelstein. 2004. Human participants challenges in youth tobacco cessation research: Researchers' perspectives. *Ethics & Behavior* 14(4): 321–334.

Dixon, M., V. J. Roscigno, & R. Hodson. 2004. Unions, solidarity, and striking. *Social Forces* 83: 3–33.

Dohan, D., & M. Sanchez-Jankowski. 1998. Using computers to analyze ethnographic field data: Theoretical and practical considerations. *Annual Review of Sociology* 24: 477–499.

Donohue, J. J., & S. D. Levitt. 2001. The impact of legalized abortion on crime. *Quarterly Journal of Economics* 116: 379–420.

Donohue, J. J., & S. D. Levitt. 2003. Further evidence that legalized abortion lowered crime: A reply to Joyce. NBER Working Paper No. 9532. February. 1–30.

Douglas, J. D. 1976. *Investigating social research: Individual and team field research.* Beverly Hill, CA: Sage.

Dovring, K. 1973. Communication, dissenters and popular culture in eighteenth-century Europe. *Journal of Popular Culture* 7: 559–568.

Dunn, M. 1997. *Black Miami in the twentieth century.* Gainesville: University Press of Florida.

Durkheim, É. 1938. *The rules of sociological method.* New York: Free Press.

Durkheim, É. [1897] 1964. *Suicide.* Reprint, Glencoe, IL: Free Press.

Durst, D., & A. Lange. 2002. Politicization and social integration of El Salvador Refugees in Regina, Canada. *Migration* 36–38: 135–150.

Dyer, J. 2001a. *Ethics and orphans.* Published by *San Jose Mercury News.* Retrieved on June 10, 2001, from www.mercurycenter.com/resources/search/

Dyer, J. 2001b. *An experiment leaves a lifetime of anguish.* Published by San Jose Mercury News. Retrieved on June 11, 2001, from www0.mercurycenter.com/resources/search/.

The Economist. 1997. Don't let it happen again. *Economist* 343: 27–28.

Eliassen, A. H., J. Taylor, & D. A. Lloyd. 2005. Subjective religiosity and depression in the transition to adulthood. *Journal for the Scientific Study of Religion* 44: 187–199.

Ellison, C. G., & D. A. Gay. 1989. Black political participation revisited: A test of compensatory, ethnic community, and public arena models. *Social Science Quarterly* 70: 101–119.

Ennett, S. T., N. S. Tobler, C. L. Ringwalt, & R. L. Flewelling. 1994. How effective is drug abuse resistance education? A meta-analysis of Project DARE outcome evaluations. *American Journal of Public Health* 84: 1394–1401.

Erikson, K. T. 1976. *Everything in its path: Destruction of community in the Buffalo Creek flood.* New York: Simon & Schuster.

Esposito, L., & J. W. Murphy. 2000. Another step in the study of race relations. *The Sociological Quarterly* 41: 171–187.

Farrell, G., & K. Clark. 2004. What does the world spend on criminal justice? HEUNI Paper No. 20. Helinski, Finland: HEUNI. http://www.heuni.fi/uploads/qjy0ay2w7l.pdf; downloaded on 1/10/06.

FCC. 2005. http://www.fcc.gov/Bureaus/Common_Carrier/Reports/FCC-State_Link/IAD/subs0305.pdf.

Feagin, J. R., A. M. Orum, & G. Sjoberg. 1991. Conclusion: The present crisis in U.S. sociology. In *A case for the case study,* 269–278, edited by J. R. Feagin, A. M. Orum, & G. Sjoberg. Chapel Hill: University of North Carolina Press.

Feagin, J. R., & H. Vera. 2001. *Liberation sociology.* Boulder, CO: Westview.

Federal Bureau of Investigation. 1995. *Uniform crime reports for the United States.* Washington, DC: U.S. Government Printing Office.

Ferraro, F. R., E. Szigeti, K. Dawes, & S. Pan. 1999. A survey regarding the University of North Dakota institutional review board: Data, attitudes and perceptions. *Journal of Psychology* 133: 272–381.

Fetterman, D., & M. Eiler. 2000. Empowerment evaluation: A model for building evaluation and program capacity. Presented at the American Evaluation Association Meetings, Honolulu, Hawaii, Nov. 1–4, 2000. Retrieved on August 2, 2001, from http://www.stanford.edu/~davidf/Capacitybld.pdf.

Fields, J. 2005. "Children having children": Race, innocence, and sexuality education. *Social Problems* 52: 549–571.

Finch, J., & L. Wallis. 1993. Death, inheritance and the life course. *Sociological Review Monograph,* 50–68.

Fine, G. A., & K. D. Elsbach. 2000. Ethnography and experiment in social psychological theory building: Tactics for integrating qualitative field data with quantitative lab data. *Journal of Experimental Social Psychology* 36: 51–76.

Fine, M., N. Freudenberg, Y. Payne, T, Perkins, K. Smith, & K. Wanzer. 2003. "Anything can happen with police around": Urban youth evaluate strategies of surveillance in public places. *Journal of Social Issues* 59(1): 141–158.

Fine, M., L. Weis, S. Wesson, & L. Wong. 2000. For whom? Qualitative research, representations, and social responsibilities. In *Handbook of qualitative research,* 2nd edition, 107–132, edited by N. K. Denzin & Y. S. Lincoln. Thousand Oaks, CA: Sage.

Finkel, S. E., T. M. Gutterbok, & M. J. Borg. 1991. Race-of-interviewer effects in a preelection poll: Virginia 1989. *Public Opinion Quarterly* 55: 313–330.

Fischman, W., B. Solomon, D. Greenspan, & H. Gardner. 2004. Making good: How young people cope with moral dilemmas at work. Cambridge, MA: Harvard University Press.

Flay, B. R., & L. M. Collins. 2005. Historical review of school-based randomized trials for evaluating problem behavior prevention programs. The ANNALS of the American Academy of Political and Social Science, 599(May): 115–146.

Fontana, A., & J. H. Frey. 1994. Interviewing: The art of science. In *Handbook of qualitative research,* 361–376, edited by N. K. Denzin & Y. S. Lincoln. Thousand Oaks, CA: Sage.

Fontana, A., & J. H. Frey. 2000. The interview: From structured question to negotiated text. In *Handbook of qualitative research,* 2nd edition, 645–672, edited by N. K. Denzin & Y. S. Lincoln. Thousand Oaks, CA: Sage.

Fox, J., C. Murray, & A. Warm. 2003. Conducting research using web-based questionnaires: Practical, methodological, and ethical considerations. *International Journal of Social Research Methodology* 6(2): 167–180.

Frank, K. 2005. Exploring the motivations and fantasies of strip club customers in relations to legal regulations. *Archives of Sexual Behavior* 34: 487–506.

Frankenberg, R. 1993. *White women, race matters: The social construction of whiteness.* Minneapolis: University of Minnesota Press.

Frankfort-Nachmias, C., & D. Nachmias. 2000. *Research methods in the social sciences,* 6th edition. New York: Worth & St. Martin's.

Frey, J. H. 1983. *Survey research by telephone.* Beverly Hills, CA: Sage.

Friedan, B. 1963. *The feminine mystique.* New York: Norton.

Friedman, J., & M. Alicea. 1995. Women and heroin: The path of resistance and its consequences. *Gender & Society* 9: 432–449.

Frith, H., & C. Kitzinger. 1998. "Emotion work" as participant resource; A feminist analysis of young women's talk-in-interaction. *Sociology* 32: 299–321.

Fritsch, E. J., T. J. Caeti, & R. W. Taylor. 1999. Gang suppression through saturation patrol, aggressive curfew, and truancy enforcement: A quasi-experimental test of the Dallas anti-gang initiative. *Crime & Delinquency* 45: 122–139.

Gallup Organization. 1998. Americans' satisfaction and well-being at all-time high levels. Retrieved on August 12, 2001, from www.gallup.com/poll/releases/pr981023.asp.

Gano-Phillips, S., & F. D. Fincham. 1992. Assessing marriage via telephone interviews and written questionnaires: A methodological note. *Journal of Marriage and the Family* 54: 630–635.

Gans, H. J. 1982. *The urban villagers: Group and class in the life of Italian Americans.* New York: Free Press.

Garza, C., & M. Landeck. 2004. College freshman at risk—social problems at issue: An exploratory study of a Texas/Mexico border community college. *Social Science Quarterly* 85(5): 1390–1400.

Gauthier, J. B. 2002. *Measuring America: The decennial censuses from 1790–2000.* (US Bureau of the Census). Washington: US Government Printing Office.

Geer, J. G. 1988. What do open-ended questions measure? *Public Opinion Quarterly* 52: 365–371.

Geer, J. G. 1991. Do open-ended questions measure "salient" issues? *Public Opinion Quarterly* 55: 360–370.

Gibson, W., P. Callery, M. Campbell, A. Hall, & D. Richards. 2005. The digital revolution in qualitative research: Working with digital audio data through Atlas.Ti. *Sociological Research Online,* 10: np. http://www.socresonline.org.uk/home.html.

Glantz, L. H. 1996. Conducting research with children: Legal and ethical issues. *Journal of the American Academy of Child and Adolescent Psychiatry* 35: 1283–1292.

Glaser, B. G., ed. 1993. Examples of grounded theory: A reader. Mill Valley, CA: Sociology Press.

Glaser, B. G. 2005. The grounded theory perspective III: Theoretical coding. Mill Valley, CA: Sociology Press.

Glaser, B. G., & A. L. Strauss. 1967. The discovery of grounded theory: Strategies for qualitative research. New York: Aldine.

Glazer, N. 1977. General perspectives. In *Woman in a man-made world,* 1–4, edited by Nona Glazer & Helen Youngelson. Chicago: Rand McNally.

Glenn, N. D. 1998. The course of marital success and failure in five American 10-year marriage cohorts. *Journal of Marriage and the Family* 60(3): 569–576.

Gold, R. L. 1958. Roles in sociological field observations. *Social Forces* 36: 217–223.

Gold, R. L. 1997. The ethnographic method in sociology. *Qualitative Inquiry* 3: 388–402.

Goode, E. 1996. The ethics of deception in social research: A case study. *Qualitative Sociology* 19: 11–33.

Gorden, R. L. 1975. *Interviewing, strategy, techniques and tactics.* Homewood, IL: Dorsey.

Gormly, E. 2005. Evangelical solidarity among the Jews: A veiled agenda? A qualitative content analysis of Pat Robertson's 700 Club program. *Review of Religious Research* 46: 255–268.

Gottman, J. M., & R. W. Levenson. 2000. The timing of divorce: Predicting when a couple will divorce over a 14-year period. *Journal of Marriage and Family* 62: 737–745.

Gould, S. J. 1995. *Dinosaur in a haystack: Reflections in natural history.* New York: Harmony.

Gould, S. J. 1997. *Full house. The spread of excellence from Plato to Darwin.* New York: Three Rivers Press.

Gray, N. B., G. J. Palileo, & G. D. Johnson. 1993. Explaining rape victim blame: A test of attribution theory. *Sociological Spectrum* 13: 377–392.

Grenz, S. 2005. Intersections of sex and power in research on prostitution: A female researcher interviewing male heterosexual clients. *Signs: Journal of Women in Culture and Society* 30(4): 2091–2113.

Gronning, T. 1997. Accessing large corporations: Research ethics and gatekeeper-relations in the case of researching a Japanese-invested factory. *Sociological Research Online* 2. www.socresonline.org.uk/socresonline/2/4/9.html.

Groves, R. M. 1989. *Survey errors and survey costs.* New York: Wiley.

Groves, R. M., S. Presser, & A. Dipko. 2004. The role of topic interest in survey participation Decisions. *Public Opinion Quarterly* 68(1): 2–31.

Gubrium, J. F., & J. A. Holstein. 2001. From the individual interview to the interview society. In *Handbook of Interview Research,* 3–32, edited by J. F. Gubrium & J. A. Holstein. Thousand Oaks, CA: Sage Publications.

Hacker, D. 2002. *Research and documentation in the electronic age,* 2nd edition. Boston: Bedford Books.

Hadaway, C. K., & P. L. Marler. 1993. What the polls don't show: A closer look at U.S. church attendance. *American Sociological Review* 58: 741–752.

Hagan, J., & A. Palloni. 1998. Immigration and crime in the United States. In *The immigration debate,* 367–387, edited by J. P. Smith and B. Edmonton. Washington, DC: National Academy Press.

Hagestad, G. O. & P. Uhlenberg. 2005. The social separation of old and young: A root of ageism. *Journal of Social Issues* 61(2): 343–360.

Hall, R. A. 2004. Inside out: Some notes on carrying out feminist research in cross-cultural interviews with South Asian women immigration applicants. *International Journal of Social Research Methodology* 7(2): 127–141.

Halse, C., & A. Honey. 2005. Unraveling ethics: Illuminating moral dilemmas of research ethics. *Signs: Journal of Women in Culture and Society* 30(4): 2149–2162.

Hamer, J. 2001. *What it means to be Daddy: Fatherhood for black men living away from their children.* New York: Columbia University Press.

Hamarat, E., D. Thompson, K. M. Sabrucky, D. Steele, K. B. Matheny, & F. Aysan. 2001. Perceived stress and coping resource availability as predictors of life satisfaction in young, middle-aged and older adults. *Experimental Aging Research* 27: 181–196.

Hammersley, M. 1995. *The politics of social research.* London: Sage.

Hammersley, M., & P. Atkinson. 1983. *Ethnography: Principles in practice.* New York: Tavistock.

Hammersley, M., & P. Atkinson. 1995. *Ethnography: Principles in practice,* 2nd edition. London: Routledge.

Hamnett, M., D. Porter, A. Singh, & K. Kumar. 1984. *Ethics, politics, and international social science research: From critique to praxis.* Hawaii: University of Hawaii Press for East-West Center.

Harding, S. 1993. Rethinking standpoint epistemology: What is "strong objectivity"? In *Feminist epistemologies,* 49–82, edited by L. Alcoff & E. Potter. New York: Routledge.

Harkness, S., & C. A. B. Warren. 1993. The social relations of intensive interviewing. *Sociological Methods & Research* 21: 317–339.

Harocopos, A., & D. Dennis. 2003. Maintaining contact with drug users over an 18-month period. *International Journal of Social Research Methodology* 6(3): 261–265.

Harter, L. M., C. Berquist, B. S. Titsworth, D. Novak, & T. Brokaw. 2005. The structuring of invisibility among the hidden homeless: The politics of space, stigma, and identity construction. *Journal of Applied Communication Research* 33: 305–327.

Health and Retirement Study (HRS). 2001. Retrieved on July 14, 2001 from http://www.umich.edu/~hrswww/studydet/develop/history.html.

Hembroff, L. A., R. Rusz, A. Rafferty, H. McGee, & N. Ehrlich. 2005. The cost-effectiveness of alternative advance mailings in a telephone survey. *Public Opinion Quarterly* 69(2): 232–245.

Hennigan, K. M., C. L. Maxson, D. Sloane, & M. Ranney. 2002. Community views on crime and policing: survey modes effects on bias in community surveys. *Justice Quarterly* 19(3): 565–587.

HERO. 2006. Higher education and research opportunities in the United Kingdom webpage at http://www.hero.ac.uk/uk/research/the_uk_research_councils/economic_and_social_research_c221.cfm.

Herrera, C. D. 1997. A historical interpretation of deceptive experiments in American psychology. *History of the Human Sciences* 10: 23–36.

Herzog, A. R., & W. L. Rogers. 1988. Interviewing older adults. *Public Opinion Quarterly* 52: 84–99.

Hess, I. 1985. *Sampling for social research surveys 1947–1980.* Ann Arbor, MI: Institute for Social Research.

Hewa, S. 1993. Sociology and public policy: The debate on value-free social science. *International Journal of Sociology and Social Policy* 13: 64–82.

Hill, E. 1995. Labor market effects of women's post-school-age training. *Industrial and Labor Relations Review* 49: 138–149.

Hills, D. 1998. Engaging new social movements. *Human Relations* 51: 1457–1476.

Hirschi, T. 1969. *Causes of delinquency.* Berkeley: University of California Press.

Hobbs, F. 2005. *Examining American Household Composition: 1990 and 2000.* U.S. Census Bureau, Census 2000 Special Reports, CENSR-24. I.S. Government Printing Office, Washington, DC.

Hochschild, A. R. 1983. *The managed heart: Commercialization of human feeling.* Berkeley: University of California Press.

Hochschild, A. R., with A. Machung. 1989. *The second shift.* New York: Avon Books.

Hofferth, S. L. 2005. Secondary data analysis in family research. *Journal of Marriage and Family* 67: 891–907.

Hoffnung, M. 2001, March. *Maintaining participation in feminist longitudinal research: Studying young women's lives over time.* Paper presented at the National Conference of the Association for Women in Psychology, Los Angeles, CA.

Hogan, R. 2005. Was Wright wrong? High-class jobs and the professional earnings advantage. *Social Science Quarterly* 86: 645–661.

Holbrook, A. L., M. C. Green, & J. A. Krosnick. 2003. Telephone versus face-to-face interviewing of national probability samples with long questionnaires. *Public Opinion Quarterly* 67(1): 79–125.

Holden, G., P. Miller, & S. Harris. 1999. The instrumental side of corporal punishment: Parents' reported practices and outcome expectancies. *Journal of Marriage and the Family* 61: 908–919.

Holster, K. 2004. The business of internet egg donation: An exploration of donor and recipient experiences. Dissertation Abstracts International.

Holsti, O. R. 1969. *Content analysis for the social sciences and the humanities.* Reading, MA: Addison-Wesley.

Holyfield, L. 2002. *Moving up and out: Poverty, education and the single parent family.* Philadelphia: Temple University Press.

Hooks, Bell, & C. Raschka. 1999. *Happy to be nappy.* New York: Hyperion Books for Children.

Hoover, E. 2005. The ethics of undercover research. *Chronicle of Higher Education.* Volume 51, Issue 47, Page A36 http://chronicle.com/free/v51/i47/47a03601.htm on 8/7/05.

Hubbard, A. 1996. School board approves random drug testing. *Kokomo Tribune,* February 6, 1996: 1.

Huff, D. 1954. *How to lie with statistics.* New York: Norton.

Hughes, E. C. 1960. Introduction: The place of field work in social science. In *Field work: An introduction to the social sciences,* iii–xiii, edited by B. H. Junker. Chicago: University of Chicago.

Huisman, K., & P. Hondagneu-Sotelo. 2005. Dress matters: Change and continuity in the dress practices of Bosnia Muslim refugee women. *Gender & Society* 19: 44–65.

Humphreys, L. 1975. *Tearoom trade: Impersonal sex in public places.* Chicago: Aldine.

Hutchinson, A. 2005. Analysing audio-recorded data: using computer software applications. *Nurse Researcher* 12: 20–31.

Ibanez, G. E., N. Khatchikian, C. Buck, D. L. Weisshaar, T. Abush-Kirsh, E. A. Lavizzo, & F. H. Norris., 2003. Qualitative analysis of social support and conflict among Mexican and Mexican-American disaster survivors. *Journal of Community Psychology* 31(1): 1:23.

Inter-Parliamentary Union. (2004). Women's suffrage. Retrieved on April 19, 2003, from http://www.ipu.org/wmn-e/suffrage.htm.

Iutocovich, J., S. Hoppe, J. Kennedy, & F. Levine. 1999. Confidentiality and the 1997 ASA code of ethics: A response from COPE. *Footnotes* (February): 5.

Jackall, R. 1988. *Moral mazes: The world of corporate managers.* New York: Oxford University Press.

Jackson, D. 2005. Say no to hip-hop's excesses. *Boston Globe,* November 12, 2005. Online at http://www.boston.com/news/globe/editorial_opinion/oped/articles/2005/11/12/say_no to hip_hops_excesses?mode=PF.

Jaggar, E. 2005. Is thirty the new sixty? Dating, age and gender in postmodern, consumer society. *Sociology* 39: 89–106.

Jarvinen, J. 2000. The biographical illusion: Constructing meaning in qualitative interviewing. *Qualitative Inquiry* 6: 370–391.

Jayartane, T. E., & A. J. Sewart. 1989. Quantitative and qualitative method in the social sciences. In *Beyond methodology feminist scholarship as lived research,* 85–106, edited by M. M. Fonow, & J. A. Cook. Bloomington: Indiana University Press.

Johnson, T. P., J. Hougland, & R. Clayton. 1989. Obtaining reports of sensitive behavior: A comparison of telephone and face-to-face interviews. *Social Science Quarterly* 70: 174–183.

Jones, J. H. 1981. *Bad blood: The Tuskegee syphilis experiment.* New York: Free Press.

Jones, R. K. 2000. The unsolicited diary as a qualitative research tool for advanced research capacity in the field of health and illness. *Qualitative Health Research* 10: 555–567.

Joyce, T. 2003. Did legalized abortion lower crime? *Journal of Human Resources* 38: 1–37.

Kahneman, D., A. B. Krueger, D. A. Schkade, N. Schwarz, & A. A. Stone. 2004. A survey method for characterizing daily life experience: The day reconstruction method. *Science* 306(5702): 1776–1780.

Kaiser Family Foundation. 2002. *Sex on TV. Executive Summary.* Menlo Park, CA: Henry J. Kaiser Family Foundation. Retrieved on February 11, 2006, from http://www.kff.org/entmedia/entmedia110905pkg.cfm.

Kallgren, K., R. R. Reno, & R. B. Cialdini. 2000. A focus theory of normative conduct: When norms do and do not affect behavior. *Personality and Social Psychology Bulletin* 26: 1002–1012.

Kaplowitz, M. D., T. D. Hadlock, & R. Levine. 2004. A comparison of web and mail survey response rate. *Public Opinion Quarterly* 68(1): 94–101.

Karney, B. R., J. Davila, C. L. Cohan, K. T. Sullivan, M. D. Johnson, & T. N. Bradbury. 1995. An empirical investigation of sampling strategies in marital research. *Journal of Marriage and the Family* 57: 909–920.

Karp, J. A., & D. Brockington. 2005. Social desirability and response validity: A comparative analysis of overreporting voter turnout in five countries. *The Journal of Politics* 67(3): 825–840.

Kaufman, P., & K. Feldman. 2004. Forming identities in college: A sociological approach. *Research in Higher Education* 45(5): 463–496.

Keeter, S., C. Miller, A. Kohut, R. Groves, & S. Presser. 2000. Consequences of reducing nonresponse in a national telephone survey. *Public Opinion Quarterly* 64: 125–148.

Kemmis, S., & R. McTaggart. 2003. Participatory action research. In *Strategies of Qualitative Inquiry,* 2nd edition, 336–3961, edited by N. K. Denzin & Y. S. Lincoln. Thousand Oaks, CA: Sage.

Kneipp, S., J. Castleman & N. Gailor. 2004. Informal caregiving burden: An overlooked aspect of the lives and health of women transitioning from welfare to employment? *Public health nursing* 21(1): 24–31.

Kenschaft, L. 2005. *Reinventing marriage. The love and work of Alice Freeman Palmer and George Herbert Palmer.* Urbana and Chicago: University of Illinois Press.

Koss, M. P., C. Gidycz, & N. Wisniewski. 1987. The scope of rape: Incidence and prevalence of sexual aggression and victimization in a national sample of higher education students. *Journal of Consulting and Clinical Psychology* 55: 162–171.

Kotlowitz, A. 1991. *There are no children here.* New York: Doubleday.

Krafchick, J. L., T. S. Zimmerman, S. A. Haddock, & J. H. Banning. 2005. Best-selling books advising parents about gender: A feminist analysis. *Family Relations* 54: 84–100.

Krimerman, L. 2001. Participatory action research: Should social inquiry be conducted democratically? *Philosophy of Social Sciences* 31: 60–82.

Krippendorff, K. 2003. *Content analysis: An introduction to its methodology. 2nd edition.* Beverly Hills, CA: Sage.

Krysan, M., H. Schuman, L. J. Scott, & P. Beatty. 1994. Response rates and response content in mail versus face-to-face surveys. *Public Opinion Quarterly* 58: 381–400.

KSL. 2000. Mayor dumps DARE program. KSL-TV website. Retrieved on August 2, 2001, at www.ksl.com/dump/news/cc/dare.htm.

Lackey, J. F., D. P. Moberg, & M. Balistrieri. 1997. By whose standards? Reflections and empowerment evaluation and grassroots groups. *Evaluation Practice* 18: 137–146.

Ladd, E. C., & K. Bowman. 1998. *What's wrong: A survey of American satisfaction and complaint.* Washington, DC: AEI Press.

Lang, A. 2001. *Economic inequality: A quantitative study of opinions related to the income gap and government involvement in its reduction.* Honors thesis at Rhode Island College.

Lasswell, H. D. 1965. Detection: Propaganda detection and the courts. In *The language of politics: Studies in quantitative semantics,* edited by Harold Lasswell et al., Cambridge, MA: MIT Press.

Lauder, M. A. 2003. Covert participation observation of a deviant community: Justifying the use of deception. *Journal of Contemporary Religion* 18(2): 185–196.

Laumann, E. O., J. H. Gagnon, R. T. Michael, & S. Michaels. 1994. *The social organization of sexuality.* Chicago: University of Chicago Press.

Lavrakas, P. J. 1987. *Telephone survey methods sampling, selection and supervision.* Newbury Park, CA: Sage.

Lazarsfeld, P. F. 1944. The controversy over detailed interviews: An offer for negotiation. *Public Opinion Quarterly* 8: 38–60.

Lee, J. 2002. From civil relations to racial conflict: Merchant-customer interactions in urban America. *American Sociological Review* 67(1): 77–98.

Lee, M. R., & G. C. Ousey. 2005. Institutional access, residential segregation, and urban black homicide. *Sociological Inquiry* 75: 3–54.

Lee, V. E., & J. B. Smith. 1995. Effects of high school restructuring and size on early gains in achievement and engagement. *Sociology of Education* 68: 241–270.

Lemert, C. 1993. *Social theory: The multicultural and classic readings*. Boulder, CO: Westview.

Leo, R. A. 1995. Trial and tribulations: Courts, ethnography, and the need for an evidentiary privilege for academic researchers. *American Sociologist* 26: 113–134.

Leslie, P. L. 2005. The presentation of college professors in popular film. Southern Sociological Society. Charlotte, NC.

Lester, D., & D. Lester, Jr., 2006. *Writing research papers in the social sciences*. New York: Longman.

Lever, J. 1981. Multiple methods of data collection. *Urban Life* 10: 199–213.

Levitt, S. D. 2004. Understanding why crime fell in the 1990s: Four factors that explain the decline and six that do not. *Journal of Economic Perspectives* 18: 163–190.

Lewis, C. A., M. J. Dorahy, & J. F. Schumaker. 1999. Depression and life satisfaction among Northern Irish adults. *Journal of Social Psychology* 139: 533–535.

Lewis, T. L., & C. R. Humphrey. 2005. Sociology and the environment; An analysis of coverage in introductory sociology textbooks. *Teaching Sociology* 33: 154–169.

Lichter, D. T., Z. Qian, & M. L. Crowley. 2005. Child poverty among racial minorities and immigrants: explaining trends and differentials. *Social Science Quarterly* 86: 1037–1051.

Liebow, E. 1967. *Tally's corner: A study of Negro streetcorner men*. Boston: Little, Brown.

Liebow, E. 1993. *Tell them who I am: The lives of homeless women*. New York: Free Press.

Lindsey, E. W. 1997. Feminist issues in qualitative research with formerly homeless mothers. *Affilia: Journal of Women and Social Work* 12: 57–76.

Lindsay, J. (2005) Getting the numbers: The unacknowledged work in recruiting for survey research. *Field Methods* 17(1): 119–128.

Link, M. L., & A. Mokdad. 2005. Advance letters as a means of improving respondent cooperation in random digit dial studies: A multistate experiment. *Public Opinion Quarterly* 69(4): 572–587.

Lo, C. 2000. Timing of drinking initiation: A trend study predicting drug use among high school seniors. *Journal of Drug Issues* 30: 525–534.

Loe, M. 2004. Sex and the senior woman: Pleasure and danger in the Viagra era. *Sexualities* 7: 303–326.

Lofland, J. 1984. *Analyzing social situations*. Belmont, CA: Wadsworth.

Lofland, J., & L. H. Lofland. 1995. *Analyzing social settings: A guide to qualitative observation and analysis,* 3rd ed. Belmont, CA: Wadsworth.

Lofland, J., D. Snow, L. Anderson, L. H. Lofland. 2005. *Analyzing social settings: A guide to qualitative observation and analysis*, 4th ed. Belmont, CA: Wadsworth.

Loftus, E. F., M. R. Klinger, K. D. Smith, & J. Fielder. 1990. A tale of two questions: Benefits of asking more than one question. *Public Opinion Quarterly* 54: 330–345.

Long, J. S., & E. K. Pavalko. 2004. The life course of activity limitations: Exploring indicators of functional limitations over time. *Journal of Aging and Health* 16(4): 490–516.

Lopata, H. Z. 1980. Interviewing American widows. In *Fieldwork experience qualitative approaches to social research,* edited by W. B. Shaffir, R. A. Stebbins, & A. Turowetz. New York: St. Martin's.

Lord, M. 2001. Mortality goes to school. *U.S. News & World Report* (June 4, 2001) 130: 50–51.

Louie, M. C. Y. 2001. *Sweatshop warriors. Immigrant women workers take on the global factory*. Cambridge, MA: South End Press.

Lowman, J., & T. Palys. 1999. Confidentiality and the 1997 ASA code of ethics: A query. *Footnotes* (February): 5.

Lynd, R. S., & H. M. Lynd. 1956. *Middletown: A study in American culture*. New York: Harcourt, Brace, Jovanovich.

Lystra, K. 1989. *Searching the heart: Women, men and romantic love in nineteenth-century America*. New York: Oxford University Press.

Macaulay, A. P., K. W. Griffin, E. Gronewold, C. Williams, & G. J. Botvin. 2005. Parenting practices and adolescent drug-related knowledge, attitudes, norms and behavior. *Journal of Alcohol & Drug Education.* 49(2): 67–83.

Madriz, E. 2000. Focus groups in feminist research. In *Handbook of qualitative research,* 2nd edition, 835–850, edited by N. K. Denzin & Y. S. Lincoln. Thousand Oaks, CA: Sage.

Maher, F. A., & M. K. Tetreault. 1994. *The feminist classroom.* New York: Basic.

Maher, K. 2005. Practicing sociology as a vocation. In *Explorations in theology and vocation,* 31–40, edited by W. E. Rogers. Greenville, NC: Furman University's Center for Theological Explorations of Vocation.

Malinowski, B. 1922. *Argonauts of the Western Pacific.* New York: Dutton.

Manning, P. K. 2002. FATHETHICS: Response to Erich Goode. *Qualitative Sociology* 25(4): 541–547.

Manza, J., & C. Brooks. 1998. The gender gap in U.S. presidential elections: When? Why? Implications? *American Journal of Sociology* 103: 1235–1266.

Marecek, J., M. Fine, & L. Kidder. 1997. Working between worlds: Qualitative methods and social psychology. *Journal of Social Issues* 53: 631–645.

Martin, L. L., T. Abend, C. Sedikides, & J. D. Green. 1997. How would I feel if . . . ? Mood as input to a role fulfillment evaluation process. *Journal of Personality and Social Psychology* 73: 242–254.

Martineau, H. 1962. *Society in America.* New York: Anchor.

Martinez, R., & M. T. Lee. 2000. Comparing the context of immigrant homicides in Miami: Haitians, Jamaicans and Mariel. *International Migration Review* 34: 794–812.

Marx, Karl. [1867] 1967. *Das Kapital.* Reprint. New York: International.

Massey, S., A. Cameron, S. Ouellette, & M. Fine. 1998. Qualitative approaches to the study of thriving: What can be learned? *Journal of Social Issues* 54: 337–355.

Maycock, B., & P. Howat. 2005. The barriers to illegal anabolic steroid use. *Drugs: Education, Prevention & Policy* 12: 317–325.

McBride, D., & A. Gienapp. 2000. Using randomized designs to evaluate client-centered programs to prevent adolescent pregnancy. *Family Planning Perspectives* 32: 227–235.

McCall, G. J. 1984. Systematic field observation. *Annual Review of Sociology* 10: 263–282.

McCorkel, J., & K. Myers. 2003. What difference does difference make? Position and privilege in the field. *Qualitative Sociology* 26(2): 199–231.

McGraw, L., A. Zvonkovic, & A. Walker. 2000. Studying postmodern families: A feminist analysis of ethical tensions in work and family research. *Journal of Marriage and the Family* 62 (Feb.): 68–77.

McNabb, S. 1995. Social research and litigation: Good intentions versus good ethics. *Human Organization* 54: 331–335.

Mead, M. 1933. *Sex and temperament in three primitive societies.* New York: Morrow.

Mears, A., & W. Finlay. 2005. Not just a paper doll: How models manage bodily capital and why they perform emotional labor. *Journal of Contemporary Ethnography* 34: 317–343.

Melrose, M. 2002. Labour pains: Some considerations on the difficulties of researching juvenile prostitution. *International Journal of Social Research Methodology* 5(4): 333–351.

Menard, S. 1991. *Longitudinal research.* Newbury Park, CA: Sage.

Merton, R. K., M. Fiske, & P. L. Kendall. 1956. *The focused interview.* Glencoe, IL: Free Press.

Messner, S. F. 1999. Television violence and violent crime: An aggregate analysis. In *How it's done: An invitation to social research,* 308–316, by E. Stier Adler & R. Clark. Belmont, CA: Wadsworth.

Michael, R. T., J. H. Gagnon, E. O. Laumann, & G. Kolata. 1994. *Sex in America.* Boston: Little, Brown.

Michell, L. 1998. Combining focus groups and interviews: Telling how it is; telling how it feels. In *Developing focus group research,* 36–46, edited by R. S. Barbour & J. Kitzinger. Thousand Oaks, CA: Sage.

Midanik, L. T., T. K. Greenfield, & J. D. Rogers. 2001. Reports of alcohol-related harm: Telephone versus face to face interviews. *Journal of Studies on Alcohol* 62: 74–78.

Mies, M. 1991. Women's research or feminist research? In *Beyond methodology: Feminist scholarship as lived research,* 60–84, edited by M. M. Fonow & J. A. Cook. Bloomington: Indiana University Press.

Miles, M. B., & A. M. Huberman. 1994. *Qualitative data analysis,* 2nd edition. Thousand Oaks, CA: Sage.

Milgram, S. 1974. *Obedience to authority: An experimental view.* New York: Harper & Row.

Miller, D. C., & N. J. Salkind., 2002. *Handbook of research design and social measurement.* Thousand Oaks, CA: Sage.

Miller, E. M. 1986. *Street woman.* Philadelphia: Temple University Press.

Miller, J. M., & J. A. Krosnick. 2004. Threat as a motivator of political activism: A field experiment. *Political Psychology* 25(4); 507–523.

Miller, W. E., & S. Traugott. 1989. *American National Election Studies Data Sourcebook, 1952–1986.* Cambridge, MA: Harvard University Press.

Mishler, E. G. 1986. *Research interviewing context and narrative.* Cambridge, MA: Harvard University Press.

Mishler, E. G. 2000. *Storylines: Craftartists' narratives of identity.* Cambridge, MA: Harvard University Press.

Mitchell, A. 1997. Survivors of Tuskegee study get apology from Clinton. *New York Times* 146: 9–10.

Morgan, D. L. 1996. Focus groups. *Annual Review of Sociology* 22: 129–153.

Morgan, D. L. 2001. Focus group interviewing. In *Handbook of Interview Research,* 141–159, edited by J. F. Gubrium & J. A. Holstein. Thousand Oaks, CA: Sage Publications.

Morris, M., & L. Jacobs. 2000. You got a problem with that? Exploring evaluators' disagreements about ethics. *Evaluation Review* 24: 384–406.

Morrow, R. 1994. *Critical theory and methodology.* Thousand Oaks, CA: Sage.

Moskowitz, J. M. 2004. Assessment of cigarette smoking and smoking susceptibility among youth. *Public Opinion Quarterly* 68(4): 565–587.

Mosteller, F. 1955. Use as evidenced by an examination of wear and tear on selected sets of ESS. In *A study of the need for a new encyclopedic treatment of the social sciences,* edited by K. Davis. (unpublished).

Muir, K. B., & T. Seitz. 2004. Machismo, misogyny, and homophobia in a male athletic subculture: Participant-observation study of deviant rituals in collegiate rugby. *Deviant Behavior* 25: 303–327.

Nathan, Rebekah. 2005. *My freshman year: What a professor learned by becoming a student.* Ithaca: Cornell University Press.

National Academy of Sciences. 1993. Methods and values in science. In *The "Racial" Economy of Science,* 341–343, edited by S. Harding. Bloomington: Indiana University Press.

National Alliance for Caregiving. 2005. Young caregivers in the U.S. findings from a national study. At http://www.uhfnyc.org/usr_doc/Young_Caregivers_Study_083105.pdf on 1/22/06.

National Center for Education Statistics. 1982. *Conferences, critiques and references on the subject of public and private schools: The Coleman report.* Washington, DC.

Neto, F., & J. Barros. 2000. Psychosocial concomitants of loneliness among students of Cape Verde and Portugal. *Journal of Psychology* 134: 503–514.

Newman, F. L., & M. J. Tejeda. 1996. The need for research that is designed to support decisions in the delivery of mental health services. *American Psychologist* 51(Oct.): 1040–1049.

NHS News. 2005. The Nurses' Health Study Annual Newsletter 12. Boston, MA. http://www.channing.harvard.edu/nhs/history/index.shtml.

NORC (National Opinion Research Center). 1995. PHDCN Project 4709. Systematic social observation coding manual, June 1995. Chicago: NORC.

NORC. 2006. General Social Survey: Study FAQs. Retrieved at http://www.norc.uchicago.edu/projects/gensoc3.asp on February 25, 2006.

Norgaard, K., & R. York. 2005. Gender equality and state environmentalism. *Gender & Society* 19: 506–522.

Norusis, M. J. 2005. *The SPSS guide to data analysis.* Chicago: SPSS Inc.

Nyden, P. A., A. Figert, M. Shibley, & D. Burrows. 1997. *Building community.* Thousand Oaks, CA: Pine Forge.

Oakley, A. 1981. Interviewing women: A contradiction in terms. In *Doing feminist research,* 30–61, edited by H. Roberts, London: Routledge & Kegan Paul.

Ogden, P. E., & R. Hall. 2000. Households, reurbanisation and the rise of living alone in the principal French cities 1975–90. *Urban Studies* 37 (InfoTrac #A61862671): 1–23.

Olson, S. 1976. *Ideas and data: The process and practice of social research.* Homewood, IL: Dorsey.

Oriel, K., M. B. Plane, & M. Mundt. 2004. Family medicine residents and the impostor phenomenon. *Family Medicine* 36: 248–252.

Orrange, R. M. 2002. Aspiring law and business professionals' orientations to work and family life. *Journal of Family Issues* 23: 287–317.

Orrange, R. M. 2003. The emerging mutable self: Gender dynamics and creative adaptations in defining work, family, and the future. *Social Forces* 82: 1–34.

Orum, A. M., J. R. Feagin, & G. Sjoberg. 1991. Introduction: The nature of the case study. In *A case for the case study,* 1–26, edited by J. R. Feagin, A. M. Orum, & G. Sjoberg. Chapel Hill: University of North Carolina Press.

Palys, T., & J. Lowman. 2001. Social research with eyes wide shut: The limited confidentiality dilemma. *Canadian Journal of Criminology* 43: 255–267.

Paradis, E. K. 2000. Feminist and community psychology ethics in research with homeless women. *American Journal of Community Psychology* 28: 839–854.

Park, P. 1993. What is participatory research? A theoretical and methodological perspective. In *Voices of change, participatory research in the United States and Canada,* edited by P. Park, M. Brydon-Miller, B. Hall, & T. Jackson. Westport, CT: Bergin & Garvey.

Pattman, R., S. Frosh, & A. Phoenix. 2005. Constructing and experiencing boyhoods in research in London. *Gender and Education* 17(5): 555–561.

Peirce, K., & E. Edwards. 1988. Children's construction of fantasy stories: Gender differences in conflict resolution strategies. *Sex Roles* 18: 393–404.

Perrin, A. J. 2005. National threat and political culture: Authoritarianism, antiauthoritarianism, and the September 11 attacks. *Political Psychology* 26: 167–194.

Perry, C. L., K. A. Komro, S. Veblen-Mortenson, & L. M. Bosma. 2000. The Minnesota DARE PLUS Project: Creating community partnerships to prevent drug use and violence. *Journal of School Health* 70: 84–88.

Pillemer, K., & D. Finkelhor. 1988. The prevalence of elder abuse: A random sample survey. *Gerontologist* 28: 51–57.

Plutzer, D., & M. Berkman. 2005. The graying of America and support for funding of the nation's schools. *Public Opinion Quarterly* 69(1): 66–86.

Portes, A., J. M. Clarke, & R. Manning. 1985. After Mariel: A survey of the resettlement experiences of 1980 Cuban refugees in Miami. *Cuban Studies* 15: 37–59.

Powell, R. A., H. M. Single, & K. R. Lloyd. 1996. Focus groups in mental health research: Enhancing the validity of user and provider questionnaires. *International Journal of Social Psychiatry* 42: 193–206.

Pyrczak, F., & R. R. Bruce. 2005. *Writing empirical research reports: A basic guide for students of the social and behavioral sciences,* 5th edition. Los Angeles, CA: Pyrczak.

Ptacek, J. 1988. Why do men batter their wives? In *Feminist perspectives on wife abuse,* edited by K. Yllo & Michele Bograd. Newbury Park, CA: Sage.

Public Agenda. 2002. Right to die: Major proposals. Public Agenda Online. www.publicagenda.org/issues/major-proposals_details.dfm? . . . Retrieved on April 18, 2002.

Punch, M. 1994. Politics and ethics of qualitative research. In *Handbook of qualitative research,* 83–97, edited by N. Denzin & Y. Lincoln. Thousand Oaks, CA: Sage.

Putnam, R. D. 2000. *Bowling alone. The collapse and revival of American community.* New York: Simon & Schuster.

Puwar, N. 1997. Reflections on interviewing women MPs. *Sociological Research Online* 2. www.socresonline.org.uk/socresonline/2/1/4.html.

Racher, F. E., J. M. Kaufert, & B. Havens. 2000. Conjoint research interviews with frail elderly couples: Methodological implications. *Journal of Nursing Research* 6: 367–379.

Rathje, W. L., & C. Murphy. 1992. *Rubbish! The archaeology of garbage.* New York: HarperCollins.

Reger, J. 2004. Organizational "emotion work" through consciousness-raising: An analysis of a feminist organization. *Qualitative Sociology* 27(2): 205–222.

Reinharz, S. 1992. *Feminist methods in social research.* New York: Oxford University Press.

Reisman, C. K. 1987. When gender is not enough: Women interviewing women. *Gender & Society* 1: 172–207.

Rennison, C. 2000. *Criminal victimization 1990.* Washington, DC: Bureau of Justice Statistics.

Rew, L., S. D. Horner, L. Riesch, & R. Cauvin. 2004. Computer-assisted survey interviewing of school-age children. *Advances in Nursing Science* 27(2): 129–137.

Rey, D., & J. Barkdull. 2005. Why do some democratic countries join more intergovernmental organizations than others? *Social Science Quarterly* 86: 386–401.

Richards, T. J., & L. Richards. 1994. Using computers in qualitative research. In *Handbook of qualitative research,* 445–462, edited by N. K. Denzin & Y. S. Lincoln. Thousand Oaks, CA: Sage.

Richman, W., S. Kiesler, S. Weisband, & F. Drasgow. 1999. A meta-analytic study of social desirability distortion in computer-administered questionnaires, traditional questionnaires and interviews. *Journal of Applied Psychology* 84: 754–775.

Riemer, J. W. 1977. Varieties of opportunistic research. *Urban Life* 5: 467–477.

Ringheim, K. 1995. Ethical issues in social science research within special reference to sexual behavior research. *Social Science and Medicine* 40: 1691–1697.

Robb, M. 2004. Exploring Fatherhood: Masculinity in the Research Process. *Journal of Social Work Practice* 18(3): 395–406.

Rocky Mountain Behavioral Science Institute (RMBSI). 1995. A model for evaluating DARE & other prevention programs. *News & Views* 22: 1–2.

Rosengren, K. E. 1981. Advances in Scandinavian content analysis. In *Advances in content analysis,* 9–19, edited by K. E. Rosengren. Beverly Hills, CA: Sage.

Rosenhan, D. L. 1971. On being sane in insane places. *Science* 179: 250–258.

Rossi, P. H., & H. E. Freeman. 1993. *Evaluation: A systematic approach.* Newbury Park, CA: Sage.

Rossi, P. H., & J. D. Wright. 1984. Evaluation research: An assessment. In *Annual review of sociology,* vol. 10, 332–352, edited by R. H. Turner & J. F. Short. Palo Alto: Annual Reviews.

Roy, K. M., C. Y. Tubbs, & L. M. Burton. 2004. Don't have no time: Daily rhythms and the organization of time for low-income families. *Family Relations* 53(2): 168–178.

Rubin, L. B. 1976. *Worlds of pain: Life in the working-class family.* New York: Basic.

Rubin, L. B. 1983. *Intimate strangers: Men and women together.* New York: Harper & Row.

Rubin, L. B. 1994. *Families on the fault line: America's working class speaks about the family, the economy, race, and ethnicity.* New York: HarperCollins.

Runcie, J. F. 1980. *Experiencing social research.* Homewood, IL: Dorsey.

Rutherford, M. B. 2004. Authority, autonomy, and ambivalence: Moral choice in twentieth-century commencement speeches. *Sociological Forum* 19: 583–609.

Sampson, R. J., & S. W. Raudenbush. 1999. Systematic social observation of public spaces: A new look at disorder in urban neighborhoods. *American Journal of Sociology* 105: 603–651.

Scarce, R. 1995. Scholarly ethics and courtroom antics: Where researchers stand in the eyes of the law. *American Sociologist* 26(1): 87–112.

Scarce, R. 2005. *Contempt of court. A scholar's battle for free speech from behind bars.* Lanham, MD: AltaMira Press.

Schaeffer, N. C., & S. Presser. 2003. The science of asking questions. *Annual Review of Sociology* 29: 65–88.

Schumaker, J. F., & J. D. Shea. 1993. Loneliness and life satisfaction in Japan and Australia. *Journal of Psychology* 127: 65–71.

Schuman, H., & S. Presser. 1981. *Questions and answers in attitude surveys.* New York: Academic Press.

Schwarz, N., & H. Hippler. 1987. What responses may tell your respondents: Informative functions of response alternatives. In *Social information processing and survey method,* 163–178, edited by H. Hippler, N. Schwarz, & S. Sudman. New York: Springer.

SDA Archive. GSS Cumulative Datafile 1972–2004. Retrieved October 20, 2005. http://sda.berkeley.edu/archive.htm.

Seidman, I. 1991. *Interviewing as qualitative research.* New York: Teachers College, Columbia University.

Seidman, I. 1998. *Interviewing as qualitative research,* 2nd edition. New York: Teachers College, Columbia University.

Seiter, E. 1993. *Sold separately: Children and parents in consumer culture.* New Brunswick, NJ: Rutgers University Press.

Serrano-Garcia, I. 1990. Implementing research: Putting our values to work. In *Researching community psychology,* 171–182, edited by P. Tolan, C. Keys, F. Chertock, & L. Jason. Washington, DC: American Psychological Association.

Shadish, W. R., Jr., T. D. Cook, & L. C. Leviton. 1991. *Foundations of program evaluation.* Newbury Park, CA: Sage.

Shapiro, G., & P. Dawson. 1998. *Revolutionary demands: A content analysis of the Cahiers de doleances of 1789.* Stanford, CA: Stanford University Press.

Shaw, C. R., & H. D. McKay. 1931. Social factors in juvenile delinquency. Volume II of *Reports on the Causes of Crime.* National Commission on Law Observance and Enforcement, Report No. 13. Washington, DC: U.S. Government Printing Office.

Sheirer, M. A. 1994. Designing and using process evaluation. In *Handbook of practical program evaluation,* 40–68, edited by J. P. Wholey, H. P. Hatry, & K. E. Newcomer. San Francisco: Jossey-Bass.

Sherwood, J. H. 2004. Talk about country clubs: Ideology and the reproduction of privilege. Ph.D. dissertation, North Carolina State University. http://www.lib.ncsu.edu/theses/available/etd-04062004-083555.

Silver, H. J. 2006. Science and politics: The uneasy relationship. *Footnotes* 34(2): 1 and 5.

Silverman, D. 2004. *Doing qualitative research,* 2nd edition. London: Sage Publications.

Sinclair, B., & D. Brady. 1987. Studying members of the United States Congress. In *Research methods for elite studies,* 61–71, edited by G. Moyser & M. Wagstaffe. London: Allen & Unwin.

Singer, E., & F. J. Levine. 2003. Protection of human subjects and research: Recent developments and future prospects for the social sciences. *Public Opinion Quarterly* 67(1): 148–164.

Singer, P. 2005. Anti-fraud research rules take effect. *National Journal* 37(28): 2210.

Singleton, R., Jr., B. C. Straits, M. M. Straits, & R. J. McAllister. 1988. *Approaches to social research.* New York: Oxford University Press.

Singleton, R. A., Jr., & B. C. Straits. 2001. Survey interviewing. In *Handbook of Interview Research,* 59–82, edited by J. F. Gubrium & J. A. Holstein. Thousand Oaks, CA: Sage Publications.

Sixsmith, J., & C. Murray, 2001. Ethical issues in the documentary analysis of Internet posts and archives. *Qualitative Health Research* 11: 423–432.

Slavin, S. 2004. Drugs, space, and sociality in a gay nightclub in Sydney. *Journal of Contemporary Ethnography* 33: 265–295.

Smith, A. E., J. Sim, T. Scharf, & C. Phillipson. 2004. Determinants of quality of life amongst older people in deprived neighborhoods. *Ageing and Society* 24: 793–814.

Smith, J. D., B. H. Schneider, P. K. Smith, & K. Ananiadou. 2004. The effectiveness of whole-school antibullying programs: A synthesis of evaluation research. *School Psychology Review* 33(4): 547–560.

Smith, P. M., & B. B. Torrey. 1996. The future of the behavioral and social sciences. *Science* 271: 611–612.

Smith, T. W. 1989. That which we call welfare by any other name would smell sweeter: An analysis of the impact of question wording on response patterns. In *Survey research methods,* 99–107, edited by E. Singer & S. Presser. Chicago: University of Chicago Press.

Smith, T. W. 2005. Generation gaps in attitudes and values from the 1970s to the 1990s. In *On the frontier of adulthood: Theory, research and public policy,* edited by R. A. Settersen, F. F. Furstenberg, Jr., & R. G. Rumbaut. Chicago: University of Chicago Press.

Smith, T. W., 2004. Coming of age in twenty-first century america: Public attitudes towards the importance and timing of transitions to adulthood. *Ageing International* 29(2): 136–148.

Smith, T. W. 2006. Altruism and empathy in America: Trends and correlates. National Opinion Research Center/University of Chicago. http://www-news.uchicago.edu/releases/06/060209.altruism.pdf.

Smith, T. W., J. Kim, A. Koch, & A. Park. 2005. Social-science research and the general social surveys. *ZUMA-Nachrichten* 56: 68–77.

Socolar, R. S., D. K. Runyan, & L. Amaya-Jackson. 1995. Methodological and ethical issues related to studying child maltreatment. *Journal of Family Issues* 16: 565–586.

Stake, R. E. 2003. Case studies. In *Strategies of Qualitative Inquiry,* 2nd edition, 134–164, edited by N. K. Denzin & Y. S. Lincoln. Thousand Oaks, CA: Sage.

Starr, P. 1987. The sociology of official statistics. In *The politics of numbers,* 7–60, edited by W. Alonso & P. Starr. New York: Russell Sage Foundation.

Stille, A. 2001. New attention for the idea that abortion averts crime. *New York Times,* April 14, Arts and Ideas Section.

Stouffer, S. A. 1950. Some observations on study design. *American Journal of Sociology* 55: 355–361.

Strauss, A. 1987. *Qualitative analysis for social scientists.* Cambridge: Cambridge University Press.

Suchman, L., & B. Jordan. 1992. Validity and the collaborative construction of meaning in face-to-face surveys. In *Questions about questions,* 241–270, edited by J. M. Tanur. New York: Russell Sage Foundation.

Sudman, S. 1967. *Reducing the cost of surveys.* Chicago: Aldine.

Sullivan, M. R., Bhuyan, K. Senturia, S. Shiu-Thornton, & S. Ciske. 2005. Participatory action research in practice. A case study in addressing domestic violence in nine cultural communities. *Journal of Interpersonal Violence* 20(8): 977–995.

Szinovacz, M., & A. Davey. 2006. Effects of retirement and grandchild care on depressive symptoms. *International Journal of Aging and Human Development* 62(1): 1–20.

Tan, J., & Y. Ko. 2004. Using feature films to teach observation in undergraduate research methods. *Teaching Sociology* 32: 109–118.

Taylor, L. 2005. All for him: Articles about sex in American lad magazines. *Sex Roles* 52: 152–163.

Taylor, L. 1995. *Occasions of faith: An anthropology of Irish Catholics.* Philadelphia: University of Pennsylvania Press.

Teachman, J. 2005. Military service in the Vietnam ear and educational attainment. *Sociology of Education* 78: 50–68.

Thompson, P. 2004. Pioneering the life story method. *International Journal of Social Research Methodology* 7(1): 80–84.

Thurlow, M. B. 1935. An objective analysis of family life. *Family* 16: 13–19.

Touliatos, J., B. F. Perlmutter, & M. A. Straus. 2001. *Handbook of family measurement techniques.* Thousand Oaks, CA: Sage.

Tourangeau, R., E. Singer, & S. Presser. 2003. Context effects in attitude surveys: effects on remote items and impact on predictive validity. *Sociological Methods & Research* 31(4): 486–513.

Tseng, V. 2004. Family interdependence and academic adjustment in college: Youth from immigrant and U.S. born families. *Child Development* 75(3): 966–983.

Turabian, K. L., 6th edition, revised by John Grossman and Alice Bennett. 1996. *A manual for writers of term papers, theses, and dissertations.* Chicago: University of Chicago Press.

Turner, J. A. 1991. *The structure of sociological theory,* 5th edition. Belmont, CA: Wadsworth

Turney, L., & C. Pocknee. 2005. Virtual focus groups: New frontiers in research. *International Journal of Qualitative Methods* 4(2), Article 3. Retrieved January, 14, 2006 from http:www.ualberta.ca/~ijqm/backissues/4_2/pdf/turney.pdf.

Ulrich, Laurel Thatcher. 1990. *A midwife's tale: The life of Martha Ballard, based on her diary, 1785–1812.* New York: Vintage Books.

Umaña-Taylor A. J., & M. Y. Bámaca. 2004. Conducting focus groups with Latino populations: lessons from the field. *Family Relations* 53(3): 261–272.

United Nations Statistics Division (2000a). The world's women 2000: Trends and statistics. Table 5.D: Indicators of economic activity.

Retrieved May 17, 2004, from http://unstats.un.org/unsd/demographic/ww2000/.

United Nations Development Program. (2002). *Human Development Report 2002: Deepening Democracy in a Fragmented World.* New York: Oxford University Press.

U.S. Bureau of the Census. 2000. Census bureau director says 92 percent of U.S. households accounted for. *United States Department of Commerce News.* May 31. Retrieved in October 2005, from www.census.gov/Press-Release/www/200041.html.

US Bureau of the Census Bureau. 2005. Computer and Internet use in the United States: 2003. Current Population Reports P23–208.

U.S. Commission on Immigration Reform. 1994. *Restoring credibility.* Washington, DC: U.S. Commission on Immigration Reform.

Unnithan, N. P. 1986. Research in a correctional setting: Constraints and biases. *Journal of Criminal Justice* 14: 301–412.

Urbaniak, G. C., & S. Plous. 2005. *Research randomizer.* Retrieved on November, 2005, from http://randomizer.org.

Van Leeuwen, T., & C. Jewitt. 2001. *Handbook of visual analysis.* Thousand Oaks, CA: Sage.

Vanneman, A. 1995. Editorial: Evaluating the evaluators. *Youth Today* 2: 2.

Venkatesh, S. A. 1997. The social organization of street gang activity in an urban ghetto. *American Journal of Sociology* 103: 82–111.

Verhovek, S. H. 1997. In poll, Americans reject means but not ends of racial diversity. *New York Times,* December 14: 1–32.

Verschuren, P. J. M. 2003. Case study as a research strategy: Some ambiguities and opportunities. *International Journal of Research Methodology* 6(2): 121–139.

Vidich, A. J., & J. Bensman. 1964. The Springdale case: Academic bureaucrats and sensitive townspeople. In *Reflections on community studies,* edited by A. Vidich, J. Bensman, & Maurice R. Stein. New York: Wiley.

Von Hippel, W., J. W. Schooler, K. J. Preacher, & G. A. Radvansky. 2005. Coping with stereotype threat: Denial as an impression management strategy. *Journal of Personality and Social Psychology* 89: 22–35.

Wallace, W. 1971. *The logic of science in sociology.* Chicago: Aldine.

Walter, T., & H. Waterhouse. 1999. A very private belief: Reincarnation in contemporary England. *Sociology of Religion* 60: 187–197.

Walton, J. 1992. Making the theoretical case. In *What is a case? Exploring the foundations of social inquiry,* 121–138, edited by C. C. Ragin, & H. S. Becker. New York: Cambridge University Press.

Ward, J., & Z. Henderson. 2003. Some practical and ethical issues encountered while conducing tracking research with young people leaving the 'care' system *International Journal of Social Research Methodology* 6(3): 255–259.

Warr, M., & C. G. Ellison. 2000. Rethinking social reactions to crime: Personal and altruistic fear in family households. *American Journal of Sociology* 106: 551–578.

Warren, C. A. B. 2001. Qualitative interviewing. In *Handbook of Interview Research,* 83–101, edited by J. F. Gubrium, & J. A. Holstein. Thousand Oaks, CA: Sage Publications.

Warren, C. A. B, T. Barnes-Brus, H. Burgess, L. Wiebold-Lippisch, J. Hackney, G. Harkness, V. Kennedy, R. Dingwall, P. C. Rosenblatt, A. Ryen, & R. Shuy. 2003. After the interview. *Qualitative Sociology* 26(1): 93–110.

Wasserman, I. M., & M. Richmond-Abbott. 2005. Gender and the Internet: Causes of variation in access, level and use. *Social Science Quarterly* 86: 252–270.

Wax, R. H. 1971. *Doing fieldwork: Warnings and advice.* Chicago: University of Chicago.

Weaver, D. 1994. The work and retirement decisions of older women: A literature review. *Social Security Bulletin* 57: 3–25.

Webb, E. J., D. T. Campbell, R. D. Schwartz, & L. Sechrest. 2000. *Unobtrusive measures,* revised edition. Thousand Oaks, CA: Sage.

Weber, M. 1947. *The theory of social and economic organization.* New York: Free Press.

Wechsler, H., M. Seibring, I-C. Liu, & M. Ahl. 2004. Colleges respond to student binge drinking: Reducing student demand or limiting access. *Journal of American College Health* 52(4): 159–168.

Weisberg, H. F. 2005. *The total survey error approach.* Chicago: The University of Chicago Press.

Weisberg, H. F., J. A. Krosnick, & B. D. Bowen. 1989. *An introduction to survey research and data analysis.* Glenview, IL: Scott, Foresman.

Weiss, R. S. 1994. *Learning from strangers: The art and method of qualitative interview studies.* New York: Free Press.

Weitzman, E., & M. B. Miles. 1995. *Computer programs for qualitative analysis.* Thousand Oaks, CA: Sage.

Weitzman, L., D. Eifler, E. Hokada, & C. Ross. 1972. Sex-role socialization in picture books for preschool children. *American Journal of Sociology* 77: 1125–1150.

Weitzman, L. J. 1985. *The divorce revolution.* New York: Free Press.

West, S. L., & K. K. O'Neal. 2004. Project D.A.R.E. outcome effectiveness revisited. *American Journal of Public Health* 94(6): 1027–1029.

Westkott, M. 1979. Feminist criticism of the social sciences. *Harvard Educational Review* 49: 422–430.

Whyte, W. F. 1955. *Street corner society.* Chicago: University of Chicago Press.

Whyte, W. F. 1997. *Creative problem solving in the field: Reflections on a career.* Walnut Creek, CA: AltaMira Press.

Whyte, W. F., D. J. Greenwood, & P. Lazes. 1991. Participatory action research: Through practice to science in social research. In *Participatory action research,* edited by W. F. Whyte. Newbury Park, CA: Sage.

Willcox, W. F. 1930. Census. *Encyclopedia of the social sciences,* 295–300. New York: Macmillan.

Williams, T. 1989. *The cocaine kids.* Reading, MA: Addison-Wesley.

Williamson, T., & A. Long. 2005. Qualitative data analysis using data displays. *Nurse Researcher* 12: 7–19.

Wolcott, H. F. 1982. Differing styles for on-site research, or, "If it isn't ethnography, what is it?" *Review Journal of Philosophy and Social Science* 7: 154–169.

Wolkomir, M. 2001. Emotion work, commitment, and the authentication of the self. *Journal of Contemporary Ethnography* 30: 305–334.

World almanac and book of facts. 2000. New York: Newspaper Enterprise Association.

World Health Organization. 1992. *World health statistics annual.* Geneva: World Health Organization.

Wysong, E., & D. W. Wright. 1995. A decade of DARE: Efficacy, politics and drug education. *Sociological Focus* 28: 283–311.

Zeller, R. A. 1997. Validity. In *Educational research, methodology, and measurement: An international handbook,* 822–829, edited by J. P. Keeves. New York: Pergamon.

Zernike, K. 2001. Antidrug program says it will adopt new strategy. *New York Times,* Feb. 15, 2001, p. A1 and A23.

Zipp, J. F., & J. Toth. 2002. She said, he said, they said: The impact of spousal presence in survey research. *Public Opinion Quarterly* 66(2): 177–208.

Zogby International. 2004. Young mobile voters pick Kerry over Bush, 55% to 40%, Rock the vote/Zogby poll reveals: National text-message poll breaks new ground. Available at: http://www .zogby.com/news/ReadNews.dbm?ID=919.

Index

Photo Credits

This page constitutes an extension of the copyright page. We have made every effort to trace the ownership of all copyrighted material and to secure permission from copyright holders. In the event of any question arising as to the use of any material, we will be pleased to make the necessary corrections in future printings. Thanks are due to the following authors, publishers, and agents for permission to use the material indicated.